Abstract Algebra via Numbers

Lars Tuset

Abstract Algebra via Numbers

Springer

Lars Tuset
Department of Computer Science
Oslo Metropolitan University
Oslo, Norway

ISBN 978-3-031-74622-2   ISBN 978-3-031-74623-9 (eBook)
https://doi.org/10.1007/978-3-031-74623-9

© The Editor(s) (if applicable) and The Author(s), under exclusive license to Springer Nature Switzerland AG 2025

This work is subject to copyright. All rights are solely and exclusively licensed by the Publisher, whether the whole or part of the material is concerned, specifically the rights of translation, reprinting, reuse of illustrations, recitation, broadcasting, reproduction on microfilms or in any other physical way, and transmission or information storage and retrieval, electronic adaptation, computer software, or by similar or dissimilar methodology now known or hereafter developed.
The use of general descriptive names, registered names, trademarks, service marks, etc. in this publication does not imply, even in the absence of a specific statement, that such names are exempt from the relevant protective laws and regulations and therefore free for general use.
The publisher, the authors and the editors are safe to assume that the advice and information in this book are believed to be true and accurate at the date of publication. Neither the publisher nor the authors or the editors give a warranty, expressed or implied, with respect to the material contained herein or for any errors or omissions that may have been made. The publisher remains neutral with regard to jurisdictional claims in published maps and institutional affiliations.

This Springer imprint is published by the registered company Springer Nature Switzerland AG
The registered company address is: Gewerbestrasse 11, 6330 Cham, Switzerland

If disposing of this product, please recycle the paper.

To my wife Maria and my kids Oliver and Evelyn

Preface

When I think of analysis, I tend to imagine a solitary scholar stooped over a table, perhaps with a telescope and a microscope nearby to explore a very minute or distant object, which cannot be reached, only approached by zooming. Algebra, on the other hand, I think of as a discipline dealing with day-to-day objects. The problem is, daily life quickly gets messy and complex, so there is a need to clean up and structure matters. This difference in attitude is reflected in the layout and style encountered when one opens a book in algebra as opposed to one in analysis.

Arguably, the most flexible and important notions in algebra are rings and modules. Rings are generalizations of numbers, where elements can be added and multiplied, while modules are generalizations of vector spaces with scalars replaced by ring elements. Rings and modules occur frequently at all levels in mathematics and are often constructed from numbers. Fields are special cases of rings, and modules over fields are actually considered as vector spaces with a ground field more general than just the fields of rational $\mathbb{Q}$, real $\mathbb{R}$, or complex numbers $\mathbb{C}$. Extensions like $\mathbb{Q} \subset \mathbb{R} \subset \mathbb{C}$ can be studied with greater sensitivity in the context of fields, and this turns out to be very fruitful. At the heart of this study, you find Galois theory, a powerful tool to study algebraic equations.

Another important notion in algebra is that of a group, which from the outset seems a bit unrelated to numbers, though most examples of groups are constructed as matrix groups with numbers as entries. Further, every group leads to an important ring, namely the group ring. Groups act on vector spaces via representations, turning the vector spaces into modules over the corresponding group rings. Yet another way of producing rings is to consider certain classes of number-valued functions on any set, then with pointwise addition and multiplication of functions. Modules over such function rings occur in geometry (including topology) as sections of vector bundles, rendering algebraic tools available to these disciplines. The interplay between geometry and algebra became strong already when coordinate systems were introduced to represent points in space, thus creating then an external bridge between our ability to visualize space and to manipulate symbols. The interplay between algebra and geometry is as strong today as it ever was.

A crucial strategy in algebra is to decompose more complicated objects into simpler ones, and then to try to understand these objects, perhaps even classify them. This principle runs as a thread through this book. The decomposition of natural numbers into primes uses arguments that pass on to decomposing polynomials into irreducible ones, and the method generalizes to Euclidean domains and PIDs. This, in turn, provides Jordan decomposition of matrices, decomposition of representations into irreducible ones, and of modules into simple ones. In group theory decomposition of groups via descending, or ascending, series into simple groups is important, a concept that reoccurs when we talk of Noetherian and Artinian modules.

Strictly speaking, no mathematical background is required to read this book. However, I have included a preliminary section with some basic notions from set theory. Exceptions to this are the introductions to each chapter, where prior mathematical knowledge is an asset, though it does no harm to try to understand them. One of my senior collaborators, John Roberts, who was a former Ph.D. student of the famous Paul Dirac, once said that when writing an article, you start with the introduction, only to find yourself rewriting the introduction when finishing the article. The same thing can be said about reading the introductions to the chapters in this book. You are probably served reading them once more when you have finished the chapters, or for those of you who are in for the long haul, when you have read the entire book. In this sense, the introductions are not entirely self-contained at a first reading, but could perhaps clarify thoughts you might have after reading the chapters. The introductions are also quite personal. I single out some material and ideas in the chapters that I find worth commenting on more informally. I should also say that the last section in the chapter on groups is not fully self-contained. I have included it based on the suggestion of the referee to provide more examples of groups coming from geometry and combinatorics. For this advice and otherwise useful comments, I am grateful.

Of course, any book will necessarily be subjective, especially when it comes to what one chooses to include as there is no end of material available. I have nevertheless tried to stick to mainstream topics that mathematical students are expected to have learned by the time they graduate. The first few chapters can be used for teaching at the first-year bachelor level, the middle section could serve for courses at a medium level, while the last chapters are probably best suited for master courses. It all depends on the appetite of the reader.

I have adapted the convention of naming a mathematical result or defining a notion whenever words or phrases are emphasized in the text.

Oslo, Norway
December 2023

Lars Tuset

Set-Theoretic Preliminaries

We recall a few things from naive set theory.

A *set* X is given in terms of its members, and we write $x \in X$ to indicate that x is a *member* or an *element* of X. Two sets are equal if their members are the same.

Sets can be indicated by listing their members in brackets, like $\{x, y, z\}$. We write $\{x \mid P\}$ for the set of all elements x with property P. Attention should be made to avoid self-referring statements, like the set of all sets, which is meaningless, and the set that is not a member of itself; a notorious statement known as Russel's paradox.

We are allowed to form various sets from other ones. The *union* $\cup_i X_i$ and *intersection* $\cap_i X_i$ of any collection of sets consists of those elements that belong to at least one of the sets, respectively, to each one of them. A *subset* $Y \subset X$ of a set X is a set Y with members only from X, and its *complement* $X \setminus Y$ consists of those elements in X that do not belong to Y.

The useful *de Morgan's laws* are the easily proved statements that

$$\cap_i (X \setminus X_i) = X \setminus \cup_i X_i \text{ and } \cup_i (X \setminus X_i) = X \setminus \cap_i X_i$$

for subsets X_i of a set X.

The *(Cartesian) product* $X \times Y$ of two sets X and Y consists of all *ordered pairs* (x, y) with $x \in X$ and $y \in Y$. By construction (x, y) is the subset $\{\{x\}, \{x, y\}\}$ of $X \cup Y$. Two ordered pairs coincide $(x, y) = (x', y')$ if and only if $x = x'$ and $y = y'$. To see this, suppose $\{\{x\}, \{x, y\}\} = \{\{x'\}, \{x', y'\}\}$. Either $\{x\} = \{x, y\} = \{x'\}$, and then all elements are equal, or $\{x\} = \{x'\}$ and $\{x, y\} = \{x', y'\}$, and then $x = x'$ and $y = y'$, or there are two other similar alternatives, and in these cases also $x = x'$ and $y = y'$.

By a *relation on a set* X, we mean any subset R of $X \times X$, and we write $x \sim y$ for $(x, y) \in R$.

A *function* or a *map* $f : X \to Y$ from X to Y is a relation $X \times Z$ on $X \cup Y$ with $Z \subset Y$ such that there is exactly one element $(x, y) \in X \times Z$ for each $x \in X$, and we then write $y = f(x)$. If there is only one such $x \in X$ to each $y \in Z$, then we say that f is *injective*, and f is *surjective* if $Z = Y$, and if it is both injective and surjective, then it is *bijective*. The relation $X \times Z$ is also called the *graph* of f with *domain* X and *image* Z.

The *composition* of $f : X \to Y$ and $g : Y \to Z$ is the function $g \circ f : X \to Z$ given by $g \circ f(x) = g(f(x))$. We sometimes write gf for $g \circ f$. Note that $h(gf) = (hg)f$ for $h : Z \to W$, so we often skip parentheses and write hgf for $(hg)f$.

If $Z = X$ and gf equals the identity map $\iota : X \to X$, then clearly g is surjective and f is injective, so if also $fg = \iota$, now with ι the identity map on Y, then both f and g are bijective. Also, one map is uniquely determined by the other because if also $g'f = \iota$ and $fg' = \iota$, then

$$g' = g'\iota = g'(fg) = (g'f)g = \iota g = g$$

and vice versa. We say that g is the *inverse map* of f and write f^{-1} for g. Thus $(f^{-1})^{-1} = f$. In this uniqueness argument, we only used $fg = \iota$ and $g'f = \iota$.

If f is bijective, then it has an inverse, namely the map g given by $g(f(x)) = x$. This definition makes sense firstly because any element of Y is of the form $f(x)$ for some $x \in X$ as f is surjective, and secondly because if $f(x) = f(y)$ then $x = y$ by injectivity of f, and thus $g(f(x)) = g(f(y))$. Hence $gf = \iota$, and g is obviously bijective, so by the same argument with the roles of f and g swopped, we also get $gf = \iota$. So f is invertible with inverse g.

A map on a finite set is injective if and only if it is surjective. Indeed, let f be a map on a finite set, and assume f is injective. To hit an element x, apply f to x repeatedly till repetitions $f^m(x) = f^n(x)$ occur. Then peal off f's till $x = f(f^k(x))$ for some k. Conversely, if f is not injective, then its image will contain too few elements for it to be surjective.

A *binary operation on a set* X is a map $X \times X \to X$, and one often writes xy or $x \cdot y$ or $x + y$, etc., for the image of (x, y) depending on context and what further properties the binary operation might have.

If a bijective map $f : X \to Y$ preserves binary operations on X and Y, say $f(x+y) = f(x) \cdot f(y)$, then so will its inverse because

$$f^{-1}(f(x) \cdot f(y)) = f^{-1}f(x+y) = x+y = f^{-1}(f(x)) + f^{-1}(f(y)).$$

Any bijective map that preserves all the relevant operations on the collection of sets under consideration is called an *isomorphism*. Often one specifies with an adjective under what operations the map is an isomorphism. Maps that preserve the operations without being necessary bijective are often called *morphisms* or *homomorphisms*.

The *product* $\prod_{i \in I} X_i$ of sets $\{X_i\}$ over any (index) set I consists of all functions $f : I \to \cup X_i$ with $f(i) \in X_i$ for all $i \in I$. We write X^I for $\prod_{i \in I} X_i$ when all $X_i = X$. When $I = \{1, \ldots, n\}$, we write X^n for X^I, so X^n consists of all *n-tuples* $(x_1, \ldots, x_n)$ with $x_i \in X$.

We denote the set $\{1, 2, 3, \ldots\}$ of natural numbers by $\mathbb{N}$.

A *sequence* $\{x_n\}$ of elements $x_n \in X$ is a function $f : \mathbb{N} \to X$ with $x_n = f(n)$, or in other words, we have $\{x_n\} = f \in \prod_{n \in \mathbb{N}} X_n$ with $X_n = X$ for all n.

Say we have a function $f : X \to Y$ and subsets $A \subset X$ and $B \subset Y$. Then the *image* $f(A)$ of A and the *inverse image* $f^{-1}(B)$ of B are defined as

$$f(A) = f(x) | x \in A \text{ and } f^{-1}(B) = x \in X | f(x) \in B$$

The *power set* $P(X)$ of X is the set of all subsets of X. The *characteristic function* of $Y \subset X$ is the function $\chi_Y : X \to \{0, 1\}$ such that $\chi_Y(x) = 1$ if $x \in Y$ and $\chi_Y(x) = 0$ if $x \notin Y$.

Note that the map which sends a subset of X to its characteristic function is a bijection from $P(X)$ to $\{0, 1\}^X$. So the number of elements in $P(\{1, \ldots, n\})$ equals 2^n, hence the terminology 'power set'.

A relation $\sim$ on a set X is called an *equivalence relation* if it is *reflexive*, $x \sim x$, *symmetric*, $x \sim y \iff y \sim x$, and *transitive*, $(x \sim y) \wedge (y \sim z) \Rightarrow x \sim z$. One can then form the *quotient set* $X/\sim$ of equivalence classes, and the *equivalence class* of $y \in X$ is the subset $\{x \in X \mid x \sim y\}$. Because $\sim$ is an equivalence relation, the quotient set is a *partition* of X, meaning that X is a disjoint (e.g. pairwise non-intersecting) union of equivalence classes.

To explain why every element of X belongs to exactly one equivalence class, first notice that due to reflexivity, any $x \in X$ belongs to its own equivalence class. And if x belongs to another equivalence class, say to that of an element y, then $x \sim y$, and any element z in the equivalence class of x, will because of transitivity, also belong to the equivalence class of y, so the equivalence class of x will be contained in the one of y. But by symmetry, we see that we also have inclusion the other way, so x belongs to only one class.

In fact, it is easy to see that to any partition of a set X, there is a unique equivalence relation $\sim$ having $X/\sim$ as the partition; just define $\sim$ by $x \sim y$ if x and y belong to the same block of the partition.

By an *order* $>$ on a set X we mean any relation that is transitive and such that for any $x, y \in X$ exactly one of the statements $x > y, x = y, y > x$ holds. Then $x \geq y$, meaning $x > y$ or $x = y$, defines a *partial order* $\geq$ on X, that is, a relation which is transitive, reflexive, and *antisymmetric*, $(x \geq y) \wedge (y \geq x) \Rightarrow x = y$. Conversely, any partial order where all elements are pairwise comparable, i.e. either $x \geq y$ or $y \geq x$, defines an order $>$ with $x > y$ if $x \geq y$ and $x \neq y$.

A *chain in a partially ordered set* is any subset of pairwise comparable elements. This is a notion that plays an important role in *Zorn's lemma*:

Axiom 2.1 *If every chain in a partially ordered non-empty set S has an upper bound, then S has a maximal element, i.e. with no elements superseding it.*

This axiom is equivalent to the *axiom of choice*:

Axiom 2.2 *To every collection of non-empty sets, there is a function that chooses exactly one element from each set.*

Or equivalently, the product of any collection of non-empty sets is non-empty, containing at least one *choice function*.

The axiom of choice is again equivalent to Cantor's *well-ordering principle*:

Axiom 2.3 *Every non-empty set can be endowed with a partial order for which it is well-ordered.*

Clearly, a choice function would then be one that picks out the minimal element in each non-empty set. The converse direction is much harder to prove and normally goes via Zorn's lemma.

Despite the controversy around the axiom of choice, due to, e.g. the not very intuitive requirement that the set $\mathbb{R}$ of real numbers (to be introduced carefully later) can be well-ordered, we accept it as a set-theoretic axiom along with the others, and these ones are commonly agreed upon to be those formulated by Zermelo-Frankel. The axiom of choice is independent of the *ZF*-axioms provided these are consistent, so neither the claim nor its negation can be proved from potentially consistent *ZF*-axioms.

Contents

1 **Number Theory** 1
    1.1   Decomposition of Natural Numbers 4
    1.2   Mathematical Induction 5
    1.3   The Well-Ordering Principle 7
    1.4   Relative Primeness and Greatest Common Divisors 8
    1.5   The Euclidean Algorithm 10
    1.6   Newton's Binomial Formula 11
    1.7   Infinity of Primes 13
    1.8   Primes in Arithmetic Progression 16
    1.9   The Function $\pi(x)$ 17
    1.10   Congruence 19
    1.11   Arithmetic Functions 22
    1.12   Primitive Roots 25
    1.13   Quadratic Reciprocity 30
    1.14   Certain Classes of Numbers 35
    1.15   Diophantine Equations 38
    1.16   Sums of Squares 41
    1.17   Fibonacci Numbers 44

2 **Construction of Numbers** 47
    2.1   Peano's Axioms 49
    2.2   The Integers Constructed from the Natural Numbers 52
    2.3   From the Integers to the Rational Numbers 55
    2.4   Finite Simple Continued Fractions 57
    2.5   Construction of the Real Numbers 59
    2.6   The Least Upper Bound Property 63
    2.7   Decimal Expansions 64
    2.8   Infinite Continued Fractions 66
    2.9   Pell's Equation 71
    2.10   Complex Numbers 74

	2.11	Absolute Values and p-Adic Numbers	76
	2.12	Cardinality	81
3	**Linear Algebra**		87
	3.1	Vector Spaces	89
	3.2	Linear Basis	93
	3.3	Linear Transformations	97
	3.4	Matrices	99
	3.5	Systems of Linear Equations	103
	3.6	Permutations	109
	3.7	Determinants	111
	3.8	Eigenvalues and Eigenvectors	118
	3.9	Jordan Canonical Form	121
	3.10	Dual Spaces, Inner Products and Tensor Products	127
4	**Groups**		131
	4.1	Groups and Semigroups	134
	4.2	Subgroups	135
	4.3	Generators	136
	4.4	Cosets and Lagrange's Theorem	137
	4.5	Morphisms	138
	4.6	Normal Subgroups	139
	4.7	Cyclic Groups	140
	4.8	Normalizers and Centralizers	142
	4.9	Correspondences	143
	4.10	More Isomorphism Theorems	143
	4.11	Permutation Groups	144
	4.12	Symmetries	146
	4.13	Automorphisms	147
	4.14	Semidirect Products	148
	4.15	The General and Special Linear Group	150
	4.16	Inner Products and Linear Subgroups	151
	4.17	Actions	154
	4.18	Spheres, Projective Spaces and Grassmannians	157
	4.19	Groups of Prime Power Orders	160
	4.20	Cauchy's Theorems and Sylow's First Theorem	161
	4.21	Sylow's Second and Third Theorems	163
	4.22	Some Examples	165
	4.23	Groups with Order the Product of Two Primes	166
	4.24	Normal Series	167
	4.25	The Theorem of Schreier	170
	4.26	Solvable Groups	171
	4.27	Nilpotent Groups	172
	4.28	Simplicity of the Alternating Group	175

	4.29	Transfer Homomorphisms	176
	4.30	Finitely Generated Abelian Groups	178
	4.31	Free Groups	182
	4.32	An Example of a Free Product	185
	4.33	Generators and Relations	187
	4.34	Ordered Groups	189
	4.35	Groups in Algebraic Topology	190
5	**Representations of Finite Groups**		**197**
	5.1	Basic Definitions	199
	5.2	Regular Functions	200
	5.3	New Representations from Old Ones	201
	5.4	Decomposition Into Irreducibles	202
	5.5	Haar Integral	203
	5.6	Regular Representation	204
	5.7	Schur's Lemma	204
	5.8	Characters of Abelian Groups	205
	5.9	Fourier Analysis	207
	5.10	Orthogonality Relations	209
	5.11	Three Auxiliary Representations	211
	5.12	Characters of Representations	212
	5.13	Group Algebra	214
	5.14	Quadratic Reciprocity from Fourier Analysis	216
	5.15	The Character Table for S_3	220
	5.16	Induced Representations	221
	5.17	Reciprocity	223
	5.18	Mackey Theory	224
	5.19	Characters of Induced Representations	227
6	**Rings**		**231**
	6.1	Basic Definitions	232
	6.2	Prime Subfields and Characteristics	234
	6.3	Examples of Non-commutative Rings	235
	6.4	Group Rings	236
	6.5	Polynomial Rings	237
	6.6	Laurent Series and Power Series	239
	6.7	Ideals	240
	6.8	Quotient Rings and Homomorphisms	241
	6.9	Rings with Generators and Relations	244
	6.10	Twisted Group Rings	245
	6.11	Simple Rings and Maximal Ideals	246
	6.12	Euclidean Domains and Principal Ideal Domains	248
	6.13	Prime Ideals and Irreducible Ideals	250
	6.14	Unique Factorization in a PID	251
	6.15	Unique Factorization Domains	254

7 Field Extensions ... 257
- 7.1 Roots and Reducible Polynomials ... 258
- 7.2 Algebraic Extensions ... 260
- 7.3 Algebraic Closures ... 264
- 7.4 Ruler and Compass ... 268
- 7.5 Splitting Fields and Normal Extensions ... 271
- 7.6 Multiple Roots ... 274
- 7.7 Finite Fields ... 276
- 7.8 Separable Extensions ... 278

8 Galois Theory ... 281
- 8.1 Automorphisms and Fixed Fields ... 282
- 8.2 The Galois Group of a Polynomial ... 286
- 8.3 The Fundamental Theorem in Galois Theory ... 287
- 8.4 Proof of the Fundamental Theorem of Algebra ... 289
- 8.5 Primitive Roots and Cyclotomic Polynomials ... 291
- 8.6 Constructable Polygons ... 293
- 8.7 Cyclic Extensions ... 294
- 8.8 Polynomials Solvable by Radicals ... 296
- 8.9 Symmetric Functions ... 300
- 8.10 Cubic and Quartic Equations ... 301

9 Modules ... 307
- 9.1 Basics ... 309
- 9.2 Exactness ... 312
- 9.3 Projectivity ... 316
- 9.4 Injectivity ... 320
- 9.5 Tensor Products and Bimodules ... 324
- 9.6 Diagram Chase ... 330
- 9.7 Flatness ... 335
- 9.8 Duals ... 341
- 9.9 Modules over PID's ... 344
- 9.10 Torsion Modules over PID's ... 346
- 9.11 Smith Normal Form ... 350
- 9.12 Applications to Linear Algebra ... 358
- 9.13 Generalized Jordan Blocks ... 361
- 9.14 The Jordan–Chevalley Decomposition ... 365
- 9.15 Semisimple Modules ... 370
- 9.16 Density ... 373
- 9.17 Semisimple Rings ... 376
- 9.18 Noetherian and Artinian Modules ... 383
- 9.19 Nilpotence ... 392
- 9.20 The Jacobson Radical ... 396
- 9.21 The Wedderburn Radical ... 402
- 9.22 Radicals Under Change of Rings ... 406

	9.23	Radicals of Polynomial Rings	410
	9.24	Radicals of Groups Rings	414
	9.25	Units in Group Rings	418
	9.26	Division Rings	422
10	**Appendix**		**429**
	10.1	The Function $\pi(x)$ for Large x	429
	10.2	The Riemann Zeta Function	433
	10.3	Bernoulli Numbers	434
	10.4	Transcendentality of e and π	437
	10.5	Proof of Liouville's Theorem	440
	10.6	Thue's Theorem	441

Bibliography 443

Index 445

About the Author

Lars Tuset has been a Full Professor of Mathematics at the Oslo Metropolitan University since 2005. He took his formal education at the Norwegian University of Science and Technology and spent several years as postdoc at University II in Rome and at University College Cork in Ireland.

He has published extensively within the areas of quantum groups, operator algebras, and noncommutative geometry, and has had many international collaborators. In 2022, he published a comprehensive Springer volume on analysis and locally compact quantum groups. He wrote another monograph (with a colleague) in 2013 for the French Mathematical Society, expanding then on quantum groups and their representation categories.

Chapter 1
Number Theory

Number theory deals with natural numbers and their properties. Prime numbers play a crucial role here. They are numbers that cannot be written as a product of two numbers, if we rule out the number one. Any number other than one, is a product of prime numbers, and this decomposition is unique up to reordering of factors. This theorem is a corner stone in number theory, and can be proved rigorously from the well-ordering principle, stating that any non-empty subset of the natural numbers has a least element. While the existence of a decomposition follows easily from this principle, the uniqueness part of the theorem is more subtle. This part of the proof resorts to the property of relative primeness; two numbers are relatively prime if they both cannot be divided by a common number other than one. They behave thus with respect to each other as if they are prime numbers without necessarily being so. When two natural numbers are not relatively prime, a pertinent notion to study is that of their greatest common divisor, which is then obviously larger than one.

Having now a decomposition result at our disposal, it seems natural to find all the prime numbers, so that we know what our building blocks under multiplication are. However, this turns out to be extremely difficult, to say the least. Starting to list them, say by sieving away the composite numbers as one proceeds, it becomes clear that they occur very irregularly, escaping any possible general formula. Moreover, one cannot hope to exhaust the list, as there happens to be infinitely many of them, a fact that can easily be established using the decomposition result above. Having no clue how to find all the prime numbers, it seems remarkable that one can prove that there are infinitely many of them. We can't even find samples of arbitrary large ones. As a matter of fact, some of the largest known prime numbers are national secrets, used in code theory.

One can nevertheless say something about the distribution of prime numbers. In doing so, one applies techniques from complex function theory, a branch nowadays called analytic function theory. One result in this direction is the prime number theorem, which gives an estimate of how many prime numbers less than any given number there are. Another result is the infinity of prime numbers along any arithmetic progression, that is, numbers with a fixed distance to consecutive members in the progression, or sequence. And then there is the notorious Riemann Hypothesis, which relates the distribution of primes to an analytic function, known as the zeta-function, a relation discovered already in the eighteenth century by Euler. Related to this are the Bernoulli numbers. We leave the bulk of this discussion to the appendix, and reduce our study in the main text to a few short sections, just to give an idea of the fascinating mystery of the occurrence of prime numbers.

More relevant for us are some of the basic techniques used to study prime numbers. Many of these notions and results might seem a bit ad hoc in the context of numbers, but are better understood in the more general context of groups, which we will study in a later chapter. For our purposes it makes sense to anticipate events. So let us just recall that a group is a set with a binary associative operation, which has a unit, and for which each element has an inverse element. Now fix a natural number n. We say that two integers are congruent modulo n if their difference is an integer multiple of n. This is an equivalence relation, and looking at its equivalence classes (see our preliminaries from set theory), we obtain a set consisting of n elements. By adding, subtracting and multiplying representatives from these so called residue classes, we get well-defined corresponding operations on this finite set $\mathbb{Z}_n$, turning it into what we later on call a commutative ring. If we consider the representatives $0, 1, 2, \ldots, n-1$ from each of the n classes, written $[0], [1], [2], \ldots, [n-1]$, it turns out that the classes with representatives that are relatively prime to n form a group under multiplication. The number of elements in this group is by definition Euler's phi-function of n, or $\phi(n)$. For a prime number p, clearly $\phi(p) = p - 1$, saying that all the classes in $\mathbb{Z}_p$, except the one containing 0, have inverse elements. In this case the ring is what we later call a field, bearing much the same properties as we have for the rational, real or complex numbers, only that it is finite. For general n, the multiplicative group in $\mathbb{Z}_n$ does not fill out almost the whole ring. But in nice cases the group will have a generator, that is, a class whose powers exhaust the whole group, and we say then that the group is cyclic. A representative for such a class is called a primitive root. We provide a theorem telling us exactly for which n this happens. As soon as n has a primitive root, there will be $\phi(\phi(n))$ of them; you get the other ones from a primitive root, by taking the powers of it with the exponents that are relatively prime to $\phi(n)$, hence the formula with two ϕ's. It is clear that if you take a generator of the group, then the $\phi(n)$th power of it is the first power that reproduces the unit, otherwise you get either too many or too few elements in the group. The phenomena of recovering the unit happens in the cyclic case if you take the $\phi(n)$th power of any element. In fact, the same holds also in the non-cyclic case, because if you haven't recovered the element before, you get too many elements in the group. And if you get it before, the power m you have reached must divide $\phi(n)$ as you partition the group into so called cosets, all having m elements. As a result you get Euler's theorem, which states that the $\phi(n)$th power of any class gives you the class back, or $a^{\phi(n)} \equiv 1 \pmod{n}$, using the congruence relation for any number a relatively prime to n.

1 Number Theory 3

For the sake of concreteness, let's consider some simple examples. In $\mathbb{Z}_{10}$ the multiplicative group consist of the four classes [1], [3], [7], [9] since 1, 3, 7, 9 are relatively prime to 10, and there are only $\phi(10) = 4$ of them. Euler's ϕ-function is an example of an arithmetic function that is multiplicative on any two numbers that are relatively prime. So using prime number decomposition, we can quickly calculate its values. For example, in the above case $\phi(2 \cdot 5) = \phi(2)\phi(5) = 1 \cdot 4 = 4$. The four classes listed above form indeed a group. For instance, we have $[3] \cdot [7] = [3 \cdot 7] = [21] = [1]$, which is the unit in the group, and this incidentally also shows that [3] and [7] are inverses of each other. They are also the $\phi(4) = 2$ generators of the group, since $[3]^2 = [9]$ and $[7]^2 = [49] = [9]$ and 3 does not divide 4. So we have a cyclic group here with two primitive roots 3, 7 of 10. They can be gotten from each other by taking cubes, as 1 and 3 are the numbers less than four that are relatively prime to 4. Indeed, we have $[3]^3 = [27] = [7]$ and $[7]^3 = [9] \cdot [7] = [63] = [3]$. Note also that [2], which does not belong to the group, has order 5, as $2^5 = 32$, and not order 4, as $2^4 = 16$, whereas in $\mathbb{Z}_8$, we have $[2]^3 = [0]$. In fact, the multiplicative group in the latter ring consists of the residue classes [1], [3], [5], [7], and all these have square [1], so 8 has no primitive root, and the multiplicative group is not cyclic here.

One might also want to solve certain equations working within the ring $\mathbb{Z}_n$. For instance, there is the question of whether square roots of any ring element exist. For example, in $\mathbb{Z}_8$ we saw that [1] had 4 square roots, namely [1], [3], [5], [7], and [4] and [0] have the square roots [2], [6] and [0], [4], respectively, while [2] and [6] have none. In general, say we have two distinct prime numbers p, q both different from 2. Thanks to a deep reciprocity result by Gauss, the question whether $[q]$ has a square root in $\mathbb{Z}_p$, is equivalent to knowing whether $[p]$ has a square root in $\mathbb{Z}_q$. Indeed, if at least one of $(p-1)/2$ or $(q-1)/2$ is even, then the answers are both the same, otherwise, the two questions have always opposite answers. Together with some easier results, this allows one to check effectively if a square root of an element in $\mathbb{Z}_n$ exists or not. A simple illustration of this is furnished by the following example. We wonder if [3] has a square root in $\mathbb{Z}_{101}$. Naively, it seems we must check if any of the 101 classes has square [3], and this is pretty cumbersome. Now 3, 101 are prime numbers and $(101-1)/2$ is even, so we can equivalently ask if [101] has a square root in $\mathbb{Z}_3$. But in the latter ring $[101] = [2]$, which is not a square, as $[0]^2 = [0], [1]^2 = [1]$ and $[2]^2 = [1]$.

Along the same lines cubic roots and so forth can be studied. We include a section on integer solutions of polynomial equations, with polynomials in several variables. We discuss the most famous of these, namely Fermat's theorem, from a purely elementary point of view. The fairly recent proof of this theorem by Wiles uses a great deal of general mathematics beyond the scope of this volume.

For cultural and historical reasons we also devote a section to certain classes of numbers, and in a separate section we scrutinize the Fibonacci numbers. We study another peculiar number theoretical problem, namely to what extend a natural number can be written as sums of squares.

For further reading, we suggest the references [9, 16].

1.1 Decomposition of Natural Numbers

The natural numbers $1, 2, 3, \ldots$ with their hideous and awinspiring complexity present themselves to us through counting. They come with two operations: addition and multiplication. The number 1 is the building block for addition in that every natural number can be obtained by adding 1 to itself sufficiently many times, e.g. $3 = 1 + 1 + 1$.

Now what are the fundamental building blocks for multiplication? The number 1 certainly won't do. Neither does 2 and 3 since we cannot obtain 5 as a product of 2's and 3's. We would have to add 5 to the list, and then 7, and 11, and so on. Being fundamental building blocks they cannot be divided further, that is, they cannot be decomposed into products of smaller natural numbers. We call them prime numbers.

Definition 1.1.1 A *prime number* is a natural number that can only be divided by itself or by 1, and we won't consider 1 as a prime number.

The prime numbers are therefore $2, 3, 5, 7, 11, 13, 17, \ldots$.

A *composite number* is any natural number larger than 1 that is not prime. So the composite numbers are $4, 6, 8, 9, 10, 12, \ldots$.

Any number except 1 can be decomposed into a product of prime numbers. Keep dividing into smaller natural numbers till you cannot divide any more. Then all the factors are prime numbers. Moreover, modulo rearrangements of factors such a decomposition is unique.

Example 1.1.2 To wit $120 = 2 \cdot 60 = 2 \cdot 2 \cdot 30 = 2 \cdot 2 \cdot 2 \cdot 15 = 2 \cdot 2 \cdot 2 \cdot 3 \cdot 5$. We could also have decomposed like this $120 = 5 \cdot 24 = 5 \cdot 8 \cdot 3 = 5 \cdot 4 \cdot 2 \cdot 3 = 5 \cdot 2 \cdot 2 \cdot 2 \cdot 3$. The two final decompositions into smallest components are the same in the sense that 5 appears once, the number 3 appears once, and 2 appears thrice, whereas 7 and 11 etc. appear zero times. So apart from factors been written in different orders, the decomposition is unique.

Since any number is unaltered under multiplication by any power of 1, we could say that 1 appears zero times or once or e.g. 13 times in the decomposition of 120, violating uniqueness. That is why 1, although it is only divisible by itself and 1, is not considered a prime number. $\diamond$

Let us record our result on the decomposition of a natural number as a theorem. It is known as the *fundamental theorem of arithmetic*:

Theorem 1.1.3 *Any whole number larger than 1 can be written as a product of primes. Such a decomposition is unique if we e.g. write the prime factors with increasing magnitude towards the right.*

In this sense $2^3 \cdot 3 \cdot 5$ is the unique decomposition of 120. Why is in general such a decomposition unique? We can convince ourselves of this by checking various cases. For instance, $21 = 3 \cdot 7$, and there is no other way of writing 21 as a product of primes. To check that $2^2 \cdot 5 \cdot 7$ is the unique decomposition of 140 requires more work, but it can be done. However, we can't keep checking case by case. After finitely many cases, we might have infinitely many left to do, so even the aid of a very powerful computer is futile. It seems we need something out of this world. What we need is a mathematical proof.

So imagine that we have two prime number decompositions of the same number. If the same primes occur as factors in both decompositions, then we can mutually cancel them out. At the end of this finite procedure we are left with an equality $p_1 \cdots p_n = q_1 \cdots q_m$, where none of the primes on the left hand side equals any of those on the right hand side. Now p_1 does divide the left hand side, since it sits there as the first factor, and therefore it must also divide the right hand side. But this is only possible if p_1 equals one of the primes $q_1, \ldots, q_m$, which should not happen, since it then ought to have been canceled. We have reached a contradiction. Since we cannot accept any result that contradicts established results, the original decomposition must be unique.

All this seems fine, except that we have used the fact that if p_1 divides $q_1 \cdots g_n$, then it must simply be one of these q's. Is this an established truth? A skeptical mind might even question the argument that allowed us to decompose a number into its smallest constituents. We need to agree upon what should be regarded as obvious truths. Over the next two sections we will do so, and then we will prove the fundamental theorem of arithmetic from this common basis.

1.2 Mathematical Induction

Some of you have perhaps encountered the principle of induction. If we believe in this principle, can we then deduce what we need? Let us first refresh what this principle says.

Say S is a subset of the set $\mathbb{N}$ of natural numbers with the property that if n belongs to S, so does its successor $n + 1$. If in addition S contains the number 1, then S must contain $2 = 1 + 1$, and $3 = 2 + 1$, and $4 = 3 + 1$, and so forth. We see that S eventually absorbs all natural numbers, and this is what the *first principle of induction* says.

Axiom 1.2.1 If S is any subset of $\mathbb{N}$ with $1 \in S$ and such that $n + 1 \in S$ whenever $n \in S$, then $S = \mathbb{N}$.

Example 1.2.2 Suppose we arrived at the formula

$$1 + \cdots + n = \frac{n(n+1)}{2}$$

as something which ought to be true for all natural numbers n, although we might have checked it only in a few special cases. To deduce it from the principle of induction, let S be the set of natural numbers n for which the formula holds. Since

$n = 1$ satisfies the formula, as $1 = \frac{1(1+1)}{2}$, the number 1 belongs to S. Next suppose that n is a member of S. This means that

$$1 + \cdots + n = \frac{n(n+1)}{2}.$$

We must show that $n + 1$ is a member of S, or in other words, that

$$1 + \cdots + (n+1) = \frac{(n+1)((n+1)+1)}{2}.$$

But this latter expression can be gotten by adding $n + 1$ to both sides of the previous formula, and then use the identity

$$\frac{n(n+1)}{2} + (n+1) = \frac{(n+1)(n+2)}{2}.$$

By the principle of induction the set S therefore consists of all natural numbers, so the initially stated formula holds for all natural numbers n. ◇

Here is a related version of this principle, known as the *second principle of induction*.

Axiom 1.2.3 Suppose S is a set of natural numbers with the property that $n \in S$ whenever $k \in S$ for all $k < n$. Then $S = \mathbb{N}$.

Since there are no natural numbers less than 1, we are assuming that $1 \in S$.

We can now show that any natural number, say n, decomposes into prime factors.

Proof of the existence part of the fundamental theorem of arithmetic: Let S be the set of natural numbers that have a prime number decomposition. Clearly 2 belongs to S; here our induction starts at 2 rather than 1. Suppose n is a natural number such that any number less than it and larger than 1 belongs to S. Now either n is already a prime number, and we are done, or it can be written as a product of two other natural numbers smaller than n. Since these two numbers belong to S by the property of n, and every number in S has a prime number decomposition, each of these two numbers can be decomposed into prime factors. But then also n, being the product of these two numbers, is itself a product of prime numbers, and belongs to S. By the second induction principle, we therefore conclude that $S = \{2, 3, 4, \ldots\}$, or in other words, that any natural number larger than 1 can be decomposed into factors of primes.

We are here taking the second principle of induction as something obvious, and we shall presently see that both induction principles follow from an even more fundamental statement.

1.3 The Well-Ordering Principle

According to the standards of rigor in contemporary mathematics, number theorists accept as valid any result that can be deduced from the *well-ordering principle*:

Axiom 1.3.1 Any non-void subset T of non-negative integers has a least member, that is, a number $n \in T$ such that $n \leq m$ for all other $m \in T$.

Experience with numbers favors this principle. For instance, the set $T = \{5, 8, 3, 11\}$ has 3 as its least element, and 2 is the least element in the set of even positive numbers. It is a deceptively obvious principle because in many cases it is not clear what the least number actually is.

Example 1.3.2 Consider the set E of all primes larger than 100000000000. What is its smallest element? Using computers one can show that the set E is indeed nonempty. In Sect. 1.7 we shall even prove that this is so by a very short argument. We shall in fact prove something much more, namely that there exist infinitely many prime numbers. However, this proof relies on the well-ordering principle, so to say that E has a least element, we have then used this principle at least twice; once to show that there are infinitely many primes, so that E is nonempty, and then once more to conclude that E has a least element. ◇

We just want to stress that the well-ordering principle is an existence result, it does not tell us how to find least elements. We shall nevertheless regard this seemingly naive principle as true. A more pedantic approach to natural numbers is to start with Peano's axioms, an approach we will discuss later in Section 2.1.

To demonstrate the power of the well-ordering principle let us show that the second induction principle holds; the first principle of induction can be proved in a similar fashion.

Proof of the second principle of induction: Suppose S has the required properties for the second induction principle to kick into action, and that $S \neq \mathbb{N}$. Then its complement is non-void, and by the well-ordering principle, it must have a least element n. Since any number less than n belongs to S, so must n, which is a contradiction. Hence $S = \mathbb{N}$.

The idea of the proof was simple enough. Once a step-by-step inductive progression has started, a collision with the least element in any complement of S is unavoidable.

Every mathematical statement requires a proof, which in the case of integers means a logical deduction from the well-ordering principle. That is the rule of the game. Even a self evident statement like the *Archimedean property* requires a proof.

Proposition 1.3.3 *For any two natural numbers a and b, there exists another n such that $na \geq b$.*

Proof Assume to the contrary that $na < b$ for all natural numbers n, and let T consist of the numbers $b - na$. By the well-ordering principle it possesses a least element, say $b - ma$, which is absurd since $b - (m + 1)a$ is smaller and certainly of a form qualifying for membership in T. □

Returning to the fundamental theorem of arithmetic, we have now shown (by induction and thus by the well-ordering principle) that any natural number admits a prime number decomposition. We would also like to prove that this decomposition is unique as explained. We needed then to show that whenever p_1 divides $q_1 \cdots q_n$, it must be one of the q's. This would follow if we could prove that whenever a prime number p is a factor of ab, then it must divide either of the natural numbers a and b. For suppose this is true, then p_1 either divides q_1, and we are done since the only way a prime number can divide another is that they are equal, or p_1 divides $q_2 \cdots q_n$. In this second case we can repeat the argument, if necessary all the way till p_1 either divides q_{n-1} or q_n, and then again it simply has to be one of these two.

Suppose therefore that p is a factor of ab that does not divide a. We want to show that it then has to divide b. That p does not divide a means that these two numbers are relatively prime, a notion we shall elaborate on in the following section; enough to complete the proof of the uniqueness part of the fundamental theorem of arithmetic.

1.4 Relative Primeness and Greatest Common Divisors

Definition 1.4.1 Two integers are said to be *relatively prime* if ± 1 are their only common integer factors.

So -8 and 9 are relatively prime without any of them being prime. Relative primeness is a relative notion. Although -8 and 9 are relatively prime, none of them are relatively prime to 6 as -8 and 6 have 2 as a common factor, while 9 and 6 are both divisible by -3.

Definition 1.4.2 The largest natural number that divides two integers a and b, not both zero, is called their *greatest common divisor*, and is denoted by $\gcd(a, b)$.

So a and b are relatively prime exactly when $\gcd(a, b) = 1$. Note also that $\gcd(6, -8) = 2$ and $\gcd(6, 9) = 3$, and $\gcd(45, 60) = 15$; fifteen being the largest common divisor among $-15, -5, -3, -1, 1, 3, 5, 15$.

As we shall see shortly, we would have the fundamental theorem of arithmetic under the belt if we could prove *Euclid's lemma*:

Lemma 1.4.3 *If a and b are relatively prime, and a is a factor of bc, then a must divide c.*

The standard way of proving this lemma is to use the *division algorithm*:

Theorem 1.4.4 *Let a be an integer and b a natural number. Then $a = bq + r$ for unique integers q and $0 \leq r < b$.*

1.4 Relative Primeness and Greatest Common Divisors

The idea behind this theorem is clear; if you divide the dotted line of integers in equal lengths b, you will find a in exactly one of the slots, and r will tell, namely slot number $|q| + 1$ away from the origin 0, and r will tell you where in the slot a is located. Let us anyway furnish a proof.

Proof Let T be the non-negative integers of the form $a - bx$, where x runs over the integers. Then T is not empty as $a - b(-|a|) \geq 0$, so it has a least element, say $r = a - bq$ for some integer q. Then $r < b$. Otherwise $a - b(q+1) = r - b$ is non-negative and hence of the required form to belong to T, and it is clearly also less than r which is the least element of T; a contradiction.

As for uniqueness, if q' and r' is another prescribed pair for a and b. Then $bq + r = bq' + r'$, or $b(q - q') = r - r'$, and as $|r - r'| < b$, we get $q - q' = 0$, and then in turn $r - r' = 0$. □

Before we prove Euclid's lemma, we need another result worth recording.

Proposition 1.4.5 *Let a and b be integers. Then gcd not both zero. Then $\gcd(a, b) = 1$ if and only if $1 = ax + by$ for some integers x and y.*

Proof Any positive common divisor of a and b will also divide $ax + by$, and if this latter number is 1, this divisor has also to be 1, so a and b are relatively prime.

Conversely, suppose that $\gcd(a, b) = 1$. Let d be the least number in the non-void set T consisting of all natural numbers of the form $au + bv$, where u and v vary over the integers. Write $d = ax + by$ for integers x and y. By the division algorithm there exist integers q and $0 \leq r < d$ such that $a = dq + r$. But then $r = a - dq = a(1 - xq) + b(-yq)$ is smaller than the least number of T, and for r not to belong to T, it must be zero. Hence $a = dq$, or in other words, d divides a. Similarly one shows that d divides b. Since a and b are relatively prime, this is only possible if $d = 1$. □

Euclid's lemma follows almost immediately from this proposition. Since the greatest common divisor of a and b is 1, there are integers x and y such that $ax + by = 1$. By assumption a divides cb, and hence also $cax + cby = c$.

Let us also record its corollary, which is the last brick in the proof of the fundamental theorem of arithmetic:

Corollary 1.4.6 *If a prime number p divides the product ab of two natural numbers, then p divides either a or b.*

Proof Just to be convinced, if p does not divide a, then $\gcd(a, p) = 1$ and hence by Euclid's lemma, it has to divide b. □

Another useful result, and a corollary to the proposition above, is the following:

Corollary 1.4.7 *Suppose a, b are relatively prime and that both divide c. Then ab divides c.*

Proof Write $c = an = bm$ and $ax + by = 1$. Then $c = ab(mx + ny)$. □

Adapting the proof of the above proposition slightly, we get the following result:

Proposition 1.4.8 *If a, b are integers, not both zero, there are integers x, y such that $ax + by = \gcd(a, b)$.*

Incidentally this gives the precise condition for the existence of solutions of the *linear Diophantine equation*.

Corollary 1.4.9 *Given integers a, b and c with ab non-zero, then the equation $ax + by = c$ has an integer solution (x, y) if and only if $\gcd(a, b)$ divides c, and then all the integer solutions are given by $(x + bn/\gcd(a, b), y - an/\gcd(a, b))$ for any integer n.*

Proof The first assertion is clear from the proposition. Moreover, the proposed integer candidates for solutions will evidently satisfy the equation, so we only need to show that any integer solution (x_0, y_0) is among these candidates. Certainly $a(x_0 - x) = b(y - y_0)$, and writing $a = k \gcd(a, b)$ and $b = l \gcd(a, b)$ for relatively prime integers k and l, we get $k(x_0 - x) = l(y - y_0)$. By Euclid's lemma, we get $y - y_0 = kn$ for some integer n, which in turn gives $x_0 - x = nl$. □

1.5 The Euclidean Algorithm

For large numbers a, b it is cumbersome to work out what $\gcd(a, b)$ is. It is even less obvious how to produce concrete solutions x and y to the equation $ax + by = \gcd(a, b)$. The division algorithm offers a method, known as the *Euclidean algorithm*, for solving these problems. We illustrate by an example.

Example 1.5.1 Let us find $\gcd(12378, 3054)$. Applying the division algorithm repeatedly, we get the equations:

$$12378 = 4 \cdot 3054 + 162$$
$$3054 = 18 \cdot 162 + 138$$
$$162 = 1 \cdot 138 + 24$$
$$138 = 5 \cdot 24 + 18$$
$$24 = 1 \cdot 18 + 6$$
$$18 = 3 \cdot 6 + 0.$$

The last nonzero remainder 6, in the second last equation, is then the sought for greatest common divisor. ◇

The argument for this hinges on repeated application of the following lemma:

Lemma 1.5.2 *If $a = qb + r$, then $\gcd(a, b) = \gcd(b, r)$.*

1.6 Newton's Binomial Formula

Proof Since $r = a - qb$, any common divisor of a and b will also divide r, so $\gcd(a, b)$ must divide r, and of course also b. We argue that $\gcd(a, b)$ is the largest such divisor. Suppose c is another common divisor of r and b. Then it also divides $a = r + qb$, so c cannot be greater than $\gcd(a, b)$. □

Example 1.5.3 In our previous example we get

$$\gcd(12378, 3054) = \gcd(3054, 162) = \gcd(162, 138)$$
$$= \gcd(138, 24) = \gcd(24, 18) = \gcd(18, 6) = \gcd(6, 0) = 6.$$

To represent the greatest common divisor 6 as a linear combination of the numbers 12378 and 3054, we start with the next-to-last equation in the previous example and successively eliminate the reminders:

$$\begin{aligned}
6 &= 24 - 18 \\
&= 24 - (138 - 5 \cdot 24) \\
&= 6 \cdot 24 - 138 \\
&= 6(162 - 138) - 138 \\
&= 6 \cdot 162 - 7 \cdot 138 \\
&= 6 \cdot 162 - 7(3054 - 18 \cdot 162) \\
&= 132 \cdot 162 - 7 \cdot 3054 \\
&= 132(12378 - 4 \cdot 3054) - 7 \cdot 3054 \\
&= 132 \cdot 12378 + (-535)3054.
\end{aligned}$$

The procedure in the general case should now be clear.

1.6 Newton's Binomial Formula

We are well acquainted with the formulas

$$\begin{aligned}
(a + b) &= a + b \\
(a + b)^2 &= a^2 + 2ab + b^2 \\
(a + b)^3 &= a^3 + 3a^2b + 3ab^2 + b^3 \\
(a + b)^4 &= a^4 + 4a^3b + 6a^2b^2 + 4ab^3 + b^4
\end{aligned}$$

with coefficients providing a configuration known as *Pascal's triangle*

$$1\ 1$$

$$1\ 2\ 1$$

$$1\ 3\ 3\ 1$$

$$1\ 4\ 6\ 4\ 1$$

$$\ldots$$

boarded by 1's and with the remaining numbers being sums of the two numbers nearest in the row above.

The general pattern is captured in *Newton's binomial formula*:

Theorem 1.6.1 *Given two commuting symbols x and y, we have the following expansion formula*

$$(x+y)^n = \sum_{m=0}^{n} \binom{n}{m} x^{n-m} y^m,$$

where the bracketed symbols are the binomial coefficients

$$\binom{n}{m} = \frac{n!}{m!(n-m)!}$$

with the convention $0! = 1$ *and inductive definition* $n! = n(n-1)!$ *for n-factorial.*

A proof of this expansion formula would typically go by induction, and is left to the reader. In the induction step one would perhaps want to use *Pascal's rule*:

Lemma 1.6.2
$$\binom{n}{k} + \binom{n}{k-1} = \binom{n+1}{k}$$

for $1 \leq k \leq n$.

Proof The lemma follows readily by multiplying the identity

$$\frac{1}{k} + \frac{1}{n-k+1} = \frac{n+1}{k(n-k+1)}$$

with

$$\frac{n!}{(k-1)!(n-k)!}.$$

□

Example 1.6.3 Here is a numerical example:

$$2^5 = (1+1)^5 = \sum_{m=0}^{5} \binom{5}{m} = \binom{5}{0} + \binom{5}{1} + \binom{5}{2} + \binom{5}{3} + \binom{5}{4} + \binom{5}{5}$$
$$= 1 + 5 + \frac{5 \cdot 4}{2} + \frac{5 \cdot 4}{2} + 5 + 1 = 32.$$

1.7 Infinity of Primes

Since prime numbers are the fundamental building blocks for the natural numbers under multiplication, it is tempting to contain them or conquer them in some way or other. However, this is easier said than done; prime numbers are mysterious.

One of the reasons for this is that we know of no concrete way of producing them, that is, we have no practical formula or algorithm that spits out only primes and a significant part of them, if not all.

Example 1.7.1 In the middle ages it was widely believed that the formula

$$f(n) = n^2 + n + 41$$

assumed only primes. For $n < 40$ it does, but then $f(40) = 41^2$ and $f(41) = 41 \cdot 43$. ◇

This failure is no coincidence.

Proposition 1.7.2 *No non-constant polynomial with integral coefficients can produce only primes.*

Proof Suppose f was such a polynomial. Let $p = f(1)$. Then $f(1 + np)$ will be divisible by p, hence equals p for all natural numbers n, and this is not possible for a non-constant polynomial. □

A mindbogglingly result says that the set of all primes is the positive range of the following polynomial

$$(k+2)(1 - (wz + h + j - q)^2 - ((gk + 2g + k + 1)(h + j) + h - z)^2 -$$
$$(16(k+1)^3(k+2)(n+1)^2 + 1 - f^2)^2 - (2n + p + q + z - e)^2 -$$
$$(e^3(e+2)(a+1)^2 + 1 - o^2)^2 - ((a^2 - 1)y^2 + 1 - x^2)^2 -$$
$$(16r^2y^4(a^2 - 1) + 1 - u^2)^2 - (n + l + v - y)^2 -$$
$$((a^2 - 1)l^2 + 1 - m^2)^2 - (ai + k + 1 - l - i)^2 -$$
$$(((a + u^2(u^2 - a))^2 - 1)(n + 4dy)^2 + 1 - (x + cu)^2)^2 -$$

$$(p + l(a - n - 1) + b(2an + 2a - n^2 - 2n - 2) - m)^2 -$$
$$(q + y(a - p - 1) + s(2ap + 2a - p^2 - 2p - 2) - x)^2 -$$
$$(z + pl(a - p) + t(2ap - p^2 - 1) - pm)^2)$$

as the 26 variables $a, \ldots, z$ vary over the non-negative integers.

Another way of generating primes goes as follows. Set $a_1 = 7$ and define a_n recursively by $a_{n+1} = a_n + \gcd(n + 1, a_n)$. Then the sequence $a_{n+1} - a_n$ as n runs over the natural numbers, consists only of ones and primes: $1, 1, 1, 5, 3, 1, 1, 1, 1, \ldots$.

However, the number of elementary operations needed to factorize a natural number n into primes grows exponentially with the size of n. So it is computationally challenging to decide whether a large number is a prime or not. The advent of quantum computers might improve this because as promised by a celebrated algorithm of Shore, such computers should be able to factorize numbers in polynomial time.

At any rate large primes are very difficult to single out, and for this reason large prime numbers play a role in cryptography. As of February 2013 the largest publicly known prime number is

$$2^{57885161} - 1$$

having 17425170 digits. It is a *Mersenne prime*, that is, a prime number of the form $2^p - 1$ for a prime number p, and it is easier to check whether such numbers can be factorized or not, see Sect. 1.14.

It is perhaps surprising therefore that there are infinitely many primes, and that people figured out this more than 2000 years ago by appealing to a remarkably short argument.

First a few words about the concept of infinity. That there are infinitely many elements of something, means for a mathematician that there are not finitely many. This sounds like a tautology, but philosophically there is a profound distinction. We don't try to comprehend something we cannot grasp, rather we acknowledge the incomprehensible as something that cannot be grasped.

So how do we see that something is not finite? We assume that it is finite and produce an absurdity. We have already used the well-ordering principle to a similar effect. To see that a collection is finite, try to list the members and then exhibit a member that cannot be on the list. This idea, that goes back to the Greeks, was used a century ago in Cantor's diagonal argument. From this he proved the existence of a whole hierarchy of infinitely large sets. We will return to this in Sect. 2.12.

Example 1.7.3 Let us first apply the simple idea of exhibiting a member not on a supposedly complete list to the collection of all natural numbers. Suppose there are finitely many of them. The successor of the largest element on the list cannot be on that list, so there are infinitely many natural numbers. ◇

Now to the result on primes, where we in the proof shall use the fundamental theorem of arithmetic.

Theorem 1.7.4 *There are infinitely many primes.*

1.7 Infinity of Primes

Proof Suppose there are only finitely many of them, listed as $p_1, \ldots, p_n$. Any prime factor in the decomposition of $p_1 \cdots p_n + 1$ cannot be on that list since it would have to divide 1. □

Another way of looking at it is to say that if $p_1, \ldots, p_n$ are the first n primes (in increasing order), any prime factor of $p_1 \cdots p_n + 1$ has to be larger than p_n.

For instance $2 \cdot 3 + 1 = 7$. In fact $p_1 \cdots p_n + 1$ are all prime numbers for n less than 6, but then $2 \cdot 3 \cdot 5 \cdot 7 \cdot 11 \cdot 13 + 1 = 59 \cdot 509$. Are there nevertheless infinitely many primes of the form $p_1 \cdots p_n + 1$? Like so many simple questions about prime numbers, the answer to this is unknown.

Knowing that there are infinitely many primes, how frequently do they occur? Since any prime factor of $p_1 \cdots p_n + 1$ has to be greater than p_n, surely the prime number p_{n+1} next to p_n has to be less than or equal to $p_1 \cdots p_n + 1$. This gives us the estimate $p_{n+1} \leq p_1 \cdots p_n + 1 < p_n^n + 1$. So for instance $5 \leq 2 \cdot 3 + 1 < 3^2 + 1$ and $7 < 5^3 + 1 = 126$; not a very sharp estimate.

A better estimate is the following.

Proposition 1.7.5 $p_n \leq 2^{2^{n-1}}$.

Proof Clearly this estimate holds for $n = 1$. Assuming that it holds for all natural numbers up to n, we get

$$p_{n+1} \leq p_1 \cdots p_n + 1 \leq 2 \cdots 2^{2^{n-1}} + 1 = 2^{1+\cdots+2^{n-1}} + 1$$
$$= 2^{2^n - 1} + 1 \leq 2^{2^n - 1} + 2^{2^n - 1} = 2^{2^{(n+1)-1}}$$

and the result follows by the second principle of induction. □

Primes tend to come as *twins*, i.e. prime numbers differing only by two. Examples are 3, 5 and 41, 43. Are there infinitely many twins? We don't know. A result by Viggo Brun says that the sum of their reciprocals is finite, and is conjectured to be 2, as opposed to the sum of the reciprocals of all primes which is infinite. We also have the following result by Chen:

Theorem 1.7.6 *There are infinitely many primes p, where $p + 2$ is either a prime or a product of two primes.*

For any natural number n, one of the numbers $n, n + 2, n + 4$ is divisible by 3, so the only triplet of primes we have is $(3, 5, 7)$.

One has also shown that there are infinitely many pairs of primes with some fixed interval between them that does not exceed 246, an estimate that has shrunk from some initial 70 million, and is actively sharpened.

In the other direction, there are arbitrary large gaps between primes. Here is a list of n consecutive composites:

$$(n + 1)! + 2, (n + 1)! + 3, \ldots, (n + 1)! + (n + 1).$$

We have the following startling result by Wells.

Theorem 1.7.7 *There exists a positive real number r such that $f(n) = [r^{3^n}]$ takes only primes.*

Here the square brackets picks out the greatest integer not exceeding r^{3^n}. This is by no means a concrete formula since basically nothing is known about r, so it rather tells us how poorly we understand the real numbers. We shall study real numbers in Section 2.5.

Here is another striking result, which we also do not prove.

Theorem 1.7.8 *The last digits 1, 3, 7, 9 of large primes occur with equal probability.*

1.8 Primes in Arithmetic Progression

A natural question is how many primes you will hit if you jump with fixed lengths along the dotted line of integers. You then follow an *arithmetic progression* $a, a + b, a + 2b, a + 3b, \ldots$. Dirichlet's theorem says that if a and b are relatively prime, you will hit infinitely many. Put more succinctly:

Theorem 1.8.1 *If $\gcd(a, b) = 1$, the set $\{a + bn \mid n \in \mathbb{N}\}$ contains infinitely many primes.*

We won't prove this result, and will limit ourselves to the particular and almost trivial case when $a = 3$ and $b = 4$.

Proposition 1.8.2 *There are infinitely many primes of the form $3 + 4n$.*

Proof Suppose we have only finitely many such, say $p_1, \ldots, p_m$. Consider the prime number decomposition $q_1 \cdots q_k$ of $N = 4p_1 \cdots p_m - 1$. By the division algorithm the q's are either of the form $4s + 3$ or $4s + 1$. They cannot all be of the latter form since then their product would also be of that form as $(4s + 1)(4t + 1) = 4(4st + s + t) + 1$, and $N = 4(p_1 \cdots p_m - 1) + 3$ is not of that form. Hence at least one of the q's has to be of the form $4s + 3$, and thus will belong to the list of the p's. But such a number must divide 1, which is impossible. □

Already the case $a = 1$ and $b = 4$ is much harder to prove, and will have to await more machinery.

Proposition 1.8.3 *No arithmetic progression consists solely of prime numbers.*

Proof Any number that appears in a progression will reoccur as a factor in infinitely many composites because

$$a + b(n + m(a + bn)) = (a + bn)(1 + bm)$$

for any n and m. □

Let us also quote the following result by Green-Tao.

Theorem 1.8.4 *There exist arbitrary long sequences of consecutive primes in arithmetic progression. More precisely, for any integer n and natural number m, there are integers a and b such that an + b is a prime for $n = 0, \ldots, m - 1$.*

It is clear from what we have said about recurrences, that no infinitely long such sequence exists.

Example 1.8.5 Sequences of consecutive primes in arithmetic progression are 2, 3 and 3, 5, 7 and 5, 11, 17, 23, 29. Although we have an existence result, it is very hard to come up with long sequences. One such sequence, of length 26, and found in 2010, is

$$43142746595714191 + n \cdot 23681770 \cdot 23$$

for $n = 0, 1, \ldots, 25$. ◇

It is conjectured that any sequence $\{a_n\}$ of positive integers with divergent series $\sum 1/a_n$ contains arbitrary long arithmetic progressions.

Of the various ways of unraveling the distribution of primes via addition, you find the *Goldbach conjecture*, which says that every even number larger than 2 can be written as a sum of two primes.

For example $4 = 2 + 2$, $8 = 5 + 3$ and $28 = 11 + 17 = 5 + 23$. The Goldbach conjecture, still being a famous conjecture is almost true, awaits a proof. It is, however, almost true in the following precise sense.

Theorem 1.8.6 *If $A(n)$ is the number of cases less than n that fail to split into a sum of two primes, then*

$$\lim_{n \to \infty} \frac{A(n)}{n} = 0.$$

This result, due to Vinigradov, does not, of course, prevent infinitely many exceptions for which the conjecture is false.

1.9 The Function $\pi(x)$

Let us consider the primes among the first one hundred natural numbers. To single out these, it is easiest to remove the composites. We can always write a composite as $n = ab$ with $1 < a \leq b$. So $a \leq \sqrt{n}$, which means that any composite has a prime factor less than or equal to $\sqrt{n}$.

Example 1.9.1 All composites less than 100 must therefore have prime factors not exceeding $\sqrt{100} = 10$. Hence, removing all numbers with 2, 3, 5, 7 as proper factors, we are left with the prime numbers less than 100. These are the 25 numbers:

2, 3, 5, 7, 11, 13, 17, 19, 23, 29, 31, 37, 41, 43, 47, 53, 59, 61, 67, 71, 73, 79, 83, 89, 97.

◇

In practice, in finding these numbers you typically list the first one hundred natural numbers, and then cross out every second (but keep the first one, i.e. 2), every third (but keep 3), every fifth (after 5) and then every seventh number after 7. The remaining unmarked numbers are the primes. This method of singling out primes is known as the *sieve of Eratosthenes*.

One way to understand the distribution of primes is to introduce functions of real or complex variables associated to prime numbers and study these functions. Especially important is the following one.

Definition 1.9.2 The function $\pi(x)$ counts the primes not exceeding x.

For instance, we see that $\pi(10) = 4$ as the primes not exceeding 10 are the four numbers 2, 3, 5, 7. Or $\pi(\sqrt{2}) = 0$.

The estimate $p_n \leq 2^{2^{n-1}}$ we found for the nth prime number immediately gives the lower bound $\pi(2^{2^{n-1}}) \geq n$.

With the sieve of Eratosthenes in mind Legendre extracted an exact formula for $\pi(x)$.

Theorem 1.9.3 *For any real number x*

$$\pi(x) = [x] - 1 + \pi(x^{1/2}) + \sum_{n=1}^{\pi(x^{1/2})} \sum_{m \in P_n} (-1)^n \left[\frac{x}{m}\right],$$

where P_n is the set of products of n distinct primes each not greater than $\sqrt{x}$.

Before we prove this theorem, let us illustrate how the formula works in a concrete case.

Example 1.9.4 If $x = 25$, then the primes not greater than $\sqrt{x}$ are 2, 3, 5, so

$$\pi(25) = 25 - 1 + 3 - \left[\frac{25}{2}\right] - \left[\frac{25}{3}\right] - \left[\frac{25}{5}\right] + \left[\frac{25}{2 \cdot 3}\right] + \left[\frac{25}{2 \cdot 5}\right] + \left[\frac{25}{3 \cdot 5}\right] - \left[\frac{25}{2 \cdot 3 \cdot 5}\right] = 9.$$

◇

Proof To see how to obtain this formula remember that we get the primes not exceeding x by taking all numbers $[x]$, subtract 1, which is not a prime, and then all the composites. As we know these have to be divisible by some of the primes $p_1, \ldots, p_u$

1.10 Congruence

not greater than $\sqrt{x}$. So you start deleting all numbers divisible by these, making for a total of $\sum_i \left[\frac{x}{p_i}\right]$ deletions, except that the primes themselves are not composites, so you better add $\pi(x^{1/2})$ first. However, this way you will have removed too much because those integers divisible by two distinct primes have been deleted twice, so you remove $\sum_{i \neq j} \left[\frac{x}{p_i p_j}\right]$ deletions. But now you have removed too many deletions since those integers divisible by three distinct primes have been removed twice, so you better add $\sum_{i \neq j \neq k} \left[\frac{x}{p_i p_j p_k}\right]$, and so forth, accounting for the alternating series. □

This inclusion-exclusion argument, although intuitive, is half-baked as it is not entirely clear at the end whether every element which should be removed is deleted exactly once. In Sect. 1.11 we will provide another proof.

1.10 Congruence

Gauss introduced the following powerful concept.

Definition 1.10.1 Two integers a and b are said to be congruent modulo a natural number n, in symbols $a \equiv b \pmod{n}$, if n divides $a - b$.

So $3 \equiv 24 \pmod 7$ and $25 \not\equiv 12 \pmod 7$.

It is easily checked that $a \equiv b \pmod n$ is an equivalence relation on the set of integers, and the equivalence classes are called the *congruence (or residue) classes modulo n*. They form a partition of the integers, with the odd and even integers being the congruence classes modulo 2. Using the fact that every integer is either even or odd, one quickly rules out any integer solution to for instance the Diophantine equation $x^2 + 11x - 39 = 0$. Congruence classes are in general parametrized by the remainders in the division algorithm, and there can't be more of them than the number one mods out with.

The following straightforward result shows that congruence respect multiplication and addition.

Proposition 1.10.2 *If $a \equiv b \pmod n$ and $c \equiv d \pmod n$, then $ac \equiv bd \pmod n$ and $a + c \equiv b + d \pmod n$.*

Let us demonstrate the usefulness of this result.

Example 1.10.3 To say that 41 divides $2^{20} - 1$, means $2^{20} \equiv 1 \pmod{41}$, and this is true as

$$2^{20} \equiv (2^5 \pmod{41})^4 \equiv ((-9)^2 \pmod{41})^2 \equiv (-1 \pmod{41})^2 \equiv (-1)^2 \pmod{41}.$$

Whether an integer can be divided by 9 or 11 is easily checked:

Corollary 1.10.4 *A natural number is divisible by 9 or 11 if and only if the sum, respectively, the alternating sum, of its digits is.*

Proof Let $n = a_m 10^m + \cdots + a_0$ and $k = a_0 + \cdots + a_m$ for a non-negative integer m and $a_i \in \{0, \ldots, 9\}$. Then $n \equiv k \pmod 9$ by the proposition.

Similarly we get $n \equiv l \pmod{11}$ with $l = a_0 - a_1 + \cdots + (-1)^m a_m$. □

So 8001 is divisible by 9, whereas 7214 is not.

According to Euclid's lemma, we also have cancellation, at least partly.

Proposition 1.10.5 *If $ca \equiv cb \pmod n$ for relatively prime c and n, then $a \equiv b \pmod n$.*

We can also consider equations modulo natural numbers.

Proposition 1.10.6 *The linear congruence $ax \equiv b \pmod n$ has an integer solution x if and only if $\gcd(a, n)$ divides b, and then we have $\gcd(a, n)$ pairwise incongruent integer solutions $x + mn/\gcd(a, n)$ for $m \in \{1, \ldots, \gcd(a, n)\}$.*

Proof Combine Corollary 1.4.9 with the division algorithm and the proposition above. □

If a and n in this proposition are relatively prime, then $ax \equiv b \pmod n$ has exactly one solution modulo n.

The following result is known as *Fermat's little theorem*.

Theorem 1.10.7 *If p is a prime number that does not divide an integer a, then $a^{p-1} \equiv 1 \pmod p$.*

Proof The $p - 1$ numbers $a, 2a, \ldots, (p-1)a$ are obviously pairwise incongruent, so they must be pairwise incongruent to the possible remainders $1, \ldots, p - 1$ under division by p. Hence by Proposition 1.10.2, the products of each family coincide modulo p, so $a^{p-1}(1-p)! \equiv (1-p)! \pmod p$. We can cancel $(p-1)!$ by Euclid's lemma since p cannot divide $(p-1)!$ by the fundamental theorem of arithmetic. □

We conclude that $a^p \equiv a \pmod p$ for any integer a and prime p.

Fermat's little theorem is clearly efficient in reducing large numbers in congruence calculations. It can also be used to check whether a natural number is composite. For instance, the number 117 is composite since $2^{117} \not\equiv 2 \pmod{117}$.

Here is another consequence.

Corollary 1.10.8 *If $a^p \equiv a \pmod q$ and $a^q \equiv a \pmod p$ for distinct primes p and q, then $a^{pq} \equiv a \pmod{pq}$.*

Proof By the theorem we have $(a^q)^p \equiv a^q \pmod p \equiv a \pmod p$, and similarly $(a^p)^q \equiv a \pmod q$, so p and q are factors of $a^{pq} - a$, and the result is clear from the fundamental theorem of arithmetic. □

1.10 Congruence

Wilson's theorem is proved along the same lines as Fermat's little theorem.

Theorem 1.10.9 *The identity* $(p-1)! \equiv -1 \pmod{p}$ *holds if and only if p is a prime.*

Proof If p is not a prime, it has a divisor $a \in \{2, \ldots, p-1\}$. So if the identity holds, then a will divide both p and $(p-1)!$, and thus also 1, which is a contradiction.

Conversely, if p is a prime, say larger than 3, then $ax \equiv 1 \pmod{p}$ for $a \in \{1, \ldots, p-1\}$ has a unique solution x in the same set by Proposition 1.10.6. Now $x = a$ if and only if $(a-1)(a+1)$ is divisible by p if and only if $a = 1$ or $a = p-1$. The remaining incongruent numbers $2, \ldots, p-2$ can be paired in distinct a and x with $ax \equiv 1 \pmod{p}$. Multiplying these congruences together and rearranging factors, gives
$$2 \cdots (p-2) \equiv 1 \pmod{p}$$
which upon multiplying with $p-1$ gives the desired result. □

We can use this theorem to solve a *quadratic congruence*.

Proposition 1.10.10 *Let p be an odd prime. Then $x^2 + 1 \equiv 0 \pmod{p}$ has a solution if and only if $p \equiv 1 \pmod{4}$.*

Proof If we have a solution x, then by Fermat's little theorem, we have
$$1 \equiv x^{p-1} \pmod{p} \equiv (x^2)^{(p-1)/2} \pmod{p} \equiv (-1)^{(p-1)/2} \pmod{p}.$$
This gives the absurdity $1 \pmod{p} \equiv -1 \pmod{p}$ if $p = 4n+3$ for some integer n.

Conversely, say p is of the form $p = 4n+1$. Then by Wilson's theorem we get
$$-1 \equiv 1 \cdot 2 \cdots \frac{p-1}{2} \cdot \frac{p+1}{2} \cdots (p-2)(p-1) \pmod{p}$$
$$\equiv 1 \cdot (-1) \cdot 2 \cdot (-2) \cdots \frac{p-1}{2} \cdot (-\frac{p-1}{2}) \pmod{p} \equiv (\frac{p-1}{2}!)^2 \pmod{p}$$
as there are $(p-1)/2 = 2n$ signs. □

The proof tells us that a solution of the quadratic congruence is $x = ((p-1)/2)!$.

The next crucial result, known as the *Chinese remainder theorem*, deals with a system of linear congruences.

Theorem 1.10.11 *Let $n_1, \ldots, n_m$ be pairwise relatively prime natural numbers. Then the congruences $x \equiv a_i \pmod{n_i}$ for integers a_i have a simultaneous solution x which is unique modulo $n_1 \cdots n_m$.*

Proof Let N_i be the product of the (n_j)'s with n_i omitted. By Proposition 1.10.6 there is an integer x_i such that $N_i x_i \equiv 1 \pmod{n_i}$, and then

$$x = a_1 N_1 x_1 + \cdots + a_m N_m x_m$$

is easily checked to be a solution of the desired congruences.

If x_0 is another solution of these congruences, then n_k must divide $x - x_0$, and by Corollary 1.4.7 we get $x_0 \equiv x \pmod{n_1 \cdots n_m}$. □

Note that the proof provides a formula for the solution provided some individual linear congruences are solved.

1.11 Arithmetic Functions

An *arithmetic function* is a function defined on the natural numbers. We study here a few basic ones that are particularly useful.

Let $\sum_{d|a} f(d)$ and $\prod_{d|a} f(d)$ denote the sum and product, respectively, of $f(d)$ over the positive divisors d of a natural number a.

Definition 1.11.1 An arithmetic function f is *multiplicative* if $f(ab) = f(a)f(b)$ whenever $\gcd(a,b) = 1$.

Proposition 1.11.2 *Given an arithmetic function f, and define another by $g(a) = \sum_{d|a} f(d)$. Then g is multiplicative if f is multiplicative.*

Proof Every divisor of ab is of the form cd for unique divisors c of a and d of b with $\gcd(c,d) = 1$. Hence $g(ab) = \sum_{c|a} \sum_{d|b} f(cd) = g(a)g(b)$ if f is multiplicative. □

The number of positive divisors of a natural number a is $\tau(a) = \sum_{d|a} 1$ and their sum is $\sigma(a) = \sum_{d|a} d$. By the proposition this defines multiplicative arithmetic functions τ and σ. In fact, if $a = p_1^{n_1} \cdots p_m^{n_m}$ for prime numbers p_i, then obviously

$$\tau(a) = \prod_i (n_i + 1) \quad \text{and} \quad \sigma(a) = \prod_i (1 + p_i + \cdots + p_i^{n_i}) = \prod_i \frac{p_i^{n_i+1} - 1}{p_i - 1}.$$

Here is another arithmetic function, which is also evidently multiplicative.

Definition 1.11.3 The *Möbius function* $\mu(m)$ is defined to be 1 if $m = 1$, and $(-1)^n$ if m is a product of n distinct primes, and otherwise set to be zero.

Lemma 1.11.4 *We have $\sum_{d|a} \mu(d) = \delta_{a1}$.*

Proof This is clear from the proposition above and the fact that the sum vanishes when a is a positive power of a prime by definition of μ. □

1.11 Arithmetic Functions

The following result is known as the *Möbius inversion formula*.

Proposition 1.11.5 *If f and g are arithmetic functions related by $g(a) = \sum_{d|a} f(d)$, then $f(a) = \sum_{d|a} \mu(d)g(a/d) = \sum_{d|a} \mu(a/d)g(d)$.*

Proof The last identity is obvious. To prove the first one, we calculate

$$\sum_{d|a} \mu(d)g(a/d) = \sum_{d|a} \sum_{c|(a/d)} \mu(d)f(c) = \sum_{c|a} \sum_{d|(a/c)} \mu(d)f(c)$$

as d divides a and c divides a/d if and only if c divides a and d divides a/c. At this stage the lemma gives the desired result. □

In particular, we get $1 = \sum_{d|a} \mu(a/d)\tau(d)$ and $a = \sum_{d|a} \mu(a/d)\sigma(d)$.

The converse of Proposition 1.11.2 is also true. Indeed, using once more the observation made in the first line of that proof, we get by the inversion formula above that

$$f(ab) = \sum_{c|a} \sum_{e|b} \mu(ce)g(ab/ce) = \sum_{c|a} \sum_{e|b} \mu(c)\mu(e)g(a/c)g(b/e) = f(a)f(b)$$

for relatively prime natural numbers a and b.

Recall that the greatest integer function $[x]$ of any real number x is the greatest integer not exceeding x.

Theorem 1.11.6 *Given a natural number a and a prime number p. Then the exponent of the highest power of p that divides $a!$ is $\sum_{i=1}^{\infty} [a/p^i]$.*

The sum is actually finite.

Proof The members among $1, \ldots, a$ divisible by p are $p, 2p, \ldots, [a/p]p$, and those divisible by p^2 are $p^2, 2p^2, \ldots, [a/p^2]p^2$, and so forth. □

This gives *Legendre's formula* $a! = \prod p^{\sum_{i=1}^{\infty} [a/p^i]}$, where we are taking the product over all primes $p \leq a$.

By Newton's binomial formula, we know that the binomial coefficients are natural numbers, so the product of a consecutive integers is divisible by $a!$.

Proposition 1.11.7 *If $g(a) = \sum_{d|a} f(d)$ for arithmetic functions f and g, then $\sum_{i=1}^{n} g(i) = \sum_{i=1}^{n} f(i)[n/i]$ for any natural number n. In particular, we have $\sum_{i=1}^{n} (\tau(i) - [n/i]) = 0$ and $\sum_{i=1}^{n} (\sigma(i) - i[n/i]) = 0$.*

Proof The first assertion holds since $i, 2i, \ldots, [n/i]i$ are the integers among $1, \ldots, n$ that are divisible by i. □

We introduce now the all important *Euler's phi-function*.

Definition 1.11.8 Let $\phi(n)$ denote the number of members among $1, \ldots, n$ that are relatively prime to n.

So $\pi(6) = 2$ and $\phi(p) = p - 1$ for any prime number p.

The following result is straightforward.

Lemma 1.11.9 *For any integers a, b, c we have that $\gcd(a, bc) = 1$ if and only if $\gcd(a, b) = 1 = \gcd(a, c)$.*

Theorem 1.11.10 *The arithmetic function ϕ is multiplicative.*

Proof Say a and b are natural numbers that are relatively prime. Write the integers from 1 to ab as $ia + j$ for $i \in \{0, \ldots, b - 1\}$ and $j \in \{1, \ldots, a\}$. By the lemma we know that $\phi(ab)$ equals the number of members that are relatively prime to both a and b. By Lemma 1.5.2 there are $\phi(a)$ numbers j with $\gcd(ia + j, a) = 1$ for all i. For fixed such j, there are $\phi(b)$ numbers $ia + j$ that are relatively prime to b because the b numbers $j, a + j, \cdots, (b - 1)a + j$ are pairwise incongruent modulo b, and are thus incongruent to $0, \ldots, b - 1$. □

Corollary 1.11.11 *We have*

$$\phi(p_1^{n_1} \cdots p_m^{n_m}) = p_1^{n_1} \cdots p_m^{n_m}(1 - 1/p_1) \cdots (1 - 1/p_m)$$

for pairwise distinct prime numbers p_i and non-negative integers n_i.

Proof By the theorem it suffices to show $\phi(p^n) = p^n - p^{n-1}$ for a prime number p and a natural number n. The integers between 1 and p^n divisible by p are $p, 2p, \ldots, p^{n-1}p$, so there are $p^n - p^{n-1}$ that are not divisible by p, and they are the ones that are relatively prime to p^n. □

From this corollary it is clear that $\phi(a)$ is even for a larger than 2.

We have the following result by Gauss.

Corollary 1.11.12 *For every natural number a we have $a = \sum_{d|a} \phi(d)$.*

Proof By the theorem, the right hand side is a multiplicative arithmetic function, so we need only check the identity for a a positive power of a prime number, and in this case it is straightforward. □

Proposition 1.11.13 *The sum of the integers between 1 and the natural number a that are relatively prime to a is $a\phi(a)/2$.*

Proof Let $b_1, \ldots, b_{\phi(a)}$ be the integers between 1 and a that are relatively prime to a. As $\gcd(b_i, a) = 1$ if and only if $\gcd(a - b_i, a) = 1$, we get $\sum b_i = \sum (a - b_i)$. □

Proposition 1.11.14 *For any natural number a we have $\phi(a) = a \sum_{d|a} \mu(d)/d$.*

Proof Apply the inversion formula to the formula in the last corollary above. □

Since $\phi(p) = p - 1$ for a prime number, the following result by Euler generalizes Fermat's little theorem.

1.12 Primitive Roots

Theorem 1.11.15 *If an integer a and a natural number n are relatively prime, then* $a^{\phi(n)} \equiv 1 \pmod{n}$.

Proof Say $b_1, \ldots, b_{\phi(n)}$ are the integers between 1 and n that are relatively prime to n. Then $ab_1, \ldots, ab_{\phi(n)}$ are pairwise incongruent modulo n, and by the lemma above they are relatively prime to n, so modulo n they correspond to the first sequence. Hence
$$(ab_1) \cdots (ab_{\phi(n)}) \equiv b_1 \cdots b_{\phi(n)} \pmod{n}$$
and by Proposition 1.10.5 we can cancel $b_1 \cdots b_{\phi(n)}$ on both sides. □

The former result by Legendre can now be formulated more succinctly.

Proposition 1.11.16 *Let n be the product of all primes not grater than* $x^{1/2}$. *Then*
$$\pi(x) = \pi(x^{1/2}) + \sum_{d|n} \mu(d)[x/d] - 1.$$

Proof For any real number x and natural number n, let $\phi(x, n)$ denote the number of natural numbers $a \leq x$ with $\gcd(a, n) = 1$. By Lemma 1.11.4 we see that $\sum_{d|a, d|n} \mu(d)$ is one if $\gcd(a, n) = 1$, and is otherwise zero. Hence as $\sum_{d|a, a \leq x} 1 = [x/d]$, we get
$$\phi(x, n) = \sum_{a \leq x} \sum_{d|a, d|n} \mu(d) = \sum_{d|n, d \leq x} \mu(d) \sum_{d|a, a \leq x} 1 = \sum_{d|n} \mu(d)[x/d].$$

In this formula let n be the product of all primes not greater than $x^{1/2}$, and observe that a natural number $a \leq x$ satisfies $\gcd(a, n) = 1$ if and only if $a = 1$ or a is a prime and $x^{1/2} < a \leq x$. □

1.12 Primitive Roots

Definition 1.12.1 Let a and b be relatively prime integers with $b \geq 2$. The *order of a modulo b* is the least natural number n such that $a^n \equiv 1 \pmod{b}$.

It follows easily from the division algorithm that if $a^k \equiv 1 \pmod{b}$ for an integer k, then n divides k. The converse statement is even easier. In particular, the order of a modulo b must divide $\phi(b)$ by Euler's theorem.

Also note that if a has order n modulo b, then $a^i \equiv a^j \pmod{b}$ if and only if $i \equiv j \pmod{n}$, so the n integers $a, \ldots, a^n$ are pairwise incongruent modulo b.

The following result is also straightforward.

Proposition 1.12.2 *If a has order n modulo b, then* a^k *for a natural number k has order* $n/\gcd(n, k)$ *modulo b. In particular, the numbers a and* a^k *have the same order modulo b if n and k are relatively prime.*

The condition for maximal order is important.

Definition 1.12.3 If the order of a modulo b is $\phi(b)$, we say that a is a *primitive root* of b.

Not every integer has a primitive root. Our aim in this section is to classify those that do.

Note that if a is a primitive root of b, then $a, \ldots, a^{\phi(b)}$ are congruent modulo b to the integers between 1 and b that are relatively prime to b. Since any primitive root of a must be found among these powers, and since the powers a^n with order $\phi(b)$ correspond to those n that are relatively prime to $\phi(b)$, we get the following result.

Proposition 1.12.4 *An integer b has $\phi(\phi(b))$ primitive roots provided it has any at all.*

Returning to existence, let us consider primes first, starting with some preliminary results.

Lemma 1.12.5 *Any nth degree integer coefficient equation modulo a prime number p that does not divide the coefficient of the highest power, has at most n pairwise incongruent solutions modulo p.*

Proof The case $n = 1$ is a linear congruence, which has at most one solution by Proposition 1.10.6. Assuming the lemma holds for n, we show that it holds for $n + 1$. Say $f(x) \equiv 0 \pmod{p}$ has degree $n + 1$, and say it has at least one solution a. By high-school polynomial division we may write $f(x) = (x - a)q(x) + r$, and $r \equiv 0 \pmod{p}$ as $f(a) \equiv 0 \pmod{p}$. If b was another solution incongruent to a, then $(b - a)q(b) \equiv 0 \pmod{p}$ so $q(b) \equiv 0 \pmod{p}$, and there can be only n such b's by our induction hypothesis. □

We will be more precise about polynomials and their properties in later chapters.

Corollary 1.12.6 *If n is a divisor of $p - 1$ for a prime number p, then $x^n - 1 \equiv 0 \pmod{p}$ has exactly n pairwise incongruent solutions.*

Proof Write $p - 1 = nk$ for some k. Then $x^{p-1} - 1 = (x^n - 1)f(x)$ with $f(x) = x^{n(k-1)} + x^{n(k-2)} + \cdots + x^n + 1$. By Fermat's little theorem the natural numbers less than p are pairwise incongruent solutions of $x^{p-1} - 1 \equiv 0 \pmod{p}$, and by the lemma at most $n(k - 1) = p - 1 - n$ of these solve $f(x) \equiv 0 \pmod{p}$. So there has to be at least $p - 1 - (p - 1 - n) = n$ solutions of $x^n - 1 \equiv 0 \pmod{p}$. Again by the lemma, there cannot be more than this. □

Proposition 1.12.7 *If n is a divisor of $p - 1$ for a prime number p, there are $\phi(n)$ pairwise incongruent integers with order n modulo p. In particular, there are $\phi(p - 1)$ pairwise incongruent primitive roots of p.*

1.12 Primitive Roots

Proof Let $\psi(n)$ be the number of natural numbers less than p with order n modulo p. By Fermat's little theorem every natural number less than p has order n for some divisor of $p - 1$, so $p - 1 = \sum_{n|p-1} \psi(n)$. By Corollary 1.11.12 it suffices to show that $\psi(n) \leq \phi(n)$. Since we are only proving an inequality, we can assume that $\psi(n) \geq 1$. If an integer a has order n modulo p, then by the corollary above, any integer with order n modulo p must be congruent to one of the pairwise incongruent numbers $a, \ldots, a^n$. Among these only $\phi(n)$ have order n, so $\psi(n) = \phi(n)$. □

Example 1.12.8 By the proposition above there are $\phi(6) = 2$ pairwise incongruent integers that have order 6 modulo the prime number 31. We aim to find these.

The same proposition tells us that there are $\phi(31 - 1) = 8$ primitive roots of 31. We find them by trial and error among the integers $2, \ldots, 30$. Now $2^5 \equiv 1 \pmod{31}$ rules out 2, but $3^{15} \not\equiv 1 \pmod{31}$, so 3 is a primitive root.

Thus any integer relatively prime to 31 is congruent to 3^n for some integer n between 1 and 30. By Proposition 1.12.2 the order of 3^n is $30/\gcd(30, n)$, and this is 6 if and only if $\gcd(n, 30) = 5$. Hence $n = 5$ or $n = 25$. But $3^5 \equiv 26 \pmod{31}$ and $3^{25} \equiv 6 \pmod{31}$, so 6 and 26 are the only integers having order 6 modulo 31. ◇

Next we exclude integers that do not have primitive roots.

Proposition 1.12.9 *None of the integers 2^n for $n \geq 3$ has primitive roots.*

Proof The integers that are relatively prime to 2^n are the odd numbers, and if x is odd, then $x^{\phi(2^n)/2} \equiv x^{2^{n-2}} \equiv 1 \pmod{2^n}$, where we prove the latter equality by induction: It holds for $n = 3$ as $1^2 \equiv 3^2 \equiv 5^2 \equiv 7^2 \equiv 1 \pmod{2^n}$, and assuming it holds for n, so $x^{2^{n-2}} = 1 + b2^n$ for an integer b, we get

$$x^{2^{n-1}} = (1 + b2^n)^2 = 1 + 2^{n+1}(b + b^2 2^{n-1}) \equiv 1 \pmod{2^{n+1}}.$$

□

Proposition 1.12.10 *If n and m are relatively prime integers greater than two, then mn has no primitive roots.*

Proof Since ϕ is multiplicative and is even on integers greater than two, we see that the natural number $b = \phi(m)\phi(n)/\gcd(\phi(m), \phi(n))$ is not greater than $\phi(mn)/2$. Any integer a relatively prime to mn is relatively prime to both m and n, so by Euler's theorem, we get $a^b \equiv 1 \pmod{m}$ and $a^b \equiv 1 \pmod{n}$. Hence $a^b \equiv 1 \pmod{mn}$. In other words, the order of any integer relatively prime to mn does not exceed $\phi(mn)/2$, so we have no primitive roots of mn. □

The previous two propositions limit the search for integers greater than one with primitive roots to 2, 4, p^n and $2p^n$ for odd primes p and natural numbers n. We will show that all these numbers indeed have primitive roots. Obviously 2 and 4 have the primitive roots 1 and 3, respectively. The two remaining cases are more involved.

Lemma 1.12.11 *Any odd prime p has a primitive root x with $x^{p-1} \not\equiv 1 \pmod{p^2}$.*

Proof By Proposition 1.12.7 we may pick a primitive root x of p. If it does not have the required property, replace it by the primitive root $x + p$. The binomial formula then gives

$$(x+p)^{p-1} \equiv x^{p-1} + (p-1)px^{p-2} \pmod{p^2} \equiv 1 - px^{p-2} \pmod{p^2} \not\equiv 1 \pmod{p^2}$$

as $\gcd(x, p) = 1$ so p does not divide x^{p-2}. $\square$

Lemma 1.12.12 *Let x be as in the previous lemma. Then*

$$x^{p^{n-2}(p-1)} \not\equiv 1 \pmod{p^n}$$

for every integer n greater than one.

Proof The lemma holds for $n = 2$. Assuming it holds for n, we proceed by induction. By Euler's theorem we may write

$$x^{p^{n-2}(p-1)} = x^{\phi(p^{n-1})} = 1 + bp^{n-1}$$

for an integer b not divisible by p. By the binomial formula we then get

$$x^{p^{n-1}(p-1)} = (1 + bp^{n-1})^p \equiv 1 + bp^n \pmod{p^{n+1}} \not\equiv 1 \pmod{p^{n+1}},$$

which completes the induction step. $\square$

Theorem 1.12.13 *The integers greater than one that have primitive roots are $2, 4, p^n$ and $2p^n$ for any odd prime p and integer n greater than one.*

Proof It remains to show that the last two type of numbers have primitive roots, starting with p^n.

By the lemmas there is a primitive root x of p with $x^{p^{n-2}(p-1)} \not\equiv 1 \pmod{p^n}$. We show that x will also be a primitive root of p^n. The order k of x modulo p^n must certainly divide $\phi(p^n) = p^{n-1}(p-1)$, and evidently $x^k \equiv 1 \pmod{p}$, so $p - 1 = \phi(p)$ must divide k, and we may write $k = p^m(p-1)$ for some non-negative integer $m \leq n - 1$. If m was less than $n - 1$, then $p^{n-2}(p-1)$ would be divisible by k, and we get the absurdity $x^{p^{n-2}(p-1)} \equiv 1 \pmod{p^n}$.

As for the second case $2p^n$, let y be a primitive root of p^n. Replacing y by $y + p^n$ if necessary, we may assume that y is odd, so $\gcd(y, 2p^n) = 1$. The order l of y modulo $2p^n$ must divide $\phi(2p^n) = \phi(p^n)$. On the other hand, we evidently have $y^l \equiv 1 \pmod{p^n}$, and as y is a primitive root of p^n, the number l is also divisible by $\phi(p^n)$, so $l = \phi(2p^n)$. $\square$

Definition 1.12.14 Say a and n are relatively prime integers. Then the *index of a relative to a primitive root b of n* is the least natural number $m \in \{1, \ldots, \phi(n)\}$ such that $a \equiv b^m \pmod{n}$. We denote m by $\text{ind}_b a$, or simply $\text{ind } a$ when the root is understood.

1.12 Primitive Roots

The index is the same for all elements congruent to a modulo n. Note that $\text{ind}_b \, 1 \equiv 0 \pmod{\phi(n)}$ and that $\text{ind}_b \, b \equiv 1 \pmod{\phi(n)}$ with notation as in the definition above. We also have the following result.

Proposition 1.12.15 *If b is a primitive root of n, and a and c are relatively prime to n, then $\text{ind}_b \, (ac) \equiv \text{ind}_b \, a + \text{ind}_b \, c \pmod{\phi(n)}$.*

Proof This is clear as $b^{\text{ind} \, a + \text{ind} \, c} \equiv ac \equiv b^{\text{ind}(ac)} \pmod{n}$. □

We may thus convert the congruence $ax^n \equiv b \pmod{n}$, where $\gcd(a, n) = 1$ and n has a primitive root, to the linear one $\text{ind} \, a + n \, \text{ind} \, x \equiv \text{ind} \, b \pmod{\phi(n)}$ in the unknown $\text{ind} \, x$.

Example 1.12.16 Consider the prime number 13. Modulo 13 the powers $2^1, \ldots, 2^{12}$ of 2 are the $13 - 1$ pairwise incongruent numbers

$$2, 4, 8, 3, 6, 12, 11, 9, 5, 10, 7, 1$$

listed in corresponding order. So 2 is one of the $\phi(\phi(13)) = 4$ primitive roots of 13. The remaining three are 2^n with $n \geq 2$ such that $\gcd(\phi(13), n) = 1$, so $n = 5, 7, 11$, giving 6, 11 and 7 modulo 13, respectively. From the lists of powers of 2 given above, we get for the numbers $a = 1, 2, \ldots, 12$, the corresponding values $\text{ind}_2 \, a = 12, 1, 4, 2, 9, 5, 11, 3, 8, 10, 7, 6$ for the index.

If we wish to solve $4x^9 \equiv 7 \pmod{13}$, we can work with the primitive root 2 of 13. The congruence is then equivalent to $\text{ind} \, 4 + 9 \, \text{ind} \, x \equiv \text{ind} \, 7 \pmod{12}$, or $9 \, \text{ind} \, x \equiv 9 \pmod{12}$, which in turn is equivalent to $\text{ind} \, x \equiv 1 \pmod{4}$. Thus $\text{ind} \, x = 1, 5$ or 9. From the list of indices we then get the three possible solutions $x = 2, 5$ and 6 modulo 13.

We could of course have worked with any other primitive root of 13, then taken the index relative to that root. ◇

The following criterion for solvability of congruences is sometimes useful.

Proposition 1.12.17 *Suppose n has a primitive root and is relatively prime to an integer a. Then $x^k \equiv a \pmod{n}$ has $d = \gcd(\phi(n), k)$ solutions if and only if $a^{\phi(n)/d} \equiv 1 \pmod{n}$.*

Proof The latter congruence is equivalent to $(\phi(n)/d) \, \text{ind} \, a \equiv 0 \pmod{\phi(n)}$, which holds if and only if d divides $\text{ind} \, a$. Now the former congruence is equivalent to $k \, \text{ind} \, x \equiv \text{ind} \, a \pmod{\phi(n)}$, and this has d solutions $\text{ind} \, x$ if and only if d divides $\text{ind} \, a$. These are then in one-to-one correspondence with the pairwise incongruent solutions of $x^k \equiv a \pmod{n}$. □

The case when n is a prime was first proved by Euler.

Example 1.12.18 By the proposition above the congruence $x^3 \equiv 4 \pmod{13}$ has no solution because $4^{\phi(13)/\gcd(\phi(13),3)} = 4^4 \equiv 9 \pmod{13}$.

The same proposition tells us also that $x^3 \equiv 5 \pmod{13}$ has solutions since $5^4 \equiv 1 \pmod{13}$, and these are found by the method outlined in the previous example. ◇

1.13 Quadratic Reciprocity

Here we focus on quadratic congruences of the form $ax^2 + bx + c \equiv 0 \pmod{p}$ for an odd prime p relatively prime to a. Any solutions x can be found by first solving $y^2 \equiv (b^2 - 4ac) \pmod{p}$ for y and then solving the linear congruence $2ax \equiv (y - b) \pmod{p}$ for x.

So it suffices to study congruences of the form $x^2 \equiv d \pmod{p}$, and to avoid trivialities, we assume that the integer d is relatively prime to p. Note that if x is a solution of this congruence, then so is $p - x$, which is clearly incongruent to x, and by Lemma 1.12.5, these are all solutions modulo p.

Definition 1.13.1 An integer a is a *quadratic residue (non-residue) of an odd prime p* if it is relatively prime to p and if $x^2 \equiv a \pmod{p}$ has (not) a solution x.

Example 1.13.2 The quadratic residues of 7 are 1, 2 and 4 because modulo 7 we have $1^2 \equiv 6^2 \equiv 1$ and $2^2 \equiv 5^2 \equiv 4$ and $3^2 \equiv 4^2 \equiv 2$. The quadratic non-residues are 3, 5 and 6. Note that the residues and the non-residues partition the numbers $1, \ldots, 6$ in two equally large parts. The reason for this is Proposition 1.13.6 below. ◇

If an odd prime p is relatively prime to an integer a, then by Fermat's little theorem, we have $(a^{(p-1)/2} - 1)(a^{(p-1)/2} + 1) = (a^p - 1) \equiv 0 \pmod{p}$, so either $a^{(p-1)/2} \equiv 1 \pmod{p}$ or $a^{(p-1)/2} \equiv -1 \pmod{p}$, and clearly both cannot hold at the same time.

The following result known as *Euler's criterion* shows that these two conditions distinguish quadratic residues from non-residues.

Proposition 1.13.3 *If p is an odd prime relatively prime to an integer a. Then a is a quadratic residue of p if and only if $a^{(p-1)/2} \equiv 1 \pmod{p}$.*

Proof The forward implication is clear since we can use Fermat's little theorem to any solution of the quadratic congruence.

For the opposite implication pick a primitive root b of p. Then $a \equiv b^n \pmod{p}$ for some integer $1 \leq n \leq p - 1$. The order $p - 1$ of b must divide $n(p - 1)/2$, so $n = 2m$ for some integer m, and a solution of the quadratic congruence is b^m. □

Since $2^{(13-1)/2} \equiv -1 \pmod{13}$ and $3^{(13-1)/2} \equiv 1 \pmod{13}$, we know that 2 is a quadratic non-residue of 13, whereas 3 is a quadratic residue. In fact, the two incongruent solutions of $x^2 \equiv 3 \pmod{13}$ are 4 and 9.

Definition 1.13.4 Given an integer a relatively prime to an odd prime p, then the *Legendre symbol* (a/p) is 1 if a is a quadratic residue of p, and is -1 if a is a quadratic non-residue of p.

Thus $(3/13) = 1$ and $(7/13) = -1$.

The nominator in the Legendre symbol clearly respects congruence modulo the prime number in question.

1.13 Quadratic Reciprocity

By the proposition above $(a/p) \equiv a^{(p-1)/2} \pmod{p}$. Hence, if b is relatively prime to p, then $(ab/p) = (a/p)(b/p)$ since this holds modulo p. So squares in nominators can typically be removed, and $(1/p) = 1$. Also $(-1/p) = (-1)^{(p-1)/2}$ since again both identities hold modulo p. Thus $(-1/p) = 1$ if $p \equiv 1 \pmod 4$ and $(-1/p) = -1$ if $p \equiv 3 \pmod 4$.

Example 1.13.5 The congruence $x^2 \equiv -38 \pmod{13}$ admits a solution because $(-38/13) = 1$ as

$$(-38/13) = (-1/13)(38/13) = (3 \cdot 2^2/13) = (3/13) \equiv 3^{(13-1)/2} \equiv 1 \pmod{13}.$$

Proposition 1.13.6 *We have $\sum_{a=1}^{p-1}(a/p) = 0$ for any odd prime p, so there are $(p-1)/2$ quadratic residues of p and $(p-1)/2$ non-residues.*

Proof Let b be a primitive root of p. Then for any $a \in \{1, \ldots, p-1\}$ there exists a unique member n among the same numbers such that $a \equiv b^n \pmod p$. Thus $(a/p) \equiv (-1)^n \pmod p$, so $(a/p) = (-1)^n$, which add up to zero. □

From the proof we see that the quadratic residues of p are congruent modulo p to the even powers of any primitive root b of p, whereas the quadratic non-residues are congruent to the odd powers of b.

For instance, the quadratic residues 1, 2, 4 of 7 are congruent to 3^2, 3^4, 3^6 modulo 7, whereas the odd powers 3^1, 3^3, 3^5 of the single primitive root 3 of 7 are congruent to the quadratic non-residues 3, 5, 6.

The following crucial result is known as *Gauss' lemma*.

Lemma 1.13.7 *Let a be an integer relatively prime to an odd prime p. If n is the number of members in $\{a, 2a, \ldots, ((p-1)/2)a\}$ whose remainders upon division by p exceed $p/2$, then $(a/p) = (-1)^n$.*

Proof Let $b_1, \ldots, b_n$ and $c_1, \ldots, c_m$ be those remainders upon division by p that exceed and succeed $p/2$, respectively. Then the $(p-1)/2$ integers

$$c_1, \ldots, c_m, p - b_1, \ldots, p - b_n$$

lie between 1 and $p/2$. They are also pairwise incongruent because if $c_i = p - b_j$, then as $c_i \equiv ak \pmod p$ and $b_j \equiv al \pmod p$ for integers k, l between 1 and $(p-1)/2$, we get $(k+l)a \equiv c_i + b_j = p \equiv 0 \pmod p$, so $k+l \equiv 0 \pmod p$ as $\gcd(a, p) = 1$, which is impossible. Hence $c_1, \ldots, c_m, p - b_1, \ldots, p - b_n$ is just a reordering of $1, \ldots, (p-1)/2$. Thus

$$((p-1)/2)! \equiv c_1 \cdots c_m (p - b_1) \cdots (p - b_n) \equiv (-1)^n c_1 \cdots c_m b_1 \cdots b_n$$
$$\equiv (-1)^n a \cdot pa \cdots ((p-1)/2)a \equiv (-1)^n ((p-1)/2)! a^{(p-1)/2}$$
$$\equiv (-1)^n ((p-1)/2)!(a/p) \pmod p,$$

which by cancelling $((p-1)/2)!$ gives the desired result. □

Example 1.13.8 The $(13-1)/2$ numbers $5, 10, 15, 20, 25, 30$ have reminders $5, 10, 2, 7, 12, 4$ upon division by 13, and of these three numbers are greater than $13/2$, so $(5/13) = (-1)^3 = -1$ by the lemma above. ◇

Proposition 1.13.9 *For any odd prime p, we have $(2/p) = 1$ if $p \equiv \pm 1 \pmod{8}$ and $(2/p) = -1$ if $p \equiv \pm 3 \pmod{8}$. Hence $(2/p) = (-1)^{(p^2-1)/8}$.*

Proof Let n be as in the lemma with $a = 2$. Then $n = (p-1)/2 - [p/4]$, and it is easily checked that n is even when $p \equiv \pm 1 \pmod 8$ and that n is odd when $p \equiv \pm 3 \pmod 8$. □

That 107 and 179 have 2 as a primitive root, is clear from the following result, which is another application of Gauss' lemma.

Corollary 1.13.10 *If p and $2p + 1$ are odd primes, then $2(-1)^{(p-1)/2}$ is a primitive root of $2p + 1$.*

Proof If $p \equiv 1 \pmod 4$, then the order n of $2(-1)^{(p-1)/2} = 2$ modulo $2p+1$ must divide $\phi(2p+1) = 2p$, and it cannot be p since $2p+1 \equiv \pm 3 \pmod 8$ which by the proposition, means that $-1 = (2/(2p+1)) \equiv 2^p \pmod{(2p+1)}$. And n certainly cannot be 1 or 2, so $n = 2p$.
The case $p \equiv 3 \pmod 4$ is proved similarly. □

So 179 has 2 as a primitive root, whereas 167 has -2 as a primitive root.
Here is another application of Proposition 1.13.9.

Corollary 1.13.11 *There are infinitely many primes of the form $8n - 1$.*

Proof Say we had only $p_1, \ldots, p_k$ such primes. Then there must be an odd prime divisor p of $b = (4p_1 \cdots p_k)^2 - 2$. So $(2/p) = 1$, and by the proposition above p must be of the form $8l \pm 1$. If all the odd prime divisors of b were of the form $8l + 1$, then b would be of the form $16c + 2$; an impossibility. So we may pick a p of the form $8l - 1$. Then p must divide both b and $p_1 \cdots, p_k$, which is also impossible. □

Lemma 1.13.12 *If a and p are relatively prime odd integers with p prime, then*

$$(a/p) = (-1)^{\sum_{i=1}^{(p-1)/2}[ia/p]}.$$

Proof By the division algorithm $ia = [ia/p] + r_i$ for integers r_i between 1 and $p-1$. Thus if $r_i < p/2$, then it is one of the integers $c_1, \ldots, c_m$ in the proof of Gauss' lemma, and it $r_i > p/2$, then is one of $b_1, \ldots, b_n$. Hence

$$\sum_{i=1}^{(p-1)/2} ia = \sum_{i=1}^{(p-1)/2}[ia/p] + \sum_{i=1}^{m} c_i + \sum_{i=1}^{n} b_i.$$

Using $\sum_{i=1}^{(p-1)/2} i = \sum_{i=1}^{m} c_i + \sum_{i=1}^{n}(p - b_i) = pn + \sum_{i=1}^{m} c_i - \sum_{i=1}^{n} b_i$ to eliminate $\sum_{i=1}^{m} c_i$, we get

1.13 Quadratic Reciprocity

$$n \equiv \sum_{i=1}^{(p-1)/2} [ia/p] \pmod 2,$$

and the result follows by Gauss' lemma. □

Thus $(5/13) = (-1)^{\sum_{i=1}^{6}[i5/13]} = (-1)^{0+0+1+1+1+2} = -1$.

Our main result is the following *Gauss' quadratic reciprocity law*.

Theorem 1.13.13 *We have*

$$(p/q)(q/p) = (-1)^{\frac{p-1}{2}\frac{q-1}{2}}$$

for distinct odd primes p and q.

Proof Consider the rectangle R in the xy-plane with vertices $(0, 0)$, $(p/2, 0)$, $(0, q/2)$, $(p/2, q/2)$. The number of lattice points, i.e. points (n, m) with integers $1 \le n \le (p-1)/2$ and $1 \le m \le (q-1)/2$, inside R is $(p-1)(q-1)/4$. None of these points lie on the diagonal $y = (q/p)x$.

The number of lattice points in R below this diagonal is $\sum_{i=1}^{(p-1)/2}[iq/p]$, whereas the number of those above is $\sum_{i=1}^{(q-1)/2}[ip/q]$. Hence

$$\frac{p-1}{2}\frac{q-1}{2} = \sum_{i=1}^{(p-1)/2}[iq/p] + \sum_{i=1}^{(q-1)/2}[ip/q]$$

and the result is clear from the lemma. □

Corollary 1.13.14 *For distinct primes p and q, we have $(p/q)(q/p) = 1$ and $(p/q) = (q/p)$ if $p \equiv 1 \pmod 4$ or $q \equiv 1 \pmod 4$, and $(p/q)(q/p) = -1$ and $(p/q) = -(q/p)$ if $p \equiv q \equiv 3 \pmod 4$.*

Proof This follows from the theorem since the number $(p-1)(q-1)/4$ is even in the first case and odd in the second case. □

We can now effectively calculate (a/p) for an odd prime p and an integer a relatively prime to p, by prime factorizing a, removing higher order powers of primes and using multiplicativity, till we are left with Legendre symbols of the type $(\pm 1/p)$ and $(2/p)$ and (q/p). The first two types we already know how to calculate, and to calculate the last type, we use the quadratic reciprocity law repeatedly together with modulo simplifications to reduce to the first two cases.

Example 1.13.15 We have

$$(29/53) = (53/29) = (24/29) = (2/29)(3/29) = -(29/3) = -(2/3) = 1.$$

By the corollary and the obvious facts that $(p/3) = 1$ if $p \equiv 1 \pmod{3}$ and $(p/3) = -1$ if $p \equiv 2 \pmod{3}$ for an odd prime $p \neq 3$, we get the following result.

Corollary 1.13.16 *For an odd prime $p \neq 3$, the Legendre symbol $(3/p)$ is 1 if $p \equiv \pm 1 \pmod{12}$ and it is -1 if $p \equiv \pm 5 \pmod{12}$.*

Example 1.13.17 Since $1357 = 23 \cdot 59$ the congruence $x^2 \equiv 196 \pmod{1257}$ has a solution if and only if both $x^2 \equiv 196 \pmod{23}$ and $x^2 \equiv 196 \pmod{59}$ are solvable. By the corollary $(196/23) = (12/23) = (3/23) = 1$, so the first of these last two congruences is solvable. The second one is also solvable because $(196/59) = (19/59) = -(59/19) = -(2/9) = 1$. ◇

We consider now composite moduli.

Lemma 1.13.18 *Say p is an odd prime relatively prime to an integer a. Then $x^2 \equiv a \pmod{p^n}$ has a solution if and only if $(a/p) = 1$.*

Proof Any solution of $x^2 \equiv a \pmod{p^n}$ solves $x^2 \equiv a \pmod{p}$, so $(a/p) = 1$.

In the opposite direction we proceed by induction on n, assuming $(a/p) = 1$. The case $n = 1$ holds by definition of (a/p). Assuming it holds for n, there are integers x and b such that $x^2 = a + bp^n$. We may pick an integer y such that $2xy \equiv -b \pmod{p}$. Then it is easily checked that $(x + yp^n)^2 \equiv a \pmod{p^{n+1}}$. □

Lemma 1.13.19 *If a is an odd number, then:*

1. *$x^2 \equiv a \pmod{2}$ has always a solution;*
2. *$x^2 \equiv a \pmod{4}$ has a solution if and only if $a \equiv 1 \pmod{4}$;*
3. *$x^2 \equiv a \pmod{2^n}$ for $n \geq 3$ has a solution if and only if $a \equiv 1 \pmod{8}$.*

Proof The first two enumerated statements are obvious. The forward implication of the third statement is clear since the square of an odd integer is of the form $8k + 1$.

For the opposite direction we induct on n, assuming $a \equiv 1 \pmod{8}$. The case $n = 3$ is clear. If we have solution of $x^2 \equiv a \pmod{2^n}$, there is an integer b such that $x^2 = a + b2^n$. Then pick a solution y of $xy \equiv -b \pmod{2}$, and check that $(x + y2^{n-1})^2 \equiv a \pmod{2^{n+1}}$. □

Theorem 1.13.20 *Say $2^{n_0} p_1^{n_1} \cdots p_r^{n_r}$ is the prime factorization of b and that b is relatively prime to a. Then $x^2 \equiv a \pmod{b}$ is solvable if and only if all $(a/p_i) = 1$ and $a \equiv 1 \pmod{4}$ if 4 and not 8 divide b and that finally $a \equiv 1 \pmod{8}$ whenever 8 divides b.*

Proof This is clear from the last two lemmas since $x^2 \equiv a \pmod{b}$ has a solution if and only if $x^2 \equiv a \pmod{2^{n_0}}$ and all $x^2 \equiv a \pmod{p_i^{n_i}}$ are solvable. □

1.14 Certain Classes of Numbers

Definition 1.14.1 A natural number is *perfect* if it is the sum of all its positive divisors less than itself.

Thus a natural number n is perfect if and only if $\sigma(n) = 2n$. It is not known whether there are infinitely many of them; perfect numbers are rare. The first six ones are 6, 28, 496, 8128, 33550336, 8589869056 and indeed $\sigma(6) = 1 + 2 + 3 + 6 = 2 \cdot 6$, etc.

Theorem 1.14.2 *If n is an integer greater than two with $2^n - 1$ prime, then $2^{n-1}(2^n - 1)$ is perfect, and every even perfect number is of this form.*

Proof If $2^n - 1$ is prime, then by multiplicativity of the arithmetic function σ, we have

$$\sigma(2^{n-1}(2^n - 1)) = \sigma(2^{n-1})\sigma(2^n - 1) = (2^n - 1)((2^n - 1) + 1) = 2 \cdot 2^{n-1}(2^n - 1),$$

so $2^{n-1}(2^n - 1)$ is perfect.

If m is even and perfect, then $m = 2^{n-1}k$ for an integer $n \geq 2$ and an odd integer k. Then

$$2^n k = 2m = \sigma(m) = \sigma(2^{n-1})\sigma(k) = (2^n - 1)\sigma(k),$$

so $k = a(2^n - 1)$ for an integer $a < k$ by Euclid's lemma. Since both k and a are divisors of k, we get $2^n a = \sigma(k) \geq k + a = 2^n a$, which shows that $a = 1$ and that $k = 2^n - 1$ is prime. □

Proposition 1.14.3 *If $a \geq 1$ and $n \geq 2$ are integers with $a^n - 1$ prime, then n is prime and $a = 2$.*

Proof Since

$$(a^n - 1) = (a - 1)(a^{n-1} + a^{n-2} + \cdots + a + 1)$$

and the second factor is greater than one, the first factor must be one, so $a = 2$.

If $n = mk$ for integers m and k greater than one, then

$$(a^n - 1) = ((a^m)^k - 1) = (a^m - 1)(a^{m(k-1)} + a^{m(k-2)} + \cdots + a^m + 1)$$

with both factors greater than one. □

Indeed, the first six perfect numbers correspond to $n = 2, 3, 5, 7, 13, 17$ in the theorem. Note that $2^{11} - 1 = 23 \cdot 89$, and that it is open whether or not there are infinitely many primes of the form $2^p - 1$.

Proposition 1.14.4 *Even perfect numbers end with the digits 6 or 8.*

Proof By the theorem and proposition above there is no limitation in considering an even perfect number $a = 2^{p-1}(2^p - 1)$ with $2^p - 1$ prime for a prime p. Clearly the result holds when $p = 2$.

If $p = 4n + 1$, then $a = 2 \cdot 16^{2n} - 16^n$. Utilizing the congruence $16^k \equiv 6 \pmod{10}$ which by induction holds for any natural number k, we get $a \equiv 6 \pmod{10}$.

If $p = 4n + 3$, then $a = 2 \cdot 16^{2n+1} - 4 \cdot 16^n$ and by $16^k \equiv 6 \pmod{10}$, we get $a \equiv 8 \pmod{10}$. □

In fact, a more careful study of the last case in the proof above (calculating modulo 100) shows that if the last digit of a perfect number is 8, then the last two digits are 28.

Definition 1.14.5 A *Mersenne prime* is a prime number of the form $2^n - 1$.

By Proposition 1.14.3 any Mersenne prime is of the form $2^p - 1$ for a prime p.
So $31 = 2^5 - 1$ is a Mersenne prime, and so is the enormous number $2^{31} - 1$. Here are some methods to decide when we have a Mersenne prime.

Proposition 1.14.6 *If p and $2p + 1$ are primes, then the latter either divides $2^p - 1$ or $2^p + 1$, but never both.*

Proof Fermat's little theorem gives $(2^p - 1)(2^p + 1) \equiv 0 \pmod{2p + 1}$. □

Which of these cases the prime $2p + 1$ divides is decided by the following result.

Proposition 1.14.7 *A prime number $p = 2n + 1$ divides $2^n - 1$ provided $p \equiv \pm 1 \pmod 8$. In particular, if n is a prime with $n \equiv 3 \pmod 4$, then p divides $2^n - 1$.*

Proof That p divides $2^n - 1$ means that $2^{(p-1)/2} \equiv 1 \pmod p$, in other words, that $(2/p) = 1$. Hence the result follows from Proposition 1.13.9. □

In particular, the number $2^{131} - 1$ is composite.

Proposition 1.14.8 *Divisors of $2^p - 1$ for odd primes p are of the form $2np + 1$.*

Proof If a is a divisor of $2^p - 1$, then the order m of 2 modulo a must divide p, and since $m \geq 2$, we get $m = p$. By Fermat's little theorem this order must also divide $a - 1$, so $a - 1 = kp$ for an integer k that must be even since a and p are odd. □

Proposition 1.14.9 *Any prime divisor q of $2^p - 1$ for an odd prime p satisfies $q \equiv \pm 1 \pmod 8$.*

Proof Write $q = 2n + 1$ and $a = 2^{(p+1)/2}$. Then $a^{q-1} \equiv 2^n \pmod q$, which combined with Euler's theorem $a^{q-1} \equiv 1 \pmod q$, shows that q divides $2^n - 1$. The conclusion then follows from Proposition 1.14.7. □

1.14 Certain Classes of Numbers

Example 1.14.10 Consider $a = 2^{17} - 1$. The integers of the form $34n + 1$ less than $\sqrt{a}$ are

$$35, 69, 103, 137, 171, 205, 239, 273, 307, 341.$$

Among these only $103, 137, 239, 307$ are prime. We can exclude the last one as $307 \not\equiv \pm 1 \pmod 8$. One checks that the remaining three potential prime divisors are not divisors. So a is a Mersenne prime. ◇

The greatest prime numbers found are Mersenne primes, and these produce very large even perfect numbers. It is open whether there are odd perfect numbers. Here are a couple of result in this direction of inquiry.

Theorem 1.14.11 *Every odd perfect number is of the form $p_1^{n_1} p_2^{2m_2} \cdots p_k^{2m_k}$ for distinct odd primes p_i and with $p_1 \equiv n_1 \equiv 1 \pmod 4$.*

Proof Let $p_1^{n_1} \cdots p_k^{n_k}$ be the prime number factorization of an odd perfect number a. Then

$$\sigma(p_1^{n_1}) \cdots \sigma(p_1^{n_1}) = \sigma(a) = 2a \equiv 2 \pmod 4,$$

so say $\sigma(p_1^{n_1}) \equiv 2 \pmod 4$ while the remaining $\sigma(p_i^{n_i})$ are odd.

Consider the two cases $p_i \equiv \pm 1 \pmod 4$. With the minus sign the number $\sigma(p_i^{n_i}) = 1 + p_i + \cdots + p_i^{n_i}$ is 0 modulo 4 if n_i is odd, and it is 1 modulo 4 if n_i is even. Thus $p_1 \equiv 1 \pmod 4$. It is also clear that if $p_i \equiv -1 \pmod 4$ for $i \geq 2$, then n_i must be even.

With the plus sign, we get $\sigma(p_i^{n_i}) \equiv 1 + n_i \pmod 4$ for all i. So $n_1 \equiv 1 \pmod 4$ as $\sigma(p_1^{n_1}) \equiv 2 \pmod 4$. Since $\sigma(p_i^{n_i}) \equiv \pm 1 \pmod 4$ for $i \geq 2$, the same identity shows that n_i must again be even. □

It is known that there are no odd perfect numbers below 10^{100}, but this does not rule out the possibility that there exists one.

Definition 1.14.12 A *Fermat number* is any number of the form $2^{2^n} + 1$ for non-negative n. It is called a *Fermat prime* if it is a prime number.

Proposition 1.14.13 *Any prime number of the form $2^n + 1$ is a Fermat prime.*

Proof If n is not a power of two, say $n = (2m+1)k$ for natural numbers m and k, then we arrive at the contradiction

$$2^n + 1 = (2^k + 1)(2^{2mk} - 2^{(2m-1)k} + \cdots + 2^{2k} - 2^k + 1).$$

□

The first five Fermat primes are $3, 5, 17, 257$ and 65537.

Proposition 1.14.14 *The number $2^{2^5} + 1$ is divisible by 641.*

Proof Set $a = 2^7$ and $b = 5$, so $1 + ab = 641$. Then observe that $1 + ab - b^4 = 1 + (a - b^3)b = 1 + 3b = 2^4$, so $2^{2^5} + 1 = 2^4 a^4 + 1 = (1 + ab - b^4)a^4 + 1 = (1 + ab)(a^4 + (1 - ab)(1 + a^2 b^2))$. □

Proposition 1.14.15 *Distinct Fermat numbers are relatively prime.*

Proof Let $m > n \geq 0$ be integers and set $a = 2^{2^m} + 1$ and $b = 2^{2^n} + 1$. Then

$$\frac{a-2}{b} = \frac{(2^{2^n})^{2^{m-n}} - 1}{2^{2^n} + 1} = \frac{x^k - 1}{x + 1} = x^{k-1} - x^{k-2} + \cdots - 1$$

with $x = 2^{2^n}$ and $k = 2^{m-n}$. Hence any divisor of b must also divide $a - 2$, and if it also divides a, then it must divide 2. □

Proposition 1.14.16 *The number 3 serves as a primitive root of any Fermat prime of the form $2^{2^n} + 1$ with $n \geq 2$, whereas 2 never does.*

Proof Set $a = 2^{2^n} + 1$. Since $2^{2^{n+1}} - 1 = a(2^{2^n} - 1)$ the order of 2 modulo a does not exceed 2^{n+1} and $\phi(a) = a - 1 = 2^{2^n} > 2^{n+1}$ by induction on n. So 2 cannot be a primitive root of a.

Note that a is of the form $12k + 5$ because $4^m \equiv 4 \pmod{12}$ for all natural numbers m by induction. Hence $(3/a) = -1$ by Corollary 1.13.16, and we get $3^{\phi(a)/2} \equiv -1 \pmod{a}$ by Euler's criterion, so 3 has order $\phi(a)$ modulo a. □

We will return to Fermat primes in the study of regular polygons.

1.15 Diophantine Equations

The most studied Diophantine equations are those associated with *Fermat's last theorem*; a long standing conjecture by Fermat proved by Andrew Wiles this millennium using techniques from algebraic geometry.

Theorem 1.15.1 *No triple (x, y, z) of natural numbers satisfy $x^n + y^n = z^n$ for any integer n greater than two.*

For $n = 2$ the statement is wrong.

Definition 1.15.2 A *Pythagorean triple* is a triple (x, y, z) of natural numbers that satisfy $x^2 + y^2 = z^2$. The triple is *primitive* if $\gcd(x, y, z) = 1$.

The most familiar primitive Pythagorean triple is $(4, 3, 5)$.

Lemma 1.15.3 *For a primitive Pythagorean triple (x, y, z) either x is odd and y is even, or vice verse, while z is always odd. In particular, no Pythagorean triple consists of primes only.*

1.15 Diophantine Equations

Proof If both x and y were even then 2 would be a divisor of x, y and z. If both x and y were odd, then z^2 would be 2 modulo 4, and this is impossible for the square of any integer. □

The following result is clear from the fundamental theorem of arithmetic.

Lemma 1.15.4 *If the product of two relatively prime natural numbers a and b is the nth power of a natural number, then so are a and b.*

So if $ab = c^n$, then $a = c_1^n$ and $b = c_2^n$ for natural numbers c and c_i.

Proposition 1.15.5 *All primitive Pythagorean triples (x, y, z) with x even are given by $x = 2ab$ and $y = a^2 - b^2$ and $z = a^2 + b^2$ with relative prime integers $a > b > 0$ such that $a \not\equiv b \pmod{2}$.*

Proof By the first lemma there is no restriction in assuming that x is even and that y and z are odd. Hence there are relatively prime integers m and n such that $z + y = 2m$ and $z - y = 2n$. As $(x/2)^2 = mn$, the second lemma provides relatively prime integers $a > b > 0$ with $m = a^2$ and $n = b^2$, giving the desired formulas for x, y and z. In order not to violate the first lemma, one of a and b is odd, while the other is even.

Conversely, it is easily checked that given numbers a and b subject to the conditions of the proposition, then $(2ab, a^2 - b^2, a^2 + b^2)$ is a Pythagorean triple with no prime dividing all three coordinates. □

Further primitive Pythagorean triples are $(12, 5, 13)$ and $(84, 13, 85)$. Using Fermat's little theorem it is also easy to see that 3 must divide exactly one of the first two coordinates of any Pythagorean triple.

A *Pythagorean triangle* is any right triangle with sides having integral length.

Corollary 1.15.6 *The radius of the inscribed circle of a Pythagorean triangle is always an integer.*

Proof Drawing the three lines from the vertices of the triangle to the center of the inscribed circle with radius r, one gets three triangles with areas $rx/2$, $ry/2$ and $rz/2$ adding up to $xy/2$, so $xy = r(x + y + z)$. By the proposition $x = n2ab$ and $y = n(a^2 - b^2)$ and $z = n(a^2 - b^2)$ for integers n, a, b. Plugging these into the former equation gives $r = nb(a - b)$. □

Here is another consequence of the classification of Pythagorean triples. Another important ingredient in the proof of the following result is the method of *infinite descent*.

Theorem 1.15.7 *No triple (x, y, z) of natural numbers satisfies the equation $x^4 + y^4 = z^2$.*

Proof Suppose to the contrary that we have such a solution (x, y, z). We may assume that x and y are relatively prime since $(x/d, y/d, z/d^2)$ is also a solution for any divisor d of both x and y. Then (x^2, y^2, z) is a primitive Pythagorean triple. By the previous proposition, we may write $x^2 = 2ab$ and $y^2 = a^2 - b^2$ and $z = a^2 + b^2$, where only one of the relatively prime natural numbers a and b is odd and where we have arranged so that x^2 and hence x is even. If b were odd, then $1 \equiv y^2 = a^2 - b^2 \equiv -1 \pmod 4$, so $b = 2n$ for an integer n. Applying the previous lemma to $(x/2)^2 = an$, there are natural numbers z_1 and w_1 with $a = z_1^2$ and $n = w_1^2$.

But (b, y, a) is evidently also a primitive Pythagorean triple, and by the previous proposition, there are relatively prime integers $s > t > 0$ such that $b = 2st$ and $y = s^2 - t^2$ and $a = s^2 + t^2$. Since $st = w_1^2$, then by the previous lemma we may write $s = x_1^2$ and $t = y_1^2$ for natural numbers x_1 and y_1. Observe that $x_1^4 + y_1^4 = z_1^2$ and $z_1 \le a \le a^2 < z$. Repeating the argument we can produce yet another solution (x_2, y_2, z_2) of natural numbers with $z_2 < z_1$, and we can continue this. Such an infinite descent among the natural numbers cannot happen. $\square$

This theorem proves Fermat's last theorem for $n = 4$. Factorizing general n, we therefore see that Fermat's last theorem is equivalent to the statement that no triple (x, y, z) of natural numbers satisfy $x^p + y^p = z^p$ for any odd prime p. An 'elementary' proof of this is not known to exist although extensive work by Kummer has shown the result to hold for a large class of primes.

The special case $n = 4$ says that there are no Pythagorean triangles with integer square lengths as sides. However, there are Pythagorean triangles whose lengths of sides, if increased by one, are integer squares, like $(13^2 + 1, 10^2 + 1, 14^2 + 1)$. It is not known whether there are infinitely many such triples. It is also unknown whether there are infinitely many Pythagorean triples with *triangular numbers* $n(n + 1)/2$ as coordinates, one of which is given by $n = 132, 143, 164$.

We consider yet another Diophantine equation studied by Fermat.

Proposition 1.15.8 *No triple (x, y, z) of natural numbers satisfies the equation $x^4 - y^4 = z^2$.*

Proof Say we have a solution (x, y, z) of natural numbers. By the well-ordering principle we may assume it is one with least value of x. If $x = 2n$ for an integer n, then $(n, 2y, 4z)$ is another solution, so x must be odd. Also, if $x = da$ and $y = db$ for natural numbers a, b, d, then $(a, b, d^2 z)$ is another solution, so x and y must be relatively prime.

Assume first that y is odd. By the previous proposition there are relatively prime integers $s > t > 0$ such that $z = 2st$ and $y^2 = s^2 - t^2$ and $x^2 = s^2 + t^2$. Then (s, t, xy) is another solution of natural numbers with $s < x$. So y must be even. Hence $y^2 = 2st$ and $z = s^2 - t^2$ and $x^2 = s^2 + t^2$.

Say s is even. By the last lemma, there are natural numbers u, v such that $2s = u^2$ and $t = v^2$. As u^2 is even, there is an integer w with $u = 2w$. Again by the previous proposition there are relatively prime integers $k > l > 0$ such that $2w^2 = 2kl$ and

$v^2 = k^2 - l^2$ and $x = k^2 + l^2$. The last lemma provides natural numbers i, j with $k = i^2$ and $l = j^2$. Then (i, j, v) is another solution with $i < x$. The case s odd is dealt with similarly. □

Corollary 1.15.9 *The area of a Pythagorean triangle is never an integral square.*

Proof Say a Pythagorean triple has sides with lengths $z > x \geq y$. If $xy/2 = n^2$ for an integer n, then $(x^2 - y^2)^2 = z^4 - (2n)^4$ contradicts the proposition. □

1.16 Sums of Squares

Here we will investigate to what extend a natural number can be written as the sum of integer squares, starting with sums of two squares. To this end we need a result by Axel Thue.

Lemma 1.16.1 *For an integer a relatively prime to a prime number p, the congruence $ax \equiv y \pmod{p}$ admits a non-trivial solution (x, y) with $|x|, |y|$ between zero and $[\sqrt{p}]$.*

Proof The set $\{ax - y \mid 0 \leq x, y \leq [\sqrt{p}]\}$ has cardinality greater than p, so there must be two distinct elements (x_i, y_i) with $a(x_1 - x_2) \equiv y_1 - y_2 \pmod{p}$, providing the required solution $(x_1 - x_2, y_1 - y_2)$. □

Proposition 1.16.2 *An odd prime p is the sum of two integer squares if and only if $p \equiv 1 \pmod{4}$.*

Proof No integer which is 3 modulo 4 can be the sum of two integer squares because the square of any integer is either 0 or 1 modulo 4. This proves the only if part.

Conversely, suppose $p \equiv 1 \pmod 4$. Pick an integer a such that $a^2 \equiv -1 \pmod{p}$, and then by the lemma a solution (x, y) of $ax \equiv y \pmod{p}$ with $|x|, |y|$ between zero and $[\sqrt{p}]$. Then $-x^2 \equiv (ax)^2 \equiv y^2 \pmod{p}$, so $x^2 + y^2$ is an integer multiple of p. As $0 < x^2 + y^2 < 2p$, we must therefore have $x^2 + y^2 = p$. □

It can also be shown that up to order and squares of negatives, any such representation of a prime is unique. As a numerical example we have the unique decomposition $13 = 2^2 + 3^2$.

Let us go beyond primes.

Proposition 1.16.3 *A natural number is the sum of two integer squares if and only if its prime factors that are 3 modulo 4 occur to even powers.*

Proof Any natural number can be written as $n^2 m$ with m square-free. Assume $m = p_1 \cdots p_k$ has no prime factor p_i which is 3 module 4. By the previous proposition each p_i is the sum of two integer squares. The identity

$$(a^2 + b^2)(c^2 + d^2) = (ac + bd)^2 + (ad - bc)^2$$

shows that the product of finitely many numbers that are sums of two squares is again a sum of two integer squares. Hence $n^2 m$ is the the sum of two integer squares.

Conversely, if $n^2 m = a^2 + b^2$ for integers a and b with a prime divisor p of m. Write $a = sd$ and $b = td$ for integers with s and t relatively prime. Then $s^2 + t^2 \equiv 0 \pmod{p}$ as m is square-free. Say s is relatively prime to p, so $su \equiv 1 \pmod{p}$ for some integer u. Then $1 + (tu)^2 \equiv 0 \pmod{p}$, and $p \equiv 1 \pmod 4$ since -1 is a quadratic residue of p. □

Uniqueness is lost as the example $5^2 + 0^2 = 3^2 + 4^2$ shows.

Proposition 1.16.4 *A natural number is the difference of two integer squares if and only if it is not 2 modulo 4.*

Proof For any integers a, b the number $a^2 - b^2$ will never be 2 modulo 4.

Conversely, suppose a natural number n is not 2 modulo 4. If n is 1 or 3 modulo 4, then $n = ((n+1)/2)^2 - ((n-1)/2)^2$. And if n is 0 modulo 4, then $n = (1 + n/4)^2 - (1 - n/4)^2$. □

We notice from the proof that any odd prime is the difference of two successive integer squares, and in this case one has uniqueness because $a^2 - b^2 = (a+b)(a-b)$.

Let us move to sums of three squares.

Theorem 1.16.5 *A natural number is the sum of three integer squares if and only if it is not of the form $4^n(8m+7)$.*

Proof We prove only the easy direction, namely, that no natural number of the stated form is the sum of three integer squares. The case $n = 0$ is trivial since $a^2 + b^2 + c^2$ can never be 7 modulo 8 for any integers a, b, c.

Next, if $4^n(8m+7) = a^2 + b^2 + c^2$ for $n \geq 1$, then a, b, c must all be even, so we can divide the identity by four and get one with $n - 1$ instead of n. Continuing this we finally arrive at the impossibility described in the first paragraph. □

Finally, we consider the even more liberal case of sums of four integer squares, aiming for Lagrange's theorem, which says that such a representation is always possible.

Lemma 1.16.6 *Any finite product of integers that are sums of four integer squares, is again such a sum.*

Proof This is clear form the identity

$$(a_1^2 + a_2^2 + a_3^2 + a_4^2)(b_1^2 + b_2^2 + b_3^2 + b_4^2)$$
$$= (a_1 b_1 + a_2 b_2 + a_3 b_3 + a_4 b_4)^2 + (a_1 b_2 - a_2 b_1 + a_3 b_4 - a_4 b_3)^2$$
$$+ (a_1 b_3 - a_2 b_4 - a_3 b_1 + a_4 b_2)^2 + (a_1 b_4 + a_2 b_3 - a_3 b_2 - a_4 b_1)^2.$$

□

1.16 Sums of Squares

Lemma 1.16.7 *For any odd prime p the congruence $x^2 + y^2 \equiv -1 \pmod{p}$ has a solution with $0 \le x, y < p/2$.*

Proof One sees that no two elements of $\{1 + 0^2, 1 + 1^1, \ldots, 1 + ((p-1)/2)^2\}$ are congruent, and nor is this the case for $\{-0^2, -1^2, \ldots, -((p-1)/2)^2\}$. Since there are $p + 1$ elements belonging to the union of these two sets, some element in the first set must be congruent to some element in the second set. In other words, there is $0 \le x, y < p/2$ with $1 + x^2 \equiv -y^2 \pmod{p}$. □

Corollary 1.16.8 *For any odd prime p there is an integer $n < p$ such that np is the sum of four integer squares.*

Proof By the lemma there are integers $0 \le x, y < p/2$ such that $x^2 + y^2 + 1^2 + 0^2 = np$ for some integer n with $np < p^2/4 + p^2/4 + 1 < p^2$. □

Theorem 1.16.9 *Any natural number is the sum of four integer squares.*

Proof By the lemma and corollary above together with the fundamental theorem of arithmetic, it suffices to show that for any odd prime p, the least natural number $n < p$ such that $np = x^2 + y^2 + z^2 + w^2$ for some integers x, y, z, w is actually one.

It certainly cannot be even, for by rearranging x, y, z, w, we may assume that $x \equiv y \pmod{2}$ and $z \equiv w \pmod{2}$, and then

$$(n/2)p = ((x-y)/2)^2 + ((x+y)/2)^2 + ((z-w)/2)^2 + ((z+w)/2)^2$$

violates minimality of n.

If $n \ge 3$, we can find integers a, b, c, d with absolute value less than $n/2$ such that $a \equiv x \pmod{n}$ and $b \equiv y \pmod{n}$ and $c \equiv z \pmod{n}$ and $d \equiv w \pmod{n}$. Then $a^2 + b^2 + c^2 + d^2 = kn$ for some natural number k such that $kn < n^2$. If $k = 0$, then $a = b = c = d = 0$ and n would divide p, which is impossible as $1 < n < p$. Also $k < n$.

Now $(np)(kn)$ can by the lemma be written as $r^2 + s^2 + t^2 + u^2$ and r, s, t, u are all divisible by n. Thus kp is the sum of four integer squares, and this contradicts minimality of n. □

This result can be generalized, very much so that a whole industry emerged working on *Waring's problem*: Does there exist a function $f : \mathbb{N} \to \mathbb{N}$ such that for any fixed natural number n, any natural number can be written as

$$a_1^n + \cdots + a_{f(n)}^n$$

for some integers a_i? Hilbert settled this in the affirmative. Once one function is known to exist, there must also be a least one, meaning that its value at every n is not greater than the value of any other candidate at n. Letting f be the minimal one, the two previous theorems tell us that $f(2) = 4$. Obviously $f(1) = 1$. Much less obvious are the facts that $f(3) = 9$ and $f(5) = 37$. In fact, it it believed that $f(n) = [(3/2)^n] + 2^n - 2$ for any n, and this has moreover been verified for all but finitely many n.

Another problem is to decide to what extend any nth power of a natural number is the sum of n terms of nth powers of natural numbers. For instance, we have $353^4 = 30^4 + 120^4 + 272^4 + 315^4$ and $72^5 = 19^5 + 43^5 + 46^5 + 47^5 + 67^5$, but for higher powers the situation is unclear. One can also manage with less terms, as $144^5 = 27^5 + 84^5 + 110^5 + 133^5$ shows.

1.17 Fibonacci Numbers

Studying the growth of a population of rabbits, Fibonacci came up with the following recursive sequence.

Definition 1.17.1 Let $u_1 = u_2 = 1$ and define the remaining *Fibonacci numbers* by $u_n = u_{n-1} + u_{n-2}$ for $n \geq 3$.

So the Fibonacci sequence is $1, 1, 2, 3, 5, 8, 13, \ldots$.

Proposition 1.17.2 *Successive Fibonacci numbers are relatively prime.*

Proof Any common divisor of u_{n+1} and u_n must by the recursive relation also divide u_{n-1}, till ultimately it must also divide $u_2 = 1$. □

More generally, we will see that the greatest common divisor of Fibonacci numbers is again Fibonacci. To this end we need the following result.

Lemma 1.17.3 *The identity $u_{m+n} = u_{m-1}u_n + u_m u_{n+1}$ holds for all natural numbers m, n with $m \geq 2$.*

Proof The proof goes by induction on n with m fixed; the case $n = 1$ being obvious. Assuming it holds for n and $n - 1$, it must also hold for $n + 1$ as

$$u_{m+n+1} = u_{m+n} + u_{m+n-1} = (u_{m-1}u_n + u_m u_{n+1}) + (u_{m-1}u_{n-1} + u_m u_n)$$
$$= u_{m-1}(u_n + u_{n-1}) + u_m(u_{n+1} + u_n) = u_{m-1}u_{n+1} + u_m u_{n+2}.$$

□

Corollary 1.17.4 *The number u_{mn} is divisible by u_m for all natural numbers m and n.*

Proof Fixing m, the claim certainly holds for $n = 1$, and assuming it holds for n, we see from the lemma that it also holds for $n + 1$. So the claim holds by induction. □

1.17 Fibonacci Numbers

Lemma 1.17.5 *If $m = qn + r$, then $\gcd(u_m, u_n) = \gcd(u_r, u_n)$.*

Proof By the lemma and the corollary above, we see that

$$\gcd(u_m, u_n) = \gcd(u_{qn-1}u_r, u_n) = \gcd(u_r, u_n)$$

as any common divisor of u_{qn-1} and u_n will be a common divisor of u_{qn-1} and u_{qn}, and must be one by the proposition above, and then Euclid's lemma applies. □

Theorem 1.17.6 *We have*

$$\gcd(u_m, u_n) = u_{\gcd(m,n)}$$

for all m and n.

Proof We may assume that $m \geq n$. By the Euclidean algorithm applied to finding the greatest common divisor of m and n, together with repeated application of the previous lemma, we see that $\gcd(u_m, u_n) = u_r$, where $r = \gcd(m, n)$ is the last non-zero remainder in the algorithm. □

Corollary 1.17.7 *The number u_m divides u_n if and only if m divides n.*

Proof By the previous corollary, it suffices to show the forward implication. If $u_m | u_n$, then $u_{\gcd(m,n)} = \gcd(u_m, u_n) = u_m$, so $m = \gcd(m, n)$. □

Adding the identities $u_m = u_{m+2} - u_{m+1}$ for $m = 1, \ldots, n$, and canceling terms, we get

$$u_1 + \cdots + u_n = u_{n+2} - 1.$$

Another identity is given by the following result.

Proposition 1.17.8 *The identity*

$$u_n^2 = u_{n+1}u_{n-1} + (-1)^{n-1}$$

holds for any integer $n \geq 2$.

Proof We have

$$u_n^2 - u_{n+1}u_{n-1} = u_n(u_{n-1} + u_{n+2}) - u_{n+1}u_{n-1} = (u_n - u_{n+1})u_{n-1} + u_n u_{n-2}$$
$$= (-1)(u_{n-1}^2 - u_n u_{n-2}) = \cdots = (-1)^{n-2}(u_2^2 - u_3 u_1).$$

□

Proposition 1.17.9 *Every natural number is a sum of distinct Fibonacci numbers.*

Proof We show by induction that each member of $1, 2, \ldots, u_n - 1$ for $n \geq 3$ is a sum of members from $\{u_1, u_2, \ldots, u_{n-2}\}$, none repeated. Obviously it holds for $n = 3$. Assuming it holds for n, say a is a natural number with $u_n - 1 < a < u_{n+1}$, then $a - u_{n-1}$ is less than u_n, and by hypothesis is a sum of distinct members from $\{u_1, \ldots, u_{n-2}\}$. So the induction step holds. □

Chapter 2
Construction of Numbers

In this chapter we turn to the staple diet of fractions and real and complex numbers digested in calculus courses, and regard the meal from a more fundamental point of view. Even the God given natural numbers are up for new investigation with the advent of set theory which on the threshold of the 20th-century aimed to land mathematics on a more solid footing. Peano had a decent shot at it by deducing everything we know about the natural numbers starting with a certain injective function on a certain set. We honor this attempt with a section in this book, although I am not sure whether his axiom is so much more natural than the well-ordering principle together with addition and multiplication of elements we think of as numbers.

Having gotten our hands on the natural numbers, the idea is to construct all the other numbers using set theory, and we do so in successive order; first the whole numbers, then the rational numbers, next the real numbers, and finally the complex numbers. The construction of the integers and their fractions is obtained by introducing a clever equivalence relation, first on the natural numbers, to obtain the integers as the equivalence classes, and then another equivalence relation on the integers to obtain the fractions. In the course of doing this we formalize the notions of a ring and a field, which will play an important role in this book. Their presence is reflected in the fact that the integers have a zero element and that numbers can be subtracted, while for the rational numbers, even division by a non-zero number makes sense.

The leap to the real numbers historically required much more effort, withstanding attempts for more than 2000 years. It must have stung the old Greeks that believed that everything, even music, could be described by rational numbers. Yet, their own Pythagorean theorem produced lengths with square two, and no such entity can actually be a rational number. So is there any hope in attaching some sort of number to such a length? At least there exists a sequence of rational numbers with square entries that approximate two, in that they get as close to two as one wishes if one goes far enough out in the sequence. The breakthrough was to consider so called Cauchy sequences of rational numbers, that is, sequences that behave as if they converge to something (without actually doing so), in that the members of the sequences get arbitrarily close to each other sufficiently far out in the sequence. An equivalence relation between such Cauchy sequences is introduced by declaring those having tails getting arbitrarily close to each other as equivalent. The algebraic operations from the rational numbers pass to equivalence classes by applying them term wise to the corresponding representatives. A distance can also be introduced by considering distances between the tails of the corresponding representatives. The rational numbers can be considered as those classes containing constant sequences. The essential point is that this ordered field of classes is complete in the sense that any Cauchy sequence will converge to a class consisting of a representative cleverly chosen among the representatives of the classes in the sequence. Hence numbers like the square root of two and π will belong there. We have sketched the construction of the real numbers. They are characterized as the complete ordered field containing the rational numbers as a dense ordered subfield. Completeness can also be described by a property reminiscent of well-ordering, namely that any subset bounded below has a largest lower bound, the infinum of the subset, or equivalently, the least upper bound, or supremum, of any subset bounded above. A more hands-on description of the real numbers, is to regard them as infinite decimal expansions, which we also study here.

It must have come as a shock to the community, to realize that something as sensible and rigorous as the real numbers, could not be counted, no matter how drilled, patient or ingenious one was. In fact, the real numbers cannot even be listed, that is, as an infinite list. You don't actually need to be able to count to just set up a string of elements. Imagine, you send your sheep off to the field in the morning, putting a stone in your pocket for every sheep that passes through the gate. When you take the sheep in for the night, you drop a stone on the ground for each sheep that passes back through the opening. When all the stones are gone from you pocket, you know that all the sheep have returned. You have kept track of the size of the set of sheep without counting. Similarly, you can say that two sets are of the same size, or have the same cardinality, if there exists a bijection between them. No bijection exists between the natural numbers and the real numbers, or what amounts to the same thing, the latter set cannot be listed. It was Cantor's diagonal argument that led to this disturbing fact, and it led set theorists even further astray. All of a sudden one had to handle infinities of different magnitudes. Soon sets could be spoken about that were so large that they were not even accessible. Thanks to Gödel, the mere existence of such types of sets could not be proved from the set theoretic axioms, demonstrating the limitations of any rich enough axiomatic approach. All this might seem pretty disillusioning, but our task is to push boarders to the limits of what we can say, a not so modest goal.

The last step in the construction of numbers, namely from the real to the complex ones, is surprisingly simple; they appear as ordered pairs of real numbers. There is a challenge in defining a reasonable product of such pairs, as the coordinate wise product won't do. The construction is more geometric. Consider the pairs as the endpoints of arrows in the plane which start at the origin. Two arrows can be multiplied by multiplying their lengths and adding their angles to the x-axis, obtaining this way a new arrow. The result is a (Cauchy) complete field, namely the complex numbers, which contains the real numbers as the horizontal arrows in the xy-plane. The stunning result of this simple construction, is that the complex numbers are complete in another sense, namely algebraically complete. In fact, any polynomial equation with complex coefficients always has complex solutions, and when counted with multiplicity, the number of solutions equals the degree of the polynomial. This is the fundamental theorem of algebra. It seems Gauss was so puzzled by it, that he furnished a dozen proofs, perhaps to convince himself that it was indeed true. From the point of view of equations, or dealing with limits, there is no need to construct any bigger number system. In some sense there is actually no room for further expansion.

We spice up the chapter with a couple of sections about continued fractions. These offer another way of looking at numbers, and using this, we provide a solution of Pell's equation in a separate section. We also look at p-adic numbers, a study which opens a gate to other number fields.

2.1 Peano's Axioms

Peano studied the natural numbers via the function $n \mapsto n + 1$. He realized that everything about them, including addition and multiplication and an order with required properties, could be constructed from the following data, which he regarded as axioms.

Axiom 2.1.1 A set $\mathbb{N}$ with a distinguished element 1 and an injective map $f : \mathbb{N} \to \mathbb{N}$ that does not hit 1 and has no proper invariant subsets that contain 1.

The property relating to invariant subsets echoes the induction principle; if $1 \in S$ and $f(S) \subset S$, then $S = \mathbb{N}$. Yet, there is no a priori reference to numbers in Peano's approach. Simplifying notation, we write a' for $f(a)$, and only after some effort, it will be clear that the elements $2 = 1', 3 = 2', 4 = 3', \ldots$ deserve their names.

Peano's axioms follow from the ZF-axioms in set theory. To see this requires a rather elaborate and systematic setup, which we won't enter here. From a philosophically point of view it might nevertheless be worth seeing how one imagines that such data would occur. A set theorist think about the creation (of natural numbers) like this: Say there is nothing. Then there is something, namely nothing. But then you have even more; you have nothing, and in addition the status of having nothing. But now you have nothing, and the status of having nothing, and the status of having both nothing and the status of having nothing. And the list goes on.

It looks more serious in symbols (with the empty set ϕ as nothing):

$$\{\phi\}, \{\phi, \{\phi\}\}, \{\phi, \{\phi\}, \{\phi, \{\phi\}\}\}, \ldots, A, A \cup \{A\}, \ldots$$

The set $\mathbb{N}$ consisting of these elements, with $1 = \{\phi\}$, and map $f(A) = A \cup \{A\}$, will do the job. Clearly f misses 1. To see that it is injective, first observe that for any elements A and $\tilde{A}$, either $A \subset \tilde{A}$ or $\tilde{A} \subset A$. For definiteness, say $A \subset \tilde{A}$. Now if $A \cup \{A\} = \tilde{A} \cup \{\tilde{A}\}$, then the only possibility left is $\{A\} = \{\tilde{A}\}$, which means that $A = \tilde{A}$. Finally, the only invariant subsets are those that consist of all elements to the right of some element A including A, so the one containing 1 has to be the whole set.

Let us now return to the axioms. We start by observing that f hits everything except 1.

Lemma 2.1.2 *For any $1 \neq a \in \mathbb{N}$, there is a unique $b \in \mathbb{N}$ such that $a = b'$.*

Proof The set $S = \mathbb{N}' \cup \{1\}$ is clearly an invariant subset of f that contains 1, so $S = \mathbb{N}$, and uniqueness of b is just injectivity of f. □

We define addition recursively.

Lemma 2.1.3 *There exists exactly one binary operation on $\mathbb{N}$, called addition $+$, such that (i) $a + 1 = a'$ and (ii) $a + b' = (a + b)'$ for all $a, b \in \mathbb{N}$. Moreover, this operation is associative, $(a + b) + c = a + (b + c)$, and commutative, $a + b = b + a$.*

Proof For existence, let S consist of all $a \in \mathbb{N}$ such that $a + b$ is defined for all $b \in \mathbb{N}$ and such that (i) and (ii) hold. Now $1 \in S$ because we can define $1 + b$ to be b' for all $b \in \mathbb{N}$, and then clearly $1 + 1 = 1'$ and $1 + b' = (b')' = (1 + b)'$, so (i) and (ii) hold. Next, if $a \in S$, so that $a + b$ is defined, we can define $a' + b$ to be $(a + b)'$ for all $b \in \mathbb{N}$. Then $a' \in S$ because $a' + 1 = (a + 1)' = (a')'$, so (i) holds for a', and $a' + b' = (a + b')' = ((a + b)')' = (a' + b)'$ for all $b \in \mathbb{N}$, so (ii) holds also for a'. Thus $S = \mathbb{N}$.

As for uniqueness, suppose $\oplus$ is another binary operation satisfying (i) and (ii). Fix $a \in \mathbb{N}$ and let $S = \{b \in \mathbb{N} \mid a + b = a \oplus b\}$. Then $1 \in S$ because $a + 1 = a' = a \oplus 1$. Also, if $b \in S$, then $b' \in S$ because $a + b' = (a + b)' = (a \oplus b)' = a \oplus b'$. So $S = \mathbb{N}$, and the two operations coincide.

To see that $+$ is associative, fix $a, b \in \mathbb{N}$ and verify that

$$S = \{c \in \mathbb{N} \mid (a + b) + c = a + (b + c)\}$$

is an invariant subset for f that contains 1. To check commutativity is even easier. □

Multiplication is also defined recursively.

Lemma 2.1.4 *There exists a unique binary operation on $\mathbb{N}$, called multiplication $\cdot$, such that (i) $a \cdot 1 = a$ and (ii) $a \cdot b' = a \cdot b + a$ for all $a, b \in \mathbb{N}$. Moreover, this operation is associative, $(ab)c = a(bc)$, commutative, $ab = ba$, and distributive, $a(b + c) = ab + ac$, where we have suppressed the dot.*

Proof The proof goes like that for the addition operation. To see that multiplication distributes over addition, fix $a, b \in \mathbb{N}$ and let

$$S = \{c \in \mathbb{N} \mid a(b + c) = ab + ac\}.$$

2.1 Peano's Axioms

Then $1 \in S$ because $a(b+1) = ab' = ab + a = ab + a1$. If $c \in S$, then $c' \in S$ because

$$a(b + c') = a(b + c)' = a(b + c) + a = (ab + ac) + a = ab + (ac + a) = ab + ac'.$$

Thus $S = \mathbb{N}$. □

As for the order operation, we need the following result.

Lemma 2.1.5 *Let $a, b \in \mathbb{N}$. Then exactly one of the following statements hold:*
(i) $a = b$;
(ii) $a = b + u$ for some $u \in \mathbb{N}$;
(iii) $b = a + v$ for some $v \in \mathbb{N}$.

Proof We show that (i) and (ii) cannot hold simultaneously, because if so, then $a = a + u$, which contradicts $S \equiv \{b \in \mathbb{N} \mid b \neq b + u\} = \mathbb{N}$. To see that these two sets are equal, observe first that $1 \in S$ because $1 \neq u' = 1 + u$. Secondly, observe that if $b \in S$, then $b' \in S$ because $b \neq b + u$ and by injectivity of f and commutativity of $+$, we get $b' \neq (u + b)' = u + b' = b' + u$. Therefore $S = \mathbb{N}$.

Similarly, one sees that (i) and (iii) cannot hold simultaneously, and using associativity, neither can (ii) and (iii).

It remains to see that at least one of the three statements must hold. Fix $a \in \mathbb{N}$ and let S denote the set of b's for which either (i), (ii) or (iii) hold. As usual we show that S is an invariant set under f that contains 1, so $S = \mathbb{N}$. Now $1 \in S$ because either $a = 1$, in which case (i) holds, or $a \neq 1$, and then $a = u'$ for some u by Lemma 2.1.2, so $a = 1 + u$, and (ii) holds. Next, if $b \in S$, then $b' \in S$ because if $b = a$, then $b' = a' = a + 1$, so (iii) holds for b', or if $b = a + v$, then $b' = a + v'$, so again (iii) holds for b', or finally $a = b + u$, and then one of two cases will occur: Either $u = 1$ and then $a = b'$, so (i) holds for b', or $u \neq 1$, and then $u = w'$ for some w by Lemma 2.1.2, so $a = b + w' = b + 1 + w = b' + w$ and (ii) holds for b'. In either case $b' \in S$. □

Definition 2.1.6 Let $a, b \in \mathbb{N}$. We say that a is greater than b, and write $a > b$, if $a = b + u$ for some $u \in \mathbb{N}$.

By the lemma above we see that either $a = b$, or $a > b$, or $a < b$, and that only one of these statements can hold. Clearly $<$ is transitive. So $>$ is an order on $\mathbb{N}$.

We also have the *cancellation property*: if $a + b = a + c$, then $b = c$, because the alternative $b = c + v$ for some v, means $a + c < a + c$, an impossibility, and the remaining alternative $c = b + v$, means $a + b > a + b$, another absurdity. From Lemma 2.1.2 we also see that any $a \in \mathbb{N}$ different from 1 is greater than 1.

We claim that $a < b$ if and only if $a + 1 \leq b$. To see this, note that if $a < b$, and $a + u = b$ with $u \neq 1$, then by Lemma 2.1.2, we can write $u = v'$ for some v, and then $a + 1 + v = b$, so $a + 1 < b$. The converse direction is even more obvious.

Finally, we prove the well-ordering principle.

Lemma 2.1.7 *Every non-empty subset of $\mathbb{N}$ possesses a least element.*

Proof Suppose $S \subset \mathbb{N}$ is non-empty, and let $T \subset \mathbb{N}$ consist of those elements not greater than any element of S. Then by the remarks above $1 \in T$. Also by Lemma 2.1.3, we have $a' > a$ for all $a \in \mathbb{N}$, so in particular $a' \notin T$ for $a \in S$. Hence $T \neq \mathbb{N}$, and T cannot be invariant, so there has to exist an element $c \in T$ such that $c' \notin T$. We are done if we can show that $c \in S$. Now if $c \notin S$, then since $c \in T$, it has to be smaller than all elements of S, but then by the remarks above $c' = c + 1$ will not be greater than any element of S, so $c' \in T$, which is absurd. □

At this point we regard the elements of $\mathbb{N}$ as natural numbers with $2 = 1'$, $3 = 2'$, etc.

2.2 The Integers Constructed from the Natural Numbers

Our next aim is to construct the integers from the natural numbers.

The idea behind the construction of the integers is to consider $a - b$ with $a, b \in \mathbb{N}$ as a formal pair (a, b), and avoid any reference to the minus sign. Now if $a - b$ is supposed to correspond to (a, b), then we should not distinguish between (a, b) and (c, d) whenever $a - b = c - d$. We can rewrite this equation as $a + d = b + c$, and then we have gotten rid of the minus sign, and are left with an expression that makes sense within $\mathbb{N}$ using only addition.

Definition 2.2.1 We say that (a, b) and (c, d) in $\mathbb{N} \times \mathbb{N}$ are equivalent, and write $(a, b) \sim (c, d)$, if $a + d = b + c$.

It is easy to see that our $\sim$ is an equivalence relation. For instance, we see that $(a, b) \sim (a, b)$ because $a + b = b + a$.

Definition 2.2.2 Write $\mathbb{Z}$ for the quotient set $\mathbb{N} \times \mathbb{N}/\sim$, and denote the equivalence class of $(a, b) \in \mathbb{N} \times \mathbb{N}$ by $\overline{(a, b)}$.

We proceed to show that the elements of $\mathbb{Z}$ deserve to be called integers.

Thinking about (a, b) as $a - b$ and because $(a - b) + (c - d) = (a + c) - (b + d)$ and $(a - b)(c - d) = (ac + bd) - (ad + bc)$, addition and multiplication in $\mathbb{Z}$ ought to be defined as follows:

Definition 2.2.3

$$\overline{(a, b)} + \overline{(c, d)} = \overline{(a + c, b + d)}, \quad \overline{(a, b)} \cdot \overline{(c, d)} = \overline{(ac + bd, ad + bc)}.$$

Although our guiding formulas do not make sense, since they involve a minus sign, our new definitions do, at least from the outset, since they only involve addition and multiplication already defined in $\mathbb{N}$.

2.2 The Integers Constructed from the Natural Numbers

There is however a more subtle problem that needs to be dealt with: We are adding and multiplying sets on the basis of having chosen some elements in them, and have then added and multiplied these representative elements and taken their equivalence classes. The resulting sets should not depend on these representatives, otherwise we have definitions that depend on the choices of elements picked, and we have not even specified how these choices are made, so in principle the reader has no idea of what we are defining. We will however see that the resulting sets do not depend on the representatives chosen, and when this is the case, we say that the operations are well-defined. When defining things this way, via some representatives, well-definedness should always be checked.

So how do we check that addition is well-defined? Suppose we have two other representatives, one $(\tilde{a}, \tilde{b}) \sim (a, b)$ and another $(\tilde{c}, \tilde{d}) \sim (c, d)$. Then we must show that $(\tilde{a} + \tilde{c}, \tilde{b} + \tilde{d}) \sim (a + c, b + d)$, or equivalently, that $\tilde{a} + \tilde{c} + b + d = \tilde{b} + \tilde{d} + a + c$. But this holds because by assumption, we know that $\tilde{a} + b = \tilde{b} + a$ and $\tilde{c} + d = \tilde{d} + c$. Of course, here we have skipped parentheses and swapped order since $+$ is associative and commutative. In a similar fashion one shows that multiplication in $\mathbb{Z}$ is well-defined.

With these definitions it is straightforward to check that addition and multiplication in $\mathbb{Z}$ are associative and commutative and that multiplication distributes over addition. Also $\mathbb{Z}$ has an identity $1 \equiv \overline{(2, 1)}$ for multiplication because

$$\overline{(a, b)}1 = \overline{(a2 + b1, a1 + b2)} = \overline{(a + (a + b), b + (a + b))} = \overline{(a, b)}$$

as $(a + (a + b), b + (a + b)) \sim (a, b)$.

The interesting bit is that $\mathbb{Z}$ also has an identity $\overline{(1, 1)}$ for addition, which we call zero, and denote 0, because $\overline{(a, b)} + 0 = \overline{(a + 1, b + 1)} = \overline{(a, b)}$, and what is more, we have $\overline{(a, b)} + \overline{(b, a)} = 0$, so $\overline{(b, a)}$ is to be thought of as minus that of $\overline{(a, b)}$.

Definition 2.2.4 Any set with two binary operations, say $+$ and $\cdot$ with elements 0 and 1, that is associative, commutative and distributive is called a *commutative ring with identity* 1 if each element x has an additive inverse $-x$, meaning that $x + (-x) = 0$. If the ring is not commutative for multiplication, one requires the distributive law $(r + s)t = rt + st$ on the other side to hold as well, and an identity must then be two-sided. Unless otherwise specified, a ring is not assumed to be commutative nor unital, that is, having an identity.

It is easy to see that 0 and 1 in any ring are uniquely determined by their basic properties, and that properties like $0r = 0$ and $(-r)(-r) = r^2$ automatically hold.

Returning to $\mathbb{Z}$, it is easy to see that $a \mapsto \overline{(a + 1, 1)}$ defines a unital, additive and multiplicative map $\mathbb{N} \to \mathbb{Z}$ that is injective, because if $\overline{(a + 1, 1)} = \overline{(b + 1, 1)}$, then $(a + 1, 1) \sim (b + 1, 1)$, so $a + 2 = b + 2$, and then $a = b$ by the cancellation property for $\mathbb{N}$. Using this embedding we identify the elements in $\mathbb{N}$ with their images in $\mathbb{Z}$, so we write a for $\overline{(a + 1, 1)}$ and $-a$ for its additive inverse $\overline{(1, a + 1)}$, which is consistent with the notation for the identity 1. Then $\overline{(a, b)} = \overline{(a + 1, 1)} + \overline{(1, b + 1)} = a + (-b)$, which we abbreviate as $a - b$, so any element of $\mathbb{Z}$ can be written as a difference of two elements in $\mathbb{N} \subset \mathbb{Z}$, which somehow was our starting point, but now we have constructed the integers from the natural numbers by a set theoretically legitimate procedure.

Next we would like to transfer the order from $\mathbb{N}$ to $\mathbb{Z}$. This will be done in such a way that $\mathbb{N}$ is the set of positive elements in $\mathbb{Z}$.

Definition 2.2.5 An *ordered domain* is a ring R together with a subset P which is closed under $+$ and $\cdot$ and such that R is the pairwise disjoint union of P, $\{0\}$ and $-P$. Then an order can be introduced on R by declaring that $a > b$ if $a - b \in P$.

So the elements of P are the positive elements, and those of $-P$ are the negative ones. Transitivity is true because P is closed under $+$. Clearly, also $a > b$ implies $a + c > b + c$ for any c, and if also $c > 0$, then since P is closed under $\cdot$, we get $ac > bc$.

Our discussion on integers culminates in the following result, yet to be proved.

Theorem 2.2.6 *The integers $\mathbb{Z}$ together with $\mathbb{N}$ is an ordered domain with identity 1, and $\mathbb{N}$ is well-ordered (considered with order from $\mathbb{Z}$). And any unital ordered domain whose set of positive elements is non-empty and well-ordered, is of this form.*

Proof We first show that $\mathbb{Z}$ together with $\mathbb{N} \subset \mathbb{Z}$ is an ordered domain. Clearly $\mathbb{N}$ is closed under addition and multiplication, so it suffices to prove that $-\mathbb{N} \cup \{0\} \cup \mathbb{N}$ is a partition of $\mathbb{Z}$. Suppose $\overline{(a,b)} \in \mathbb{Z}$ is non-zero, so $a \neq b$. Then by Lemma 2.1.5, either $a = b + u$ or $b = a + v$ for some $u, v \in \mathbb{N}$, and the first case happens precisely when $\overline{(a,b)} \in \mathbb{N}$, whereas the second case happens exactly when $\overline{(a,b)} \in -\mathbb{N}$. The verification of this is routine; at some point one needs the cancellation property for addition in $\mathbb{N}$. For instance, if $\overline{(a,b)} \in -\mathbb{N}$, then $(a,b) \sim (1, v+1)$ for some $v \in \mathbb{N}$, so $a + v + 1 = b + 1$ and $b = a + v$.

This also shows that the embedding $n \mapsto \overline{(n+1,1)}$ of $\mathbb{N}$ into $\mathbb{Z}$ is order preserving, because $a > b$ in $\mathbb{N}$ means that $a = b + u$ for some $u \in \mathbb{N}$, and thus

$$a - b = \overline{(a+1,1)} - \overline{(b+1,1)} = \overline{(a+1,1)} + \overline{(1,b+1)} = \overline{(a+2,b+2)} = \overline{(a,b)} \in \mathbb{N},$$

so $a > b$ also in $\mathbb{Z}$. But then $\mathbb{N} \subset \mathbb{Z}$ is well-ordered by Lemma 2.1.7.

Next assume that R is any unital ordered domain such that the set P of positive elements is well-ordered. Denote the identity in R by e, and define a map $\theta \colon \mathbb{Z} \to R$ by $\theta(n) = ne$, where ne means 0 if $n = 0$, or e added to itself $n - 1$ times if $n \in \mathbb{N}$, or $-((-n)e)$ if n is negative. It is easy to see that θ is unital and preserves addition and multiplication; a verification that strictly speaking should be checked by induction. We are done if we can show that θ is bijective and that $\theta(\mathbb{N}) = P$, because then θ preserves all relevant structure, i.e. is an isomorphism. Uniqueness (of form) is a statement about unique structure; that is all that matters, not how we name things. Since we are talking about partitions and because $\theta(-n) = -\theta(n)$ for all $n \in \mathbb{Z}$, we only need to check that θ is bijective and that $\theta(\mathbb{N}) \subset P$.

First observe that the square a^2 of any non-zero element a in R is positive, because either $a > 0$, and then $a^2 = aa > 0$, or $-a > 0$, and then $a^2 = (-a)(-a) > 0$. Since

P is non-empty, the identity e cannot be zero, for otherwise $b = be = b0 = 0$ for all $b \in R$, and $-P, \{0\}, P$ is not a partition of $R = \{0\}$. But then $e = e^2 > 0$, so $ne \in P$ for any $n \in \mathbb{N}$ because $e \in P$ and P is closed under addition. Hence $\theta(\mathbb{N}) \subset P$.

If $\theta(n) = \theta(m)$ for $n, m \in \mathbb{Z}$, and if say $n > m$, then $\theta(n - m) = 0$ and $n - m \in \mathbb{N}$, which contradicts $\theta(\mathbb{N}) \subset P$. So θ is injective.

As for surjectivity, suppose that $S \equiv R \backslash \theta(\mathbb{Z})$ is not empty. Then S has to contain at least one positive element, because if $a \in S$, then also $-a \in S$, otherwise if $-a \in \theta(\mathbb{Z})$, then also $a \in \theta(\mathbb{Z})$. Since P is well-ordered, there exists a least element a in the subset of positive elements in S. Then $a - e \neq 0$ as $e \in \theta(\mathbb{Z})$. If $a - e > 0$, then $a - e \notin \theta(\mathbb{Z})$ would be smaller than a. The only option left is that $e - a > 0$. Then $e > a = ae > a^2$, and a^2 cannot belong to $\theta(\mathbb{Z})$ as otherwise $e > a^2 = ne$ for some $n \in \mathbb{N}$ and this is impossible. So $a^2 \in S$ but this also contradicts the minimality of a. Hence θ is surjective. □

This theorem characterizes the integers and the natural numbers. We also state the following consequence.

Corollary 2.2.7 *Any unital ordered domain R whose set of positive elements P is non-empty and well-ordered, is commutative and has the property that if $rs = 0$, then either $r = 0$ or $s = 0$.*

Proof Commutativity is immediate from the theorem above.

To check the second property, by the theorem above, we may as well work with $\mathbb{Z}$. Suppose $\overline{(a, b)} \cdot \overline{(c, d)} = \overline{(1, 1)}$, so $ac + bd + 1 = ad + bc + 1$. If $\overline{(a, b)} \neq 0$, then $a \neq b$, so either $a = b + u$ or $b = a + v$. We consider only the case $a = b + u$ as the discussion for the other case is similar. Then plugging this into the previous identity and using the cancellation property in $\mathbb{N}$, we get $uc = ud$. If $c \neq d$, then either $c = d + n$ or $d = c + m$ for some $n, m \in \mathbb{N}$. In the first case $ud + un = u(d + n) = uc = ud$, so $ud > ud$, which is a contradiction. We get a similar contradiction in the second case. So $c = d$ and $\overline{(c, d)} = 0$. □

Definition 2.2.8 Any commutative unital ring such that $rs = 0$ implies either $r = 0$ or $s = 0$, is called an *integral domain*.

This means that in the equation $rs = r's$ one can cancel any non-zero s. This property will be crucial when we now form the rational numbers from the integers, since indeed $\mathbb{Z}$ is an integral domain.

2.3 From the Integers to the Rational Numbers

We think of a rational number as a quotient a/b of integers a, b where $b \neq 0$. To construct such quotients we proceed as we did when we produced the integers from the natural numbers.

Definition 2.3.1 Define a relation $\sim$ between ordered pairs of integers with non-zero second coordinates by requiring that $(a, b) \sim (c, d)$ if $ad = bc$.

Considering a/b as (a, b) this is just a rewriting of $a/b = c/d$ that makes sense within $\mathbb{Z}$.

Now $\sim$ is an equivalence relation. Reflexivity and symmetry are immediate from commutativity of $\mathbb{Z}$, and as for transitivity, if also $(c, d) \sim (c', d')$, then $(a, b) \sim (c', d')$, because $(ad')d = d'(ad) = d'(bc) = b(cd') = b(dc') = (bc')d$ and we can cancel the non-zero d.

Definition 2.3.2 Write $\mathbb{Q}$ for $(\mathbb{Z} \times (\mathbb{Z} \setminus \{0\}))/\sim$ and let $[a, b]$ be the equivalence class of the pair (a, b).

Since we think of $[a, b]$ as a/b, addition and multiplication in $\mathbb{Q}$ should be defined as follows.

Definition 2.3.3

$$[a, b] + [c, d] = [ad + bc, bd], \quad [a, b][c, d] = [ac, bd].$$

This makes sense since first of all $bd \neq 0$ as $b \neq 0$ and $d \neq 0$, and secondly because the equivalence classes on the right of $=$ do not depend on the chosen representatives in the equivalence classes on the left, i.e. the operations are well-defined. Drawing on properties from $\mathbb{Z}$ it is easy to see that $\mathbb{Q}$ is a commutative ring with zero $0 = [0, 1]$ and identity $1 = [1, 1]$, and $-[a, b] = [-a, b]$. What is more interesting is that each non-zero element $[a, b]$ has a multiplicative inverse, namely the element $[b, a]$, as $[a, b][b, a] = [ab, ba] = 1$.

Definition 2.3.4 A *field* is a commutative unital ring such that any non-zero element has a multiplicative inverse.

So $\mathbb{Q}$ is a field. Now the map $a \mapsto [a, 1]$ from $\mathbb{Z}$ to $\mathbb{Q}$ is clearly unital, additive, multiplicative and injective, and upon identifying $\mathbb{Z}$ with its image in $\mathbb{Q}$, we see that $[a, b] = [a, 1][1, b] = [a, 1][b, 1]^{-1} = a(1/b) \equiv a/b$, where we have written $1/b$ for b^{-1}. So $\mathbb{Q}$ is indeed the desired field of quotients of elements in $\mathbb{Z}$, and should be called the rational numbers. It is obviously also the smallest field containing $\mathbb{Z}$, and this characterizes $\mathbb{Q}$.

Let $\mathbb{Q}_+$ be the subset of $\mathbb{Q}$ of elements a/b where either both a and b are positive, or both are negative. Then clearly $\mathbb{Q}$ together with $\mathbb{Q}_+$ is an ordered domain, and the natural numbers will be among the positive elements $\mathbb{Q}_+$.

Definition 2.3.5 An *ordered field* is an ordered domain that is also a field.

We have proved the following result.

Theorem 2.3.6 *The rational numbers $\mathbb{Q}$ is the smallest ordered field that contains $\mathbb{Z}$.*

Clearly any field of *characteristic zero*, that is, such that $n \cdot 1 \neq 0$ for all $n \in \mathbb{N}$, has to contain $\mathbb{Q}$, which is then generated by the identity 1 of the field.

2.4 Finite Simple Continued Fractions

Here we will look at rational numbers from a different perspective.

Definition 2.4.1 A *finite simple continued fraction* is a fraction of the form

$$a_0 + \cfrac{1}{a_1 + \cfrac{1}{\cdots + \cfrac{1}{a_{n-1} + \cfrac{1}{a_n}}}},$$

where a_0 is an integer and the *partial denominators* $a_1, \ldots, a_n$ are natural numbers. We also write $[a_0; a_1, \ldots, a_n]$ for the continued fraction.

The representation $a = [a_0; a_1, \ldots, a_n]$ is not unique: If $a_n > 1$, we may also write $a = [a_0; a_1, \ldots, a_n - 1, 1]$, and if $a_n = 1$, then $a = [a_0; a_1, \ldots, a_{n-1} + 1]$. But these are the only options, so every finite simple continued fraction has a unique representation with an (even) odd number of partial denominators. For instance, we may write $2 + \frac{1}{3+\frac{1}{5}} = [2; 3, 5] = [2; 3, 4, 1]$.

Obviously, every finite simple continued fraction is a rational number. The converse is also true.

Proposition 2.4.2 *Any rational number can be written as a finite simple continued fraction.*

Proof Say we have a rational number a/b with $b > 0$. By Euclid's algorithm, there are integers n, a_i, r_i with $0 < r_1 < b$ and $0 < r_i < r_{i-1}$ such that $a = ba_0 + r_1$ and $b = r_1 a_1 + r_2$ and $r_{i-2} = r_{i-1} a_{i-1} + r_i$ for $i \in \{1, \ldots, n+1\}$ with the exception that $r_{n+1} = 0$. Then $a/b = a_0 + 1/(b/r_1)$ and $b/r_1 = a_1 + 1/(r_1/r_2)$ and $r_{i-2}/r_{i-1} = a_{i-1} + 1/(r_{i-1}/r_i)$ show that $b/a = [a_0; a_1, \ldots, a_n]$ as $r_{n-1}/r_n = a_n$. □

So we can indeed study rational numbers from the point of view of finite simple continued fractions.

Consider now the Fibonacci numbers u_n. Applying the technique in the proof above to rewrite a rational number as a continued fraction, we get $u_{n+1}/u_n = [1; 1, \ldots, 1]$ with $n + 1$ ones. So as a continued fraction the ratio of two successive Fibonacci numbers looks particularly simple.

Definition 2.4.3 The *kth convergent* C_k of $[a_0; a_1, \ldots, a_n]$ is $[a_0; a_1, \ldots, a_k]$ for $1 \leq k \leq n$. Set $C_0 = a_0$.

Note that if a_k in C_k is replaced by $a_k + 1/a_{k+1}$, then one gets C_{k+1}.

Proposition 2.4.4 *Given $[a_0; a_1, \ldots, a_n]$ and $0 \leq k \leq n$, we have $C_k = p_k/q_k$, where $p_0 = a_0, q_0 = 1$ and $p_1 = a_1 a_0 + 1, q_1 = a_1$ and $p_k = a_k p_{k-1} + p_{k-2}$ and $q_k = a_k q_{k-1} + q_{k-2}$ for $k \geq 2$.*

Proof The proposition is easily verified for $k = 0, 1, 2$. Assuming it is true for $2 \leq k < n$, we have

$$C_k = p_k/q_k = (a_k p_{k-1} + p_{k-2})/(a_k q_{k-1} + q_{k-2}).$$

Since the p's and q's in this formula do not depend on a_k, we may replace a_k by $a_k + 1/a_{k+1}$ in the formula. Hence by the remark prior to this proposition together with the recursive formulas for the p's and q's, the induction step is seen to hold. □

Proposition 2.4.5 *If $C_k = p_k/q_k$ is the kth convergent of a a simple continued fraction $[a_0; a_1, \ldots, a_n]$ with $1 \leq k \leq n$, then*

$$p_k q_{k-1} - q_k p_{k-1} = (-1)^{k-1}.$$

In particular, the integers p_k and q_k are relatively prime.

Proof The formula obviously holds for $k = 1$. Assuming it holds for $k < n$, and using the recursive formulas for p and q, we see that the formula also holds for $k + 1$.

As for the second assertion, the formula shows that any common divisor of p_k and q_k must divide $(-1)^{k-1}$. □

The second statement of this proposition shows that p_k and q_k are uniquely determined by $C_k = p_k/q_k$.

Since for Fibonacci numbers $u_{n+1}/u_n = [1; 1, \ldots, 1]$, the formula in the proposition above reproduces Proposition 1.17.8.

The formula in the proposition above may also be used to solve linear Diophantine equations. Namely, if $ax + by = 1$ for relatively prime integers a and b, then writing a/b as $[a_0; a_1, \ldots, a_n]$, we see that $a = p_n$ and $b = q_n$ as $a/b = C_n = p_n/q_n$. So

$$a q_{n-1} - b p_{n-1} = p_n q_{n-1} - q_n p_{n-1} = (-1)^{n-1},$$

which gives particular solutions $x_0 = q_{n-1}, y_0 = -p_{n-1}$ when n is odd, and $x_0 = -q_{n-1}, y_0 = p_{n-1}$ when n is even. The general solution is then $x = x_0 + bm$ and $y = y_0 - am$ for every integer m.

Example 2.4.6 We find all integers x and y such that $172x + 20y = 1000$. Division by four gives $43x + 5y = 250$, where 43 and 5 are relatively prime. We first solve the Diophantine equation $43x + 5y = 1$. We find the finite simple continued fraction of $43/5$ using Euclid's division algorithm: $43 = 8 \cdot 5 + 3, 5 = 1 \cdot 3 + 2, 3 = 1 \cdot 2 + 1$ and $2 = 2 \cdot 1$. So $43/5 = [8; 1, 1, 2]$. The convergents are therefore $C_0 = 8/1, C_1 = 9/1, C_2 = 17/2$ and $C_3 = 43/5$, so $p_2 = 17, q_2 = 2, p_3 = 43$ and $q_3 = 5$. Then

$$43 \cdot 2 - 5 \cdot 17 = p_3 q_2 - q_3 p_2 = (-1)^{3-1} = 1,$$

which when multiplied with 250 gives $43 \cdot 500 + 5(-4250) = 250$. Hence the general integer solution of $172x + 20y = 1000$ is $x = 500 + 5m$ and $y = -4250 - 43m$ for every integer m. ◇

Lemma 2.4.7 *The denominators of the convergents of a finite simple continued fraction grow with the length of the convergents, and strict inequality occurs from the second convergent onwards.*

Proof This is clear from the formula $q_{k+1} = a_{k+1}q_k + q_{k-1}$ for the denominator of the $(k+1)$th convergent C_{k+1} of a finite simple continued fraction $[a_0; a_1, \ldots, a_n]$. □

Proposition 2.4.8 *Let C_k be the kth convergent of a finite simple continued fraction. Then $C_0 < C_2 < C_4 < \cdots$ and $C_1 > C_3 > C_5 > \cdots$ and every convergent with odd subscript is greater than any one with even subscript.*

Proof The first two inequality statements are immediate from the lemma and the following identity

$$C_{k+2} - C_k = \frac{p_{k+2}}{q_{k+2}} - \frac{p_{k+1}}{q_{k+1}} + \frac{p_{k+1}}{q_{k+1}} - \frac{p_k}{q_k} = \frac{(-1)^k(q_{k+2} - q_k)}{q_k q_{k+1} q_{k+2}}.$$

As for the third statement, divide the identity $p_k q_{k-1} - q_k p_{k-1} = (-1)^{k-1}$ by $q_k q_{k-1}$ to get $C_k - C_{k-1} = (-1)^{k-1}/q_k q_{k-1}$. From this also the second inequality in

$$C_{2i} < C_{2i+2j} < C_{2i+2j-1} < C_{2j-1}$$

holds for all integers i and j for which the expression makes sense. □

2.5 Construction of the Real Numbers

The Pythagoreans knew that the diagonal d of a square of length 1 satisfies $d^2 = 2$. This puzzled them since their aim was to reduce everything to numbers, including harmonies, and rational numbers were the only numbers they knew of. And yet they were facing a perfectly reasonable length that could not be realized as a rational number.

Proposition 2.5.1 *There exists no rational number d such that $d^2 = 2$.*

Proof Say $d = a/b$ for integers a and b not both even, which we can safely assume. Then $a^2 = 2b^2$ is even, so a has to be even (otherwise a^2 would be odd), showing that b^2 and thus b is even, a contradiction. □

More generally, we have the following result.

Proposition 2.5.2 *If a natural number n is not the mth power of another, then $n^{1/m}$ will be irrational.*

Proof If we had $n^{1/m} = a/b$ for natural numbers a and b, then each prime factor in a^m will occur a multiple of m times, and this will obviously not be so for all prime factors in nb^m. □

Similar problems relate to the area and circumference of the unit disc since π also fails to be rational.

All this seems contraintuitive since between any two rational numbers there is another, e.g. their average, so they are pretty dense, suggesting that there should be few if no gaps between them. But this is very far from the truth. It took however more than 2000 years before the real numbers were constructed, and in one stroke every length, area, volume, angle, etc. could be represented by numbers. This was a dramatic event because up till that point their absence mounted to obstructions to simple answers in geometry and algebra, and where indeed considered as such by rigorous thinkers, but with the advent of the real numbers these obstructions where embodied in numbers that could be studied effectively as algebraic objects. This sort of process happens all the time in mathematics and in life; out of a thesis and an antithesis grows a synthesis.

We shall also see that there are many more real than rational numbers amounting to a certain substance that the rationals never could mobilize.

So how are the real numbers constructed? As often is the case in mathematics, the answer might well lie right in front of your nose. First of all, observe that the rational numbers approximate lengths and other geometric and physical quantities arbitrarily well.

Example 2.5.3 Rational numbers approximate any d with $d^2 = 2$ as well as we please. One way of seeing this is to note that d lies between 1 and 2, and then decide which half of this interval d belongs to. It will be the lower interval since $1^2 < 2 < (3/2)^2$. Continue this division till required accuracy is achieved. ◇

Why don't we therefore consider the approximation itself as some sort of new number? Does this make sense? And what are we asking precisely?

Pursuing this idea further, we consider new numbers as sequences of rational numbers intended to approximate something arbitrarily well. If a sequence is supposed to approximate something, the elements in the sequence should get closer to each other the further out in the sequence one goes.

Definition 2.5.4 A sequence $\{a_n\}$ of rational numbers is a *Cauchy sequence* if for any natural number k, the non-negative difference $|a_n - a_m| < 1/k$ for all n and m greater than some N.

Clearly, two (or more) Cauchy sequences might approximate the same thing, and then they ought to correspond to the same new number. Their eventual term wise distance ought then to be zero, and we don't want to distinguish between such sequences.

2.5 Construction of the Real Numbers

Definition 2.5.5 Two Cauchy sequences $\{a_n\}$ and $\{b_n\}$ are said to be equivalent if for any natural number k there exists another N such that $|a_n - b_n| < 1/k$ for every $n > N$.

This is indeed an equivalence relation as any sequence has zero eventual term wise distance to itself, and if one sequence has zero distance to another, then that other sequence has no distance to the first one, and finally it is transitive, because if a sequence has no distance to a second one, and that second one has no distance to a third, then so will there be no distance between the first and the third.

Definition 2.5.6 The *real numbers* are all the equivalence classes of Cauchy sequences of rational numbers.

Note that the definition of an equivalence class of Cauchy sequences of rational numbers involves only rational numbers, otherwise we would be cheating.

For the definition to be any good, various things need to be checked. For these equivalence classes to be called numbers, we must be able to define number-like operations on them. But this is easy.

Definition 2.5.7 We add and subtract and multiply and divide real numbers by picking representatives in the classes, do these operations term wise and then form the class of the resulting sequence.

We should convince ourselves that these operations do not depend on the particular representatives chosen. For instance, the sum $[\{a_n\}] + [\{b_n\}]$ of two classes $[\{a_n\}]$ and $[\{b_n\}]$ is by definition the class which contains the Cauchy sequence $\{a_n + b_n\}$. If we picked other representatives $\{a'_n\}$ and $\{b'_n\}$, then for any natural number k, we get

$$|(a'_n + b'_n) - (a_n + b_n)| \leq |a'_n - a_n| + |b'_n - b_n| < \frac{1}{k}$$

for all natural numbers n large enough. So $\{a'_n + b'_n\}$ will belong to the same class as $\{a_n + b_n\}$, and addition is well-defined. We therefore have a number system with an obvious identity and zero class. These operations inherit properties from the rational numbers, so we get a field.

It is also an ordered field with order defined as follows:

Definition 2.5.8 $[\{a_n\}] < [\{b_n\}]$ if $a_n < b_n$ eventually, that is, for all n greater than some N.

Well-definedness is again easy to verify.

The rational numbers correspond to those classes which contain constant sequences. This is a one-to-one correspondence which preserves order and addition and multiplication, so the real numbers, being this ordered field of classes, is an extension of the rational numbers.

Example 2.5.9 When we approximated d with $d^2 = 2$ we produced a Cauchy sequence $\{a_n\}$ of rational numbers such that eventually $|a_n^2 - 2| < 1/k$. The class of this sequence deserves the status as $\sqrt{2}$ because $[\{a_n\}]^2 - 2 = [\{a_n^2 - 2\}] = 0$. ◇

By construction we get everything that can be approximated by rationals. But what about quantities that can be approximated by these new numbers, can they also be represented by numbers of the same type, or do we need to extend further? Put differently, will Cauchy sequences of real numbers have limits as real numbers? The answer is yes; real numbers are complete in this sense. Let us first formalize what we mean.

Definition 2.5.10 A sequence of classes $[\{a(k)_n\}]$, or real numbers, converge to $[\{b_n\}]$, or has this number as a limit, if for any natural number m there is another N such that
$$|[\{a(k)_n\}] - [\{b_n\}]| < \frac{1}{m}$$
for any $k > N$, which again means that there exists a natural number M_k such that
$$|a(k)_n - b_n| < \frac{1}{m}$$
for every $k > N$ and $n > M_k$. Cauchy sequences of real numbers are defined analogously.

It is now clear that any Cauchy sequence $[\{a(k)_n\}]_k$ of classes will converge to the class of the sequence $\{a(n)_n\}$, and the latter is indeed a Cauchy sequence as
$$|a(n)_n - a(m)_m| < |a(n)_n - a(m)_n| + |a(m)_n - a(m)_m|.$$

The rational numbers are dense in the set of real numbers, because given a class of a Cauchy sequence $\{a_n\}$ of rational numbers, there is a sequence $[\{b(k)_n\}]_k$ of classes that will converge to $[\{a_n\}]$. Namely, let $b(k)_n = a_n$ for $n < k$ and $b(k)_n = a_k$ for $n \geq k$, and note that the Cauchy sequence $\{b(k)_n\}$, with k fixed, is equivalent to the constant sequence associated to the rational number a_k.

In conclusion we have constructed an ordered field $\mathbb{R}$ that is *Cauche complete*, in that every Cauchy sequence of elements in it has a limit belonging to it. Moreover, the rational numbers $\mathbb{Q}$ sits inside as a dense ordered field. In fact, these properties characterize $\mathbb{R}$, meaning that it is uniquely determined by these properties.

Theorem 2.5.11 *There exists a unique Cauchy-complete ordered field $\mathbb{R}$ that contains $\mathbb{Q}$ as a dense ordered field.*

Proof To prove uniqueness, suppose that $\tilde{\mathbb{R}}$ is another field with the same properties as $\mathbb{R}$. We want to identify $\mathbb{R}$ and $\tilde{\mathbb{R}}$ by setting up a map $a \mapsto \tilde{a}$, which should be bijective and preserve all properties. How do we define it? Well, to $a \in \mathbb{R}$ we know there exists a sequence $\{a_n\}$ of rational numbers such that $a = \lim a_n$, and being necessarily Cauchy, this sequence also has to have a limit $\tilde{a}$ in $\tilde{\mathbb{R}}$. Clearly, any other sequence of rational numbers with limit a, will produce the same limit $\tilde{a}$, so we have a well-defined map $a \mapsto \tilde{a}$ from $\mathbb{R}$ to $\tilde{\mathbb{R}}$. Using this argument once more, we can define a map in the opposite direction $\tilde{\mathbb{R}} \to \mathbb{R}$, and it is easy to see that this map is

2.6 The Least Upper Bound Property   63

the inverse of the previous one. So $a \mapsto \tilde{a}$ is bijective, and clearly it preserves order and addition and multiplication since these properties are governed by those in $\mathbb{Q}$. □

The members of $\mathbb{R}$ are the real numbers, and we don't need to think of them any more as classes of Cauchy sequences of rational numbers, which admittedly is rather cumbersome. How they are constructed is not important, it is their properties that matter, and we have singled out the essential ones, those that characterize them.

Our construction is due to Cantor. There are other constructions or realizations of the real numbers. Dedekin, another pioneer, focused on the order structure and regarded real numbers as certain subsets of $\mathbb{Q}$ called cuts. We won't repeat his approach here.

2.6 The Least Upper Bound Property

The Cauchy-completeness property of the real numbers can be replaced by a property reminiscent of the well-ordering principle for natural numbers, namely, the *least upper bound property*:

Proposition 2.6.1 *For any non-empty subset S of $\mathbb{R}$ with an upper bound, there exists a least number among the numbers greater than or equal to every element of S. Such a least upper bound is called the supremum of S, and denoted by $\sup S$. Equivalently one can talk about the greatest lower bound, or infinum, with symbol $\inf S$, of a non-empty subset S of $\mathbb{R}$ that is bounded below. Furthermore, Theorem 2.5.11 is valid with Cauchy-completeness replaced by this least upper (or greatest lower) bound property.*

To get from one property to another, note that any Cauchy sequence $\{a_n\}$ is a non-empty subset of $\mathbb{R}$ that is bounded above and below, and converges to

$$\limsup a_n \equiv \inf\nolimits_m \sup\{a_n\}_{n=m}^\infty = \sup\nolimits_m \inf\{a_n\}_{n=m}^\infty \equiv \liminf a_n$$

provided the least upper (and hence the greatest lower) bound property holds. Conversely, if a subset S of $\mathbb{R}$ has some member s and an upper bound $u \in \mathbb{R}$. Then adapt the dividing-in-half-procedure that we performed for $\sqrt{2}$. If the average $(s+u)/2$ is still an upper bound, repeat the argument with $(s+u)/2$ instead of u, or otherwise with s replaced by $(s+u)/2$. The limit of the Cauchy sequence of upper bounds obtained this way will obviously be a least upper bound for S.

Repeating the argument we used for natural numbers, with the well-ordering principle replaced by the least upper bound property, we see that the Archimedean property holds also for real numbers, so any real number will be smaller than any positive real number multiplied by a large enough natural number.

2.7 Decimal Expansions

It is customary to consider real numbers as decimal expansions, a realization we can now get to quickly and which in view of our discussion, also seems very natural. The decimal expression $364, 78$ is the number $3 \cdot 10^2 + 6 \cdot 10^1 + 4 \cdot 10^0 + 7 \cdot 10^{-1} + 8 \cdot 10^{-2}$, which is of course a rational number, as would any finite expansion be with or without the convention of adding a tail of infinitely many zeros.

Definition 2.7.1 A decimal expansion

$$n_{-r} \cdots n_{-1} n_0, n_1 n_2 \cdots$$

with digits $n_k \in \{0, 1, \ldots, 9\}$ and non-negative integer r, means the real number that is the limit of the Cauchy sequence $\{a_m\}$ of rational numbers

$$a_m = \sum_{k=-r}^{m} n_k \cdot 10^{-k}.$$

Proposition 2.7.2 *Any real number admits a decimal expansion.*

Proof The Archimedean property and the type of arguments used to establish this property, will do. Indeed, given any positive real number a pick the greatest integer r such that $10^r \leq a$, and then the largest non-negative integer n_{-r} such that $n_{-r} \cdot 10^r \leq a$. Having chosen $n_{-r}, \ldots, n_{m-1}$, let n_m be the greatest non-negative integer such that

$$a_m \equiv \sum_{k=-r}^{m} n_k \cdot 10^{-k} \leq a.$$

Then the sequence $\{a_m\}$ clearly converges to a. If a was negative, find the expansion for $-a$ and put a minus sign in front. □

Proposition 2.7.3 *Let $a \in \mathbb{R}$. Then*

$$\sum_{n=0}^{m} a^n = \frac{1 - a^{m+1}}{1 - a}.$$

Hence the geometric series $\sum_{n=0}^{\infty} a^n$ converges if and only if $|a| < 1$, and it then converges to

$$\sum_{n=0}^{\infty} a^n = \frac{1}{1 - a}.$$

Proof Let $S \equiv \sum_{n=0}^{m} a^n$. Then $(1 - a)S = S - aS = 1 - a^{m+1}$ upon telescoping terms, and we are done. □

2.7 Decimal Expansions

Infinite expansions can still be rational numbers.

Example 2.7.4 Consider the number $1/3$, which has the expansion $0, 333\cdots$. To recover $1/3$ from the expansion, we single out a geometric series:

$$0, 333\cdots = 3(1/10 + 1/100 + \cdots) = 3 \cdot (1/10)/(1 - 1/10) = 1/3.$$

◇

The rational numbers are distinguished by expansions that turn periodic, that is, with digits ultimately repeated in blocks.

Example 2.7.5 The rational number $9/11 = 0.818181\cdots$ has period 2, while $3227/555 = 5, 8144144144\cdots$ has period 3 with repeated block 144. Note that an expansion that terminates, like $21, 7 = 21.700\cdots = 21.699\cdots$, has period 1. ◇

An expedient way to recover the fraction from a periodic expansion goes as follows:

Example 2.7.6 Consider $x = 5, 8144144\cdots$. Then

$$(10000 - 10)x = 10000x - 10x = 58144, 144\cdots - 58, 144\cdots = 58086.$$

◇

So all periodic expansions are rational numbers. This means that expansions of irrational numbers will never turn periodic.

Example 2.7.7 Some people take pride in memorizing digits of $\sqrt{2}$ and π, but will ever only approximate such numbers by rational numbers, doomed to cut off the expansions at best further to the right than their competitors.

In fact, any pattern in the digits of π would come as a surprise as it is conjectured to be normal to base 10, meaning that any block of digits occur with expected frequency, so 58 is expected to occur a hundredth of the time, and 467 every thousand.

It is also unknown whether every digit occur infinitely many times in the expansion of π.

The expansion of a fraction can be obtained by long division:

Example 2.7.8 For $x = 5/74$ we get $x = 0, 06\cdots$ because $10 \cdot 10 \cdot 5/74 = 6 + 56/74$. Next $x = 0, 067\cdots$ because $10 \cdot 56/74 = 7 + 42/74$, and in the next step $x = 0, 0675\cdots$ as $10 \cdot 42/74 = 5 + 50/74$. At this stage we start on a new period because $10 \cdot 50/74 = 6 + 56/74$, just as we had before. So we get $x = 0, 0675675675\cdots$.

The remainders in the division are $56, 42, 50, 56, 42, 50, \ldots$. Clearly, all it takes for the expansion of a fraction to become periodic is that some remainder will be repeated (in this case 56 is the first such), but the possible remainders are limited by the denominator of the fraction (in this case to $0, 1, \ldots, 73$), so a periodic expansion is ultimately unavoidable. ◇

We have just seen that expansions of fractions will always turn periodic. In conclusion we have shown the following result.

Proposition 2.7.9 *The rational numbers are precisely the real numbers that admit periodic expansions.*

There are various results on the relation between fractions and the periodicity in their expansions.

Proposition 2.7.10 *If p is a prime number that is not a divisor of* 10, *the repeating block in the extension of* $1/p$ *starts just after the comma. Furthermore, the length of the block equals the order of* 10 *modulo p, that is, the smallest natural number n such that* $10^n \equiv 1 \pmod{p}$.

By Fermat's little theorem $10^{p-1} \equiv 1 \pmod{p}$, the number n has to be a factor of $p - 1$. If the order is maximal, then the first $p - 1$ digits after the comma in the expansion of any integer multiple of $1/p$ are rotated.

Example 2.7.11 The fraction $1/7 = 0, 142857 \cdots$ has period 6, and the first 6 digits after the comma in $3 \cdot 1/7 = 0, 428571 \cdots$ are obtained by rotating those of $1/7$, and so are those of $2 \cdot 1/7 = 0, 285714 \cdots$, etc. ◇

Similar types of results hold for expansions of reciprocals of composites with prime factors relatively prime to 10.

In our discussion on expansions we could have replaced the base 10 by any integer larger than one, producing different realizations of the real numbers. We happened to pick 10 because we have ten fingers, but 2 is a better choice for computers with their on-and-off modus, allowing to handle the real numbers as binary numbers with digits 0 and 1. For example, the number 73, 5 (in the decimal expansion) reads 1001001, 1 in the binary expansion which by definition equals $2^6 + 2^3 + 2^0 + 2^{-1}$.

2.8 Infinite Continued Fractions

The generalization of finite continued fractions to infinite ones is clear.

Definition 2.8.1 Given a sequence a_n of integers with $a_n \geq 1$ for $n \geq 1$, the *infinite simple continued fraction* $[a_0; a_1, a_2, \ldots]$ is the real number $\lim_{n \to \infty} C_n$, where C_n is the nth convergent $[a_0; a_1, \ldots, a_n]$.

This definition makes sense because by Proposition 2.4.8, the even-numbered convergents form a monotone increasing sequence with supremum b, while the odd-numbered convergents form a monotone decreasing sequence with infinum b', and the proof of the proposition shows that

$$|b' - b| = b' - b \leq C_{2n} - C_{2n-1} \leq 1/q_{2n}q_{2n-1},$$

2.8 Infinite Continued Fractions

so $b' = b$, which we suggestively have denoted by $[a_0; a_1, a_2, \ldots]$.

The infinite simple continued fraction $x = [1; 1, 1, \ldots]$ associated to the Fibonacci numbers u_n satisfies

$$x = \lim_n \frac{u_{n+1}}{u_n} = \lim_n \frac{u_n + u_{n-1}}{u_n} = \lim_n (1 + \frac{1}{u_n/u_{n-1}}) = 1 + \frac{1}{x},$$

or $x^2 - x - 1 = 0$, which has as positive solution the golden ratio $x = (1 + \sqrt{5})/2$.

Definition 2.8.2 If the integers in an infinite simple continued fraction $x = [a_0; a_1, a_2 \ldots]$ from a certain point n onwards repeat themselves in blocks $b_1, \ldots, b_m$ of length m, we write $[a_0; a_1, \ldots, a_n, \overline{b_1, \ldots b_m}]$ for x. In this case the shortest such block is called the period of x, and x is said to be periodic.

Example 2.8.3 To determine $x = [3; 6, \overline{1, 4}]$ write $x = [3; 6, y]$ with

$$y = [\overline{1; 4}] = [1; 4, y] = 1 + \frac{1}{4 + 1/y}.$$

Then $4y^2 - 4y - 1 = 0$ with $y = (1 + \sqrt{2})/2$ as its positive solution. This in turn gives $x = (14 - \sqrt{2})/4$. ◇

That the infinite simple continued fractions so far have been irrational numbers is no coincidence.

Lemma 2.8.4 *All infinite simple continued fractions are irrational numbers.*

Proof Say $[a_0; a_1, a_2, \ldots]$ is a rational number a/b. Then

$$0 < |C_{2n} - a/b| < C_{2n} - C_{2n+1} = 1/q_{2n}q_{2n+1},$$

so $0 < |aq_{2n} - bp_{2n}| < |b/q_{2n+1}|$, which is impossible since $aq_{2n} - bp_{2n}$ is an integer and $q_n \to \infty$ as $n \to \infty$. □

Lemma 2.8.5 *If $[a_0; a_1, a_2, \ldots] = [b_0; b_1, b_2, \ldots]$, then $a_n = b_n$ for all n. In particular, distinct infinite simple continued fractions represent distinct irrational numbers.*

Proof First observe that if $a < y < a + 1/b$ for integers a, b with $b \geq 1$, then $a < y < a + 1$, so $[y] = a$. As

$$a_0 + 1/[a_1; a_2, a_3, \ldots] = [a_0; a_1, a_2, \ldots] = x = [b_0; b_1, b_2, \ldots] = b_0 + 1/[b_1; b_2, b_3, \ldots],$$

we therefore get $a_0 = [x] = b_0$. Then $[a_1; a_2, a_3, \ldots] = [b_1; b_2, b_3, \ldots]$. By the same argument we get $a_1 = b_1$, and then $a_2 = b_2$, and so forth. □

Lemma 2.8.6 *Every irrational number is an infinite simple continued fraction.*

Proof Say x is an irrational number. Define a_n inductively by $a_n = [x_n]$, where $x_{n+1} = 1/(x_n - [x_n])$ and $x_0 = x$. Since x is irrational, so is x_n for all n, and therefore $x_{n+1} > 1$ as $0 < x_n - [x_n] < 1$. Hence we get infinitely many natural numbers $a_1, a_2, \ldots$.

Since $x_k = a_k + 1/x_{k+1}$, repeated substitution yields $x = [a_0; a_1, \ldots, a_n, x_{n+1}]$ for every n. But

$$|x - C_n| = \left|\frac{x_{n+1}p_n + p_{n-1}}{x_{n+1}q_n + q_{n-1}} - \frac{p_n}{q_n}\right| = \frac{1}{(x_{n+1}q_n + q_{n-1})q_n}$$
$$< \frac{1}{(a_{n+1}q_n + q_{n-1})q_n} = 1/q_n q_{n+1}$$

tends to zero as $n \to \infty$. So $x = [a_0; a_1, a_2, \ldots]$. □

The following result is an immediate consequence of the three previous lemmas.

Theorem 2.8.7 *The map* $\mathbb{Z} \times \mathbb{N}^\mathbb{N} \to \mathbb{R}\backslash\mathbb{Q}$ *given by*

$$(a_0, a_1, a_2, \ldots) \mapsto [a_0; a_1, a_2, \ldots]$$

is a bijection.

In other words, the irrational numbers can be considered as infinite simple continued fractions. We have also seen that the rational numbers can be considered as finite simple continued fractions; in this case we have a one-to-one correspondence between finite sequences and the rational numbers with continued fractions of say odd length.

The proof of the last lemma provides an algorithm for working out the continued fraction representation of an irrational number. For instance, with $x = \sqrt{23}$ one gets $x_0, \ldots, x_4$ with greatest integer parts $a_0 = 4, a_1 = 1, a_2 = 3, a_3 = 1$ and $a_4 = 8$. But then $x_5 = x_1$, so $\sqrt{23} = [4; \overline{1, 3, 1, 8}]$.

In fact, it is not hard to see that the periodic infinite simple continued fractions are precisely the irrational solutions of quadratic equations with integer coefficients. So infinite simple continued fractions of cube roots are not periodic.

One checks that $e = [2; 1, 2, 1, 1, 4, 1, 1, 6, 1, 1, 8, \ldots]$, which has a periodic pattern of length three with the first entry increased by two at each repetition. The standard way of showing that $e = \sum_{m=0}^\infty 1/m!$ is irrational goes as follows: Assume to the contrary that it is written as a fraction of integers, and let n be larger than the positive denominator. Then $a = n!(e - \sum_{m=0}^n 1/m!)$ must be a natural number, and yet

$$a = 1/(n+1) + 1/(n+1)(n+2) + \cdots < 1/(n+1) + 1/(n+1)^2 + \cdots = 1/n < 1.$$

Similarly, one proves that e^b and $\tan b$ are irrational for any non-zero rational number b.

2.8 Infinite Continued Fractions

Since $\tan(\pi/4) = 1$, one infers that π is also irrational. Resorting again to a pocket calculator one sees that $\pi = [3; 7, 15, 1, 292, \ldots]$, and here no pattern seems to emerge. The convergents $3/1, 22/7, 333/106, \ldots$ do indeed approach π. For instance, for the so-called *Archimedian value* $22/7$ of π, we know that $|\pi - 22/7| < 1/7^2$. This is a special case of the useful inequality from the proof of the last lemma, which we state as a separate result.

Corollary 2.8.8 *For an infinite simple continued fraction* $x = [a_0; a_1, a_2, \ldots]$, *we have*
$$|x - p_n/q_n| < 1/q_n q_{n+1} < 1/q_n^2.$$

A natural question is how well irrational numbers can be approximated by rational numbers. The next question is what one should mean by 'approximate' since in the naive sense they can be approximated arbitrarily well. One approach is to consider rational numbers with the same denominator. The next result shows that among these, the convergents of the infinite simple continued fraction of an irrational number offer the best possible approximations.

Proposition 2.8.9 *Consider an irrational number x expressed as an infinite simple continued fraction* $[a_0; a_1, a_2 \ldots]$. *For a rational number a/b with $1 \leq b \leq q_n$, we have* $|x - p_n/q_n| \leq |x - a/b|$.

Proof The system of equations $p_n u + p_{n+1} v = a$ and $q_n u + q_{n+1} v = b$ for fixed n, has the unique solution $u = (-1)^{n+1}(aq_{n+1} - bp_{n+1})$ and $v = (-1)^{n+1}(bp_n - aq_n)$.

If $u = 0$, then $aq_{n+1} = bp_{n+1}$ and as $\gcd(q_{n+1}, p_{n+1}) = 1$, the number q_{n+1} must divide b, and this contradicts $b \leq q_n < q_{n+1}$. So $u \neq 0$.

If $v = 0$, then we certainly have the required inequality, so we may assume that $v \neq 0$. In this case it is clear from the equations for u and v, that u and v will have opposite signs. And so will $q_n x - p_n$ and $q_{n+1} x - p_{n+1}$ as x lies between the convergents p_n/q_n and p_{n+1}/q_{n+1}. Hence $u(q_n x - p_n)$ and $v(q_{n+1} x - p_{n+1})$ have the same signs. Therefore

$$|bx - a| = |(q_n u + q_{n+1} v)x - (p_n u + p_{n+1} v)| = |u(q_n x - p_n) + v(q_{n+1} x - p_{n+1})|$$
$$= |u||q_n x - p_n| + |v||q_{n+1} x - p_{n+1}| > |q_n x - p_n|.$$

$\square$

So the Archimedian value $22/7$ is the best possible approximation of π among rational numbers with positive denominators not greater than seven.

The next result shows that if a rational number is sufficiently 'close' to an irrational number x, then it must be one of the convergents of x.

Corollary 2.8.10 *Let x be an irrational number. If a rational number a/b with $b \geq 1$ satisfies $|x - a/b| < 1/2b^2$, then a/b must be one of the convergents p_n/q_n of x.*

Proof Assume $a/b \neq p_n/q_n$ for all n. Then $1 \leq |bp_n - aq_n|$. Pick n such that $q_n \leq b < q_{n+1}$. Then by the proposition above $|q_n x - p_n| \leq |bx - a|$, so $|x - p_n/q_n| < 1/2bq_n$. Hence

$$\frac{1}{bq_n} \leq |\frac{bp_n - aq_n}{bq_n}| \leq |\frac{p_n}{q_n} - x| + |x - \frac{a}{b}| < \frac{1}{2bq_n} + \frac{1}{2b^2},$$

which implies the contradiction $b < q_n$.   □

Definition 2.8.11 An *algebraic real number* is any real solution x of an *algebraic equation* $a_n x^n + \cdots + a_1 x + a_0 = 0$ with integer coefficients a_i. The *transcendental real numbers* are the real numbers that are not algebraic.

Transcendental numbers are obviously irrational, and by definition, the square root of two is algebraic, and so is any algebraic combination of nth roots. Although we will see that most irrational numbers are transcendental, they are not easy to detect. In the appendix we show that the familiar numbers e and π are transcendental. On the other hand, it is not even known whether $e + \pi$ or $e\pi$ or π^e are irrational. However, the first two of these cannot both be rational since e is transcendental and satisfies $x^2 - (e + \pi)x + e\pi = 0$.

Liouville was the first to prove the existence of transcendental numbers by studying how well irrational numbers can be approximated by rational numbers. We have already seen in the second last corollary above that for any irrational number a, there are infinite many rational numbers m/n with $n > 0$ such that $|a - m/n| < 1/n^2$. However, for any rational number $k/l \neq m/n$ with $l > 0$, we have $|k/l - m/n| = |(kn - ml)/ln| \geq 1/ln$. So rational numbers cannot be approximated as well by other rational numbers as irrational numbers can.

The following result by Liouville, which is proved in the appendix, shows that in a certain sense there is also a limit to how well algebraic numbers can be approximated by rationals.

Theorem 2.8.12 *Say a is an irrational solution of an algebraic equation of degree $n > 0$. Then there is a real number $c > 0$ such that $|a - p/q| > c/q^n$ for all integers p, q with $q > 0$.*

Definition 2.8.13 A *Liouville number* is a real number a such that for any natural number n, there are integers p, q with $q \geq 2$ such that $0 < |a - p/q| < 1/q^n$.

By our previous remark no Liouville number is rational. In fact, by the previous theorem we have the following stronger result.

Corollary 2.8.14 *All Liouville numbers are transcendental.*

We will now construct a Liouville number, hence a transcendental number. Consider the irrational number $a = \sum_{k=1}^{\infty} 1/10^{k!} = 0,110001\ldots$ with non-periodic expansion. Introduce the integers $q_n = 10^{n!}$ and $p_n = q_n \sum_{k=1}^{n} 1/10^{k!}$. Then

2.9 Pell's Equation

$$0 < |a - p_n/q_n| = \sum_{k=n+1}^{\infty} 1/10^{k!} \le \sum_{k=(n+1)!}^{\infty} 1/10^k < (9/10^{(n+1)!}) \sum_{k=0}^{\infty} 1/10^k$$
$$= (9/10^{(n+1)!})(10/(10-1)) \le 10^{n!}/10^{(n+1)!} = 1/q_n^n,$$

so a is Liouville. This specific number is called *Liouvilles's constant*. This argument also holds for the abundance of numbers of the type $\sum_{k=1}^{\infty} a_k/b^{k!}$ for integers a_k and b with $b \ge 2$ and $0 \le a_k \le b-1$, with $a_k \ne 0$ for infinitely many k. All these numbers are therefore Liouville and hence transcendental.

Roth improved Liouville's result drastically by showing that for any real algebraic number a of degree $n \ge 2$ and for any real number $\varepsilon > 0$, there are at most finitely many rational numbers p/q with $q > 0$ such that $|a - p/q| > 1/q^{2+\varepsilon}$. Hence there is a constant $c > 0$ such that $|a - p/q| > c/q^{2+\varepsilon}$ for all rational numbers p/q with $q > 0$. This deep theorem, which surpassed previous improvements by Siegel and Thue, is as good as it gets since there are infinitely many rational numbers p/q with $q > 0$ such that $|a - p/q| < 1/q^2$. The estimate in Roth's theorem does not depend on n as opposed to the improvements of Thue and Siegel, who obtained $1 + n/2$ and $2\sqrt{n}$ as exponents for q. But as shown by Thue, see the appendix, any lowering of the exponent beyond Liouville's estimate have profound implications on the study of Diophantine equations.

2.9 Pell's Equation

Pell's equation $x^2 - dy^2 = 1$ has been studied since ancient times. We are interested in finding positive integer solutions x and y when d is a natural number which is not a perfect square. There is an intimate connection to continued fractions. To keep the length down, we adapt a brief style here.

Proposition 2.9.1 *For any positive integer solution (x, y) of $x^2 - dy^2 = 1$ with d a natural number which is not a perfect square, the number x/y is a convergent of the infinite simple continued fraction of $\sqrt{d}$.*

Proof The identity $(x - y\sqrt{d})(x + y\sqrt{d}) = 1$ shows that $x > y\sqrt{d}$, so we have

$$0 < x/y - \sqrt{d} = 1/y(x + y\sqrt{d}) < \sqrt{d}/y(y\sqrt{d} + y\sqrt{d}) = 1/2y^2$$

and the result follows from Corollary 2.8.10. □

The estimate in Corollary 2.8.8 tells us that $|p^2 - dq^2| < 1 + 2\sqrt{d}$ for any convergent p/q of $\sqrt{d}$, but $p^2 - dq^2$ need not be the integer 1. Finding solutions among the convergents of $\sqrt{d}$ requires more work.

Lemma 2.9.2 *If p_k/q_k are the convergents of $\sqrt{d}$ for a natural number d which is not a perfect square, and n is the length of the periode of the continued fraction expansion of $\sqrt{d}$, then*

$$p_{nk-1}^2 - dq_{kn-1}^2 = (-1)^{kn}$$

for every natural number k.

Proof First observe that the infinite simple continued fraction of $\sqrt{d}$ for any natural number d which is not a perfect square, is always of the form

$$\sqrt{d} = [a_0; \overline{a_1, a_2, a_3, \ldots, a_3, a_2, a_1, 2a_0}].$$

We leave the proof of this claim as an exercise to the reader.

Having done so, we can write $\sqrt{d} = [a_0; a_1, a_2, \ldots a_{kn-1}, b_{kn}]$, where

$$b_{kn} = [2a_0; \overline{a_1, a_2, \ldots a_{n-1}, 2a_0}] = a_0 + \sqrt{d}.$$

Substituting this into $\sqrt{d} = (b_{kn} p_{kn-1} + p_{kn-2})/(b_{kn} q_{kn-1} + q_{kn-2})$ gives

$$\sqrt{d}(a_0 q_{kn-1} + q_{kn-2} - p_{kn-1}) = a_0 p_{kn-1} + p_{kn-2} - dq_{kn-1},$$

which splits into two relations

$$a_0 q_{kn-1} + q_{kn-2} - p_{kn-1} = 0 = a_0 p_{kn-1} + p_{kn-2} - dq_{kn-1}.$$

Multiplying the first of these by p_{kn-1} and the second by $-q_{kn-1}$, and then adding, gives

$$p_{kn-1}^2 - dq_{kn-1}^2 = p_{kn-1} q_{kn-2} - q_{kn-1} p_{kn-2} = (-1)^{kn}.$$

$\square$

The following result is then immediate.

Theorem 2.9.3 *Say p_k/q_k are the convergents of $\sqrt{d}$ for a natural number d which is not a perfect square, and that n is the length of the periode of the continued fraction expansion of $\sqrt{d}$. Then $(x, y) = (p_{kn-1}, q_{kn-1})$ are solutions of $x^2 - dy^2 = 1$ for all natural numbers k, when n is even, and (p_{2kn-1}, q_{2kn-1}) are solutions for all natural numbers k, when n is odd.*

A more careful analysis shows that these are all the positive integer solutions of $x^2 - dy^2 = 1$.

Definition 2.9.4 *The fundamental solution of the Pell equation $x^2 - dy^2 = 1$ for a natural number d which is not a perfect square, is the integer solution (x, y) with least positive value for both x and y among all integer solutions.*

The results above guarantee the existence and uniqueness of a fundamental solution.

2.9 Pell's Equation

Example 2.9.5 Since $\sqrt{7} = [2; \overline{1, 1, 1, 4}]$ has periode with length four, the convergens $p_3/q_3 = 8/3$, $p_7/q_7 = 127/48$ and $p_{11}/q_{11} = 2024/765$ produce respectively the first three positive integer solutions $(x, y) = (8, 3)$, $(127, 48)$ and $(2024, 765)$ of the Pell equation $x^2 - 7y^2 = 1$.

Since $\sqrt{13} = [3; \overline{1, 1, 1, 1, 6}]$ has periode of length five, the fundamental solution $(649, 180)$ of $x^2 - 13y^2 = 1$ correspond to the convergent $p_9/q_9 = 649/180$.

The fundamental solution can some times be difficult to find. For the equation $x^2 - 1000099y^2 = 1$ it's x-value has 1118 digits due to the form of the continued fraction expansion of $\sqrt{1000099}$ which has a period of length 2174. Also, while the fundamental solution of the Pell equation with $d = 60$ is $(31, 4)$, the one for $d = 61$ is $(17663319049, 226153980)$, so small perturbations in d can cause huge variations.

The next result shows how all the positive integer solutions can be obtained from a fundamental solution.

Theorem 2.9.6 *Let d be a natural number which is not a perfect square, and let (x_1, y_1) be the fundamental solution of $x^2 - dy^2 = 1$. The condition $x_n + y_n\sqrt{d} = (x_1 + y_1\sqrt{d})^n$ uniquely determines pairs (x_n, y_n) for $n = 1, 2, \ldots$ which are positive integer solutions, and these pairs exhaust the positive integer solutions.*

Proof Clearly, the condition uniquely determines positive integers x_n and y_n for every n. It is also easy to check that the condition implies $x_n - \sqrt{d}y_n = (x_1 - y_1\sqrt{d})^n$. Then

$$x_n^2 - dy_n^2 = (x_n + y_n\sqrt{d})(x_n - y_n\sqrt{d}) = (x_1 + y_1\sqrt{d})^n(x_1 - y_1\sqrt{d})^n = (x_1^2 - dy_1^2)^n = 1$$

shows that (x_n, y_n) is a solution.

Suppose (u, v) is a positive integer solution that is not of the asserted form. Since $a \equiv x_1 + y_1\sqrt{d} > 1$, there is n such that $a^n < u + v\sqrt{d} < a^{n+1}$. Multiplying both sides of these inequalities by the positive number $x_n - y_n\sqrt{d}$, and invoking the condition in the theorem to use Pell's equation, we get

$$1 < (x_n - y_n\sqrt{d})(u + v\sqrt{d}) < x_1 + y_1\sqrt{d}.$$

Define integers r and s by $r + s\sqrt{d} = (x_n - y_n\sqrt{d})(u + v\sqrt{d})$. Then $r^2 - ds^2 = 1$, and $1 < r + s\sqrt{d} < x_1 + y_1\sqrt{d}$ shows that (r, s) is a positive integer solution with $r < x_1$ and $s < y_1$, which cannot be as (x_1, y_1) is a fundamental solution. □

Example 2.9.7 Since $(x_1, y_1) = (6, 1)$ is a fundamental solution of $x^2 - 35y^2 = 1$, the condition

$$x_2 + y_2\sqrt{35} = (6 + \sqrt{35})^2 = 71 + 12\sqrt{35}$$

gives another solution $(71, 12)$, whereas $(x_3, y_3) = (846, 143)$ is the next solution in the sequence. ◇

2.10 Complex Numbers

The quest for more general numbers is driven by the need to solve equations. The extension from natural numbers to integers arose from the need to solve e.g. the equation $x + 7 = 1$. Further extension to the rational numbers solved equations of the type $3x - 11 = 0$. The real numbers extended the rational numbers and offered for instance solutions to equations like $x^2 - p = 0$ with p prime.

Now what about the equation $x^2 + 1 = 0$? Such an equation has no solution among real numbers because $x^2 + 1 > 0$ for any real number x. To solve this equation we need more numbers, and if we insist on extending the real numbers, these are the complex numbers. A solution of the equation $x^2 + 1 = 0$ amounts to finding square roots of -1. How do we construct such numbers? Secondly, does this extension business ever stop? Do we need infinitely many number systems to cover all types of equations? Even worse, the more numbers we introduce, the more equations we generate that require solutions since nothing prevents us from picking coefficients from the new numbers, just like the equation $\sqrt{2}x^3 - \pi x^2 + ex - 3/4 = 0$ emerges from having introduced the real numbers.

The solution to all this is highly satisfactory. As soon as we have constructed the complex numbers, we do not need more general numbers to solve equations even with complex numbers as coefficients; all thanks to the *fundamental theorem of algebra*, which says that the complex numbers are closed in this respect.

Theorem 2.10.1 *Any equation of the form*

$$a_n x^n + \cdots a_1 x + a_0 = 0$$

with complex coefficients a_i, has exactly n complex solutions counted with multiplicity.

Gauss was the first to prove this result, he gave many proofs. Since then several other proofs have occurred. One of the easiest and shortest uses complex function theory, and goes as follows: If a non-trivial polynomial f has no zeroes, then $1/f$ is a holomorphic and bounded function, so it must be a constant by Liouville's theorem in the appendix, and this is absurd. Having gotten one root, we can perform polynom division, obtaining a polynomial of one grade lower, which by induction hypothesis has the remaining roots of f. We shall provide an algebraic proof in Section 8.4.

The construction of complex numbers is very geometric and surprisingly simple. Consider the vectors in the plane, and think of a vector as starting at the origin and ending at the point (x, y) with $x, y \in \mathbb{R}$. Such a point and a vector amount to the same thing, but in what follows it is easier to think in terms of arrows than points. We know how to add (and hence subtract) vectors both geometrically (via parallelograms) and algebraically (coordinate wise addition). But is there a way to multiply vectors, producing a new vector in the plane? Indeed there is. First specify a vector by indicating its length r and its angle θ to the x-axis.

2.10 Complex Numbers

Definition 2.10.2 We multiply two vectors in the real plane $\mathbb{R} \times \mathbb{R}$ by multiplying their lengths and adding their angles. This gives a new vector which by definition is the product of the two vectors.

The first thing we have to check is that this is a sensible operation, in that the vectors form a field. Clearly the product is commutative and associative with the unit vector along the x-axis as the unit for multiplication. The inverse of a non-zero vector is obviously the vector with reciprocal length and angle with opposite sign. A little exercise in trigonometry shows that the distributive law also holds, so we get indeed a field.

Definition 2.10.3 The *complex numbers* $\mathbb{C}$ is the field of vectors $\mathbb{R} \times \mathbb{R}$ under the previously prescribed product.

The real numbers correspond to the vectors lying on the x-axis, called the *real axis*, and we write a for $(a, 0)$, so the unit is 1.

The rest of the plane are new numbers, with the y-axis coined the *imaginary axis*.

The unit vector $(0, 1)$ on this axis, denoted by i, has the property that $i^2 = -1$ because the length remains the same upon squaring whereas the angle doubles $2(\pi/2)$, so we end up with an arrow with its tip at -1. Thus i is a square root of -1 (there are two of them) and we write $i = \sqrt{-1}$.

Since multiplication by i means adding $\pi/2$ to the angle, real numbers y are erected to vertical vectors $iy = (0, y)$.

Definition 2.10.4 Any complex number $z = (x, y)$ can be written as

$$z = x + iy,$$

called its *normal form*.

The components x and y are the *real and imaginary parts* of z, and we sometimes use the notation Rez and Imz for them. The *complex conjugate* $\bar{z}$ of z is obtained by reflecting the vector about the x-axis, so $\bar{z} = x - iy$, and the length $|z|$ of the vector z is clearly the square root of the non-negative number $z\bar{z}$.

The normal form allows for a more algebraic approach to complex numbers. When you multiply together two complex numbers in normal form just use ordinary rules for calculating with numbers together with the identity $i^2 = -1$. So on the one hand $zz' = rr'(\cos(\theta + \theta') + i \sin(\theta + \theta'))$, and on the other hand

$$\begin{aligned} zz' &= (x + iy)(x' + iy') = xx' - yy' + i(xy' + yx') \\ &= rr'(\cos\theta \cos\theta' - \sin\theta \sin\theta' + i(\cos\theta \sin\theta' + \sin\theta \cos\theta')), \end{aligned}$$

and we have recovered the trigonometric formulas for addition of angles. This can also be used to show the distributive law.

The following result, which characterizes the complex numbers, is immediate from the fundamental theorem of algebra, and involves the notion of *algebraic closure*, to be defined and studied in Section 7.3.

Proposition 2.10.5 *The field of complex numbers is the algebraic closure of the real numbers* $\mathbb{R}$.

However, there exists no order on $\mathbb{C}$ turning it into an ordered field like the situation was for the real numbers.

The distance between two complex numbers z and z' is the usual Euclidean distance
$$|z - z'| = \sqrt{(x - x')^2 + (y - y')^2}$$
between them as points in the plane.

Proposition 2.10.6 *The complex numbers are complete, in that every Cauchy sequence has a limit.*

Proof To see that every Cauchy sequence $\{z_n\}$ converges to a complex number, it suffices to observe that the corresponding real sequences $\{x_n\}$ and $\{y_n\}$ are Cauchy. □

2.11 Absolute Values and *p*-Adic Numbers

Definition 2.11.1 An *absolute value* on a field F is a function $a \mapsto |a|$ from F to the non-negative real numbers that is zero at zero, and only then, and satisfies $|ab| = |a| \cdot |b|$ and $|a + b| \leq |a| + |b|$ for all $a, b \in F$. It is called *non-archimedean* or a *valuation* if it satisfies the stronger inequality $|a + b| \leq \max\{|a|, |b|\}$ for all $a, b \in F$.

Observe that $|1| = 1$ when $1 \neq 0$, and that $|a| = |-a|$ and $|a^{-1}| = |a|^{-1}$ for any non-zero $a \in F$. The *trivial absolute value* on any field is the non-archimedean absolute value that is one for every non-zero element.

We can obviously define Cauchy sequences and convergence of sequences with elements in F just as we did for the rational numbers, but now with respect to the given absolute value on F. This allows also for the formation of a completion of F in the same way as we formed the real numbers from the rational numbers. Such a completion is clearly the unique field which contains F and has the property that any element a of it can be approximated by some sequence $\{a_n\} \subset F$ in the sense that $|a - a_n| \to 0$ as $n \to \infty$. The field F is contained in the completion as the classes having Cauchy sequences that eventually become constant with the constants being the elements in F, and the absolute value was extended to an absolute value on the completion by declaring $|a| = \lim |a_n|$, which exists as $\{|a_n|\}$ is Cauchy in $\mathbb{R}$.

Definition 2.11.2 Two absolute values on the same field are equivalent if whenever a sequence converges with respect to one absolute value, it does so for the other absolute value as well.

2.11 Absolute Values and p-Adic Numbers

The ordinary absolute value $|\cdot|$ on $\mathbb{Q}$ is an archimedean absolute value, and so is $|\cdot|^r$ for any $r \in \langle 0, 1 \rangle$. However, these are all equivalent to $|\cdot|$, and their completions is the field of real numbers.

Proposition 2.11.3 *Two non-trivial absolute values $|\cdot|_i$ are equivalent if and only if $|a|_1 < 1$ implies $|a|_2 < 1$. In this case there is a positive real number r such that $|a|_2 = |a|_1^r$ for all a.*

Proof Note that $|a|_i < 1$ if and only if $\lim a^n = 0$, which proves the first statement.

For the second assertion, first note that $|a|_1 > 1$ implies $|a|_2 > 1$ by going to inverses. Since $|\cdot|_1$ is non-trivial, there is c such that $|c|_1 > 1$. Let $r = \log(|c|_2)/\log(|c|_1)$, so $|c|_2 = |c|_1^r$.

For any non-zero element a, there is a real number s such that $|a|_1 = |c|_1^s$. For any integers m, n with $m/n > s$ and $n > 0$, we have $|a|_1 > |c|_1^{m/n}$, or $|a^n/c^m|_1 < 1$. So $|a^n/c^m|_2 < 1$, or $|a|_2 < |c|_2^{m/n}$. Thus $|a|_2 \leq |c|_2^s$. Arguing similarly with $m/n < s$, we get $|a|_2 = |c|_2^s = |c|_1^{rs} = |a|_1^r$. $\square$

In the other direction we have the following approximation result by Artin-Waples.

Theorem 2.11.4 *Let $|\cdot|_i$ be non-trivial pairwise inequivalent absolute values on a field F. For $\varepsilon > 0$ and any finite collection of elements $a_i \in F$, there is $a \in F$ such that $|a - a_i|_i < \varepsilon$.*

Proof By assumption we have $a, b \in F$ with $|a|_1, |b|_s > 1$ and $|a|_s, |b|_1 \leq 1$. With $c = a/b$, we get $|c|_1 > 1$ and $|c|_s < 1$. We show by induction that there is $d \in F$ such that $|d|_1 > 1$ and $|d|_i < 1$ for $i = 2, 3, \ldots, s$. Assume this holds with s replaced by $s - 1$. If $|d|_s \leq 1$, then replacing d by $d^n c$ for large n, the induction step is seen to hold. If $|d|_s > 1$, replace d by $d^n c/(1 + d^n)$ for large n to perform the induction step. As the case $s = 2$ was already established, we have the required element d.

Now $d^n/(1 + d^n)$ tends to 1 with respect to $|\cdot|_1$, and goes to 0 with respect to the other absolute values. So for each i we have an element $b_i \in F$ that is close to 1 with respect to $|\cdot|_i$, and which is close to 0 with respect to $|\cdot|_j$ for all $j \neq i$. Hence $a = \sum a_i b_i$ has the required property. $\square$

Pick a prime number p. By the fundamental theorem of arithmetic, write any non-zero rational number as $p^n a/b$ for a unique integer n such that neither a nor b contains factors of p. The *p-adic absolute value* $|\cdot|_p$ on $\mathbb{Q}$ is the non-archimedean absolute value given by $|p^n a/b|_p = 1/p^n$.

Example 2.11.5 Since $a = 63/550 = 2^{-1} 3^2 5^{-2} 7^1 11^{-1}$ we have

$$|a|_2 = 2, \ |a|_3 = 1/9, \ |a|_5 = 25, \ |a|_7 = 1/7, \ |a|_{11} = 11, \ |a|_p = 1$$

for all primes $p > 11$. $\diamond$

Definition 2.11.6 The *p-adic numbers* is the completion $\mathbb{Q}_p$ of $\mathbb{Q}$ with respect to the p-adic absolute value.

Notice that the completed absolute value on the field $\mathbb{Q}_p$ has range within the set $\{0, 1/p^n \mid n \in \mathbb{Z}\}$. Also note that due to non-archimedeaness, a series of p-adic numbers converge if and only if the terms of the series tend to zero. This suggests that analysis in $\mathbb{Q}_p$ is easier than in $\mathbb{R}$.

The significance of the p-adic numbers is due to the following result by Ostrowski.

Theorem 2.11.7 *Every absolute value on $\mathbb{Q}$ is either trivial, equivalent to the usual absolute value or to the p-adic absolute value for some p.*

Proof Suppose $|\cdot|$ is an absolute value on $\mathbb{Q}$. We distinguish two cases. Say first that there is some $n \in \mathbb{N}$ with $|n| > 1$. Pick the least such n, and then a real number r such that $|n| = n^r$. Write any natural number m as

$$m = a_0 + a_1 n + \cdots + a_k n^k$$

with $a_i \in [0, n)$ and $a_k \neq 0$. Then all $|a_i| \leq 1$ by definition of n, and obviously $n^k \leq m$, so

$$|m| \leq |a_0| + |a_1| n^r + \cdots + |a_k| n^{kr} = n^{kr}(1 + n^{-r} + \cdots + n^{-kr}) \leq bm^r$$

with $b = \sum_{k=0}^{\infty}(1/n^r)^k$. Replacing m by m^l for $l \in \mathbb{N}$ gives $|m| \leq b^{1/l} m^r$. Letting $l \to \infty$ we thus get $|m| \leq m^r$.

For the opposite inequality, note that $n^k \leq m < n^{k+1}$, so

$$|m| \geq |n^{k+1}| - |n^{k+1} - m| \geq n^{(k+1)r} - (n^{k+1} - m)^r \geq n^{(k+1)r} - (n^{k+1} - n^k)^r \geq cm^r$$

with $c = 1 - (1 - 1/n)^r$. Replacing again m by m^l and letting $l \to \infty$, we get $|m| \geq m^r$. Hence $|d| = d^r$ for all $d \in \mathbb{Q}$, and $|\cdot|$ is equivalent to the usual absolute value.

Next assume that $|n| \leq 1$ for all $n \in \mathbb{N}$, and that $|\cdot|$ is non-trivial. Pick the least $p \in \mathbb{N}$ with $|p| < 1$. Then p must be prime because if $p = n_1 n_2$ for $n_i \in \{2, 3, \ldots\}$, then $|n_1| = |n_2| = 1$ as $n_i < p$, so $|p| = 1$, which is absurd.

Consider another prime $q \neq p$. If $|q| < 1$, then $|q^u| < 1/2$ and $|p^v| < 1/2$ for large $u, v \in \mathbb{N}$. Pick integers s, t with $1 = sq^u + tp^v$, so $1 \leq |s||q^u| + |t||p^v| < 1/2 + 1/2$, which is absurd. So $|q| = 1$. Hence for any $d \in \mathbb{Q}$, we have $|d| = |p|^w$, where $d = p^w e/f$ for the integer w such that p does not divide e nor f. So $|\cdot|$ is equivalent to $|\cdot|_p$. $\square$

Let us be more concrete about the p-adic numbers.

Lemma 2.11.8 *For any rational number a with $|a|_p \leq 1$ and any natural number i, there is an integer $b \in [0, p^i)$ such that $|a - b|_p \leq 1/p^i$.*

Proof Write $a = c/d$ for relative prime integers c and d. Since $|a|_p \leq 1$, the prime number p cannot divide d, so there are integers m, n such that $md + np^i = 1$. Then

$$|mc - a|_p \leq |c/d|_p |md - 1|_p \leq |np^i| \leq 1/p^i.$$

2.11 Absolute Values and p-Adic Numbers

Let $b \in [0, p^i)$ be mc plus an appropriate integer multiple of p^i. Then $|b - a|_p \le 1/p^i$ by non-archimedeaness. □

Theorem 2.11.9 *Every equivalence class $a \in \mathbb{Q}_p$ with $|a|_p \le 1$ has exactly one representative Cauchy sequence $\{a_i\}$ such that $a_i \in \{0, 1, \ldots, p^i - 1\}$ and $a_{i+1} \equiv a_i \pmod{p^i}$ for all $i \in \mathbb{N}$.*

Proof Say we have a Cauchy sequence $\{b_i\}$ in the class a. For $j \in \mathbb{N}$ pick $N(j) \in \mathbb{N}$ such that $|b_i - b_k|_p \le 1/p^j$ for every $i, k \ge N(j)$. We may assume that $\{N(j)\}$ increases strictly with j.

Note that $|b_i|_p \le 1$ for $i \ge N(1)$ since

$$|b_i|_p \le \max\{|b_k|_p, |b_i - b_k|_p\} \le \max\{|b_k|_p, 1/p\}$$

for $k \ge n(1)$, and $|b_k|_p \to |a|_p \le 1$ as $k \to \infty$. Thus $|b_{N(j)}|_p \le 1$ for all j. By the lemma we may therefore pick integers $a_j \in [0, p^j)$ such that $|a_j - b_{N(j)}|_p \le 1/p^j$. Then

$$|a_{j+1} - a_j|_p \le \max\{|a_{j+1} - b_{N(j+1)}|_p, |b_{N(j+1)} - b_{N(j)}|_p, |a_j - b_{N(j)}|_p\} \le 1/p^j,$$

so $a_{j+1} \equiv a_j \pmod{p^j}$. Also, for $i \ge N(j) \ge j$, we have

$$|a_i - b_i|_p \le \max\{|a_i - a_j|_p, |a_j - b_{N(j)}|_p, |b_i - b_{N(j)}|_p\} \le 1/p^j,$$

so $|a_i - b_i|_p \to 0$ as $i \to \infty$, and $\{a_i\}$ therefore belongs to the class a. □

If $|a|_p > 1$ for $a \in \mathbb{Q}_p$, then $|p^n a| \le 1$ for some $n \in \mathbb{N}$, so if $\{a_i\}$ represents $p^n a$ as in the theorem above, then $\{p^{-n} a_i\}$ represents a.

Writing $a_i = b_0 + b_1 p + \cdots + b_{i-1} p^{i-1}$ for digits $b_i \in \{0, 1, \ldots, p - 1\}$, the condition $a_{i+1} \equiv a_i \pmod{p^i}$ means that the first i digits for a_i coincide with the corresponding first i digits of a_{i+1}.

Hence we have a unique *p-adic expansion*

$$a = b_0 p^{-n} + b_1 p^{-n+1} + \cdots + b_n + b_{n+1} p + b_{n+2} p^2 + \cdots$$

of a. The partial sums s_m obtained by cutting off this expansion after m terms form a Cauchy sequence in $\mathbb{Q}$ with respect to $|\cdot|_p$ that converges to a. We write the expansion of a in base p as

$$\cdots b_{n+2} b_{n+1} b_n . b_{n-1} \cdots b_1 b_0$$

with powers of p increasing from right to left. Sometimes we put p as an index to indicate the base. Note that the ambiguity that occurs in any base expansion of the real numbers, like $0, 99 \cdots = 1.00 \cdots$, does not occur here; the expansion is unique. We can think of p-adic numbers as formal Laurent series over the variable p, where we define what we mean by a formal Laurent series over a variable in the chapter on rings. Pointwise addition and multiplication of Cauchy sequences convert to ordinary addition and convolution product of series, but the coefficients are shifted to fall in the correct range; just like we handle addition and multiplication of real decimal numbers.

We illustrate the arithmetic's with an example.

Example 2.11.10 Since

$$2 + 2 \cdot 3 + 2 \cdot 3^2 + \cdots + 2 \cdot 3^n = 2(1 - 3^{n+1})/(1 - 3) \equiv -1 \ (\mathrm{mod}\, 3^{n+1})$$

for all $n \in \mathbb{N}$, we get $\cdots 222_3 = -1$ in base 3.

In base 5 we have $(5^4 - 1)/3 = 4444_5/3 = 1313_5$, so $-1/3 = \cdots 1313_5$ and $-2/3 = \cdots 1313_5 \cdot 2 = \cdots 3131_5$ and $1/3 = 1 - 2/3 = \cdots 3132_5$, which upon multiplication with 3 gives $\cdots 001_5$. ◇

Definition 2.11.11 The *p-adic integers* are those $a \in \mathbb{Q}_p$ with $|a|_p \leq 1$, which means that the p-adic expansion of a has no negative powers of p. The p-adic integers form a unital subring $\mathbb{Z}_p$ of $\mathbb{Q}_p$, and we refer to the subring of ordinary integers $\mathbb{Z} \subset \mathbb{Z}_p$ as the *rational integers*.

Clearly $\mathbb{Z}_p$ is an integral domain, but not a field, and should not be confused with the field of equivalence classes in $\mathbb{Z}$ modulo p. The p-adic numbers in the example above are all p-adic integers, but most of them are not rational integers, although the lemma above tells us that p-adic integers can be approximated arbitrary well by rational integers. In fact, most of the p-adic integers in the example above are rational numbers. It is easy to see that the rational numbers are the p-adic numbers with p-adic expansions that eventually become periodic.

The units in $\mathbb{Z}_p$ are the invertible elements under multiplication, so $a \in \mathbb{Z}_p$ is a unit if and only if $|a|_p = 1$, alternatively, if and only if the first digit in its p-adic expansion is non-zero. Clearly every $a \in \mathbb{Q}_p$ can be written uniquely as $a = p^n u$ for a unit u in $\mathbb{Z}_p$ and a rational integer n. Also, given a natural number m, the rational integers are seen to be congruent modulo p^m in $\mathbb{Z}$ if and only if they are congruent modulo p^m in $\mathbb{Z}_p$. So there are p^m residue classes modulo p^m in $\mathbb{Z}_p$.

We could of course have worked with other digits than $\{0, 1, \ldots, p-1\}$ in the expansion of a p-adic number. Any collection $\{a_1, \ldots, a_{p-1}\}$ of p-adic integers with $a_i \equiv i \ (\mathrm{mod}\, p)$ will do as digits.

Remark 2.11.12 The field of p-adic numbers is not algebraically closed. Its algebraic closure $\overline{\mathbb{Q}_p}$ has infinite degree over $\mathbb{Q}_p$ as opposed to the field of complex numbers which has only degree two over $\mathbb{R}$. And while $\mathbb{C}$ is metrically complete, this is not so for $\overline{\mathbb{Q}_p}$ with respect to an extension of $|\cdot|_p$, which exists and is unique. The Cauchy completeness Ω_p of $\overline{\mathbb{Q}_p}$ is however an algebraically closed field, and should therefore be thought of as the p-adic analogue of the field of complex numbers.

Here is a remark for the analyst who is familiar with topology and locally compact groups.

Remark 2.11.13 A *local field* is a locally compact topological field with a non-discrete topology. As an additive group it has a Haar measure μ which is unique up to a positive real number. Then $a \mapsto \mu(aX)/\mu(X)$ for any measurable set X with finite non-zero measure, defines an absolute value on the field which obviously does not depend on the chosen set X.

One shows then that the archimedean local fields are isomorphic to $\mathbb{R}$ or $\mathbb{C}$, whereas any non-archimedean one is isomorphic to a finite extension of $\mathbb{Q}_p$ for some p, or to the field of formal Laurent series over a finite field of characteristic p. ◇

2.12 Cardinality

We shall see that in some sense there are many more real numbers than rational numbers. Of course, there are infinitely many in both cases.

Definition 2.12.1 We say that two sets X and Y have the same number of elements, or the same *cardinality*, if there exists a bijection between them. We then write $|X| = |Y|$.

Any set with the same cardinality as $\{1, \cdots, n\}$ is *finite* and is said to consist of $n \in \mathbb{N}$ elements.

Definition 2.12.2 A set X is *countable* if $|X| = |\mathbb{N}|$.

The members of X can then be counted because having a bijection $f \colon \mathbb{N} \to X$ means that you can list up the distinct members $\{f(1), f(2), \ldots\}$. If f was only injective, you wouldn't find all the members on the list, and if f was only surjective, the list would be complete but there would be repetitions. Removing these, you could form a new list of the distinct members of X which is at most countable:

Definition 2.12.3 A set is *at most countable* if it is either countable or finite.

Proposition 2.12.4 *Any subset of a countable set X is at most countable.*

Proof From a list of all the members of X one can form a new list by systematically weeding out the members not in the subset and replacing them by a preselected element of the subset. □

So the even numbers are countable (we can also list them directly $2, 4, \ldots$), and so are the odd ones, although in both cases the feeling is that there are only half as many because $\mathbb{N}$ is the union of these two subsets.

Peculiarly enough, the following result holds.

Proposition 2.12.5 *A countable union $\cup_{n=1}^{\infty} X_n$ of at most countable sets X_n is at most countable.*

Proof To list the members of this union, first write the members up in an array with the list of the members of the first set in the first row, the list of the second set in the second row, and so forth. Then the list of the union is obtained by starting in the upper left corner and proceeding by a zig-zag movement mopping up all the members as you move towards the lower right corner at infinity. □

Corollary 2.12.6 *The integers are countable.*

Proof Write $\mathbb{Z} = -\mathbb{N} \cup \{0\} \cup \mathbb{N}$. Alternatively we can list the integers directly: $0, 1, -1, 2, -2, \ldots$. □

Corollary 2.12.7 *If X and Y are at most countable, so is $X \times Y$.*

Proof Write $X \times Y = \cup_n \{(x_n, y) | y \in Y\}$, where $x_1, x_2, \ldots$ are the members of X. □

Since the map $\mathbb{Z} \times \mathbb{N} \to \mathbb{Q}: (m, n) \mapsto m/n$ is surjective, it follows from what we have said that $\mathbb{Q}$ is countable.

Corollary 2.12.8 *The rational numbers are countable.*

By induction any finite Cartesian product of sets that are at most countable is at most countable, but this is not so for infinite products.

It suffices to consider the countable Cartesian product $\{0, 1\}^\mathbb{N}$ with members being sequences of 0's and 1's. No vertical list of such sequences will be exhaustive because the sequence with the nth entry that is one minus the nth entry of the nth sequence cannot be on the list since it will differ from every sequence on the list at least one place.

This is known as Cantor's diagonal argument, and can easily be generalized to an arbitrary set X, using $|\{0, 1\}^X| = |P(X)|$.

Proposition 2.12.9 *There exists no surjection from a set X to its power set $P(X)$.*

Proof We must show that there is no surjection from X to $\{0, 1\}^X$. Suppose we had one, say $x \mapsto f_x$, where $f_x: X \to \{0, 1\}$. Then the function $g: X \to \{0, 1\}$ defined by $g(x) = 1 - f_x(x)$ is not of the form f_x for any x. □

Furthermore, the map $x \mapsto \{x\}$ is clearly an injection from X to $P(X)$. This suggests the notation $|X| < |P(X)|$.

Definition 2.12.10 We write $|X| < |Y|$ for sets X and Y if there exists an injection $X \to Y$ and $|X| \neq |Y|$.

Forming power sets increases the cardinality, and we get a hierarchy of sets

$$\mathbb{N} \subset P(\mathbb{N}) \subset P(P(\mathbb{N})) \subset \cdots,$$

which breaks radically with the idea that every set is at most countable; there is an infinite string of uncountable sets, each with cardinality larger than that of their predecessor.

2.12 Cardinality

Cantor originally used his argument to show that the real numbers are uncountable. Viewing real numbers as equivalence classes of Cauchy sequences of rational numbers, they are members of $P(P(\mathbb{Q}))$, so it is perhaps not surprising that they are not countable.

Proposition 2.12.11 $|\mathbb{N}| < |P(\mathbb{N})| = |\mathbb{R}|$.

Proof Perhaps the easiest way to see this is to look at real numbers as binary expansions, and consider the obvious surjection $\{n_k\} \mapsto 0, n_1 n_2 \cdots$ from $\{0, 1\}^{\mathbb{N}}$ to $[0, 1]$. This map is not injective because as we know e.g. the two expansions $0, 1000 \cdots$ and $0, 0111 \cdots$ represent the same number. To get rid of this problem remove the rational entities in both camps, i.e. those sequences and expansions that turn periodic. Then the map is bijective between the remaining sets and both parts removed are countable.

To see directly that the sequences that turn periodic are countable, categories them in lengths of periods and where the periods start, and you have contained them in a countable union of countables.

Composing the bijection from the non-periodic sequences to the irrational numbers in $\langle 0, 1 \rangle$ with any bijection from $\langle 0, 1 \rangle$ to $\mathbb{R}$, like $x \mapsto \tan(\pi(x - 1/2))$, we get a bijection from the non-periodic sequences to a subset of $\mathbb{R}$, and we have left out countable sets in both $\{0, 1\}^{\mathbb{N}}$ and $\mathbb{R}$ that will be taken care of by some bijection we know exists. □

We saw that the rational numbers, being those expansions that turn periodic, are countable, and this seems reasonable as there should be many more expansions of the other kind. This multitude is what gives the real numbers substance as opposed to the scanty rational ones. Note also that the p-adic numbers and the p-adic integers are both uncountable and have the same cardinality as $\mathbb{R}$, since their members can be expressed uniquely as infinite expansions with digits taking arbitrary values in $\{0, 1, \ldots, p - 1\}$.

We end this discussion on cardinality with some more general results that will be needed later. We will derive them using a mixture of Zorn's lemma and the axiom of choice.

Proposition 2.12.12 *If there is a surjection from one set to another, then there exists an injection in the other direction.*

Proof Say $f \colon X \to Y$ is a surjection. Then any choice function $g \colon Y \to X$ that picks an element $g(y) \in f^{-1}(\{y\})$ for every $y \in Y$ will obviously be injective. □

The converse direction also holds and does not even use the axiom of choice. The following result is known as the Shröder-Bernstein theorem.

Proposition 2.12.13 *Two sets have the same cardinality if there are injections both ways between them.*

Proof To injections $f\colon X \to Y$ and $g\colon Y \to X$, let $Z = \bigcup_{n=0}^{\infty}(fg)^n(f(X)^c)$. Define a map $h\colon X \to Y$ by sending $x \in g(Z)$ to $g^{-1}(x)$ and $x \in g(Z)^c$ to $f(x)$. It is easy to check that $x \in g(Z)$ if and only if $f(x) \in Z$. Using this ones sees that h is a bijection. $\square$

A host of injections are provided by the following result.

Proposition 2.12.14 *Given any two non-empty sets, there is at least one injection between them.*

Proof Consider non-empty sets X and Y. Let $\mathcal{F}$ be the family of all injections from any subset of X to Y. Then $\mathcal{F}$ is non-empty because for any elements $x \in X$ and $y \in Y$, it contains the function $\{x\} \to Y$ which sends x to y. Now partially order $\mathcal{F}$ by saying that $f \le g$ for injections $f\colon A \to Y$ and $g\colon B \to Y$ with A, B subsets of X, if $A \subset B$ and $g(x) = f(x)$ for $x \in A$, so g is an extension of f.

Every chain $\{f_i\colon A_i \to Y\}$ in $\mathcal{F}$ has an upper bound, namely the function $f\colon \bigcup A_i \to Y$ defined to be $f(x) = f_i(x)$ for $x \in A_i$. It is well-defined because if also $x \in A_j$ for some j, then either $A_i \subset A_j$ or $A_j \subset A_i$, and in both cases $f_i(x) = f_j(x)$. Also f is injective because if $x \ne x'$, with $x \in A_i$ and $x' \in A_j$, then again both $x, x' \in A_i$, in which case $f_i(x) \ne f_i(x')$ as f_i is injective, or both $x, x' \in A_j$, and then $f_j(x) \ne f_j(x')$. So $f \in \mathcal{F}$ and it is obviously an upper bound for the chain.

By Zorn's lemma there is a maximal element $g\colon B \to Y$ in $\mathcal{F}$. Now either $B = X$, and we are done, or g is surjective, and then by Proposition 2.12.12, we are done. The last option is that g is not surjective and $B \ne X$. But then we can pick $x \in X \setminus B$ and $y \in Y \setminus g(B)$, and define a function $h\colon B \cup \{x\} \to Y$ by $h(x) = y$ and $h = g$ on B. Clearly $h \in \mathcal{F}$ and $g < h$, which contradicts maximality of g.

So we are left with the first two options, saying that either $g\colon X \to Y$ is an injection, or there is an injection $Y \to X$. $\square$

This means that the cardinality of any two non-empty sets X and Y is comparable, that is, either $|X| < |Y|$, or $|X| = |Y|$, or $|Y| < |X|$.

A natural question then is whether there are subsets of $\mathbb{R}$ with cardinality strictly between $|\mathbb{N}|$ and $|\mathbb{R}|$. The *Continuum Hypothesis* claims that this is not the case. However, it has been shown that this claim is independent of the set theoretic axioms of Zermelo-Frankel, including the axiom of choice, provided these are consistent, leaving us free to add the hypothesis to the axioms or cutting it out.

The following result says that you can partition an infinite set in blocks that are countable, and not just at most countable.

Lemma 2.12.15 *Any infinite set can be written as a disjoint union of countable subsets.*

Proof Say X is an infinite set. By Proposition 2.12.14 either there is an injection from $\mathbb{N}$ to X, or there is an injection the other way, but then as X is infinite, it has to be countable. So in either case there is an injection $\mathbb{N} \to X$.

2.12 Cardinality

This shows that the family $\mathcal{F}$ of all disjoint countable subsets of X is non-empty, and we can partially order $\mathcal{F}$ by inclusion. Also every chain in $\mathcal{F}$ has an upper bound, namely the collection of all subsets of the chain, because any pair of subsets in this collection will belong to a common subcollection which requires them to be disjoint.

By Zorn's lemma $\mathcal{F}$ has a maximal element C of disjoint countable subsets of X. If the union Y of the members of the collection C is not the whole of X, there are two options. Either $X \setminus Y$ is finite, in which case these finitely many elements can be joined to a member of C, and we are done. Or $X \setminus Y$ is infinite. But then as in the first paragraph of this proof, we can get a copy Z of $\mathbb{N}$ inside $X \setminus Y$, and the collection $C \cup \{Z\}$ will be disjoint and strictly greater than C, which is impossible. $\square$

The following result generalizes $|\mathbb{N} \times \mathbb{N}| = |\mathbb{N}|$ from Corollary 2.12.7, and also the result $|\mathbb{R} \times \mathbb{N}| = |\mathbb{R}|$, which is easily gotten by noting that the map $\langle 0, 1 \rangle \times \mathbb{N} \to \mathbb{R}$ which sends (a, n) to $n + a$ is an injection.

Proposition 2.12.16 *For any infinite set X we have $|X \times \mathbb{N}| = |X|$.*

Proof By Lemma 2.12.15 we can write X as a disjoint union $\cup X_i$ of countable subsets X_i of X. Then $X \times \mathbb{N}$ equals the disjoint union $\cup (X_i \times \mathbb{N})$.

Corollary 2.12.7 tells us that $|X_i \times \mathbb{N}| = |X_i|$, so by the axiom of choice there is a family $\{f_i \colon X_i \to X_i \times \mathbb{N}\}$ of bijections. Using this family we can define a function $f \colon X \to X \times \mathbb{N}$ between the disjoint union of the X_i's and that of the $X_i \times \mathbb{N}$'s by $f(x) = f_i(x)$ for $x \in X_i$, and this f is obviously bijective. $\square$

It can also be shown that $|X \times X| = |X|$ if and only if the set X consists of one element or is infinite.

One could say that the 'cardinal number' $|X|$ of a set X should be the equivalence class of all sets having a bijection to X. Thanks to Proposition 2.12.14 and the Shröder-Bernstein theorem, our relation $|X| < |Y|$ would then be an order on the 'cardinal numbers', and one could also prove within naive set theory that this would be a well-ordering.

However, the collection of all sets is not a set, so this definition does not work within the ZF-axioms including the axiom of choice. If one limits to 'allowed' sets, one can manage. But a better and more conventional way to proceed is the one suggested by von Neumann, who defined the *cardinal number* of a set to be the least ordinal with a bijection to the set.

We won't be more precise here as we have not considered ordinals. Let us just mention that von Neumann's definition does require the axiom of choice to guarantee the existence of such an ordinal; without it the situation becomes more complicated and adjustments have to be made. It should also be said that the cardinal number of a set with $n \in \mathbb{N}$ elements will be n under the appropriate identification.

Chapter 3
Linear Algebra

Euclidean space with its linear structure of vectors, or arrows, that can be added and scaled seems to be deeply rooted in our way of thinking. Some philosophers even claim that it is an a priory prerequisite for our ability to reason. No wonder then that we try to build up so much mathematics around linear structures, and equally important, transfer or reduce complicated problems to a linear setting.

There is no good way of multiplying arrows beyond the plane, but in space we can still add them together and scale them, and (regarded as elements rather than arrows) this can also be done in multi-parameter spaces way beyond ordinary space. One adapts an axiomatic approach to vector spaces, as sets where we can add elements, coined vectors, together, and where vectors can be scaled sensibly by elements in any field. This is basically all we require. Out of thin air one can then prove the existence of a linear basis, that is, a subset of elements such that any vector can be written as a unique linear combination of the basis vectors. Having at least one basis for the vector space, one can produce other bases by forming appropriate linear combinations of the basis vectors one started with. This way one gets more than countably many bases for non-trivial real vector spaces, and each basis may contain more than countably many vectors. Yet, one can prove the astonishing fact that in any fixed vector space any two bases will always have the same cardinality, defined as the dimension of the vector space. From an abstract point of view, one basis is equally good as another. There is democracy, no canonical basis exists, and one should keep this subtle point in mind. However, in applications certain candidates might point themselves out as more convenient than others as one already can easily have picked a basis unknowingly. For instance, in the finite dimensional case, when dealing with matrices, that is, arrays with scalars as entries, a choice is already made.

A matrix represents the more fundamental notion of a linear transformation between vector spaces, which is basically a map that preserves the linear structures in the two spaces. Having fixed a basis in each one of two finite dimensional vector spaces, one can set up a complete correspondence between the linear transformations between the two spaces and matrices of a certain size. The linear transformations can be pointwise added and scaled, forming thus a new vector space. This is reflected in the fact that matrices can be added and scaled in a coordinate wise sense. If there is a third vector space with a chosen finite basis, then the given linear transformation can be composed with any linear transformation leaving the target space and entering the third space. This composition is reflected in a peculiar product of the two corresponding matrices. Whenever the three vector spaces are one and the same, then we get quadratic matrices. They form a unital ring (even an algebra) under this type of product, where the identity I corresponds to the identity transformation.

Beyond dimension one, this ring will never be a field since it will not be commutative, and only non-zero matrices corresponding to bijective transformations have invertible matrices. A quadratic matrix A has an inverse A^{-1}, which by definition satisfies $AA^{-1} = I = A^{-1}A$, if and only if the determinant $|A|$ of A is non-zero. The determinant of a quadratic matrix is a scalar attached to the matrix which respects multiplication of matrices and is one on the identity matrix I. The scalar is defined by adding together with appropriate signs products of the entries of the matrix. Convenient inductive procedures are available for computing determinants.

Matrices are useful in solving systems of linear equations. For instance, the system

$$a_{11}x_1 + a_{12}x_2 = v_1$$
$$a_{21}x_1 + a_{21}x_2 = v_2$$

with variables x_1, x_2 can be written as the matrix equation $Ax = v$, where

$$A = \begin{pmatrix} a_{11} & a_{12} \\ a_{21} & a_{22} \end{pmatrix} \quad \text{and} \quad v = \begin{pmatrix} v_1 \\ v_2 \end{pmatrix} \quad \text{and} \quad x = \begin{pmatrix} x_1 \\ x_2 \end{pmatrix}.$$

The system can then be solved by 'dividing by A', or more precisely, if A is invertible, the solution is obtained as follows $x = Ix = (A^{-1}A)x = A^{-1}(Ax) = A^{-1}v$.

The inverse of a matrix can be found by invoking Cramer's rule involving determinants, which is not a very efficient method, but is of theoretical importance. A more efficient way is by Gauss-Jordan elimination, which amounts to a systematic way of eliminating variables in systems of linear equations. When there are more or less variables than equations, say when the variables are under- or over-determined, then the coefficient matrix will no longer be quadratic. The image of the corresponding linear transformation might easily fail to have the same dimension as the domain space, and the discrepancy is exactly measured by the dimension of the space of vectors killed by the transformation.

Given a finite dimensional complex vectors space V and a linear transformation $A \colon V \to V$. We are interested in scalars λ such that $A - \lambda I$ is not invertible. Fix such a scalar λ. A generalized eigenvector of A with eigenvalue λ is a non-zero vector $v \in V$ such that $(A - \lambda I)^n(v) = 0$ for some natural number n. In the final section of this chapter we prove that such vectors with respect to some eigenvalue form a linear basis for V, and when they are properly organized, the corresponding matrix becomes particularly simple, namely on Jordan canonical form. All its entries are then zero except on the diagonal (where only eigenvalues appear) and just off below the diagonal (where only ones or zeros appear). In the real case the form obtained is just slightly more complicated.

The generalized eigenvectors corresponding to $n = 1$ are simply the eigenvectors of A, as $A(v) = \lambda v$ says that A reproduces the vector up to a scalar. If they already form a basis for V, the Jordan canonical form becomes perfectly diagonal.

Crucial in the entire investigation is the so called characteristic polynomial $p(x) = |A - xI|$, as the eigenvalues are found by solving the algebraic equation $p(x) = 0$. The complex numbers are algebraically complete, meaning that p can be completely factorized into first order polynomials, while in the real case one might also need some second order polynomials, which cannot be further factorized as polynomials with real coefficients. The decomposition into such irreducible polynomials, resembles the decomposition of an integer into prime numbers, and this turns out to be more than an analogy, as we shall see in Chap. 6. What irreducible polynomials one has to deal with depends on the properties of the ground field, and this will be studied in Chaps. 7 and 8. And finally in Chap. 9, we will see that this polynomial decomposition leads to the Jordan canonical form. In that context we consider V as a module over the polynomial ring by letting a polynomial $q(x)$ act on vector $u \in V$ via the matrix A as $q(A)(u)$.

We end this chapter with a section about dual spaces, inner products and tensor products, notions which we need in the next two chapters. Inner products are generalizations of scalar products, and they lead to the familiar notion of distance between points in a vector space. Tensor products are a vehicle which turns multilinear maps into linear ones. The notion is a bit abstract, but it is introduced at this stage because we need tensor products when discussing finite dimensional representations before we embark on tensor products of rings and modules.

3.1 Vector Spaces

Having introduced real numbers, it is natural to consider the Euclidean space $\mathbb{R}^n$ of all n-tuples of real numbers. This set comes with two natural algebraic operations: addition of n-tuples, performed coordinate wise

$$x + y = (x_1 + y_1, \ldots, x_n + y_n),$$

and multiplication of an n-tuple by a scalar $a \in \mathbb{R}$, again performed coordinate wise

$$ax = (ax_1, \ldots ax_n).$$

We considered $\mathbb{R}^2$ with this sort of addition when we constructed the complex numbers. Then we also defined multiplication of two 2-tuples and produced $\mathbb{C}$ as a field, but this is not possible in the general case.

It turns out to be extremely useful to settle for addition of tuples and multiplication of such by scalars. This way some of the field properties of $\mathbb{R}$ are carried into $\mathbb{R}^n$.

The geometric picture for $\mathbb{R}^3$ is then of arrows, or vectors, starting at the origin and ending at a point $x \in \mathbb{R}^3$, with addition according to the well known parallelogram-rule, and with ax amounting to a rescaling of the arrow corresponding to x by a magnitude a when $a \geq 0$, and in addition with a rotation of the arrow so that it points in the opposite direction when a is negative. This is the geometric intuition we bring with us also when dealing with $\mathbb{R}^n$ for general n.

We could allow only rational rescaling of arrows, or more generally, consider the set $\mathbb{R}^n$ with the same addition, but with multiplication only by scalars from $\mathbb{Q} \subset \mathbb{R}$, and $\mathbb{Q}$ is after all a field.

We could also restrict to rational n-tuples, and replaced $\mathbb{R}^n$ by $\mathbb{Q}^n$ with addition as before and with multiplication by scalars from $\mathbb{Q}$. Of course, then we could not use $\mathbb{R}$ as scalars since $\mathbb{R}\mathbb{Q}$ is not a subset of $\mathbb{Q}$.

Perhaps we also would want to consider $\mathbb{C}^n$ instead of $\mathbb{R}^n$, again allowing for multiplication by scalars belonging to a field, which in this case could be $\mathbb{C}$, $\mathbb{R}$, $\mathbb{Q}$, or any other subfield of $\mathbb{C}$.

There are various combinations, and we are well-served with a notion that is sufficiently flexible and yet embodies the essential properties. Instead of talking about a set of n-tuples, we work with an arbitrary set V having two basic operations to be though of as addition and multiplication by a scalar of a field F.

Definition 3.1.1 A *vector space* V over a field F is a set with:

1. an associative, commutative binary operation $+$, called addition, that has a zero element 0 and an inverse element $-x \in V$ for each $x \in V$,

2. a map $F \times V \to V$; $(a, x) \mapsto ax$, called multiplication by scalars, such that

$$a(bx) = (ab)x, \quad (a+b)x = ax + bx, \quad a(x+y) = ax + ay, \quad 1x = x$$

for all $x, y \in V$ and $a, b \in F$, where 1 is the identity in F.

When it is clear from the context what field we have in mind, we simply speak of a vector space. Elements of vector spaces are usually called *vectors*.

Example 3.1.2 We leave it as an exercise to check that properties like

$$0x = 0 = a0, \quad (-a)x = a(-x) = -(ax)$$

automatically hold in any vector space. ◇

3.1 Vector Spaces

One can clearly restrict the scalars, so that any vector space V over a field F is a vector space over a subfield E of F. To distinguish the two cases we sometimes write V_F and V_E, respectively, for V.

Example 3.1.3 Given any field F, the set $V = F^n$ is a vector space over F with coordinate wise addition and with coordinate wise multiplication by scalars. One checks that the compatibility conditions hold, these being inherited from the field as the operations are coordinate wise.

Thus the set $\mathbb{C}^n$ under coordinate wise addition is a vector space over $\mathbb{C}$ or $\mathbb{R}$ or $\mathbb{Q}$ under coordinate wise multiplication by scalars.

A special case is obtained by considering any field F as the vector space V with addition from the field and with multiplication by scalars from a subfield E of F, turning F into a vector space over E. Briefly $V = F_E$. Such examples will be important when we in Chap. 7 study algebraic equations from the point of view of Galois theory. ◇

We state the following important observation as a trivially verified result.

Proposition 3.1.4 *Any subset W of a vector space V over a field F such that*

$$x + y \in W \quad \text{and} \quad ax \in W$$

for all $x, y \in W$ and $a \in F$, is again a vector field over F with operations given by restriction to $W \times W$ and $F \times W$. Such subsets are called subspaces of the vector field.

When $W + W \subset W$, we say that W is closed under addition, and that W is closed under multiplication by scalars if $FW \subset W$, with e.g. the understood convention that $W + W = \{x + y \mid x, y \in W\}$.

Example 3.1.5 Any subset of $\mathbb{C}^n$ that is closed under addition and multiplication by scalars is a subspace of $\mathbb{C}^n$. So for instance, the subset

$$W = \{x \in \mathbb{C}^n \mid \sum_i x_i = 0\}$$

of $\mathbb{C}^n$ is a vector space over $\mathbb{C}$ because $\sum_i (ax)_i = \sum_i ax_i = a \sum_i x_i = a0 = 0$ and $\sum_i (x+y)_i = \sum_i (x_i + y_i) = \sum_i x_i + \sum_i y_i = 0 + 0 = 0$ for any $x, y \in W$ and $a \in \mathbb{C}$, showing that $ax, x + y \in W$.

The set

$$W' = \{(x, y, z) \in \mathbb{R}^3 \mid y = 0\}$$

is a subspace of the vector space $\mathbb{R}^3$ over $\mathbb{R}$ because $0 + 0 = 0$ and $a0 = 0$ in the second coordinate. So it is a vector space over $\mathbb{R}$, and hence also over $\mathbb{Q}$. As for the geometric picture, the vector space W' over $\mathbb{R}$ is just the xz-plane in $\mathbb{R}^3$.

The sphere is not a subspace of $\mathbb{R}^3$ since it does not allow arbitrary rescaling. Neither is any line that does not contain the origin, because any subspace will contain the same zero element 0 as $0 = 0x$.

In fact, the subspaces of $\mathbb{R}^3$ are $\{0\}$, $\mathbb{R}^3$, any straight line that goes through the origin, and any plane that contains the origin.

We will realize that all subspaces of $\mathbb{R}^n$ over $\mathbb{R}$ are essentially of the form $\mathbb{R}^m$ for some $m \leq n$, but the proof of this requires some preparation. Geometrically, in $\mathbb{R}^3$ the tilted planes are just copies of the xy-plane, which can be thought of as $\mathbb{R}^2$. The sloped lines through the origin are rotated versions of the x-axis, which should be thought of as $\mathbb{R}$. ◇

We have seen that there are many vector spaces over a field F that are not subspaces of F^n; the simplest example being $\mathbb{R}_\mathbb{Q}$, which cannot be of the form $\mathbb{Q}^n$ as the latter is countable whereas $\mathbb{R}$ is uncountable.

But there are many other examples, where instead of going to subfields, one presents large sets V associated to a field F, most notably various function spaces.

Example 3.1.6 The set V of all functions from a set X to a field F is a vector space over F under pointwise addition and multiplication by scalars.

The subset

$$\{f \in V \mid f(y) = 0 \text{ for all } y \in Y\}$$

is a subspace of V for any subset $Y \subset X$. In fact, this vector space is essentially the same as the one given by all functions from $X \backslash Y$ to F under pointwise operations.

Restricting to bounded functions produces another subspace of V. More subspaces of V can be obtained by appropriately restricting the class of functions on X suggested by structures imposed on X, like that of a topology or a σ-algebra, or simply by letting X itself be a vector space. ◇

We shall see that the vector space V from the example above is infinite dimensional (whatever that means) when X is infinite, in contrast to the vector space F^n over any field F, which has finite dimension n.

A more general way of forming new vector spaces from old ones, with additional subspaces coming from various potential restrictions on the elements, is as follows:

Definition 3.1.7 The product $\prod V_i$ of vector spaces V_i over a field F is a vector space over F with coordinate wise operations, called the *algebraic product of the vector spaces* V_i. The *direct sum* $\oplus V_i$ of the vector spaces V_i is the subspace of $\prod V_i$ consisting of all elements x such that $x_i = 0$ for all but finitely many i's.

Note that for a finite family of vector spaces the direct sum and the algebraic product coincide.

Curiously enough we will see that all these constructions of vector spaces come down to one basic example.

Example 3.1.8 Let V consist of all functions from a set I to a field F that are non-zero only at finitely many points of X. This is clearly a vector spaces over F with pointwise operations. In fact, it is just $\oplus_i V_i$ with $V_i = F$ for all $i \in I$, which we denote by $\oplus_i F$. ◇

3.2 Linear Basis

We introduce some fundamental notions for vector spaces.

Definition 3.2.1 Suppose S is a subset of a vector space V over a field F. Then S *spans* V if every element of V is a *linear combination*

$$a_1 x_1 + \cdots + a_n x_n$$

of some $x_i \in S$ and $a_i \in F$. The elements of S are *linear independent* if for any $x_i \in S$, the equation

$$a_1 x_1 + \cdots + a_n x_n = 0$$

only holds when all $a_i = 0$. Otherwise the elements of S are called *linear dependent*. The collection S is a *linear basis* of V if S spans V and its elements are linear independent.

Note that S need not be finite for this definition to make sense.

The *span of a subset of a vector space* is the smallest subspace that contains the subset. It clearly consists of all linear combinations of elements of the subset.

The notion of linear independence is based on the idea that two non-zero vectors are linear independent if they are not parallel. More generally, we see that a collection of vectors are linear independent precisely when none of them is a linear combination of the others, so the picture we have in mind is that each arrow points out of the subspace spanned by any subcollection excluding the arrow, spanning together with this subcollection a strictly larger subspace. We are onto an inductive argument here. Together with Zorn's lemma we can take this further.

Theorem 3.2.2 *Every non-trivial vector space has a linear basis. In fact, any linear independent subset S of a vector space V can be enlarged to a basis S' for V, and by enlarged we mean $S \subset S'$.*

Proof The first assertion follows from the second because any subset consisting of a non-zero element x of V is linear independent; otherwise $ax = 0$ for $a \neq 0$, and then

$$0 = a^{-1} 0 = a^{-1}(ax) = (a^{-1}a)x = 1x = x \neq 0,$$

which is an absurdity.

To prove the second claim, let $\mathcal{F}$ be the family of all linear independent subsets of V that contain S. This family is non-empty because S belongs to it.

Order $\mathcal{F}$ by inclusion. Then every chain has an upper bound, namely the union of all members of the chain. By Zorn's lemma $\mathcal{F}$ has a maximal element S', which we claim is a basis for V.

We need only show that S' spans V. If not, pick any member x of V outside the span. Then $S' \cup \{x\}$ is linear independent. To convince ourselves of this crucial claim, say $S' = \{x_i\}$, and suppose that a linear combination $\sum a_i x_i + ax$ is 0 for some scalars a_i and a. If $a \neq 0$, then

$$x = \sum (-a^{-1} a_i) x_i$$

is in the span of S', which is impossible. So $a = 0$, and then $\sum a_i x_i = 0$ forces all the a_i's to be 0 as the x_i's are linear independent.

So $S' \cup \{x\}$ is linear independent and obviously contains S, so it belongs to $\mathcal{F}$. But it is strictly larger than S' as $x \notin S'$, contradicting the maximality of S'. $\square$

It can be shown that the axiom of choice follows from the other axioms of Zermelo-Frankel if every non-trivial vector space admits a basis.

The most important property of a basis is the following, which geometrically means that the vectors of a basis form a coordinate system for the vector space with axes in the direction of the basis elements and unit lengths along the axes set to be the length of the corresponding basis element.

Proposition 3.2.3 *If S is a linear basis for a vector space V over F, then any $x \in V$ can be written as*

$$x = a_1 x_1 + \cdots + a_n x_n$$

for unique vectors $x_i \in S$ and scalars $a_i \in F$. The unique scalars $\{a_i\}$ are called the coordinates of x with respect to the basis S, with a_i corresponding to x_i.

Proof Existence of such a linear combination is saying that S spans V, and uniqueness is immediate from linear independence. $\square$

Any subset S of a vector space with the property stated in this proposition is automatically a basis.

There are in general many coordinate systems or bases for a given vector space.

Example 3.2.4 The vectors $\{(1, 0), (0, 1)\}$ form a basis for $\mathbb{R}^2$ because $(a, b) = a(1, 0) + b(0, 1)$. The vectors $\{(k, 0), (1, 1)\}$ with $k \neq 0$ also form a basis for $\mathbb{R}^2$ because $a(k, 0) + b(1, 1) = (ak + b, b)$, which is $(0, 0)$ only if $b = 0$ and $ak + b = 0$, so $a = 0$, showing linear independence. And $\frac{(a-b)}{k}(k, 0) + b(1, 1) = (a, b)$ shows that they span $\mathbb{R}^2$. $\diamond$

The fact that there are infinitely many bases for any non-trivial vector space, and that none of them are natural, or point themselves out as unique in some way or

3.2 Linear Basis

other, does generate various problems, since the concept of a vector space is defined without any reference to a specific basis. And yet many vector spaces seem to come with a preselected basis, like for instance $\mathbb{R}^n$, which comes with the *standard basis* $\{e_1, \ldots, e_n\}$, where

$$e_1 = (1, 0, \ldots, 0), \ e_2 = (0, 1, 0, \ldots, 0), \ldots, e_n = (0, \ldots, 0, 1).$$

From a theoretical point of view there is nothing special about this basis, it is as good as any other basis.

Amazingly though, any two bases have the same number of elements, which is n for $\mathbb{R}^n$. The verification of such a statement in full generality is actually a bit tricky, and calls for the axiom of choice or some weaker version of it.

Theorem 3.2.5 *Any two bases in a vector space have the same cardinality.*

Proof Say we have bases $\{x_i\}_{i \in I}$ and $\{y_j\}_{j \in J}$ of a vector space V. Aiming for a contradiction, we may assume by Proposition 2.12.14 that $|J| < |I|$. We have to distinguish two cases.

Suppose first that I is infinite. By assumption any y_j is a linear combination

$$y_j = \sum_{i \in F_j} a_{ij} x_i$$

of the x's, where F_j is a finite subset of I.

By the axiom of choice we have a family $\{f_j\}_{j \in J}$ of injections $f_j : F_j \to \mathbb{N}$ as each F_j is finite. Hence we have an injection

$$\bigcup_{j \in J} F_j \to J \times \mathbb{N},$$

which sends $i \in F_j$ to $(j, f_j(i))$. Together with Proposition 2.12.16, this shows that the cardinality of $\cup F_j$ is less than that of I. So there is a $k \in I$ that does not belong to F_j for any $j \in J$.

But by assumption it must be possible to write x_k as a linear combination of the y_j's, and each of these y_j's can be written as a sum of the form

$$\sum_{i \in F_j} a_{ij} x_i$$

and k does not belong to F_j for any $j \in J$. So x_k is a linear combination of the x_i's, none of which can be x_k, and this shows that $\{x_i\}_{i \in I}$ is linear dependent; a contradiction.

Next suppose that I is finite, say with $|I| = n$ and $|J| = m$, so $m < n$. Then

$$\{x_1, y_1, y_2, \ldots, y_m\}$$

spans V as $\{y_i\}$ already spans V. But $x_1 \neq 0$ because $\{x_i\}$ is linear independent. Writing x_1 as a linear combination of the y's, then at least one of the coefficients is non-zero, and the corresponding y can therefore be written as a linear combination of x_1 and the other y's. Removing this y we still get vectors that span V. Upon renumbering the indices of the y's, we therefore get a list

$$x_1, y_1, \cdots, y_{m-1}$$

of m vectors spanning V.

Next, the list with x_2 included will certainly also span V. Writing $x_2 \neq 0$ as a non-trivial linear combination of x_1 and the $m-1$ new y's, then not all the coefficients of the y's can be 0 as otherwise x_2 would be a rescaling of x_1, and $\{x_i\}$ is linear independent. Removing any y with non-trivial coefficient, we get m vectors that still span V, say

$$x_1, x_2, y_1, \cdots, y_{m-2},$$

where again we have renumbered the indices of the y's.

We can continue this way and inductively remove y's, till we get a list

$$x_1, \ldots, x_n, y_1, \cdots, y_{m-n}$$

of m vectors that still span V. In particular, we see that $n \leq m$, which is a contradiction. □

We now know that a basis exists, and that any two of them have the same cardinality, so the following definition makes sense.

Definition 3.2.6 The *dimension* dim V of a vector space V is the cardinal number of any basis for it. If this cardinal number is finite n, we talk about a *finite dimensional vector space* of dimension n. Otherwise we say that the vector space is *infinite dimensional*.

Example 3.2.7 The vector space F^n for any field F is finite dimensional with dimension n, whereas $\mathbb{R}_\mathbb{Q}$ is an infinite dimensional vector space. We consider F^0 as the *trivial vector space* $\{0\}$, which has dimension 0 as there are no linear independent vectors in $\{0\}$. Note that every vector space contains the trivial vector space as a subspace, and that the empty set is not a vector space. ◇

Corollary 3.2.8 *Say V is a vector space of finite dimension n. Then any collection of vectors in V that spans V has to have at least n elements. And any linearly independent collection of elements in V cannot contain more than n elements.*

Proof From the spanning collection keep removing members, one by one, that are linear combinations of others till they are linear independent. Then you have a basis counting n members by Theorem 3.2.5.

Alternatively, one can get the result from the last inductive part of the proof of the same theorem, where we only used that $\{y_i\}$ spanned V.

The last claim is immediate from Theorems 3.2.2 and 3.2.5. □

3.3 Linear Transformations

All vector spaces considered in this section are over a fixed field F.

In relating vector spaces to each other, the following notion is very natural.

Definition 3.3.1 A *linear transformation* or a *linear operator* A is a map from a vector space V to a vector space W satisfying

$$A(ax + by) = aA(x) + bA(y)$$

for all $x, y \in V$ and scalars a, b. Denote the set of linear transformations from V to W by $\text{End}(V, W)$, and use $\text{End}(V)$ for $\text{End}(V, V)$. Elements of $\text{End}(V)$ are sometimes referred to as linear operators on V.

Note that any linear transformation takes 0 to 0.

A linear transformation $A: V \to W$ is completely determined by its action on any basis $\{x_i\}$ of V. To see this, write any $x \in V$ as a unique linear combination

$$x = \sum a_i x_i$$

and then

$$A(x) = \sum a_i A(x_i),$$

which shows that $A(x)$ is uniquely determined by the vectors $A(x_i)$.

Conversely, any map that ascribes arbitrary vectors $A(x_i)$ in W to the basis elements x_i, defines a linear $A: V \to W$ according to the formula

$$A(x) = \sum a_i A(x_i),$$

for $x \in V$ written in the form $x = \sum a_i x_i$.

Definition 3.3.2 A *linear isomorphism* is a bijective linear transformation. Two vector spaces V and W are said to be *isomorphic vector spaces*, with symbolic notation $V \cong W$, if there exists a bijection between them.

Linear isomorphisms preserve the two linear operations on a vector space, and identify the vector spaces as sets, so isomorphic vector spaces are essentially the same, and are often identified, sometimes with the isomorphism suppressed.

Example 3.3.3 Given a family $\{V_i\}$ of vector subspaces of a vector space V such that every element of V is a finite sum of elements from the family, and such that $\sum v_i = 0$ for $v_i \in V_i$ implies $v_i = 0$. Then evidently the map $\oplus V_i \to V$ which sends $\{x_i\}$ to $\sum x_i$ is a linear isomorphism. The image of this map is sometimes called the *internal direct sum* of the family. ◇

If $A \in \text{End}(V, W)$ and $B \in \text{End}(U, V)$, then clearly the composition AB belongs to $\text{End}(U, W)$, and it will be an isomorphism if both A and B are. So the relation of being isomorphic is transitive. Also it is reflexive with the identity map $I \in \text{End}(V)$ as an isomorphism. That the relation is also symmetric, is clear as the inverse map A^{-1} of a linear isomorphism $A \colon V \to W$ is also a linear isomorphism. So the relation of being linearly isomorphic is an equivalence relation.

Theorem 3.3.4 *A vector space is uniquely determined by its dimension, that is, if V and W are two vector spaces with $\dim V = \dim W$, then $V \cong W$. Thus every vector space V over a field F is isomorphic to $\oplus_{i \in I} F$ for some set I with $|I| = \dim V$.*

Proof Pick bases $\{x_i\}_{i \in I}$ and $\{y_j\}_{j \in J}$ for V and W, respectively, and pick a bijection $f \colon I \to J$. Then the linear map $A \colon V \to W$ uniquely determined by $A(x_i) = y_{f(i)}$ is a linear isomorphism with inverse given by $A^{-1}(y_j) = x_{f^{-1}(j)}$.

For the second assertion, define for each $i \in I$, functions $f_i \colon I \to F$ by $f_i(i) = 1$ and $f_i(j) = 0$ for $j \ne i$, and note that $\{f_i\}_{i \in I}$ is a basis for $\oplus_{i \in I} F$, which shows that $\dim(\oplus_{i \in I} F) = |I|$. □

One can think of dim as a bijection from the class of equivalence classes of vector spaces to the class of cardinal numbers, except that the class of equivalence classes of vector spaces is not a set in the ZF-axioms, and neither is the class of all cardinal numbers. But we have vector spaces of every dimension and up to isomorphisms they are distinguished by their dimension. In this sense dim is a complete invariant for all vector spaces, classifying them in terms of cardinal numbers, which for finite dimensional vector spaces are the non-negative integers.

Definition 3.3.5 We define the *image* and *kernel* of a linear transformation $A \colon V \to W$ to be $\text{im} A \equiv A(V)$ and $\ker A \equiv \{x \in V \mid A(x) = 0\}$, respectively.

Note that $\text{im} A$ and $\ker A$ are subspaces of W and V, respectively, so the following result known as the *rank-nullity theorem* makes sense.

Proposition 3.3.6 *Suppose $A \colon V \to W$ is a linear transformation and that V finite dimensional. Then*
$$\dim V = \dim \ker A + \dim \text{im} A.$$

Proof Clearly $\dim \text{im} A < \infty$. Let $\{x_i\}$ be a basis for $\ker A$, and pick vectors y_j in V so that $\{A(y_j)\}$ is a basis for $\text{im} A$. It is enough to show that $\{x_i\} \cup \{y_j\}$ is a basis for V. They span V because if $x \in V$, then $A(x) = \sum b_j A(y_j)$ for some b's, and $x - \sum b_j y_j \in \ker A$ shows that $x = \sum a_i x_i + \sum b_j y_j$ for some a's. They are also linear independent because if $\sum_i a_i x_i + \sum_j b_j y_j = 0$, then $\sum_j b_j A(y_j) = 0$ as all $A(x_i) = 0$. But $\{A(y_j)\}$ is linear independent, so all $b_j = 0$, and then in turn all $a_i = 0$ as $\{x_i\}$ is linear independent. □

Corollary 3.3.7 *A linear operator on a finite dimensional vector space is injective if and only if it is surjective. So it is an isomorphism if it is either injective or surjective.*

Proof Note that $A \in \text{End}(V)$ is injective if and only if $\ker A = \{0\}$, because $A(x) = A(y)$ if and only if $x - y \in \ker A$.

Hence by Proposition 3.3.6 we see that $\dim A(V) = \dim V$ if A is injective, so any basis for $A(V)$ will also be a basis for V, and therefore $A(V) = V$. If A is surjective, then by Proposition 3.3.6, we see that $\dim \ker A = 0$, so $\ker A = \{0\}$ and A is injective. □

Given $A, B \in \text{End}(V, W)$ and scalars $a, b \in F$, we define $aA + bB \in \text{End}(V, W)$ by $(aA + bB)(x) = aA(x) + bB(x)$, turning $\text{End}(V, W)$ into a vector space over F for which $aA + bB$ is a linear combination of A and B.

When $V = W$ the composition AB belongs to $\text{End}(V)$ making $\text{End}(V)$ a unital algebra over F with composition as multiplication.

Definition 3.3.8 An *algebra $\mathcal{A}$ over a field F* is a vector space over F together with multiplication xy turning $(\mathcal{A}, +, \cdot)$ into a ring, and such that

$$a(xy) = (ax)y = x(ay)$$

for $x, y \in \mathcal{A}$ and $a \in F$. The algebra $\mathcal{A}$ is said to be a *unital algebra* if the ring $(\mathcal{A}, +, \cdot)$ has an identity, and $\mathcal{A}$ is a *commutative algebra* if the ring is. An element in the algebra is *invertible* if it is invertible in the ring.

The following example shows that the algebra $\text{End}(V)$ is only commutative when $\dim V \leq 1$.

Example 3.3.9 Let $V = F^2$ with standard basis $\{e_1, e_2\}$. Define $A, B \in \text{End}(F^2)$ by $A(e_1) = e_2, A(e_2) = 0$ and $B(e_1) = 0, B(e_2) = e_1$. Then $AB(e_1) = 0$ and $AB(e_2) = e_2$, whereas $BA(e_1) = e_1$ and $BA(e_2) = 0$. So $AB \neq BA$. ◇

3.4 Matrices

In this section all vector spaces are assumed to be finite dimensional.

Linear transformations between finite dimensional vector spaces can conveniently be represented by matrices of scalars.

Definition 3.4.1 An $m \times n$ *matrix* (a_{ij}) over a field F is a map

$$(i, j) \mapsto a_{ij} \in F,$$

where $i \in \{1, \ldots, m\}$ and $j \in \{1, \ldots, n\}$. The scalar a_{ij} is the ij-*entry* of (a_{ij}).

It is common to visualize a matrix $A = (a_{ij})$ as a rectangular array of m rows and n columns:

$$A = \begin{pmatrix} a_{11} & a_{12} & \cdots & a_{1n} \\ a_{21} & a_{22} & \cdots & a_{2n} \\ \cdot & \cdot & \cdots & \cdot \\ \cdot & \cdot & \cdots & \cdot \\ a_{m1} & a_{m2} & \cdots & a_{mn} \end{pmatrix}.$$

Definition 3.4.2 The *matrix of* $A \in \text{End}(V, W)$ *with respect to bases* $\{x_j\}_{j=1}^n$ and $\{y_i\}_{i=1}^m$ of V and W, respectively, is the $m \times n$-matrix $\tilde{A}$ with entries a_{ij} uniquely determined by

$$A(x_j) = \sum_{i=1}^m a_{ij} y_i.$$

Note that the coordinates of the vector $A(x_j)$ is the jth column of the matrix. More generally, if $x = \sum c_j x_j$, then

$$A(x) = \sum_i (\sum_j a_{ij} c_j) y_i \qquad (3.1)$$

has coordinates $\sum_j a_{ij} c_j$ with respect to $\{y_i\}$.

Example 3.4.3 Consider $A \in \text{End}(F^2, F^3)$ with $A(e_1) = e_2$ and $A(e_2) = e_3$ in standard bases for F^2 and F^3. Then the 3×2-matrix (a_{ij}) of A with respect to these bases is

$$\begin{pmatrix} 0 & 0 \\ 1 & 0 \\ 0 & 1 \end{pmatrix}$$

$\diamond$

Proposition 3.4.4 *The map* $: A \mapsto \tilde{A}$ *from* $\text{End}(V, W)$ *to the set of all* $m \times n$-*matrices over* F *is a bijection.*

Proof Injectivity is clear. Surjectivity follows from the fact that the formula for $A(x)$ in Eq. 3.1 defines $A \in \text{End}(V, W)$ for any matrix (a_{ij}) over F, and this matrix will by definition be the matrix $\tilde{A}$ with respect to the bases $\{x_j\}$ and $\{y_i\}$ of V and W, respectively. $\square$

Using the bijection $A \mapsto \tilde{A}$ we can transfer operations on $\text{End}(V, W)$ to the set of all $m \times n$-matrices over F. The linear combination $aA + bB$ of $A, B \in \text{End}(V, W)$ corresponds then to the matrix $(aa_{ij} + bb_{ij})$, turning the set of $m \times n$-matrices over F into a vector space with coordinate wise addition and multiplication by scalars.

3.4 Matrices

Suppose U is a finite dimensional vector space with basis $\{z_k\}_{k=1}^p$, and say (b_{jk}) is the $n \times p$-matrix of $B \in \mathrm{End}(U, V)$ with respect to bases $\{z_k\}$ and $\{x_j\}$, respectively, so

$$B(z_k) = \sum_j b_{jk} x_j.$$

Then the $m \times p$-matrix (c_{ik}) of $AB \in \mathrm{End}(U, W)$ with respect to bases $\{z_k\}$ and $\{y_i\}$, respectively, is given by

$$c_{ik} = \sum_j a_{ij} b_{jk}$$

because

$$(AB)(z_k) = A(B(z_k)) = \sum_j b_{jk} A(x_j) = \sum_j b_{jk} \sum_i a_{ij} y_i = \sum_i (\sum_j a_{ij} b_{jk}) y_i.$$

We define matrix multiplication accordingly.

Definition 3.4.5 The *(matrix) product* of an $m \times n$-matrix $A = (a_{ij})$ and an $n \times p$-matrix $B = (b_{ij})$ is the $m \times p$-matrix AB with ik-entry

$$\sum_j a_{ij} b_{jk}$$

gotten by taking the *scalar product* of the ith row of A with the kth column of B.

Definition 3.4.6 Let $M(n, F)$ be the set of $n \times n$-matrices over a field F.

We consider $M(n, F)$ an algebra over F under matrix multiplication and coordinate wise linear combinations, so by definition we have the following result.

Proposition 3.4.7 *Let* $n = \dim V$. *Then* $\mathrm{End}(V)$ *and* $M(n, F)$ *are isomorphic as unital algebras.*

The isomorphism $A \mapsto \tilde{A}$ is not canonical since it depends on some chosen basis $\{x_j\}$ for V.

Let $\delta_{ij} = 1$ if $i = j$ and $\delta_{ij} = 0$ for $i \neq j$. The *identity matrix* is the $n \times n$-matrix $I_n = (\delta_{ij})$ with 1's along the diagonal and 0's everywhere else. Obviously I_n is the unity $\tilde{I}$ of the algebra $M(n, F)$. We sometimes write I for I_n.

Any linear isomorphism $A \in \mathrm{End}(V)$ produces an invertible matrix $\tilde{A} \in M(n, F)$ with inverse $(\tilde{A})^{-1} = \widetilde{A^{-1}}$.

If we consider another basis $\{y_j\}$ for V and define $C \in \mathrm{End}(V)$ by $C(x_j) = y_j$, then C will be invertible. Moreover, it is easy to see that $\tilde{C}\tilde{A}\tilde{C}^{-1}$ will be the matrix of $A \in \mathrm{End}(V)$ with respect to the basis $\{y_j\}$.

Definition 3.4.8 Two matrices $A, B \in M(n, F)$ are *similar matrices*, and we write $A \sim B$, if there is an invertible matrix $C \in M(n, F)$ such that $B = CAC^{-1}$.

Clearly $\sim$ is an equivalence relation on $M(n, F)$, and we have seen that matrices of operators on V with respect to different bases are always similar. Also it is clear that similarity accounts exactly for the discrepancy of representing a linear operator in terms of matrices associated to different bases.

Example 3.4.9 The linear operators $A, B \in \text{End}(F^2)$ from Example 3.3.9 can be represented by matrices

$$A = \begin{pmatrix} 0 & 0 \\ 1 & 0 \end{pmatrix}, \quad B = \begin{pmatrix} 0 & 1 \\ 0 & 0 \end{pmatrix}$$

with respect to the standard basis $\{e_1, e_2\}$ for F^2. Matrix multiplication then gives

$$AB = \begin{pmatrix} 0 & 0 \\ 0 & 1 \end{pmatrix} \neq BA = \begin{pmatrix} 1 & 0 \\ 0 & 0 \end{pmatrix}$$

and again we see that $AB(e_1) = BA(e_2) = 0$ and $AB(e_2) = e_2$ and $BA(e_1) = e_1$. Yet $AB \sim BA$ as $AB = C(BA)C^{-1}$ with

$$C = \begin{pmatrix} 0 & 1 \\ 1 & 0 \end{pmatrix} = C^{-1}.$$

The linear isomorphism of the matrix C sends e_1 to e_2 and e_2 to e_1. So AB, seen as a matrix with respect to bases $\{e_2, e_1\}$ will equal BA, seen as a matrix with respect to bases $\{e_1, e_2\}$. ◇

Definition 3.4.10 The $n \times n$-matrix $E_{kl} = (\delta_{ik}\delta_{jl})$ is called a *matrix unit* for the algebra $M(n, F)$.

Note that the identity matrix I_n for $M(n, F)$ equals $\sum_i E_{ii}$. The following result is straightforward.

Proposition 3.4.11 *The matrix units* $\{E_{ij}\}$ *form a basis for* $M(n, F)$ *and*

$$A = \sum_{ij} a_{ij} E_{ij}$$

for $A = (a_{ij})$. *Hence the product in* $M(n, F)$ *is completely specified by the formulas:*

$$E_{ij} E_{kl} = \delta_{jk} E_{il}.$$

Example 3.4.12 With A and B from Example 3.4.9, we see that $A = E_{21}$ and $B = E_{12}$, so $AB = E_{22}$ and $BA = E_{11}$, which is consistent with Proposition 3.4.11. ◇

Definition 3.4.13 The *transpose* of an $m \times n$-matrix A with ij-entry a_{ij} is the $n \times m$-matrix A^T with ij-entry a_{ji}.

The operation of taking transposes is a linear map from the vector space of $m \times n$-matrices to the vector space $n \times m$-matrices, and if you take the transpose twice, you obviously get the identity map. If B is an $n \times p$-matrix, so AB makes sense, then clearly $(AB)^T = B^T A^T$. Hence if A is invertible, its transpose is also invertible and $(A^T)^{-1} = (A^{-1})^T$.

Definition 3.4.14 The *complex conjugate* $\bar{A}$ of a complex matrix A is gotten by taking the complex conjugate of each entry of A. The *adjoint* A^* of A is $\bar{A}^T$.

Clearly $\bar{A}^T = \overline{A^T}$ and $A^{**} = A$ and $\overline{(AB)} = \bar{A}\bar{B}$, so $(AB)^* = B^*A^*$ and $(A^*)^{-1} = (A^{-1})^*$ whenever this makes sense.

Definition 3.4.15 The *trace* $\mathrm{Tr}(A)$ of an $m \times n$-matrix A with ij-entry a_{ij} in a field is the element of the field given by $\mathrm{Tr}(A) = \sum_i a_{ii}$.

If $A, B \in M(n, F)$, then $\mathrm{Tr}(A^T) = \mathrm{Tr}(A)$ and $\mathrm{Tr}(AB) = \sum a_{ij}b_{ji} = \mathrm{Tr}(BA)$, so $\mathrm{Tr}(BAB^{-1}) = \mathrm{Tr}(A)$ whenever B is invertible. Hence we can speak of the trace of an endomorphism A, then being the trace of any associated matrix. The trace of the identity endomorphism on a vector space V is the dimension of V. Note that $\mathrm{Tr}(A^*) = \overline{\mathrm{Tr}(A)}$ when A is complex, and that $\mathrm{Tr}(A^*A) = \sum_{ij} |a_{ij}|^2 > 0$ when $A = (a_{ij}) \neq 0$.

3.5 Systems of Linear Equations

Linear algebra grew out of studying linear equations. The following assembly

$$a_{11}x_1 + \cdots + a_{1n}x_n = b_1$$
$$a_{21}x_1 + \cdots + a_{2n}x_n = b_2$$
$$\dots\dots\dots\dots$$
$$a_{m1}x_1 + \cdots + a_{mn}x_n = b_m$$

is a *system of m linear equations in n unknowns x_i with coefficients a_{ij} in a field F*. When the scalars $b_i \in F$ are all 0, we talk about a *homogeneous system* as opposed to an *inhomogeneous system*.

Using standard bases, introduce vectors $x = \sum x_j e_j \in F^n$ and $b = \sum b_i e_i \in F^m$ and a linear transformation $A \in \mathrm{End}(F^n, F^m)$ given by $A(e_j) = \sum_i a_{ij} e_i$. Then the system of linear equations above is equivalent to the single equation

$$A(x) = b.$$

The same would be true if we worked with other bases than the standard bases.

In this setting it is customary to write n-tuples as column vectors, or as $n \times 1$-matrices, with coordinates ordered from top to bottom, and then $A(x)$ is the column vector obtained by matrix multiplication

$$\begin{pmatrix} a_{11} & a_{12} & \cdots & a_{1n} \\ a_{21} & a_{22} & \cdots & a_{2n} \\ \cdot & \cdot & \cdots & \cdot \\ \cdot & \cdot & \cdots & \cdot \\ a_{m1} & a_{m2} & \cdots & a_{mn} \end{pmatrix} \begin{pmatrix} x_1 \\ x_2 \\ \cdot \\ \cdot \\ x_n \end{pmatrix}.$$

The study of the system of linear equations have quickly become an issue about linear transformations. Finding solutions of the system of equations simply amounts to finding vectors $x \in F^n$ that satisfy $A(x) = b$. Such vectors exist exactly when b belongs to the image of the linear transformation A, and there is only one solution exactly when A is injective. So a unique solution x of the system of linear equations exists precisely when

$$\ker A = \{0\} \quad \text{and} \quad b \in \operatorname{im} A.$$

In the case $m = n$, this happens exactly when A is invertible, and then the solution is $x = A^{-1}(b)$. So it becomes imperative to find the inverse of a linear transformation, or in other words, to find the inverse of a matrix. Hence we need algorithms for inverting matrices.

Before we enter this, let us look at the more general case where m might be different from n. How do we solve such systems?

First let us remove the surplus x. The *augmented matrix* of the system $A(x) = b$ is the $m \times (n+1)$-matrix

$$(a_1 \ \cdots \ a_n \ b),$$

where $a_i = \sum_m a_{mi} e_m$ is the ith *column vector* of A.

We 'simplify' this matrix by successively performing three types of operations, so called *elementary row operations*:

(O_1) interchanging two rows;

(O_2) multiplying a row by a non-zero scalar;

(O_3) multiplying a row by a scalar and then adding this row to another row.

None of these operations performed on the corresponding system $A(x) = b$, with row replaced by equation, will alter its solution set.

Definition 3.5.1 If a matrix A is obtained from another matrix B by a (perhaps empty) sequence of elementary row operations, then A and B are *row equivalent*, and we write $A \sim_r B$.

The relation $\sim_r$ is an equivalence relation. Reflexivity follows because each elementary row operation of type O_i is obviously reversible by an operation of the same type.

3.5 Systems of Linear Equations

Example 3.5.2 Applying row operations of the type O_1, then O_2 and then O_3, we get
$$\begin{pmatrix} 1 & 0 \\ 0 & 0 \end{pmatrix} \sim_r \begin{pmatrix} 0 & 0 \\ 1 & 0 \end{pmatrix} \sim_r \begin{pmatrix} 0 & 0 \\ 5 & 0 \end{pmatrix} \sim_r \begin{pmatrix} 7 & 0 \\ 5 & 0 \end{pmatrix}.$$

◇

Note that $A(x) = a_1 x_1 + \cdots + a_n x_n$, so $\operatorname{im} A = \operatorname{span}\{a_i\}$. Also, when $A \sim_r B$, we know that $A(x) = 0$ if and only if $B(x) = 0$, so linear dependence relations of column vectors of row equivalent matrices are the same.

By a 'simplification' of the augmented matrix, we mean the following.

Definition 3.5.3 A matrix is in *reduced echelon form* if:
(i) all zero-rows are at the bottom;
(ii) every *leading entry in a row* i.e., the leftmost non-zero entry in the row, is 1;
(iii) the leading entry is the only non-zero entry in its column;
(iv) every leading entry is to the right of a leading entry in the row above.

Example 3.5.4 The matrices
$$\begin{pmatrix} 1 & 0 & 4 & 3 & 0 & 0 & 0 \\ 0 & 1 & 2 & 2 & 1 & 0 & 0 \\ 0 & 0 & 0 & 0 & 0 & 1 & 0 \end{pmatrix} \quad \text{and} \quad \begin{pmatrix} 0 & 1 & 0 \\ 0 & 0 & 0 \\ 0 & 0 & 0 \end{pmatrix}$$

are in reduced echelon form.

Note that the *leading columns*, that is, those columns with a leading entry in them, form a basis for the vector space spanned by all column vectors of these matrices. This is true in general. In fact, we see that the first leading columns form a basis for the vector space of all columns not to the right of any of these leading columns. Hence we get the following result.

Proposition 3.5.5 *The columns in a matrix A corresponding to the leading columns of a row equivalent matrix on reduced echelon form is a basis for* $\operatorname{im} A$.

Gauss-Jordan elimination is an algorithm that brings a matrix A to reduced echelon form by elementary row operations, and goes as follows:

(i) Take any leftmost leading entry, rescale it to 1 by an operation of the type O_2, then move it to the first row by an operation of the type O_1;

(ii) Apply operations of the type O_3 to provide only 0's as entries below the leading entry in the same column;

(iii) Repeat (i) and (ii) to the matrix with the first row of the previous matrix covered (or ignored), and continue with additional covering till there are no more non-zero rows to modify.

At this stage you are finished with the forward phase, and have now transformed the matrix to echelon form. To get it in reduced echelon form, you must enter the backward phase:

(iv) Beginning with the rightmost leading entry, and working upward and to the left, create 0's above each leading entry by applying operations of the type O_3.

Theorem 3.5.6 *Each matrix is row equivalent to one and only one reduced echelon form.*

Proof That a matrix A is row equivalent to some matrix B in reduced echelon form, is guaranteed by Gauss-Jordan eliminating A.

Suppose C is another candidate, say obtained by Gauss-Jordan eliminating A differently. We claim that $C = B$.

As opposed to any other column, a leading column of a matrix in reduced echelon form is one that is not a linear combination of any column to its left. The position of such a column can be read straight off the linear dependence relations between the columns. As these relations are the same for row equivalent matrices, the matrices B and C have their leading columns at the same positions. But then these columns are the same because the first leading column of a matrix in reduced echelon form looks like e_1, and the second leading column looks like e_2, and so on.

The other columns of B and C must also coincide because they are the same unique linear combinations of leading columns. □

To solve $A(x) = b$, Gauss-Jordan eliminate the augmented matrix to reduced echelon form, and consider this new matrix as an augmented matrix for a new system of equations with the same solution set. In writing out this system, consider any x_i that does not have a leading entry as coefficient, as a free parameter, and find the other x's with respect to these parameters by back substitution, working upwards towards the left.

This way of solving a system of linear equations is quite efficient. It can be shown that for a system of n equations in n unknowns, the number of arithmetic operations needed is less than some constant times n^3.

Example 3.5.7 We want to solve the linear system

$$2x_2 + 2x_3 = 8$$
$$3x_1 - 15x_3 = 3$$
$$x_2 + x_3 = 4.$$

This can of course be solved directly by some non-systematic way, but let us follow the recipe to illustrate the general method. Here $A(x) = b$ with

$$A = \begin{pmatrix} 0 & 2 & 2 \\ 3 & 0 & -15 \\ 0 & 1 & 1 \end{pmatrix}, \quad b = \begin{pmatrix} 8 \\ 3 \\ 4 \end{pmatrix} \quad \text{and} \quad x = \begin{pmatrix} x_1 \\ x_2 \\ x_3 \end{pmatrix}.$$

3.5 Systems of Linear Equations

We then Gauss-Jordan eliminate the augmented matrix

$$(A\ b) = \begin{pmatrix} 0 & 2 & 2 & 8 \\ 3 & 0 & -15 & 3 \\ 0 & 1 & 1 & 4 \end{pmatrix} \sim_r \begin{pmatrix} 3 & 0 & -15 & 3 \\ 0 & 1 & 1 & 4 \\ 0 & 2 & 2 & 8 \end{pmatrix}$$

$$\sim_r \begin{pmatrix} 1 & 0 & -5 & 1 \\ 0 & 1 & 1 & 4 \\ 0 & 2 & 2 & 8 \end{pmatrix} \sim_r \begin{pmatrix} 1 & 0 & -5 & 1 \\ 0 & 1 & 1 & 4 \\ 0 & 0 & 0 & 0 \end{pmatrix},$$

where we in the first step interchanged rows, and in the second step we rescaled the first row, and in the last step we multiplied the second row by -2 and added this row to the third row. The final result is a matrix in reduced echelon form. The leading columns are in the first and second place, and they have leading coefficients 1 as required. According to Proposition 3.5.5 one basis for im A consists of the first two columns of A.

To solve $A(x) = b$ we write down the system corresponding to the augmented matrix in reduced echelon form:

$$x_1 - 5x_3 = 1$$
$$x_2 + x_3 = 4.$$

The free parameter is x_3, say $x_3 = t$. Solving with respect to t gives

$$x_1 = 5t + 1$$
$$x_2 = -t + 4,$$

so we have infinitely many solutions; one for each choice of t. This is typical for solutions of systems of linear equations; either there is no solution, or there is exactly one, or there are infinitely many, and in the latter case there can be one or several free parameters.

Geometrically each equation corresponds in this case to a plane in space, so the solution of all 3 equations is the intersections of these planes, being the intersection of the solution sets of each single equation. For a solution to exist no pairs of planes can be parallel unless they coalesh. Here there are three possibilities; either none of them coalesh, and then there is only one solution, or two of them coalesh and intersect the third plane in a straight line parametrized by one parameter, or finally all three planes coincide, so one gets a plane described by two free parameters.

Note that one basis for ker A consists of

$$\begin{pmatrix} 5 \\ -1 \end{pmatrix},$$

gotten by putting $b = 0$ in the calculation above and looking at the solution x. ◇

We can also use Gauss-Jordan elimination to find the inverse of a matrix. To this end, first notice that the effect of an elementary row operation O_E on an $n \times n$-matrix A is to produce the matrix EA, where E is the *elementary matrix* obtained by applying the same operation O_E to the identity matrix I_n. This is straightforward to check.

Example 3.5.8 Consider
$$A = \begin{pmatrix} a & b & c \\ d & e & f \\ g & h & i \end{pmatrix}.$$

Swapping O_E of the first two rows of A produces EA, because
$$EA = \begin{pmatrix} 0 & 1 & 0 \\ 1 & 0 & 0 \\ 0 & 0 & 1 \end{pmatrix} \begin{pmatrix} a & b & c \\ d & e & f \\ g & h & i \end{pmatrix} = \begin{pmatrix} d & e & f \\ a & b & c \\ g & h & i \end{pmatrix}.$$

◇

Moreover, the elementary matrices E obtained this way are all invertible as the elementary row operations are reversible.

Now suppose we have Gauss-Jordan eliminated A and gotten I_n, say after k steps. Then $E_k \cdots E_1 A = I_n$, where E_i is the elementary matrix corresponding to the ith operation. But then $A^{-1} = E_k \cdots E_1$ is the matrix obtained by applying the same elementary row operations to I_n. To keep track of the operations applied, we consider the enlarged matrix $(A|I_n)$ and Gauss-Jordan eliminate this one to obtain $(I_n|A^{-1})$, where A^{-1} can be read off directly.

If we do not obtain I_n by Gauss-Jordan eliminating A, then the system $A(x) = 0$ has more than one solution, and therefore A is not invertible.

Example 3.5.9 We have
$$\begin{pmatrix} 1 & 2 & | & 1 & 0 \\ 1 & 1 & | & 0 & 1 \end{pmatrix} \sim_r \begin{pmatrix} 1 & 2 & | & 1 & 0 \\ 0 & -1 & | & -1 & 1 \end{pmatrix}$$
$$\sim_r \begin{pmatrix} 1 & 2 & | & 1 & 0 \\ 0 & 1 & | & 1 & -1 \end{pmatrix} \sim_r \begin{pmatrix} 1 & 0 & | & -1 & 2 \\ 0 & 1 & | & 1 & -1 \end{pmatrix},$$

so
$$\begin{pmatrix} 1 & 2 \\ 1 & 1 \end{pmatrix}^{-1} = \begin{pmatrix} -1 & 2 \\ 1 & -1 \end{pmatrix},$$

which can readily be verified.

◇

3.6 Permutations

Denote the set of *permutations*, or bijections, on $\{1, \ldots, n\}$ by S_n. A permutation $\sigma \in S_n$ can be exhibited in the form

$$\begin{pmatrix} 1 & 2 & \cdots & n \\ \sigma(1) & \sigma(2) & \cdots & \sigma(n) \end{pmatrix}.$$

Definition 3.6.1 A *cycle* $(i_1, \ldots, i_m)$ is a permutation $\sigma \in S_n$ such that $\sigma(i_k) = i_{k+1}$ for $k < m$ and $\sigma(i_m) = i_1$ and which keeps all other elements i fixed, meaning that $\sigma(i) = i$. Cycles with no numbers in common are *disjoint cycles*. Any cycle of length 2 is called a *transposition*.

So a cycle permutes some of the numbers cyclically when spaced on a circle as an orbit. The order in which disjoint cycles are composed is obviously irrelevant, and e.g. the cycles $(2, 3, 5, 1)$ and $(5, 1, 2, 3)$ are identical whether they both act on $\{1, \ldots, 5\}$ or both act on $\{1, \cdots, 8\}$.

Proposition 3.6.2 *Any permutation can be written as a composition of disjoint nontrivial cycles, and modulo order, such a composition is unique.*

Proof Pick $i \in \{1, \ldots, n\}$ and apply the permutation σ repeatedly to i just till $\sigma^m(i) = i$. Then $(i, \sigma(i), \ldots, \sigma^{m-1}(i))$ is a cycle that represents the action of σ on the numbers $\{i, \sigma(i), \ldots, \sigma^{m-1}(i)\}$. Next pick an element in the complement of these numbers, and repeat the argument to get another cycle that represents the permutation on another subset of $\{1, \ldots, n\}$. Continue this till all elements of the set have fallen into orbits, and you have the desired disjoint cycles.

Clearly every element of $\{1, \ldots, n\}$ will fall into exactly one of these orbits, so the decomposition is unique as described. □

A further decomposition into transpositions is also possible.

Corollary 3.6.3 *Every permutation can be written as a composition of transpositions.*

Proof It is easily verified that any cycle $(i_1, \ldots, i_m)$ can be written as the decomposition

$$(i_1, i_m)(i_1, i_{m-1}) \cdots (i_1, i_2)$$

of $m - 1$ transpositions. Now the result follows from the proposition. □

Example 3.6.4 We have the decomposition

$$\begin{pmatrix} 1 & 2 & 3 & 4 & 5 & 6 \\ 4 & 6 & 2 & 1 & 3 & 5 \end{pmatrix} = (1\ 4)(2\ 6\ 5\ 3)$$

and
$$(2\ 6\ 5\ 3) = (2\ 3)(2\ 5)(2\ 6) = (3\ 2\ 6\ 5) = (3\ 5)(3\ 6)(3\ 2).$$

◇

This example shows that the decomposition into transpositions is not unique, not even the total number of them occurring; we could have composed further with $(5\ 6)(6\ 5)$ without altering the permutation.

However, we will see that the number of transpositions in any decomposition of a permutation into transpositions is either always even or always odd, and this allows us to define the *sign of a permutation*.

Theorem 3.6.5 *There exists a unique map* sign: $S_n \to \{\pm 1\}$ *that is* -1 *on transpositions and satisfies* $\text{sign}(\sigma\tau) = \text{sign}(\sigma)\text{sign}(\tau)$ *for any* $\sigma, \tau \in S_n$.

Proof Consider the function $f: \mathbb{Z}^n \to \mathbb{Z}$ given by

$$f(x_1, \ldots, x_n) = \prod_{i<j}(x_j - x_i),$$

where the product ranges over all pairs (i, j) of integers with $1 \le i < j \le n$.

Let $\sigma \in S_n$ act on f giving another function $\sigma(f): \mathbb{Z}^n \to \mathbb{Z}$ defined according to the formula

$$\sigma(f)(x_1, \ldots, x_n) = \prod_{i<j}(x_{\sigma(j)} - x_{\sigma(i)}).$$

Then clearly $(\sigma\tau)(f) = \sigma(\tau(f))$ for any $\tau \in S_n$.

Moreover, if τ is a transposition, then $\tau(f) = -f$. To see this, say τ interchanges r and s, and say $r < s$. Then τ changes the factor $x_s - x_r$ in f to $x_r - x_s$. All other factors in f involving r and s can be paired as follows:

$$(x_k - x_s)(x_k - x_r) \text{ if } k > s,$$
$$(x_s - x_k)(x_k - x_r) \text{ if } r < k < s,$$
$$(x_s - x_k)(x_r - x_k) \text{ if } k < r,$$

and each one of these pairs remains unchanged under the action of τ on f. The overall effect is indeed that $\tau(f) = -f$.

Since by Corollary 3.6.3 each permutation can be decomposed into a composition of transpositions, each producing a sign change when acting successively, we can define sign: $S_n \to \{\pm 1\}$ by

$$\text{sign}(\sigma) f = \sigma(f),$$

and sign clearly has all the required properties. Obviously there can be only one map satisfying these properties. □

Note that $\text{sign}(\sigma) = 1$ precisely when there is an even number of transpositions in the decomposition of $\sigma \in S_n$, and we talk then of an *even permutation*. If $\text{sign}(\sigma) = -1$, we say that σ is an *odd permutation*.

Also note that $\text{sign}(\sigma^{-1}) = \text{sign}(\sigma)$ for any $\sigma \in S_n$ as sign is multiplicative and $\text{sign}(\iota) = 1$.

In conclusion we can perform any permutation by swapping two and two numbers repeatedly, and no matter how we arrange the swapping, we will either always need an even number of swaps, or always need an odd number of swaps.

3.7 Determinants

Determinants offer a fancy and theoretically interesting way of deciding whether a matrix is invertible, and when it is, they provide a formula for the inverse.

Definition 3.7.1 Let V be a vector space over a field F and let $m \in \mathbb{N}$. A map from V^m to another vector space over F is *multilinear or m-linear* on V if it is linear in each coordinate, that is, while the other ones are kept fixed. A map $V^m \to F$ is *alternating* if it changes sign whenever two coordinates are interchanged.

Note that a multilinear map is alternating if and only if it is zero whenever two variables are equal. This is evident from the fact that

$$f(x+y, x+y) = f(x,x) + f(x,y) + f(y,x) + f(y,y)$$

for a 2-linear, or bilinear, map f on any vector space.

Proposition 3.7.2 *Let $n \in \mathbb{N}$. Then there exists exactly one alternating multilinear map Δ on F^n that satisfies $\Delta(e_1, \ldots, e_n) = 1$.*

Proof Clearly such a map is uniquely defined. In fact, if $x_i = \sum a_{ij} e_j$ is a general vector of V, then expanding $\Delta(x_1, \ldots, x_n)$ using multilinearity, we get a sum of all possible terms of type

$$a_{1,\sigma(1)} \cdots a_{n,\sigma(n)} \Delta(e_{\sigma(1)}, \ldots, e_{\sigma(n)}),$$

where $\sigma \in S_n$. We refer to Sect. 3.6 for the definition of S_n and of fundamental properties of permutations σ.

Rearranging $(e_{\sigma(1)}, \ldots, e_{\sigma(n)})$ by repeated swaps till we get the standard orientering, we thus get

$$\Delta(x_1, \ldots, x_n) = \sum_{\sigma \in S_n} \text{sign}(\sigma) a_{1,\sigma(1)} \cdots a_{n,\sigma(n)} \Delta(e_1, \ldots, e_n)$$

$$= \sum_{\sigma \in S_n} \text{sign}(\sigma) a_{1,\sigma(1)} \cdots a_{n,\sigma(n)}.$$

On the other hand, we can take this formula as the definition of $\Delta(x_1, \ldots, x_n)$. Clearly it is multilinear and satisfies $\Delta(e_1, \ldots, e_n) = 1$. It only remains to check that it is alternating.

Suppose $\tau \in S_n$. Then

$$\Delta(x_{\tau(1)}, \ldots, x_{\tau(n)}) = \sum_{\sigma \in S_n} \text{sign}(\sigma) a_{\tau(1), \sigma(1)} \cdots a_{\tau(n), \sigma(n)}$$

$$= \sum_{\sigma \in S_n} \text{sign}(\sigma \tau) a_{\tau(1), \sigma \tau(1)} \cdots a_{\tau(n), \sigma \tau(n)}$$

$$= \sum_{\sigma \in S_n} \text{sign}(\sigma \tau) a_{1, \sigma(1)} \cdots a_{n, \sigma(n)}$$

$$= \text{sign}(\tau) \sum_{\sigma \in S_n} \text{sign}(\sigma) a_{1, \sigma(1)} \cdots a_{n, \sigma(n)}$$

$$= \text{sign}(\tau) \Delta(x_1, \ldots, x_n),$$

where in the second step we used that $\sigma \mapsto \sigma \tau$ is a bijection on S_n, and in the third step we reordered the product of a's in each term, and in the fourth step we used the multiplicative property of the function sign.

The result now follows since in the particular case of a transposition τ, which amounts to a swap of two variables, we have $\text{sign}(\tau) = -1$. □

Definition 3.7.3 The *determinant* of an $n \times n$-matrix A is the scalar

$$\det(A) = \Delta(a_1, \ldots, a_n),$$

where a_i is the ith column vector of A. We sometimes write $|A|$ for $\det(A)$.

By definition $\det: M(n, F) \to F$ is the only map that is alternating multilinear on the column vectors of the input matrices, and that satisfies $\det(I_n) = 1$. Also, from the proof of Proposition 3.7.2, together with the proposition below, we see that

$$\det(A) = \sum_{\sigma \in S_n} \text{sign}(\sigma) a_{\sigma(1), 1} \cdots a_{\sigma(n), n}$$

as $a_i = \sum a_{ji} e_j$.

Example 3.7.4 Consider a 3×3-matrix A with entries a_{ij}. Then S_3 has the $3! = 6$ elements:

$$\iota = \begin{pmatrix} 1 & 2 & 3 \\ 1 & 2 & 3 \end{pmatrix}, \quad \sigma_1 = \begin{pmatrix} 1 & 2 & 3 \\ 2 & 3 & 1 \end{pmatrix}, \quad \sigma_2 = \begin{pmatrix} 1 & 2 & 3 \\ 3 & 1 & 2 \end{pmatrix},$$

$$\sigma_3 = \begin{pmatrix} 1 & 2 & 3 \\ 2 & 1 & 3 \end{pmatrix}, \quad \sigma_4 = \begin{pmatrix} 1 & 2 & 3 \\ 3 & 2 & 1 \end{pmatrix}, \quad \sigma_5 = \begin{pmatrix} 1 & 2 & 3 \\ 1 & 3 & 2 \end{pmatrix}.$$

3.7 Determinants

Of these, the first three are even, and the others are odd. Thus

$$\det(A) = a_{11}a_{22}a_{33} + a_{21}a_{32}a_{13} + a_{31}a_{12}a_{23}$$
$$- a_{21}a_{12}a_{33} - a_{31}a_{22}a_{13} - a_{11}a_{32}a_{23}.$$

◇

Proposition 3.7.5 *Let $A, B \in M(n, F)$. Then*

(i) $\det(A^T) = \det(A)$;
(ii) $\det(AB) = \det(A)\det(B)$.

Proof We have

$$\det(A) = \sum_{\sigma \in S_n} \text{sign}(\sigma) a_{1,\sigma(1)} \cdots a_{n,\sigma(n)}$$
$$= \sum_{\sigma \in S_n} \text{sign}(\sigma) a_{\sigma^{-1}(1), \sigma\sigma^{-1}(1)} \cdots a_{\sigma^{-1}(n), \sigma\sigma^{-1}(n)}$$
$$= \sum_{\sigma \in S_n} \text{sign}(\sigma^{-1}) a_{\sigma^{-1}(1), 1} \cdots a_{\sigma^{-1}(n), n}$$
$$= \sum_{\sigma \in S_n} \text{sign}(\sigma) a_{\sigma(1), 1} \cdots a_{\sigma(n), n}$$
$$= \det(A^T),$$

where we in the first step rearranged the product of a's in each term, simplified the expression and used that $\text{sign}(\sigma^{-1}) = \text{sign}(\sigma)$ in the second step, and in the third step, we used that $\sigma \mapsto \sigma^{-1}$ is a bijection on S_n.

To prove the second claim, observe first that by replacing the e_i's by general vectors, say y_i's, in the first part of the proof of Proposition 3.7.2, we get

$$\Delta(x_1, \ldots, x_n) = \sum_{\sigma \in S_n} \text{sign}(\sigma) a_{1,\sigma(1)} \cdots a_{n,\sigma(n)} \Delta(y_1, \ldots, y_n)$$
$$= \det(A^T) \Delta(y_1, \ldots, y_n),$$

where $x_i = \sum a_{ij} y_j$. Now let

$$x_i = (AB)_i = \sum (AB)_{ji} e_j = \sum a_{jk} b_{ki} e_j = \sum b_{ki} y_k,$$

where $y_k = \sum a_{jk} e_j = a_k$. Then

$$\det(AB) = \Delta(x_1, \ldots, x_n) = \det(B) \Delta(y_1, \ldots, y_n) = \det(B)\det(A).$$

□

So det: $M(n, F) \to F$ is unital and multiplicative, but it is not linear; it is alternating multilinear on the rows of any input matrix. Hence $\det(aA) = a^n \det(A)$ for $a \in F$, whereas the operation of multiplying a row in A by a scalar and adding this to another row does not alter the determinant.

Definition 3.7.6 The *determinant of a linear operator A* on a finite dimensional vector space V is the determinant of its associated matrix $\tilde{A}$.

This definition makes sense because any two matrices of a linear operator with respect to different bases of V are similar, and if $A \sim B$, say $B = CAC^{-1}$, then

$$|B| = |CAC^{-1}| = |C||A||C^{-1}| = |CC^{-1}||A| = |I||A| = |A|.$$

This argument also shows that the determinant of an invertible matrix is non-zero. To see that the converse also holds, we invoke *Cramer's rule*, which is a method to solve system of linear equations.

Proposition 3.7.7 *Say $A(x) = b$, where $A \in M(n, F)$. Then the ith coordinate of x is given by*

$$x_i \det(A) = \Delta(a_1, \ldots, b, \ldots, a_n),$$

where the column vector b is placed at the ith place.

Proof The proof is shockingly simple. Note that $A(x) = b$ can be written as $\sum a_j x_j = b$. Plug in this expression for b on the right hand side above. All terms will obviously be zero, except for $j = i$, which gives the left hand side above. □

Now suppose $\det(A) \neq 0$. Then one can divide by $\det(A)$ in Cramer's rule and get a unique solution x with coordinates

$$x_i = \det(A)^{-1} \Delta(a_1, \ldots, b, \ldots, a_n).$$

In particular, when $b = 0$ the only solution is $x = 0$. So A is injective, and therefore by Corollary 3.3.7 it is invertible.

To find a formula for the inverse, consider all terms in the sum

$$\det(A) = \sum_\sigma \text{sign}(\sigma) a_{1,\sigma(1)} \cdots a_{n,\sigma(n)}$$

that contain a given entry a_{ij} of A as a factor. They are associated to those σ's that satisfy $\sigma(i) = j$. Hence, the sum of all such terms is

$$\sum_{\sigma(i)=j} \text{sign}(\sigma) a_{1,\sigma(1)} \cdots a_{n,\sigma(n)} = a_{ij} C_{ij},$$

3.7 Determinants

where
$$C_{ij} = \sum_{\sigma(i)=j} \text{sign}(\sigma) a_{1,\sigma(1)} \cdots a_{i-1,\sigma(i-1)} a_{i+1,\sigma(i+1)} \cdots a_{n,\sigma(n)}$$

is the *cofactor* of a_{ij} in $\det(A)$.

Now
$$\det(A) = \sum_{\sigma} \text{sign}(\sigma) a_{1,\sigma(1)} \cdots a_{n,\sigma(n)} = \sum_{j=1}^{n} \sum_{\sigma(i)=j} \text{sign}(\sigma) a_{1,\sigma(1)} \cdots a_{n,\sigma(n)},$$

leading to the following *expansion of* $\det(A)$ *according to the ith row*:

$$\det(A) = a_{i1} C_{i1} + \cdots + a_{in} C_{in}.$$

Let $k \neq i$. Then

$$\sum_{j=1}^{n} a_{kj} C_{ij} = \sum_{\sigma} \text{sign}(\sigma) a_{1,\sigma(1)} \cdots a_{i-1,\sigma(i-1)} a_{k,\sigma(i)} a_{i+1,\sigma(i+1)} \cdots a_{n,\sigma(n)} = 0,$$

as this is the determinant of a matrix with two identical rows, namely the ith and kth row.

All in all we have
$$\sum_{j=1}^{n} a_{kj} C_{ij} = \delta_{ik} \det(A). \tag{3.2}$$

This tells us that if $\det(A) \neq 0$, then A^{-1} exists and its ij-entry is the scalar $\det(A)^{-1} C_{ji}$.

But we are still not entirely happy. We would like a more hands on way of computing cofactors.

Definition 3.7.8 Let A_{ij} be the matrix obtained by removing the ith row and jth column of A.

Lemma 3.7.9 *Let A be an $n \times n$-matrix. Then $C_{ij} = (-1)^{i+j} \det(A_{ij})$.*

Proof Now
$$C_{11} = \sum_{\sigma} \text{sign}(\sigma) a_{2,\sigma(2)} \cdots a_{n,\sigma(n)},$$

where the summation is over all permutations σ of $\{2, \ldots, n\}$. Hence C_{11} is the determinant of A_{11}.

The general case can be reduced to this situation by first bringing the ij-entry a_{ij} of A to the upper left corner by $i - 1$ obvious interchanges of rows and then $j - 1$ interchanges of columns, a procedure which results in the sign factor $(-1)^{i+j}$ when computing the determinant of this new matrix. By the first paragraph, the cofactor of a_{ij} in this new determinant is the determinant of the matrix with the first row and column omitted, and this matrix is A_{ij}. Hence $C_{ij} = (-1)^{i+j} \det(A_{ij})$. □

In conclusion we have the following result.

Proposition 3.7.10 *Let $A \in M(n, F)$. Then*

$$\det(A) = (-1)^{i+1} a_{i1} \det(A_{i1}) + \cdots + (-1)^{i+n} a_{in} \det(A_{in}),$$

where A_{ij} is the submatrix of A obtained by omitting the ith row and jth column. Moreover, if $\det(A) \neq 0$, then A^{-1} exists and its ij-entry is

$$\det(A)^{-1}(-1)^{i+j} \det(A_{ji}).$$

Considering the transpose one can obviously also expand the determinant according to columns.

This proposition offers a way to compute determinants inductively.

Example 3.7.11 We have

$$\begin{vmatrix} a & b \\ c & d \end{vmatrix} = ad - bc,$$

which also happens to be the area of the parallelogram spanned out by the vectors (a, c) and (b, d). If $ad - bc \neq 0$, then

$$\begin{pmatrix} a & b \\ c & d \end{pmatrix}^{-1} = \frac{1}{ad - bc} \begin{pmatrix} d & -b \\ -c & a \end{pmatrix}.$$

Using this we can then calculate determinantes of 3×3-matrices.

Example 3.7.12 Expanding according to the first row, we get

$$\begin{vmatrix} 1 & 2 & 4 \\ 0 & 2 & 1 \\ 0 & 1 & 2 \end{vmatrix} = 1 \begin{vmatrix} 2 & 1 \\ 1 & 2 \end{vmatrix} - 2 \begin{vmatrix} 0 & 1 \\ 0 & 2 \end{vmatrix} + 4 \begin{vmatrix} 0 & 2 \\ 0 & 1 \end{vmatrix} = 3,$$

but it would have been easier to expand according to the first column. In general it is a good idea to pick a row or column with many zeros. As the determinant is non-zero, we know that the corresponding matrix A is invertible with inverse

$$A^{-1} = \frac{1}{\det(A)} \begin{pmatrix} |A_{11}| & -|A_{21}| & |A_{31}| \\ -|A_{12}| & |A_{22}| & -|A_{32}| \\ |A_{13}| & -|A_{23}| & |A_{33}| \end{pmatrix} = \frac{1}{3} \begin{pmatrix} 3 & 0 & -6 \\ 0 & 2 & -1 \\ 0 & -1 & 2 \end{pmatrix}.$$

3.7 Determinants

A geometric interpretation can also be given for 3×3-determinants: The determinant equals the volume of the parallelepiped spanned out by the column vectors of the matrix.

In computing the determinant of an $n \times n$-matrix by expanding according to some row or column in general one first has to compute n determinants of $(n-1) \times (n-1)$-matrices, and expanding these, one must compute $n(n-1)$ determinants of $(n-2) \times (n-2)$-matrices, and so on, yielding in the end $n!$ terms with alternating signs to add up. Not a very pleasant job.

It is much more efficient to first simplify the matrix by performing elementary row or column operations, keeping track of signs and factors that occur in the process. The determinant of a matrix in reduced echelon form is obviously very easy to calculate.

Example 3.7.13 Say $x_i \in F$. The *Vandermonde determinant* V is defined to be

$$V = \begin{vmatrix} 1 & \cdots & 1 \\ x_1 & \cdots & x_n \\ \cdot & \cdots & \cdot \\ \cdot & \cdots & \cdot \\ x_1^{n-1} & \cdots & x_n^{n-1} \end{vmatrix}$$

with explicit value

$$V = \prod_{i<j}(x_j - x_i).$$

The reader might recall that we used such an expression to define the sign of a permutation, which was then used to introduce determinants, so the circle is completed. Let us illustrate by considering $n = 3$, how one finds this expression:

$$\begin{vmatrix} 1 & 1 & 1 \\ a & b & c \\ a^2 & b^2 & c^2 \end{vmatrix} = \begin{vmatrix} 1 & 1 & 1 \\ a & b & c \\ 0 & (b-a)b & (c-a)c \end{vmatrix} = \begin{vmatrix} 1 & 1 & 1 \\ 0 & b-a & c-a \\ 0 & (b-a)b & (c-a)c \end{vmatrix}$$

$$= \begin{vmatrix} b-a & c-a \\ (b-a)b & (c-a)c \end{vmatrix} = (b-a)(c-a)\begin{vmatrix} 1 & 1 \\ b & c \end{vmatrix}$$

$$= (b-a)(c-a)(c-b),$$

where in the first step we multiplied the second row by $-a$ and added this row to the third, and in the next step we multiplied the first row by $-a$ and added that to the second row, and then we expanded according to the first column, and then we used multilinearity of the columns to extract the factor $(b-a)(c-a)$, leaving us with a smaller matrix of the same type to compute the determinant of. ◇

Let us finally gather various settled conditions for invertibility.

Proposition 3.7.14 *Let $A \in M(n, F)$. Then the following conditions are equivalent:*

(i) *A is invertible;*
(ii) *$A(x) = 0$ forces $x = 0$;*
(iii) *$\operatorname{im} A = F^n$;*
(iv) *$\det(A) \neq 0$.*

3.8 Eigenvalues and Eigenvectors

Let us start with some motivation.

Definition 3.8.1 An $n \times n$-matrix D is *diagonal* if it has only zeros off the diagonal, i.e. if it is of the form $D = (\lambda_i \delta_{ij})$ for scalars $\lambda_i \in F$.

Diagonal matrices are particularly easy to deal with. Say we are dealing with complex scalars. The power D^m of a diagonal matrix $D = (\lambda_i \delta_{ij})$ is again diagonal with entries λ_i^m along the diagonal. The determinant $\det(D)$ is the product $\lambda_1 \cdots \lambda_n$ of the diagonal entries, so D is invertible exactly when none of these entries vanish, and then D^{-1} is diagonal with diagonal entries λ_i^{-1}. The linear system $D(x) = b$ has then a unique solution x with coordinates $x_i = \lambda_i^{-1} b_i$.

Wanting to reduce to such a simple situation, the question is whether an arbitrary $n \times n$-matrix A by any chance is similar to a diagonal matrix D. Any matrix with such a property is called *diagonalizable*.

Powers of $A = PDP^{-1}$ with D diagonal is still easy to compute. Actually,

$$A^m = PD^m P^{-1}$$

because a P between the D's cancel its inverse P^{-1}.

Now $A \sim D$ means that A is diagonal with respect to some basis. Concretely, if $A = PDP^{-1}$, then with respect to $\{v_i\}$, where $v_i = P(e_i)$, we have

$$A(v_i) = AP(e_i) = PD(e_i) = P(\lambda_i e_i) = \lambda_i v_i.$$

Definition 3.8.2 An *eigenvector* of $A \in M(n, F)$ with *associated eigenvalue* $\lambda \in F$ is any non-zero vector $v \in F^n$ such that $A(v) = \lambda v$.

Note that eigenvalues of similar matrices are the same, so eigenvalues are intrinsic to linear operators on finite dimensional vector spaces.

Proposition 3.8.3 *Suppose there are n eigenvectors v_i for $A \in M(n, F)$ with eigenvalues λ_i. Then $AP = PD$, where $D = (\lambda_i \delta_{ij})$ and $P = (v_1, \ldots, v_n)$. Hence if $\{v_i\}$ is linear independent, then P is invertible and $A = PDP^{-1}$.*

3.8 Eigenvalues and Eigenvectors

Proof We have

$$AP(e_i) = A(v_i) = \lambda_i v_i = \lambda_i P(e_i) = P(\lambda_i e_i) = PD(e_i)$$

for all i, so $AP = PD$.

The second statement is clear as an $n \times n$-matrix is invertible exactly when its column vectors are linear independent. □

We focus therefore on eigenvectors. Geometrically we are seeking vectors that under the linear transformation A remain parallel to themselves. It is not at all clear that such vectors exist, except for diagonalizable matrices, of course.

Since we hope to build a whole basis of eigenvectors, we require them to be non-zero. The zero vector will always satisfy $A(0) = \lambda 0$ for any λ. In fact, the set consisting of 0 and all the eigenvectors associated to a fixed eigenvalue is obviously a subspace of F^n.

For example, this set associated to the eigenvalue 1 of the identity matrix is the whole space as $I(v) = v$ for all $v \in F^n$.

Proposition 3.8.4 *Eigenvectors of an $n \times n$-matrix associated to distinct eigenvalues are linear independent.*

Proof We proceed by induction on the number of eigenvectors with distinct eigenvalues. Any single eigenvector is non-zero and thus linear independent. Suppose the proposition holds for $m - 1$ eigenvectors with distinct eigenvalues, and say

$$c_1 v_1 + \cdots + c_m v_m = 0,$$

where $A(v_i) = \lambda_i v_i$ with $v_i \neq 0$ and distinct λ's. Applying $A - \lambda_1 I$ to the left hand side, we get

$$c_2(\lambda_2 - \lambda_1)v_2 + \cdots + c_m(\lambda_m - \lambda_1)v_m = 0.$$

Thus $c_2 = \cdots = c_m = 0$, and $c_1 v_1 = 0$ forces also $c_1 = 0$. □

The following result is now immediate.

Corollary 3.8.5 *Any $n \times n$-matrix with n distinct eigenvalues is diagonizable.*

Example 3.8.6 Consider $(x, y)^T$ as a vector in F^2 with coordinates x and y. Then the matrix

$$A = \begin{pmatrix} 0 & 1 \\ 1 & 0 \end{pmatrix}$$

sends $(x, y)^T$ to $(y, x)^T$, so it reflects arrows in the xy-plane across the line $y = x$. Clearly the vector $v_1 = (1, 1)^T$ remains fixed under such a reflection, so it is an eigenvector of A with eigenvalue 1; as is any rescaling of v_1 to another non-zero vector. Whereas the vector $v_2 = (1, -1)^T$ perpendicular to the reflection line, will when subject to the action of A, end up pointing in the opposite direction and is otherwise unaltered, so it is an eigenvector with eigenvalue -1. Since we have two distinct eigenvalues, we know that A is diagonalizable and $A = PDP^{-1}$ with $P = (v_1, v_2)$. We also see directly that $\{v_1, v_2\}$ is a basis for F^2. ◇

This example made it possible either by geometric intuition or by simple algebraic considerations to find a basis of eigenvectors. But how do we proceed for a general $n \times n$-matrix A? To solve the system $A(v) = \lambda v$ involves $n + 1$ unknown quantities, and a possible solution is going to involve at least one free parameter since we are flexible within eigenvector spaces. Here determinants come to our rescue.

We want to find $v \neq 0$ such that $(A - \lambda I)(v) = 0$ for some scalar λ. Such a v exists precisely when $A - \lambda I$ is not invertible, or in other words, when

$$\det(A - \lambda I) = 0.$$

This is an nth degree algebraic equation in the unknown λ, and it does not involve the eigenvectors at all. We have separated the unknown coordinates of v from λ, which we can find by solving an algebraic equation, known as the *characteristic equation* of A. Of course, this is easier said than done, and existence of solutions depends on the field F. However, as soon as we have a solution λ, we can easily find an associated eigenvector v by solving $(A - \lambda I)(v) = 0$ using Gauss-Jordan elimination.

Example 3.8.7 Let

$$A = \begin{pmatrix} 1 & 2 & 2 \\ 0 & 3 & 2 \\ 0 & 0 & 4 \end{pmatrix}.$$

Then $|A - \lambda I| = 0$ gives $(1 - \lambda)(3 - \lambda)(4 - \lambda) = 0$, so the eigenvalues are $\lambda_1 = 1$, $\lambda_2 = 3$ and $\lambda_3 = 4$. Gauss-Jordan eliminating the augmented matrix of $A - I$ to reduced echelon form gives

$$\begin{pmatrix} 0 & 2 & 2 & 0 \\ 0 & 2 & 2 & 0 \\ 0 & 0 & 3 & 0 \end{pmatrix} \sim_r \begin{pmatrix} 0 & 1 & 0 & 0 \\ 0 & 0 & 1 & 0 \\ 0 & 0 & 0 & 0 \end{pmatrix}$$

with solutions $\{(t, 0, 0)^T \mid t \in F\}$. Thus one eigenvector for λ_1 is $v_1 = (1, 0, 0)^T$. Similarly, an eigenvector for λ_2 is $v_2 = (1, 1, 0)^T$, and an eigenvector with eigenvalue λ_3 is $v_3 = (2, 2, 1)^T$. ◇

Here are two examples that illustrate the relevance of the field F in questions about eigenvalues.

Example 3.8.8 The characteristic equation of the matrix

$$A = \begin{pmatrix} 0 & 2 \\ 1 & 0 \end{pmatrix}$$

is $\lambda^2 = 2$, which has no solutions in the field $F = \mathbb{Q}$. But it has two real solutions $\pm\sqrt{2}$. So A is diagonalizable as a matrix over $F = \mathbb{R}$, but not as a matrix over the field of rational numbers. ◇

Example 3.8.9 The linear transformation of the matrix

$$A = \begin{pmatrix} 0 & -1 \\ 1 & 0 \end{pmatrix}$$

rotates a vector $(x, y)^T$ in the real plane 90 degrees counter clockwise. So as a matrix over $\mathbb{R}$ it cannot have any eigenvectors. This is also manifest in its characteristic equation $\lambda^2 = -1$, which has no real solutions. But this equation has complex solutions $\pm i$, and A is therefore diagonazible over $F = \mathbb{C}$. ◇

We should stress that the eigenvectors of a matrix do not in general form a basis. In many cases there are simply not enough of them.

Example 3.8.10 The matrix

$$A = \begin{pmatrix} 0 & 1 \\ 0 & 0 \end{pmatrix}$$

over any field is not diagonizable. To see this consider its characteristic equation $\lambda^2 = 0$, which only produces the eigenvalue 0. This gives the eigenvector $v = (1, 0)^T$, and every other eigenvector is clearly proportional to v. ◇

3.9 Jordan Canonical Form

In general we cannot diagonalize a quadratic matrix. For matrices over the complex numbers the closest we get is to write it in what we shall call its Jordan canonical (or normal) form.

Given a complex quadratic matrix A. Suppose we have a non-zero vector v_0 that satisfies $(A - \lambda I)^n v_0 = 0$ for some $\lambda \in \mathbb{C}$ and $n \in \mathbb{N}$, but that $(A - \lambda I)^{n-1} v_0 \neq 0$. Set $v_k = (A - \lambda I)^k v_0$. Then v_{n-1} is an eigenvector of A with eigenvalue λ, and we call $v_0, \ldots, v_{n-1}$ a *chain of generalized eigenvectors of A of length n*. An easy induction argument shows that the elements v_i are linear independent.

We have the following fundamental result.

Theorem 3.9.1 *Let A be a complex quadratic matrix with m eigenvalues λ_i and corresponding chains $v_0^i, v_1^i, \ldots, v_{n_i-1}^i$ of generalized eigenvectors such that $v_{n_1-1}^1, \ldots, v_{n_m-1}^m$ are linear independent and span the subspace of 0 and all the eigenvectors of A. This can always be arranged, and then the complex matrix*

$$P = (v_0^1, v_1^1, \ldots, v_{n_1-1}^1, v_0^2, v_1^2, \ldots, v_{n_2-1}^2, \ldots, v_0^m, v_1^m, \ldots, v_{n_m-1}^m)$$

is quadratic and invertible, and $A = PJP^{-1}$, where $J = \text{diag}(J_1, \ldots, J_m)$ is called the Jordan canonical form of A with Jordan blocks *defined as the quadratic matrices*

$$J_i = \begin{pmatrix} \lambda_i & 0 & \cdots & \cdot & 0 \\ 1 & \lambda_i & \cdots & \cdot & 0 \\ \cdot & \cdot & \cdots & \cdot & \cdot \\ 0 & \cdot & \cdots & \cdot & 0 \\ 0 & 0 & \cdots & 1 & \lambda_i \end{pmatrix}$$

of size n_i. The Jordan canonical form is unique up to similarity and reordering of Jordan blocks.

Proof All we need to show is that the n-dimensional vector space V which the matrix A acts on has a basis consisting of generalized eigenvectors. With respect to such a basis, say with P as in the theorem, the equality $AP = PJ$ is then easily verified on the standard basis. We assume the theorem holds for all matrices of size less than n, the case $n = 1$ being trivially true. Let λ be an eigenvalue of A, which exists in the complex numbers. Replacing A by $A - \lambda I$, we may assume that $\lambda = 0$. Since $\ker A$ is non-trivial, we know by the rank-nullity theorem that $\dim(\text{im} A) < n$. By the induction hypothesis, the matrix obtained by restricting the action of A to the invariant subspace $\text{im} A$ has the appropriate chains $\{v_i^j\}$ numbered by j. Let $u_j \in V$ be preimages of v_0^j belonging to the, say k, chains with associated eigenvalue 0. Let w_l be any basis of the subspace of $\ker A$ that meets $\text{im} A$ only at $\{0\}$. All in all we have then $\dim(\text{im} A) + k + \dim(\ker A) - k = n$ generalized eigenvectors v_i^j and u_j and w_l of A.

We claim that these vectors are linear independent. Assume there are scalars a_i^j and b_j and c_l such that

$$\sum_{i,j} a_i^j v_i^j + \sum_j b_j u_j + \sum_l c_l w_l = 0.$$

Applying A to both sides yields an equation where only the vectors v_i^j occur, and such that the coefficient of v_0^j with associated eigenvalue 0 is b_j. Hence all $b_j = 0$ as the vectors v_i^j for all i and j are linear independent. But then all $a_i^j = 0 = c_l$ as $\text{im} A$ and the vector space spanned by the w_l's intersect only at $\{0\}$.

Concerning the uniqueness statement, it is easily checked that two Jordan canonical forms of A with unmatched sizes of Jordan blocks all having the same scalar along the diagonal cannot possibly be similar. □

The statement that a matrix A is similar to a Jordan canonical form J in essentially a unique way is true over any field provided it contains all the roots of the characteristic equation of A. The proof of the above theorem evidently works also in this case. This low tech direct proof will be replaced by a more refined and natural one in the context of polynomials, see Corollary 9.13.2.

3.9 Jordan Canonical Form

Notice that A^T is similar to J^T, and that the latter has Jordan type blocks with 1's on the superdiagonal rather than on the subdiagonal. Both conventions occur in the literature.

With A and J as in the theorem we see that

$$\det(A - \lambda I) = \det(J - \lambda I) = (\lambda_1 - \lambda)^{n_1} \cdots (\lambda_m - \lambda)^{n_m}.$$

Hence the multiplicity of an eigenvalue in the characteristic equation of A is the sum of the sizes of the Jordan blocks with that eigenvalue along the diagonal. In particular, no chain of generalized eigenvectors associated to an eigenvalue is longer than the multiplicity of that eigenvalue in the characteristic equation.

Example 3.9.2 The matrix

$$A = \begin{pmatrix} 0 & 0 & 1 & 0 \\ 0 & 0 & 0 & 1 \\ -2 & 2 & -3 & 1 \\ 2 & -2 & 1 & -3 \end{pmatrix}$$

has characteristic equation $\lambda(\lambda+2)^3 = 0$, so $\lambda_1 = 0$ and $\lambda_2 = -2$ are eigenvalues of multiplicity 1 and 3, respectively. Then $v_0^1 = (1, 1, 0, 0)^T$ is an eigenvector for λ_1, whereas the eigenvector space V, with 0 included, of λ_2 is spanned by $(1, 0, -2, 0)^T$ and $(0, 1, 0, -2)^T$, which brings us one dimension short; one says that λ_2 has *defect* one. So there must be exactly one chain of generalized eigenvectors for λ_2 of length two. Solving $(A + 2I)^2 v_0^2 = 0$ gives $v_0^2 = (0, 0, 1, -1)^T$, which is determined up to a scalar. Set $v_1^2 = (A + 2I)v_0^2 = (1, -1, -2, 2)^T$. Let v_0^3 be any vector in V that is linear independent of $v_1^2 \in V$. Then $\{v_0^1, v_0^2, v_1^2, v_0^3\}$ is a basis of generalized eigenvectors for A. Letting P have these as column vectors in the order listed, we get $A = PJP^{-1}$ with $J = \text{diag}(J_1, J_2, J_3)$, where $J_1 = (0)$ and $J_3 = (-2)$ and

$$J_2 = \begin{pmatrix} -2 & 0 \\ 1 & -2 \end{pmatrix}.$$

One can also directly verify that

$$J = P^{-1}AP = \begin{pmatrix} 0 & 0 & 0 & 0 \\ 0 & -2 & 0 & 0 \\ 0 & 1 & -2 & 0 \\ 0 & 0 & 0 & -2 \end{pmatrix}.$$

The following important result is known as the *Cayley-Hamilton theorem*, and will be reproved later in a more general context.

Corollary 3.9.3 *Let A be a complex quadratic matrix. In the characteristic equation* $\det(A - \lambda I) = 0$ *we may replace λ by A and each scalar by I times the scalar to obtain an algebraic equation of the same type.*

Proof Let A and J be as in the theorem. Then it suffices to check that

$$(J - \lambda_1 I)^{n_1} \cdots (J - \lambda_m I)^{n_m} = 0$$

as $A = PJP^{-1}$. But this is clear since the ith block of the matrix on the left hand side involves the factor

$$(J_i - \lambda_i I)^{n_i} = \begin{pmatrix} 0 & 0 & \cdots & \cdot & 0 \\ 1 & 0 & \cdots & \cdot & 0 \\ \cdot & \cdot & \cdots & \cdot & \cdot \\ 0 & \cdot & \cdots & \cdot & 0 \\ 0 & 0 & \cdots & 1 & 0 \end{pmatrix}^{n_i} = 0.$$

$\square$

Example 3.9.4 Since

$$A = \begin{pmatrix} 1 & 1 \\ 1 & 2 \end{pmatrix}$$

has characteristic equation $\lambda^2 - 3\lambda + 1 = 0$, the Cayley-Hamilton theorem tells us that $A^2 - 3A + I = 0$, so $A^{-1} = 3I - A$. $\diamond$

When A is a quadratic matrix with real entries, one seeks perhaps a Jordan type decomposition into matrices with real entries. To this end note that if v is a complex eigenvector of A with complex eigenvalue λ, then $\bar{v}$ is an eigenvector of A with eigenvalue $\bar{\lambda}$, as is seen by taking complex conjugation on both sides of $Av = \lambda v$. So eigenvalues and eigenvectors come in complex conjugate pairs. Let $\mathrm{Re}\,v$ and $\mathrm{Im}\,v$ denote the vectors obtained by taking real and imaginary parts, respectively, of the entries of v.

Corollary 3.9.5 *Suppose A is a quadratic matrix with real entries. Then $A = PJP^{-1}$ for matrices P and J with real entries, where J is block diagonal with blocks described as follows: For each real eigenvalue of A we use an ordinary Jordan block with that eigenvalue along the diagonal, whereas for each non-real complex conjugate pair $a \pm ib$ of eigenvalues the block is double the size of the Jordan block of $a + ib$, and is of the form*

$$\begin{pmatrix} C & 0 & \cdots & \cdot & 0 \\ I & C & \cdots & \cdot & 0 \\ \cdot & \cdot & \cdots & \cdot & \cdot \\ 0 & \cdot & \cdots & \cdot & 0 \\ 0 & 0 & \cdots & I & C \end{pmatrix} \quad \text{with} \quad C = \begin{pmatrix} a & -b \\ b & a \end{pmatrix}.$$

3.9 Jordan Canonical Form

Proof Take the basis of generalized complex eigenvectors of A described in the theorem. Replace the vectors from two chains corresponding to $a \pm ib$ by the sequence $\operatorname{Re} v_0, \operatorname{Im} v_0, \ldots, \operatorname{Re} v_k, \operatorname{Im} v_k$, where $\{v_0, \ldots v_k\}$ is the chain associated to $a + ib$. This way one gets a new basis forming the columns in P, and one gets blocks of the form C associated to $a \pm ib$ since the linear map that sends a complex number $z = x + iy = (x, y)^T$ to $((a + ib)z)^T = (ax - by, bx + ay)^T$ is given by the matrix C in the standard basis. □

Later we shall see that matrices over any field are similar to Jordan canonical forms with so called generalized Jordan blocks.

The following cumbersome result is straightforward.

Proposition 3.9.6 *Let $V = \oplus_{i=1}^m V_i$ be a finite direct sum of vector spaces. Given $B_{ij} \in \operatorname{End}(V_j, V_i)$, define $\oplus B_{ij} \in \operatorname{End}(V)$ by*

$$(\oplus B_{ij})(v_1, \ldots, v_m) = (\sum_j B_{1j} v_j, \ldots, \sum_j B_{mj} v_j).$$

The canonical inclusions $\iota_i : V_i \to V$ and projections $\pi_i : V \to V_i$ are linear maps that satisfy $\sum \iota_i \pi_i = \iota$ and $\pi_i \iota_j = \delta_{ij} \iota$. For any linear maps satisfying these relations for any vector space V and any finite family $\{V_i\}$ of vector spaces, the map

$$\operatorname{End}(V) \to \oplus_{i,j} \operatorname{End}(V_i, V_j)$$

which sends A to $(\pi_j A \iota_i)$ is a linear isomorphism, and under the isomorphism $V \cong \oplus V_i$ the map A becomes $\oplus \pi_j A \iota_i \in \operatorname{End}(\oplus V_n)$.

If V is a vector space and $\{E_i\}$ is a finite subset of $\operatorname{End}(V)$ such that $\sum_i E_i = \iota$ and $E_i E_j = \delta_{ij} \iota$, then $E_i^2 = E_i$, and $\pi_i = E_i : V \to V_i \equiv E_i V$ together with the inclusions $\iota_i : V_i \to V$ satisfy $\sum \iota_i \pi_i = \iota$ and $\pi_i \iota_j = \delta_{ij} \iota$. Any such family $\{E_i\}$ is called a complete family of orthogonal idempotents.

In the theorem above let V be the linear span of the basis

$$v_0^1, v_1^1, \ldots, v_{n_1-1}^1, v_0^2, v_1^2, \ldots, v_{n_2-1}^2, \ldots, v_0^m, v_1^m, \ldots, v_{n_m-1}^m$$

and define $E_k \in \operatorname{End}(V)$ by $E_k(\sum c_i^j v_i^j) = \sum_i c_i^k v_i^k$, so

$$E_i = \operatorname{diag}(0, \ldots, 0, I, 0, \ldots, 0)$$

with the $(n_i \times n_i)$-identity matrix I at the ith diagonal block and with zeroes elsewhere. Obviously $\{E_i\}$ is a complete family of orthogonal projections that satisfy

$$E_i J = J E_i = \operatorname{diag}(0, \ldots, 0, J_i, 0, \ldots, 0)$$

with J_i at the ith diagonal block. Under the isomorphism $V \cong \oplus E_i V$ the map J sends $(v_1, \ldots, v_m)$ to $(J_1 v_1, \ldots, J_m v_m)$, so $J_i = \pi_i J \iota_i$ and $J = \oplus J_i$.

As for the matrix A in the theorem, let $N_i = (0, \ldots, 0, J_i - \lambda_i I, 0, \ldots, 0)$ be the matrix with a block at the ith place along the diagonal which has only ones on the subdiagonal and otherwise only zeroes. Then clearly $A = A_s + A_n$, where $A_s = \sum_{i=1}^{m} \lambda_i \tilde{E}_i$ and $A_n = \sum_{i=1}^{m} \tilde{N}_i$ with $\tilde{E}_i = PE_i P^{-1}$ and $\tilde{N}_i = PN_i P^{-1}$. The family $\{\tilde{E}_i\}$ is a complete set of orthogonal idempotents with members that commute with the mutually commuting matrices N_i.

The matrix A_s is certainly diagonalizable and commutes with A_n, which is *nilpotent*, meaning that some power of it vanish. The decomposition $A = A_s + A_n$ is known as the *Jordan-Chevalley decomposition of A*, and A_s and A_n are uniquely determined by these properties. We will return to this later, where we also work over more general fields.

We consider now a decomposition of A that is in general different from the one above. Say A has all in all q distinct eigenvalues $\lambda_1, \ldots, \lambda_q$. This gives the decomposition $V = \oplus \ker(A - \lambda_i I)^{k_i}$, where the *index* k_i is the size of the largest Jordan block with eigenvalue λ_i. In other words, the subspace $\ker(A - \lambda_i I)^{k_i}$ is spanned by all generalized eigenvectors associated to λ_i, so k_i is the least natural number such that

$$\ker(A - \lambda_i I)^{k_i} = \ker(A - \lambda_i I)^l$$

for all $l \geq k_i$. Glob together the Jordan blocks of A that have the same eigenvalue, say λ_i, along the diagonal, and form P_i just as we formed E_i with respect to these in general larger blocks. This gives a complete orthogonal family $\{P_i\}$ of idempotents that commute with A and where P_i projects onto the component $\ker(A - \lambda_i I)^{k_i}$ in the above decomposition of V.

We can describe these projections in a different way. Namely, write the characteristic polynomial $f(x) = \det(xI - A)$ of A as $f(x) = \prod_i (x - \lambda_i)^{m_i}$, where for the precise definition of a polynomial over a field, we refer to later chapters. Performing a partial fraction decomposition of $1/f$, we get polynomials a_i such that

$$\frac{1}{f(x)} = \frac{a_1(x)}{(x - \lambda_1)^{m_1}} + \cdots + \frac{a_q(x)}{(x - \lambda_q)^{m_q}}.$$

Multiplying up with $f(x)$, we get $1 = \sum p_i$, where p_i is the polynomial

$$p_i(x) = a_i(x) \prod_{j \neq i} (x - \lambda_j)^{m_j}.$$

Since $p_i p_j$ for $i \neq j$ contains f as a factor, it is immediate from Cayley's theorem that $\{p_i(A)\}$ is a complete family of orthogonal idempotents that commute with A. In the same vain we see that the matrices $p_i(A)(A - \lambda_i I)$ mutually commute and are nilpotent with vanishing m_ith power, so the Jordan-Chevalley decomposition of A is clear from

$$A = \sum (\lambda_i p_i(A) + p_i(A)(A - \lambda_i I)).$$

In fact, as $p_i(A)$ contains the factor $(A - \lambda_j I)^{m_j}$ when $i \neq j$, we see that $p_i(A)v_j = 0$ for $v_j \in \ker(A - \lambda_j I)^{m_j}$. Since $I = \sum p_i(A)$, we therefore also get $p_i(A)v_i = v_i$, so we get $p_i(A) = P_i$.

3.10 Dual Spaces, Inner Products and Tensor Products

For a vector space V over a field F, the *dual space* V^* of V consists of all linear maps $V \to F$. We regarded it as a vector space over F under pointwise operations.

Proposition 3.10.1 *Suppose V is finite dimensional with a linear basis $\{v_i\}$. Define $x_i \in V^*$ by $x_i(v_j) = \delta_{ij}$. Then $\{x_i\}$ is a basis for V^*, called the* dual basis *of $\{v_i\}$, so $\dim V^* = \dim V$. We have a linear isomorphism $V \to V^{**} \equiv (V^*)^*$ given by $v \mapsto f_v$, where $f_v(x) = x(v)$ for $v \in V$ and $x \in V^*$.*

Proof Any $x \in V^*$ can be written uniquely as a finite sum $x = \sum x(v_i)x_i$, which is checked by evaluating at each v_j.

Clearly f_v is linear on V^* by definition of the vector spaces operations on V^*. It is also evident that $v \mapsto f_v$ is linear by definition of the vector spaces operations on V^{**}. To see that the map is injective, say $x(v) = 0$ for all $x \in V^*$, and observe that any $v \in V$ can be written as $v = \sum x_i(v)v_i$. As $\dim V^{**} = \dim V$, the map is also surjective. □

The trace Tr on $V = M(n, F)$ belongs to V^*. When $A \in V$ is diagonalizable, its trace is the sum of the eigenvalues of A counted with multiplicities.

Recall that a *metric* on a set X is a map $d \colon X \times X \to [0, \infty)$ such that $d(x, y) = d(y, x)$ and $d(x, z) \leq d(x, y) + d(y, z)$ for $x, y, z \in X$, and which satisfies $d(x, y) = 0$ if and only if $x = y$. The *distance* between x and y is then $d(x, y)$.

For instance, the arc length between two points on a circle defines a metric on the circle, and so does the length of the straight line between any two points. Another metric on any set, for that sake, is the function that is always one on distinct points. We won't be discussing metric spaces in great detail here, we will only focus on metrics that come from norms on vector spaces. Say V is a vector space over a field F with an absolute value. Then a *norm* on V is a map $\|\cdot\| \colon V \to [0, \infty)$ such that $\|av\| = |a|\|v\|$ and $\|u + v\| \leq \|u\| + \|v\|$, and finally such that $\|w\| = 0$ implies $w = 0$. Then clearly $d(u, v) = \|u - v\|$ defines a metric on V. For example, the usual Euclidean distance between points in the complex plane is the metric coming from the norm defined as the absolute value $|a + ib| = (a^2 + b^2)^{1/2}$, or length, of an arrow $a + ib \in \mathbb{C}$ written in normal form.

In fact, we will limit ourselves to consider norms coming from inner products, of which the latter case above is an example. An *inner product* on a complex vector space V is a map $(\cdot|\cdot) \colon V \times V \to \mathbb{C}$ that is linear in the first variable, satisfies $(u|v) = \overline{(v|u)}$ for $u, v \in V$, and such that $(w|w) = 0$ implies $w = 0$. This means that the inner product is conjugate linear in the second variable, often referred to as being a sesquilinear form. The *associated norm* of an inner product is defined to be $\|v\| \equiv (v|v)^{1/2}$.

The inequality

$$|a|^2 \|v\|^2 + 2\operatorname{Re} a(v|w) + \|w\|^2 = \|av + w\|^2 \geq 0$$

for any vectors v, w and numbers a shows that the *Cauchy-Schwarz inequality*

$$|(v|w)| \leq \|v\| \|w\|$$

and the triangle inequality $\|v + w\| \leq \|v\| + \|w\|$ hold, so the associated norm of an inner product is indeed a norm. Note that equality in the Cauchy-Schwarz inequality holds if and only if v and w are linear dependent. Also, the proof of this inequality did not use the property that $(w|w) = 0$ only holds for $w = 0$.

Example 3.10.2 The complex vector space $\mathbb{C}^n$ has an inner product, called the *standard inner product*, defined as $(v|w) = \sum_{i=1}^{n} v_i \bar{w}_i$ for $v = \{v_i\}$, $w = \{w_j\} \in \mathbb{C}^n$. The associated norm is given by $\|v\| = (\sum_{i=1}^{n} v_i \bar{v}_i)^{1/2} = (\sum_{i=1}^{n} |v_i|^2)^{1/2}$. Note that the standard inner product on $\mathbb{C}^n$ restricts to an inner product on $\mathbb{R}^n$, and the distance given by the associated norm is the usual Euclidean distance. ◇

For any inner product on a complex vector space the *polarization identity*

$$4(v|w) = \sum_{n=0}^{3} i^n \|v + i^n w\|^2$$

and the *parallelogram law*

$$\|v + w\|^2 + \|v - w\|^2 = 2\|v\|^2 + 2\|w\|^2$$

are easily seen to hold for the associated norm. Conversely, if the parallelogram law holds for a norm, then one checks that the first of these formulas defines an inner product having the norm as the associated one. This holds also for real vector spaces except that the polarization identity must be replaced by $4(v|w) = \|v + w\|^2 - \|v - w\|^2$. The parallelogram law has an obvious geometric significance in the Euclidean case. The polarization identities obviously hold without requiring that $(w|w) = 0$ only for $w = 0$.

Two vectors v, w are *orthogonal* with respect to an inner product if $(v|w) = 0$. For a subset X of an inner product space denote by $X^\perp$ the subset of those vectors that are orthogonal to all the vectors of X. We also write $X \perp Y$ to say that every element of X is orthogonal to every element of Y.

Example 3.10.3 Given a finite family of inner product spaces V_i, we define an inner product on their algebraic direct sum by $(v|w) = \sum (\pi_i(v)|\pi_i(w))$. Each V_i is a subspace of $\oplus V_i$, and they are mutually orthogonal.

3.10 Dual Spaces, Inner Products and Tensor Products

For norms associated to inner products *Pythagoras' identity*

$$\|v + w\|^2 = \|v\|^2 + \|w\|^2$$

holds for mutually orthogonal vectors v and w.

Definition 3.10.4 An *orthonormal basis* in a finite dimensional complex vector space with an inner product is a family of orthogonal vectors of norm one that span the whole space.

Note that an orthonormal basis is automatically a linear basis. In fact, if $\{v_i\}$ is an orthonormal basis in V, then V is the finite orthogonal sum of the one-dimensional subspaces spanned by each v_i. Any $v \in V$ can obviously be expanded as $v = \sum (v|v_i) v_i$ in terms of the *Fourier coefficients* $(v|v_i)$ of v. Then the *Parseval identity*

$$\|v\|^2 = \sum |(v|v_i)|^2$$

holds by an inductive application of Pythagoras' identity.

Every finite dimensional complex vector space V with an inner product has an orthonormal basis. Indeed, a concrete way of producing an orthonormal basis $\{v_n\}$ from a linear basis $\{w_n\}$ is the following *Gram-Schmidt orthonormalization* process, which sets $v_1 = w_1/\|w_1\|$ and defines inductively v_n to be the normalization of the component $w_n - \sum_{m=1}^{n-1} (w_n|v_m) v_m$ of w_n that is orthogonal to $\mathrm{span}\{v_1, \ldots, v_{n-1}\} = \mathrm{span}\{w_1, \ldots, w_{n-1}\}$. So if W is a subspace of V, then V is isomorphic to $W \oplus W^\perp$. Note also that $x_n = (\cdot|v_n)$ form a dual basis for V^*, so maps of the form $(\cdot|v)$ for $v \in V$ exhaust the dual space of V.

The *tensor product of two vector spaces* V, W over the same field F is a vector space $V \otimes W$ and a bilinear map $h \colon V \times W \to V \otimes W$ such that any bilinear map $f \colon V \times W \to U$ into another vector space U factorizes as $f = gh$ for a unique linear map $g \colon V \otimes W \to U$. The tensor product is unique up to isomorphism of vector spaces because if U was another one we would get maps $V \otimes W \to U$ and $U \to V \otimes W$ that by uniqueness of the factorization would be inverses to each other. We won't prove existence carefully here, and it is not important exactly how the tensor product is constructed, as long as we know its characterizing properties.

Let us nevertheless say a few words about existence. To this end we need to use two notions that will be define later. Consider the semigroup algebra over F of the free semigroup generated by $V \times W$. Divide out by the ideal generated by elements of the type

$$(a_1 v_1 + b_1 w_1)(a_2 v_2 + b_2 w_2) - a_1 a_2 v_1 v_2 - a_1 b_2 v_1 w_2 - b_1 a_2 w_1 v_2 - b_1 b_2 w_1 w_2$$

for scalars a_i, b_i and $v_i \in V$ and $w_i \in W$, to get $V \otimes W$ with h as the quotient map restricted to $V \times W$.

We denote $h(v, w)$ by $v \otimes w$ and call the latter an *elementary tensor*. Thus every element of $V \otimes W$ is a finite linear combination of elementary tensors. Such a linear combination is in general not unique.

Given linear maps $f_i \colon V_i \to W_i$ between vector spaces over F, we form their tensor product $f_1 \otimes f_2 \colon V_1 \otimes V_2 \to W_1 \otimes W_2$ as the unique map in the factorization of the bilinear map $V_1 \times V_2 \to W_1 \otimes W_2$ which sends (v_1, v_2) to $f_1(v_1) \otimes f_2(v_2)$ for $v_i \in V_i$. In particular, when all W_i are 1-dimensional, we can use this to show that $\{v_i \otimes w_j\}$ is a linear basis of $V \otimes W$ whenever $\{v_i\}$ and $\{w_j\}$ are linear bases of V and W, respectively. Thus every element of $V \otimes W$ can be written as a finite sum $\sum u_j \otimes w_j$ for unique $u_j \in V$.

The *flip* is the linear isomorphism $V \otimes W \to W \otimes V$ given by the factorization of the bilinear map $(v, w) \mapsto w \otimes v$ for $v \in V$ and $w \in W$. In the same vain we have an isomorphism $V_1 \otimes (V_2 \otimes V_3) \to (V_1 \otimes V_2) \otimes V_3$ for vector spaces V_i.

If V, W are finite dimensional complex vector spaces that come with inner products, we get a well-defined inner product on $V \otimes W$ by setting $(v_1 \otimes w_1 | v_2 \otimes w_2) = (v_1|v_2)(w_1|w_2)$ for $v_i \in V$ and $w_i \in W$. We certainly get a sesquilinear map, and if $(u|u) = 0$ then $u = 0$ because if $u = \sum v_i \otimes w_i$ with the w_i's orthonormal, then $\|u\|^2 = \sum \|v_i\|^2$, so all $v_i = 0$. Note that $\|v \otimes w\| = \|v\| \|w\|$. If $\{v_i\}$ is an orthonormal basis of V and $\{w_j\}$ is an orthonormal basis of W, then $\{v_i \otimes w_j\}$ is an orthonormal basis of $V \otimes W$.

Chapter 4
Groups

A semigroup is the simplest algebraic object one can think of in the sense that it is a non-empty set with only one binary associative operation called multiplication. If in addition you require that there should be a unit element and an inverse to each element, then you have a group. The fact that groups are easy to define does not mean that they are easy to study. On the contrary, to understand something well, the more structure you are handed from the outset, the better. In any case, groups are versatile and extremely important in mathematics.

Their axioms were set down at the beginning of the twentieth century, although they were used prolifically centuries before. Of course, one can say that even the Pythagoreans used groups in that they dealt with numbers, but their focus was on the numbers themselves and not on the entire collection of them with its multiplicative or additive structure.

Then again, the important thing about groups is not the groups themselves but the way they act as bijective maps on other sets. This is perhaps one of the reasons why the notion of a group was not pinned down very accurately earlier. Philosophically, one can say that groups rule aesthetics in mathematics. They tell us what we should focus on, or what beauty in a mathematical landscape really is. Loosely speaking, beauty is manifested through the invariants associated to the group. Invariants are quantities that are left unchanged under the action of a group, or in other words, appear the same for every element of the group. Within a group of observers these quantities represent truth, something all the observers can agree on. And indeed, geometry, according to the Erlangen program, is the study of invariants of groups. This is also the reason why groups are so important in physics. It suffices to mention Noether's theorem, covariance and gauge theories.

Early on therefore, we talk about actions of groups. Prior to this we introduce subgroups and show how new groups can be obtained as quotient spaces by subgroups. A subgroup of a group is just a subset which is a group under the multiplication inherited from the original group. When the order of the elements in a product does not matter, we talk about abelian groups, coined after the Norwegian mathematician

Niels Henrik Abel. A quotient set G/H of a group G with respect to a subgroup H consists of the equivalence classes in G with equivalence relation $\sim$ given by $a \sim b$ if $ab^{-1} \in H$. The quotient sets of subgroups of abelian groups are always groups with multiplication $(aH)(bH) = abH$. Simple examples of abelian groups are the cyclic ones, named so because of their action on the circle. By definition they are groups generated by only one element. We say that a group is generated by a subset if every group element is a finite product of elements from this subset.

The process of forming quotients of groups with respect to subgroups is fundamental in group theory. For quotient sets to be groups in the non-abelian setting the subgroup must be normal, that is, closed under conjugation by any element of the enveloping group, meaning that the product aba^{-1} must belong to the subgroup for any element a of the group and any element b of the subgroup. Groups are compared by means of homomorphisms, that is, maps between them that preserve the products. And two groups are considered the same if the homomorphism is bijective, then called an isomorphism. The fundamental isomorphism theorem tells us that the kernel of a homomorphism, which is the subset of those members that hit the unit, is a normal subgroup with a quotient group isomorphic to the subgroup which is the image of the homomorphism. The second and third isomorphism theorems are similar in spirit to the first one.

Understanding the structure of subgroups of a group amounts to understanding all actions of the group. This is due to a basic result which sets up a correspondence between the orbits (or minimal invariant subsets) of an action and an isotropy group, which is a subgroup whose maps fix an element of the set. There is such a subgroup for every element of the set, and elements belonging to the same orbit have conjugate isotropy groups, so there is some reduction. The orbits are then identified with the quotient sets of the group by the isotropy groups.

In this chapter we look at examples of Lie groups (named after the Norwegian mathematical Sophus Lie) and their actions on spheres and Grassmannian manifolds. This is normally done in courses on geometry and algebraic topology, but I feel it is unsatisfactory to talk about groups and actions without dealing at least superficially with these examples that are so crucial also in advanced group theory.

The formation of a quotient set partitions the group and leads to Lagrange's theorem, which tells us that the order, i.e. the number of elements, of a subgroup divides the order of the group. This is a special case of the theorem on orbits of actions, which leads to the class formula. This formula says that you can count the number of elements in a group by first counting the orbits, and then the number of members in each orbit, and the collection of orbits can be regarded as a quotient group by an isotropy group of the original group. Using this, the Norwegian mathematician Peter Sylow showed how to unravel the structure of subgroups having order equal to some power of a prime number. The techniques of Sylow, focusing mainly on the order of the groups, can be taken surprisingly far.

It is desirable to decompose a group into simpler ones. For instance, finite dimensional vector spaces are copies of the ground field, and each of these copies are additive subgroups of the vector space. For finitely generated abelian groups one has something similar. They are direct products of simpler ones, namely of cyclic groups.

But most cyclic groups can by decomposed further although not as a direct product of simpler ones. Non-trivial groups having no proper normal subgroups are called simple. Any non-trivial finite group can be decomposed into such groups, but then in terms of composition series. A normal series is a chain of descending subgroups that are only required to be normal in their immediate larger subgroups, as opposed to subproducts of factors in a direct product, which are normal in the entire group. A composition series is a normal series with simple quotient groups. Such series are gotten by filling in all possible normal subgroups till the chain is so long that the quotient groups are simple. The Jordan-Hölder theorem tells us that the quotient groups of such series are unique up to isomorphisms and rearrangements, so one is talking about a property of the group. In this sense, the theory of groups is said to be reduced to that of simple groups. Up to isomorphism, the simple abelian groups are easily shown to be either $\mathbb{Z}$ or $\mathbb{Z}_p$, where p is a prime number.

A natural thing to do then is to classify all finite simple groups. After two decades with intense research this was considered completed in the 1980's. The proof, yet to be written up in its entirety and in a coherent way, amounts to more than 10 000 pages of relentless mathematics. The classification of the finite simple groups is a landmark in mathematics. The theorem says that every finite simple non-abelian group is either an alternating group (see the paragraph below), or it is of Lie group type over certain finite fields, and these are categorized in four infinite series, plus some exceptional ones, or it is one of the 26 sporadic simple groups, and among them is the monster group consisting of more than 10^{53} elements.

At the opposite extreme are the solvable groups, being groups with normal series having abelian quotient groups. You also have the more restricted class of nilpotent groups with normal series of groups produced by successive formation of centers of quotient groups. The center of a group is the normal abelian subgroup consisting of all elements that commute with any other element. The origin of the prefix 'solvable' is due to Abel and Galois, who considered permutations associated to an algebraic equation. Both realized that such equations can be solved in terms of radicals, that is, by extracting roots, if these permutations form a solvable group. The reason why an algebraic equation of fifth order in general is not solvable by radicals, is because the even permutations on five elements form a simple non-abelian group, known as the alternating group, which prevents the group of all permutations from being solvable. We prove simplicity of alternating groups on more than four elements in this chapter, and leave the part concerning algebraic equations to Chap. 8, where we discuss the theory of Galois.

Many groups can be given in terms of generators and relations that completely specify the multiplication table of the group. To define groups this way is more subtle than it seems at first. For a start, how do we identify a relation? Such questions lead to the notion of a free group. By definition such groups possess no auxiliary restraining relations. All other groups are quotients of such groups, where the auxiliary constrains are packed into the normal subgroup the quotient group is formed with respect to. We end the chapter with an informal discussion of various groups occurring in algebraic topology. This sketchy section is not self-contained, and required preferably some background in general topology.

4.1 Groups and Semigroups

Definition 4.1.1 A *group* is a set G with a binary operation · that is *associative*, i.e. so that $a(bc) = (ab)c$ for $a, b, c \in G$, with a *unit element* e, meaning that $ae = a = ea$ for all a, and which for each element a has an *inverse element*, i.e. an element b such that $ab = e = ba$. A group is *abelian* or *commutative* if $ab = ba$ for all its elements a and b. The *order of a group* G is its cardinality $|G|$.

We wrote ab instead of $a \cdot b$. Due to associativity, we often skip parentheses.

There is only one unit element in a group because if also f is a unit, then $e = ef = f$. In this argument it suffices that e is a *left unit* ($ea = a$) and that f is a *right unit* ($af = a$). We talk about the unit, and it makes sense to have assigned a symbol to it.

Similarly, each element has only one inverse because if both a and b are inverses of c, then $a = ae = a(cb) = (ac)b = eb = b$. For this argument it suffices that a is a *left inverse* of c and that b is a *right inverse* of c.

We write a^{-1} for the inverse of an element a. Clearly, if b is an inverse of a, then a is an inverse of b, so $(a^{-1})^{-1} = a$. Also $(ab)^{-1} = b^{-1}a^{-1}$ for all elements a and b.

For abelian groups we often talk about addition $+$ rather than multiplication to stress the abelianess of the binary operation. We talk then of an *additive group*.

If we dispense with the existence of inverses all together, we are left with a *monoid*. A *semigroup* is a non-empty set with an associative binary operation, so a monoid is a semigroup with unit. We also talk about abelian semigroups and monoids. The order of a monoid or a semigroup is defined as for groups.

Proposition 4.1.2 *A semigroup is a group if it has a left unit and left inverses for all its elements.*

Proof Let a be a left inverse of a left inverse b of c, so $ab = e$ and $bc = e$. Then $cb = ab(cb) = a(bc)b = a(eb) = ab = e$, so b is also a right inverse of c. But then $ce = cbc = ec = c$, so e is also a right unit, and we have a group. □

Let $n \in \mathbb{N}$ and let a be an element of a semigroup. By a^n (or na in additive notation), we mean a multiplied with itself n-times. If the semigroup is a monoid with unit e, then $a^0 \equiv e$, and if we have a group, then a^{-n} means $(a^{-1})^n$.

Example 4.1.3 The integers $\mathbb{Z}$ is a group under addition $+$ with unit 0 and inverse $-a$ for $a \in \mathbb{Z}$. But $\mathbb{Z}$ is only a monoid under multiplication, with unit 1.

The rational numbers $\mathbb{Q}$ is a group under addition, and is a monoid under multiplication, and $\mathbb{Q} \equiv \mathbb{Q}\backslash\{0\}$ is a group under multiplication. The same is true for $\mathbb{R}$ and $\mathbb{C}$, or any field F, and we denote the multiplicative group $F\backslash\{0\}$ by F_*.

The natural numbers $\mathbb{N}$ is a monoid under addition and multiplication, but lacks inverses in both cases. The finite set $X = \{1, \ldots, n\}$ is not a semigroup under addition nor multiplication, but $\mathbb{N}\backslash X$ is a semigroup under both operations.

4.2 Subgroups

The odd integers is a monoid under multiplication, while the even integers is a group under addition and a semigroup under multiplication. All objects considered in this example are abelian, and none are finite. ◇

Example 4.1.4 Let $n \in \mathbb{N}$. Consider the set $\mathbb{Z}_n$ of *congruence (or residue) classes* in $\mathbb{Z}$ modulo n, that is, equivalence classes in $\mathbb{Z}$ with respect to the relation $a \equiv b \pmod{n}$, which as we recall, means that $a - b$ divides n. We can add such classes by saying that $[a] + [b]$ should be the equivalence class $[a + b]$ of $a + b$. This is well-defined because if $c \in [a]$ and $d \in [b]$, then $c - a = nk$ and $d - b = nl$, and therefore $(c + d) - (a + b) = n(k + l)$, so $[c + d] = [a + b]$. This additive operation turns $\mathbb{Z}_n$ into a group with unit $0 = [0] = [n]$ and with $-[a] = [-a]$. We have constructed a finite abelian group of order n. ◇

Proposition 4.1.5 *A semigroup is a group if and only if $ax = b = ya$ have solutions x, y for all a and b.*

Proof If we have a group, then $x = a^{-1}b$ and $y = ba^{-1}$ are solutions.

Conversely, suppose we always have solutions in a semigroup. Pick any a. Then there exists $y = e$ such that $ea = a$. Write any b as $b = ax$ for some x that we know exists. Then $eb = eax = ax = b$, so e is a left unit. By assumption for any b, there is a y such that $yb = e$, so y is a left inverse of b. Hence the semigroup is a group by the proposition above. □

Corollary 4.1.6 *A finite semigroup is a group if and only if the cancellation property holds, that is, if $ax = ax'$ implies $x = x'$ and if $ya = y'a$ implies $y = y'$ for all a.*

Proof The semigroup satisfies the cancellation properties if and only if the maps $x \mapsto ax$ and $y \mapsto ya$ are injective for all a. On a finite set this holds if and only if these maps are surjective, which says that $ax = b = ya$ have solutions x, y for all a and b. □

4.2 Subgroups

Definition 4.2.1 A non-empty subset H of a group G is a *subgroup*, and we write $H < G$, if the multiplication on G restricts to a multiplication on H and turns H into a group. If H is different from G and the *trivial subgroup*, which consists of the unit alone, we say that H is a *proper subgroup*.

The unit f of a subgroup is the same as the unit e of the group, because if f^{-1} is the inverse of f in the group, then $e = ff^{-1} = fff^{-1} = fe = f$. The inverse of an element in a subgroup equals the inverse of the same element in the group because both are inverses in the group and there is only one such element.

Proposition 4.2.2 *Any non-empty subset H of a group G is a subgroup if and only if $ab^{-1} \in H$ for all $a, b \in H$.*

Proof From what we have just said, the forward implication is clear.

For the opposite direction, pick any element c in the non-empty set H. Then the unit $e = cc^{-1}$ of G belongs to H. Hence if $a, b \in H$, then $b^{-1} = eb^{-1} \in H$, and thus $ab = a(b^{-1})^{-1} \in H$. □

Example 4.2.3 The circle $\mathbb{T} = \{z \in \mathbb{C} \mid |z| = 1\}$ is a subgroup of the multiplicative group $\mathbb{C}\setminus\{0\}$ because if $z, w \in \mathbb{T}$, then $|zw^{-1}| = |z||w|^{-1} = 1$. ◇

The following result shows that finite semigroups of groups are automatically subgroups.

Proposition 4.2.4 *Any non-empty finite subset H of a group is a subgroup if $ab \in H$ for all $a, b \in H$.*

Proof The semigroup H has the cancellation property thanks to the enveloping group, so the result follows from Corollary 4.1.6. □

If A, B are subsets of a semigroup, we define AB to be the set of all elements ab with $a \in A$ and $b \in B$. The following result is easy to prove.

Proposition 4.2.5 *Suppose H and K are subgroups of a group G. Then $HK < G$ if and only if $HK = KH$.*

Thus, if G is an abelian group with subgroups H and K, then HK is a subgroup of G.

4.3 Generators

By Proposition 4.2.2 the intersection of any non-empty family of subgroups of a group is again a subgroup. So the following definition makes sense.

Definition 4.3.1 Let X be a subset of a group G. The subgroup $\langle X \rangle$ of G is the smallest subgroup that contains X, and is called the *subgroup generated by X*. If G is generated by some finite subset X, it is said to be a *finitely generated group*, and we write $\langle x_1, \ldots, x_n \rangle$ for $\langle X \rangle$, where $x_1, \ldots, x_n$ are the members of X, called the *generators of G*. A *cyclic group* is a group generated by one element.

In a similar fashion we can define subsemigroups and submonoids, and such objects generated by subsets, including cyclic versions.

Note that a group generated by X consists of all finite products of elements of X and their inverses. Thus finitely (or even countably) generated groups are at most countable. In particular, a cyclic group $\langle a \rangle$ generated by an element a equals $\{a^n \mid n \in \mathbb{Z}\}$, and is clearly abelian. Any non-trivial group has a non-trivial cyclic subgroup gotten by considering $\langle a \rangle$ for any element a that is not a unit. Since cyclic groups are at most countable, the real numbers $\mathbb{R}$ have proper cyclic subgroups, and $\mathbb{Z}$ is one such subgroup, but there are many more.

Example 4.3.2 The set $n\mathbb{Z}$ is a cyclic subgroup of $\mathbb{Z}$ generated by a non-negative integer n. These are all the subgroups of $\mathbb{Z}$. To see this, suppose $H < \mathbb{Z}$ is proper, and let n be the least natural number in H. Then $H = n\mathbb{Z}$. Otherwise there is a number $a \in H$ that is not in $n\mathbb{Z}$. By the division algorithm, we can write $a = nq + r$ for integers q and $0 \leq r < n$. But $r = a - qn \in H$, so $r = 0$ as n is the least natural number in H. Thus $a = nq \in n\mathbb{Z}$, which is a contradiction. So all subgroups of $\mathbb{Z}$ are cyclic and of the form $n\mathbb{Z}$, and they are infinite, except when $n = 0$.

By the same reasoning, we see that $n\mathbb{Z} < m\mathbb{Z}$ if and only if m divides n, and these are all the subgroups of $m\mathbb{Z}$. ◇

Example 4.3.3 The group $\mathbb{Z}_n$ is cyclic with generator $[1]$. In fact, any element $[a]$ with a and n relatively prime, will serve as a generator, and these are all the single generators. This is true because $\gcd(a, n) = 1$ if and only if $ak + nl = 1$ for integers k and l if and only if $k[a] = [1]$ for some integer k.

Again by the division algorithm, the subgroups of $\mathbb{Z}_n$ are of the form $\langle [m] \rangle$, where m divides n. ◇

Definition 4.3.4 The *order of an element* a in a group, is the least natural number $o(a)$ such that $a^{o(a)} = e$, where e is the unit of the group. If no such number exists, then a is said to be of *infinite order*.

Using the division algorithm, it is easy to check that $o(a) = |\langle a \rangle|$ for any element a of finite order in a group G. Also, if $a^m = e$ for some m, then $o(a)$ divides m. If G is finite, then for the product n over G of all $o(a)$, we obviously get $b^n = e$ for all $b \in G$. But this is an overkill, evidently already $b^{|G|} = e$, so $o(b)$ divides $|G|$.

This can be generalized.

4.4 Cosets and Lagrange's Theorem

Definition 4.4.1 Suppose H is a subgroup of a group G. The *coset* of H determined by $a \in G$ is the set aH, and the collection of cosets is denoted by G/H. The *index* $[G : H]$ of $H < G$ is the order of G/H. Similarly, the set $H \backslash G$ denotes the collection of *right cosets* Ha.

The map $G/H \to H \backslash G$ given by $aH \mapsto Ha^{-1}$ is well-defined and bijective, so $[G : H] = |H \backslash G|$.

The (left) cosets of $H < G$ form a partition of G with respect to the equivalence relation $\sim$ on G given by $b \sim a$ if $b \in aH$. Also the map $H \to aH$ sending b to ab is a bijection, so $|H| = |aH|$. This proves *Lagrange's theorem*.

Proposition 4.4.2 *The order of any subgroup of a finite group divides the order of the group. In fact, if $K < H < G$ and $[G : K] < \infty$, then*

$$[G : K] = [G : H][H : K].$$

Proposition 4.4.3 *The intersection of two subgroups of finite index is of finite index.*

Proof Say H and K are subgroups of G with finite indices. Trivially

$$a(H \cap K) = aH \cap aK,$$

and there can be only finitely many such subsets. $\square$

We will need the following result later.

Proposition 4.4.4 *If H_i are subgroups of a group G and $a_{ij} \in G$ are such that*

$$G = \cup_{i=1}^{m} \cup_{j=1}^{n} a_{ij} H_i,$$

then some H_i is of finite index in G.

Proof We proceed by induction on the number m of subgroups H_i. The case $m = 1$ is obvious. Assume that the result holds for any number less than m, and consider $m \geq 2$. If all the indices are finite we are done, so by interchanging the H_i's we may assume that $[G : H_1]$ is infinite. Since all cosets form a partition of G at least one coset, say bH_1, must be disjoint from $\cup_{j=1}^{n} a_{1j} H_1$. Hence by assumption

$$bH_1 \subset \cup_{i=2}^{m} \cup_{j=1}^{n} a_{ij} H_i.$$

Left multiplication by $a_{1k}b^{-1}$ gives

$$a_{1k} H_1 \subset \cup_{i=2}^{m} \cup_{j=1}^{n} a_{1k}b^{-1}a_{ij} H_i.$$

So G is covered by cosets of $H_2, \ldots, H_m$. By the induction hypothesis one of these has finite index in G. $\square$

4.5 Morphisms

Definition 4.5.1 A *homomorphism* from a group G to another group H is a map $f : G \to H$ such that $f(ab) = f(a)f(b)$ for all $a, b \in G$. If f is injective (surjective) it is called a *monomorphism* (*epimorphism*). If it is bijective, then it is an *isomorphism*, and we write $G \cong H$. A monomorphic image of a group G in H is said to be an *embedding* of G in H. If $G = H$, we say that f is an *endomorphism*, and it is an *automorphism* if in addition it is an isomorphism. The *kernel* of f is the subgroup $\ker f \equiv \{a \in G \mid f(a) = e\}$ of G.

The relation of being isomorphic is an equivalence relation. The kernel of a homomorphism is trivial if and only if the homomorphism is a monomorphism. The image $\operatorname{im} f \equiv f(G)$ of a homomorphism $f : G \to H$ is a subgroup of H.

The *trivial homomorphism* sends all elements in a group to the unit of another group. The identity map on a group is an automorphism.

Example 4.5.2 The exponential function is an epimorphism from the additive group $\mathbb{C}$ to the multiplicative group $\mathbb{C}\setminus\{0\}$. ◇

Example 4.5.3 We have seen that $n\mathbb{Z}$ is a subgroup of $\mathbb{Z}$. But the map $n\mathbb{Z} \to \mathbb{Z}$ given by $nm \mapsto m$ is an isomorphism, so $n\mathbb{Z} \cong \mathbb{Z}$, and yet the set $\mathbb{Z}/n\mathbb{Z}$ has n members. ◇

Definition 4.5.4 The *direct product of groups* G_i is the direct product $\prod G_i$ of the sets G_i with multiplication defined pointwise.

Then the *projection map* $\pi_i \colon \prod G_j \to G_i$ given by $\pi_i(f) = f(i)$ is an epimorphism with kernel consisting of those $f \in \prod G_j$ such that $f(i) = e_i$, where e_i is the unit of G_i.

Proposition 4.5.5 *Let H and K be subgroups of a group G such that $H \cap K$ is trivial and $HK = G$ and with $hk = kh$ for all $h \in H$ and $k \in K$. Then the map $H \times K \to G$ given by $(h, k) \mapsto hk$ is an isomorphism.*

Proof The map is obviously an epimorphism. To see that it also is injective, assume that (h, k) maps to the unit e. Then $hk = e$, or $h = k^{-1}$, so both h and k belong to $H \cap K = \{e\}$. □

4.6 Normal Subgroups

Definition 4.6.1 A subgroup H of a group G is a *normal subgroup* of G, and we write $H \triangleleft G$, if $aHa^{-1} \subset H$ for $a \in G$.

It is easy to check that $H \triangleleft G$ if and only if $a^{-1}Ha = H$ for $a \in G$. So the left- and right cosets of an element in a normal subgroup coincide.

Clearly any subgroup of an abelian group is normal.

Proposition 4.6.2 *Let H be a normal subgroup of a group G. Then G/H is a group, called the quotient group of G by H, with multiplication $(aH)(bH) = abH$ for $a, b \in G$.*

Proof Multiplication is well-defined because if $x, y \in H$, then

$$(axH)(byH) = axbyH = ab(b^{-1}xb)H = abH,$$

where we twice absorbed elements of H into H. Clearly this binary operation is associative. The unit is H and $(aH)^{-1} = a^{-1}H$. □

Definition 4.6.3 Let H be a normal subgroup of a group G. The epimorphism $G \to G/H$ which sends a to aH is called the *quotient map*.

Example 4.6.4 Let $n \in \mathbb{N}$. Then $n\mathbb{Z}$ is a normal subgroup of the additive group $\mathbb{Z}$, and we can consider the quotient group $\mathbb{Z}/n\mathbb{Z}$. On the other hand we have the group $\mathbb{Z}_n$ of congruence classes $[m]$ modulo n. The map $[m] \mapsto mn\mathbb{Z}$ is an isomorphism from $\mathbb{Z}_n$ to $\mathbb{Z}/n\mathbb{Z}$. ◇

We have arrived at the *first isomorphism theorem*.

Theorem 4.6.5 *Let $f: G \to H$ be a homomorphism of groups. Then $\ker f \triangleleft G$, and the map $g: G/\ker f \to H$ given by*

$$a \ker f \mapsto f(a)$$

is a monomorphism such that $f = gh$, where $h: G \to G/\ker f$ is the quotient map. Hence $\operatorname{im} f \cong G/\ker f$.

Proof Let $a \in G$ and $b \in \ker f$. Then $aba^{-1} \in \ker f$ because $f(aba^{-1}) = f(a)f(b)f(a^{-1}) = f(a)ef(a^{-1}) = f(aa^{-1}) = e$, so $\ker f$ is a normal subgroup of G.

If $x \in \ker f$, then $ax \ker f \mapsto f(ax) = f(a)f(x) = f(a)e = f(a)$, so g is well-defined. Clearly it is a homomorphism. It is injective since $f(a) = e$ means that $a \in \ker f$, so $a \ker f = \ker f$, which is the unit in the quotient group.

Thus g is an isomorphism onto its image $\operatorname{im} f$. □

Definition 4.6.6 A *simple group* is a non-trivial group that has no proper normal subgroups.

4.7 Cyclic Groups

Let us begin with an example, where we utilize the first isomorphism theorem.

Example 4.7.1 Let $n \in \mathbb{N}$. The map

$$e_n: \theta \mapsto \exp(\frac{2\pi i \theta}{n}) \equiv \cos(\frac{2\pi i \theta}{n}) + i \sin(\frac{2\pi i \theta}{n})$$

is clearly a homomorphism from the additive group $\mathbb{R}$ to the group $\mathbb{C}_*$. Its image is the circle $\mathbb{T}$ and the kernel is $n\mathbb{Z}$. Hence $\mathbb{T} \cong \mathbb{R}/n\mathbb{Z}$.

Restrict e_n to the subgroup $\mathbb{Z}$ of $\mathbb{R}$. Then its image is the cyclic subgroup of $\mathbb{T}$ generated by $\exp(\frac{2\pi i}{n})$, and the kernel is still $n\mathbb{Z}$. Hence

$$\langle \exp(\frac{2\pi i}{n}) \rangle \cong \mathbb{Z}/n\mathbb{Z} \cong \mathbb{Z}_n.$$

4.7 Cyclic Groups

The n numbers $\exp(\frac{2\pi i m}{n})$ for $m \in \{0, 1, \ldots, n-1\}$ are placed around the circle at equal distance, and are obtained by multiplying the generator $\exp(\frac{2\pi i}{n})$, which is an n-root of 1, with itself m times, thus moving counter clockwise around the circle. Hence the terminology cyclic.

Drawing straight lines between these n points on the circle one gets a *regular n-polygon*. For instance, the group $\mathbb{Z}_4$ corresponding to $\{\pm 1, \pm i\}$ form a square with sides $\sqrt{2}$ and with one vertex at 1. ◇

We have a complete classification of cyclic groups.

Theorem 4.7.2 *Every cyclic group is either finite and is isomorphic to $\mathbb{Z}_n$ for some $n \in \mathbb{N}$, or it is infinite and isomorphic to $\mathbb{Z}$.*

Proof If $\langle a \rangle$ is infinite, then $m \mapsto a^m$ is an isomorphism from $\mathbb{Z}$ to $\langle a \rangle$. Injectivity must hold, otherwise $\langle a \rangle$ would be finite.

If $\langle a \rangle$ is finite, then $m \mapsto a^m$ is an epimorphism from $\mathbb{Z}$ to $\langle a \rangle$. Its kernel is $n\mathbb{Z}$ where $n = o(a)$, so by the first isomorphism theorem, we see that $\langle a \rangle \cong \mathbb{Z}/n\mathbb{Z}$. □

So cyclic groups of the same order are isomorphic.

Corollary 4.7.3 *Homomorphic images of cyclic groups are cyclic, and subgroups of cyclic groups are cyclic.*

Proof The first statement is trivial as the homomorphic image of a generator is a generator.

The second statement is immediate from the theorem and the two examples from Sect. 4.3. □

All proper subgroups of an infinite cyclic group G are isomorphic to G. This follows from the theorem above as the only cyclic subgroups of $\mathbb{Z}$ are $n\mathbb{Z}$ for some integer n.

Proposition 4.7.4 *There is at most one subgroup of a given order of a finite cyclic group.*

Proof Any cyclic subgroup of a finite cyclic group $\langle a \rangle$ is cyclic, and is of the form $\langle a^m \rangle$ for some m. Let d be the greatest common divisor of m and the order n of $\langle a \rangle$. Then $\langle a^m \rangle = \langle a^d \rangle$ because d divides m, so the inclusion $\subset$ holds. The opposite inclusion holds because $d = rm + sn$ for some $r, s \in \mathbb{Z}$, so $a^d = a^{rm}$.

But $\langle a^d \rangle$ has order n/d. So if two subgroups have generators with different such d's, their orders are also different. In other words, two subgroups of the same order have a common generator and must coincide. □

If G is a finite cyclic group, there is exactly one subgroup of G for each divisor m of $|G|$, and the order of this subgroup is $|G|/m$, and moreover, these are all the subgroups. This is clear from the theorem above and the proposition above.

In particular, the simple cyclic groups are those of prime order. This suggests that simple groups should be thought of as building blocks for groups.

Proposition 4.7.5 *If G and H are finite cyclic groups with relatively prime orders. Then $G \times H$ is cyclic of order $|G||H|$.*

Proof Let $|G| = m$ and $|H| = n$, so $|G \times H| = mn$. Say a and b are generators of G and H, respectively. Now $(a, b)^{mn} = (a^{mn}, b^{mn}) = (e, e)$, so $d \equiv o(a, b)$ divides mn. But as $(a, b)^d = (e, e)$, we see that both m and n divide d. Thus mn also divides d as m and n are relatively prime. So $d = mn$. Thus (a, b) is a generator for $G \times H$. □

4.8 Normalizers and Centralizers

Definition 4.8.1 The *normalizer* of a subset X of a group G is the subgroup $N(X)$ of G given by
$$N(X) = \{a \in G \mid aXa^{-1} = X\}.$$

The reason why $N(X) < G$ is that $(ab^{-1})X(ab^{-1})^{-1} = ab^{-1}bXb^{-1}ba^{-1}$ for $a, b \in N(X)$.

We write $N(a)$ when X consists of a single element a.

Proposition 4.8.2 *A subgroup H of a group G is a normal subgroup of $N(H)$, and $N(H)$ is the largest subgroup of G in which H is normal.*

If $K < N(H)$ for a subgroup H of a group G, then $KH < G$ and $H \triangleleft KH$.

Proof As $aHa^{-1} = H$ for a in a group H, we see that $H \triangleleft N(H)$, and it is also clear that $N(H)$ is the largest possible subgroup of G that contains H as a normal subgroup.

As for the second statement, notice that every element $a \in K < N(H)$ satisfies $aHa^{-1} = H$, so $KH = HK$. By Proposition 4.2.5 we conclude that KH is a group, and clearly $H \triangleleft KH$. □

We can generalize normalizers of single elements in another direction.

Definition 4.8.3 The *centralizer* of a subset X of a group G is the subgroup $Z(X)$ of G defined as
$$Z(X) = \{a \in G \mid ax = xa \text{ for all } x \in X\}.$$

The *center of a group* are the elements in the group that commute with every other element of the group, so it is the centralizer of the group as a subset of itself.

Indeed $Z(X) < G$, because $Z(X)$ is obviously a monoid, and it contains inverses of its elements since $ax = xa \Leftrightarrow xa^{-1} = a^{-1}x$ for $a, x \in G$.

The following proposition is straightforward.

Proposition 4.8.4 *An abelian subgroup H of a group G is a subgroup of $Z(H)$, and any abelian subgroup of G that contains H is a subgroup of $Z(H)$.*

The center of a group G is a normal subgroup of G. A group G is abelian if and only if $Z(G) = G$.

4.9 Correspondences

The *correspondence theorem* relates subgroups of groups connected by a homomorphism.

Theorem 4.9.1 *Let $f : G \to H$ be a homomorphism between two groups. Then the map $K \mapsto f(K)$ is a bijection between subgroups of G containing $\ker f$ and subgroups of $\operatorname{im} f$ such that normal subgroups correspond to normal subgroups.*

Proof We have seen that $f(K) < \operatorname{im} f$ for $K < G$, and clearly $f(K) \triangleleft \operatorname{im} f$ if K is normal.

If $\ker f \subset K < G$, then $K = f^{-1}(f(K))$ since trivially $K \subset f^{-1}(f(K))$, and if $a \in f^{-1}(f(K))$, then $f(a) \in f(b)$ for some $b \in K$, so $ab^{-1} \in \ker f \subset K$ and $a \in K$.

Hence, if $f(K_1) = f(K_2)$ for subgroups K_1 and K_2 of G containing $\ker f$, then $K_1 = f^{-1}(f(K_1)) = f^{-1}(f(K_2)) = K_2$.

If $L < \operatorname{im} f$, then $\ker f \subset f^{-1}(L) < G$, with $f^{-1}(L)$ normal if $L \triangleleft \operatorname{im} f$, and $f(f^{-1}(L)) = L$ as $L \subset \operatorname{im} f$. □

Here is an important application.

Corollary 4.9.2 *Suppose $H \triangleleft G$. To any $L < G/H$ there is a unique $K < G$ such that $L = K/H$, and K is normal if and only if L is normal.*

Proof Apply the theorem to the quotient map $G \to G/H$. □

Definition 4.9.3 A normal subgroup H of a group G is *maximal* if $H \neq G$ and if H is not properly contained in a proper normal subgroup.

The existence of a maximal normal subgroup in a group G is a simple application of Zorn's lemma to the family of normal subgroups of G different from G and ordered under inclusion.

Corollary 4.9.4 *Let H be a proper normal subgroup of a group G. Then H is maximal if and only if G/H is simple.*

4.10 More Isomorphism Theorems

The following result, known as the *diamond isomorphism theorem*, will prove useful.

Theorem 4.10.1 *If H and K are subgroups of a group G with $HK = KH$, then*

$$f : H \to HK/K; \quad f(a) = aK$$

is an epimorphism with $\ker f = H \cap K$, so $H/H \cap K \cong HK/K$. In particular, this holds for any H, K with $H < G$ and $K \triangleleft G$.

Proof Note that $HK < G$ and $K \triangleleft HK$, and that f is the usual quotient map. The result is then immediate from the first isomorphism theorem. □

If H, K are subgroups of a group, and $HK = KH$ is finite, then this theorem shows that
$$|HK|/|K| = |H|/|H \cap K|.$$

In fact, this holds for any subgroups H and K of a finite group.

To see this, define an equivalence relation $\sim$ on $H \times K$ by $(a, b) \sim (a', b')$ if $a' = ac$ and $b' = c^{-1}b$ for some $c \in H \cap K$. Then $c \mapsto (ac, c^{-1}b)$ is clearly a bijection from $H \cap K$ to the equivalence class $[(a, b)]$, so $|[(a, b)]| = |H \cap K|$. Thus

$$|H||K| = |H \times K| = |(H \times K)/\sim||H \cap K|,$$

so we are done if we can show that $HK \cong (H \times K)/\sim$. But $[(a, b)] \mapsto ab$ is such an isomorphism. It is well-defined as $(ac)c^{-1}b = ab$, it is obviously surjective, and it is injective because if $ab = a'b'$ with $a, a' \in H$ and $b, b' \in K$, then $c = a^{-1}a' = b(b')^{-1} \in H \cap K$ and $a' = ac$ and $b' = c^{-1}b$, so $[(a', b')] = [(a, b)]$.

The *third (or double quotient) isomorphism theorem* shows that forming quotient groups is transitive.

Theorem 4.10.2 *Let H and K be normal subgroups of a group G with $K \subset H$. Then*
$$f: G/K \to G/H; \ f(aK) = aH$$
is a well-defined epimorphism with $\ker f = H/K$, *so* $(G/K)/(H/K) \cong G/H$.

Proof The map is well-defined because if $aK = bK$ for $a, b \in G$, then $b^{-1}a \in K \subset H$, so $aH = bH$. It is obviously a homomorphism by definition of the product in quotient groups.

Surjectivity is clear, and $\ker f = \{aK \mid aH = H\} = \{aK \mid a \in H\} = H/K$, so the result follows from the first isomorphism theorem. □

4.11 Permutation Groups

In all our examples so far we have only dealt with abelian groups. Let us now consider non-abelian groups.

Definition 4.11.1 The *permutation group* $\mathrm{Perm}(X)$ of a set X consists of all bijections $X \to X$, or *permutations* of X, with multiplication given by composition of maps. Any subgroup of $\mathrm{Perm}(X)$ is also called a permutation group.

The unit of a permutation group is the identity map, and the inverse of a permutation is the inverse map.

4.11 Permutation Groups

In Sect. 3.6 we studied permutations of a finite set, and worked with the *symmetric group* $S_n \equiv \text{Perm}(\{1, \ldots, n\})$ without considering it a group at the time.

We saw that every element in S_n can be written as a product of disjoint non-trivial cycles, and that such a product is unique modulo order of factors. Also we saw that each cycle can be written as a product of transpositions.

Now S_n is a finite group of order $n!$, and each cycle generates a cyclic subgroup $\mathbb{Z}_m$, where $m \in \{1, \ldots, n\}$ is the length of the cycle. Apart from S_1, and S_2, which is isomorphic to $\mathbb{Z}_2$, the symmetric groups are non-abelian.

To see this, first notice that when $m \leq n$, the group S_m can be embedded into S_n and recognized as those elements that permute only the first m elements and leave the rest fixed. Then consider the following example, which shows that S_3 is non-abelian.

Example 4.11.2 The cycle $\sigma = (1\ 2\ 3)$ and the transposition $\tau = (2\ 3)$ generate S_3, and $S_3 = \{e, \sigma, \sigma^2, \tau, \sigma\tau, \sigma^2\tau\}$. In fact, we have the relations

$$\sigma^3 = e = \tau^2, \quad \tau\sigma = \sigma^2\tau,$$

which completely specify the multiplication. ◇

We have the following more general result.

Proposition 4.11.3 *The symmetric group S_n is generated by $(1\ 2\ \cdots\ n)$ and $(n-1\ n)$.*

Proof Let G be the subgroup of S_n generated by $\sigma \equiv (1\ 2\ \cdots\ n)$ and $\tau \equiv (n-1\ n)$. It suffices to show that G contains all transpositions since these generate S_n.

Now G contains $\sigma^m \tau \sigma^{-m}$, which equals $(m-1\ m)$ by induction. Since

$$(1\ 2)(2\ 3)(1\ 2) = (1\ 3) \quad \text{and} \quad (1\ 3)(3\ 4)(1\ 3) = (1\ 4),$$

and so forth, we see that $(1\ m) \in G$, and hence $(m\ k) = (1\ m)(1\ k)(1\ m)$ also belongs to G. □

We saw that in the symmetric group, the number of transpositions in any decomposition is either always even or always odd. This gave us the sign function, which can be regarded as an epimorphism sign: $S_n \to \{\pm 1\}$. Its kernel A_n, consisting of even permutations, is therefore a normal subgroup of S_n, and is known as the *alternating group*. By Lagrange's theorem $|A_n| = n!/2$.

The following result is known as *Cayley's theorem*.

Proposition 4.11.4 *Let G be a group, and let $a \in G$. Define $\lambda_a : G \to G$ by $\lambda_a(b) = ab$. Then the map $a \mapsto \lambda_a$ is a monomorphism from G to $\text{Perm}(G)$. Hence every group is a permutation group.*

Proof Clearly λ_e is the identity map, and $\lambda_{ab} = \lambda_a \lambda_b$, so λ_a is invertible with inverse $\lambda_{a^{-1}}$. □

4.12 Symmetries

A special kind of permutation group is particularly important in geometry.

Definition 4.12.1 A permutation σ of a space X with a metric d is a *symmetry* if $d(\sigma(x), \sigma(y)) = d(x, y)$ for all $x, y \in X$.

The set of symmetries is a subgroup, the *symmetry group*, of $\mathrm{Perm}(X)$ because if σ and τ are symmetries, then

$$d(\tau\sigma^{-1}(x), \tau\sigma^{-1}(y)) = d(\sigma^{-1}(x), \sigma^{-1}(y)) = d(\sigma\sigma^{-1}(x), \sigma\sigma^{-1}(y)) = d(x, y).$$

Let X be the points in $\mathbb{R}^2$, with the usual distance, that form the perimeter of a regular n-polygon P_n. Geometrically it is clear that any symmetry of this infinite subset X of $\mathbb{R}^2$ is uniquely determined by its effect on the vertices of P_n, which we label as $1, 2, \ldots, n$. Thus we can consider these symmetries as a subgroup of S_n.

Definition 4.12.2 The *dihedral group* D_n of degree $n \in \mathbb{N}$ is the symmetry group of P_n.

Obviously, an element of S_n is a symmetry of P_n if and only if it takes adjacent vertices of P_n to adjacent vertices. This happens if and only if the cyclic order $1, 2, \ldots, n$ is preserved or reversed, that is, if and only if the permutation is trivial or of the form σ^m, or of the form $\sigma^m \tau$, where $m \in \{1, \ldots, n-1\}$ and

$$\sigma = \begin{pmatrix} 1 \; 2 \; \cdots \; n \end{pmatrix} \text{ and } \tau = \begin{pmatrix} 1 & 2 & \cdots & n \\ 1 & n & \cdots & 2 \end{pmatrix}.$$

It is easy to check that $\tau\sigma = \sigma^{n-1}\tau$, which gives the following result.

Proposition 4.12.3 *Let n be an integer greater than one. Then the dihedral group D_n has order $2n$ with distinct elements σ^m and $\sigma^m \tau$ for $m \in \{0, 1, \ldots, n-1\}$. The multiplication is completely determined by the relations*

$$\sigma^n = e = \tau^2 \text{ and } \tau\sigma = \sigma^{n-1}\tau.$$

The permutation σ corresponds to a rotation of P_n by an angle $2\pi/n$, and τ corresponds to a reflection.

We see that D_n is non-abelian for n greater than two. We also see that $D_3 = S_3$, which is consistent with the fact that any two vertices of an equilateral triangle are adjacent. The *optic group* D_4 is however not isomorphic to S_4. We can obtain S_4 as the symmetries in $\mathbb{R}^3$ of the tetrahedron, since there any two vertices are adjacent.

4.13 Automorphisms

Another permutation group is the set $\mathrm{Aut}(G)$ of automorphisms of a group G. Recall that f is an automorphism if f is a permutation of G such that $f(ab) = f(a)f(b)$ for $a, b \in G$, and such maps obviously form a group under composition.

Definition 4.13.1 The *inner automorphism* $\mathrm{Ad}(a)$ of a group G determined by $a \in G$ is the map $\mathrm{Ad}(a)\colon G \to G$ given by

$$\mathrm{Ad}(a)(b) = aba^{-1}$$

for all b.

This is indeed an automorphism because $\mathrm{Ad}(a)(bc) = aba^{-1}aca^{-1}$, so $\mathrm{Ad}(a)$ is a homomorphism. Also $\mathrm{Ad}(a)(b) = e$ implies $b = e$, and finally $\mathrm{Ad}(a)(a^{-1}ba) = b$.

Proposition 4.13.2 *The set* $\mathrm{Inn}(G)$ *of inner automorphisms of a group G is a normal subgroup of* $\mathrm{Aut}(G)$.

Proof Since $\mathrm{Ad}(a)(\mathrm{Ad}(b))^{-1} = \mathrm{Ad}(ab^{-1})$ for any $a, b \in G$, we conclude that $\mathrm{Inn}(G)$ is a subgroup of $\mathrm{Aut}(G)$, and it is normal because $f\,\mathrm{Ad}(a)\,f^{-1} = \mathrm{Ad}(f(a))$ for $f \in \mathrm{Aut}(G)$ and $a \in G$. □

Definition 4.13.3 The quotient group $\mathrm{Out}(G) \equiv \mathrm{Aut}(G)/\mathrm{Inn}(G)$ is the *outer conjugacy classes* of the group G.

Thus two automorphisms f and g of a group G are equivalent, or *conjugate*, if $f^{-1}g \in \mathrm{Inn}(G)$, or in other words, if there is an element $a \in G$ such that $g = f\,\mathrm{Ad}(a)$. So f and g are conjugate if and only if $g = bf(\cdot)b^{-1}$ for some $b \in G$.

The following result is obvious.

Proposition 4.13.4 *The map $a \mapsto \mathrm{Ad}(a)$ from a group G to $\mathrm{Inn}(G)$ is an epimorphism with kernel $Z(G)$, so* $\mathrm{Inn}(G) \cong G/Z(G)$.

Definition 4.13.5 A group G is a *complete group* if $\mathrm{Aut}(G) \cong G$, that is, if its center is trivial and if every automorphism is inner.

Note that the order of an element in a group is the same as the order of its image under an automorphism. Also any homomorphism from a group is uniquely determined on any set of generators of the group.

Example 4.13.6 The symmetric group S_3 is complete. To see this, consider the generators σ and τ of S_3 from Example 4.11.2. The relations

$$\sigma^3 = e = \tau^2, \quad \tau\sigma = \sigma^2\tau$$

show that $Z(S_3)$ is trivial, so $|\mathrm{Inn}(S_3)| = |S_3| = 6$. They also show that σ and σ^2 are of order three, whereas τ, $\sigma\tau$ and $\sigma^2\tau$ are of order 2. So there cannot be more than 6 automorphism of S_3. Hence $\mathrm{Aut}(S_3) \cong S_3$. Of course, these are not all the permutations of S_3. The order of $\mathrm{Perm}(S_3)$ is $6! = 720$. ◇

Example 4.13.7 The order of Aut(G) for a finite cyclic group G is $\phi(|G|)$. To see this first observe that for an integer n, the map $a \mapsto a^n$ is an automorphism of a finite abelian group G if n and $|G|$ are relatively prime. This is also a necessary condition when $G = \langle a \rangle$ is a finite cyclic group, and in this case any automorphism is uniquely determined on a, and the image of a would be a^n for some n. Hence Euler's ϕ-function.

4.14 Semidirect Products

Definition 4.14.1 The *semidirect product* $K \rtimes H$ of the groups K and H with respect to a homomorphism $f \colon H \to \text{Aut}(K)$ is the group which is $K \times H$ as a set but with product given by the formula

$$(a, b) \cdot (c, d) \equiv (af(b)c, bd).$$

It is straightforward to check that the semidirect product $K \rtimes H$ is indeed a group with unit (e, e) and $(a, b)^{-1} = (f(b^{-1})a^{-1}, b^{-1})$.

Clearly, when f sends every element of H to the identity map on K, then $K \rtimes H = K \times H$ even as groups.

In general we have $(a, e)(e, b) = (a, b)$, whereas $(e, b)(a, e) = (f(b)a, b)$, so $(e, b)(a, e)(e, b)^{-1} = (f(b)a, e)$ for $a \in K$ and $b \in H$. Considering K and H as subgroups of $K \rtimes H$ under the embeddings $a \mapsto (a, e)$ and $b \mapsto (e, b)$, respectively, we see that K is normal, that $K \rtimes H = KH$ and that $K \cap H$ is trivial.

The following straightforward result tells us when a group is isomorphic to a semidirect product.

Proposition 4.14.2 *Suppose G is a group with a subgroup H and a normal subgroup K such that $KH = G$ and $H \cap K = \{e\}$. Then $K \rtimes H$ with respect to $f \colon H \to \text{Aut}(K)$ given by $f(b)a = bab^{-1}$ is isomorphic to G under $(a, b) \mapsto ab$.*

Example 4.14.3 The $(ax + b)$-group G consists of all maps $T_{a,b} \colon \mathbb{R} \to \mathbb{R}$ given by $T_{a,b}(x) = ax + b$, where $a \in \mathbb{R}_*$ and $b \in \mathbb{R}$ and with multiplication given by composition of maps. Since

$$a(cx + d) + b = (ac)x + (ad + b),$$

we do get a group this way which has $\mathbb{R}_*$ and $\mathbb{R}$ as subgroups under the embeddings $a \mapsto T_{a,0}$ and $b \mapsto T_{1,b}$. It is easy to check that the requirements of the proposition above hold with $K = \mathbb{R}$ and $H = \mathbb{R}_*$. Thus the $(ax + b)$-group is isomorphic to the semidirect product $\mathbb{R} \rtimes \mathbb{R}_*$ with $f \colon \mathbb{R}_* \to \text{Aut}(\mathbb{R})$ given by $f(a)b = ab$ as it is readily verified that $T_{a,0} T_{1,b} T_{a,0}^{-1} = T_{1,ab}$. ◇

Remark 4.14.4 In physics one thinks of the $(ax + b)$-group as the group of Galilean transformations related to a particle moving in one dimension. Its trajectory is then

4.14 Semidirect Products

described by the particles position $x = x(t)$ on the x-axis at the time t. If the particle is moving freely, then as long as one is referring to an inertial observer, it obeys Newtons law of motion $d^2x/dt^2 = 0$, saying that its acceleration is zero. This law is covariant with respect to all observers in inertial systems, that is, looks the same for everybody moving with constant speed with respect to each other, provided the coordinate transformations between these observers are Galilean transformations. This means that if the particles motion is described with respect to another inertial observer, say with coordinate $y(s)$ in his frame at absolute time $s = t$, then if the speed of the observer is constantly equal to a and his origin is translated to $-b$, the transformation is $y(s) = ax(t) + b$. As $d^2y/ds^2 = ad^2x/dt^2$ from calculus, we again get $d^2y/ds^2 = 0$. So Newtons law for free motion is covariant; none of the two observers experience that the particle is accelerating.

Insisting on covariance led Einstein to the theory of relativity [6]. In special theory of relativity the group of Galilean transformations (in three dimension) was replaced by the Poincare group of transformations between inertial observers because Maxwell's equations are covariant under these transformations, and they described electromagnetism, which Einstein sought to unify with mechanics. This implied that the speed of light should be the same with respect to all inertial observers, and came at the cost of abandoning the concepts of absolute time and rigid objects.

The general theory of relativity allows including observers in accelerated motion with respect to each other. This was achieved by including gravity; matter curves spacetime and free motion follows geodesics, or straightest possible lines in spacetime, which well could mean accelerated motion in ordinary space, an effect then attributed to a gravitational field. The appropriate mathematical language for this theory is differential geometry [18], which won't be treated in this book. With this little digression we just want to point out that groups, playing the role as symmetry objects in a physical theory, has played an important role in the development of modern physics, not to mention as gauge groups in the Standard model describing particle physics [20]. ◇

Example 4.14.5 Let G be a group. Define a homomorphism $f \colon S_n \to \mathrm{Aut}(G^n)$ by

$$f(\sigma)(a_1, \ldots, a_n) = (a_{\sigma(1)}, \ldots, a_{\sigma(n)})$$

for $\sigma \in S_n$ and $a_i \in G$. The corresponding semidirect product $G^n \rtimes S_n$ is known as the *wreath product*. ◇

Example 4.14.6 Consider the semidirect product $\mathbb{Z}_n \rtimes \mathbb{Z}_2$ with respect to the homomorphism $f \colon \mathbb{Z}_2 \to \mathrm{Aut}(\mathbb{Z}_n)$ given by $f([k])([m]) = [(-1)^k m]$. This group is isomorphic to the dihedral group D_n under the isomorphism $([m], [k]) \mapsto \sigma^m \tau^k$.

4.15 The General and Special Linear Group

Linear algebra provides many important automorphism groups. Let V be a finite dimensional vector space over a field F.

Remark 4.15.1 What we in linear algebra called the endomorphisms $\text{End}(V)$ of V, are the linear maps, and these are obviously homomorphism of V considered as an additive group. But the converse is not true, there are additive maps that do not respect multiplication by scalars.

To wit, consider $\mathbb{R}$ as a vector space over $\mathbb{Q}$. Then additive maps are uniquely defined by ascribing arbitrary values on a linear basis for $\mathbb{R}$ over $\mathbb{Q}$. This way we get additive maps $\mathbb{R} \to \mathbb{R}$ that are not continuous. Such maps cannot be linear, since any linear map from $\mathbb{R}$ to $\mathbb{R}$ is automatically continuous.

However, if F has characteristic 0, then any additive map will be linear over $\mathbb{Q}$ because
$$mf(v) = f(mn(1/n)v) = nf((m/n)v),$$
for all integers m and $n \neq 0$ and vectors $v \in V$. So if $\mathbb{Q}$ is dense in F and F is a field with continuous operations, which is the case for $\mathbb{R}$, then any continuous additive map on V will be linear. This follows because for any $a \in F$, there are rational numbers a_n such that $a = \lim a_n$, and therefore as scalar multiplication is continuous for finite dimensional vector spaces over F, we get
$$f(av) = f(\lim(a_n v)) = \lim f(a_n v) = \lim a_n f(v) = a f(v)$$
for all vectors v.

Likewise, due attention has to be made to $\text{Aut}(V)$, which in the context of linear algebra means the bijective linear maps on V. These will be automorphisms of V considered as an additive group, while the converse need not hold.

Consider now the permutation group $\text{Aut}(V)$ of V consisting of bijective linear maps. Picking a basis for V we get an isomorphism $\text{Aut}(V) \cong GL(n, F)$, where $n = \dim V$ and $GL(n, F)$ denotes the group of invertible $n \times n$-matrices with entries in F. Under this isomorphism composition of linear maps becomes multiplication of matrices with unit $e = I_n$.

Definition 4.15.2 The group $GL(n, F)$ is called the *general linear group*.

We see that two matrices in $GL(n, F)$ are similar if and only if one of them is the image of the other one under some inner automorphism.

The map $a \mapsto a I_n$ embeds the group F_* into $GL(n, F)$.

Proposition 4.15.3 *The center of $GL(n, F)$ is F_*. So $\text{Inn}(GL(n, F)) \cong GL(n, F)/F_*$, and the latter group is known as the* projective group $PGL(n, F)$.

Proof Consider the matrix units E_{ij}. Now $I_n + E_{ij}$ are all invertible with determinant 1 if $i \neq j$ and determinant 2 if $i = j$.

Let $A \in Z(GL(n, F))$. Then in particular $A(I_n + E_{ij}) = (I_n + E_{ij})A$, or equivalently $AE_{ij} = E_{ij}A$, for all i and j. But the matrix AE_{ij} has the ith column of A as its jth column, and with all other columns containing only zeros. Whereas the matrix $E_{ij}A$ has the jth row of A as its ith row, and with all other rows containing only zeros. These two matrices can only be equal if A is diagonal with $a_{ii} = a_{jj}$ for all i and j. Thus $A \in F_*$. □

The determinant is a unital homomorphism from the monoid $\mathrm{Mat}(n, F)$ to the multiplicative monoid F. The general linear group is the submonoid consisting of those matrices with non-vanishing determinant. Consider the epimorphism $\det\colon GL(n, F) \to F_*$. Let $SL(n, F)$ denote its kernel, so

$$GL(n, F)/SL(n, F) \cong F_*.$$

Definition 4.15.4 The *special linear group* $SL(n, F)$ is the normal subgroup of $GL(n, F)$ consisting of the matrices with determinant 1.

By essentially the same argument as in the proof of Proposition 4.15.3, we see that also $SL(n, F)$ has trivial center.

Note also that by Proposition 3.7.10, we observe that the matrices in $SL(n, F)$ with integer entries form a subgroup of $SL(n, F)$, which we denote by $SL(n, \mathbb{Z})$.

4.16 Inner Products and Linear Subgroups

Consider $\mathbb{C}^n$ with the usual inner product $(x|y) = \bar{y}^T x$ of columns vectors $x, y \in \mathbb{C}^n$. Here bar of a matrix means taking complex conjugates of its entries. This inner product produces the usual distance in $\mathbb{C}^n$, and we can form subgroups of $GL(n, \mathbb{C})$ consisting of symmetries, or isometries.

Definition 4.16.1 A *unitary matrix* $A \in M(n, \mathbb{C})$ is an isometry of $\mathbb{C}^n$, that is, it satisfies $(Ax|Ay) = (x|y)$ for all $x, y \in \mathbb{C}^n$. The *unitary group* is the subgroup $U(n)$ of $GL(n, \mathbb{C})$ consisting of all such isometries.

Since $\overline{(Ax)}^T Ay = \bar{x}^T \bar{A}^T Ay$ we infer that

$$U(n) = \{A \in M(n, \mathbb{C}) \mid \bar{A}^T A = I_n\}.$$

For any $A \in U(n)$, we have $1 = \det(I_n) = \det(\bar{A}^T A) = |\det(A)|^2$, so the determinant restricts to an epimorphism $\det\colon U(n) \to \mathbb{T}$. The kernel $SU(n)$ of this epimorphism is a normal subgroup of $U(n)$, so

$$U(n)/SU(n) \cong \mathbb{T}.$$

Definition 4.16.2 The group $SU(n) = U(n) \cap SL(n, \mathbb{C})$ is called the *special unitary group*.

Example 4.16.3 It is easy to check that

$$SU(2) = \left\{ \begin{pmatrix} z & -\bar{w} \\ w & \bar{z} \end{pmatrix} \mid z, w \in \mathbb{T} \text{ with } |z|^2 + |w|^2 = 1 \right\}.$$

◇

The matrices in $U(n)$ with real entries obviously form a subgroup $O(n)$. Restricting the inner product on $\mathbb{C}^n$ to $\mathbb{R}^n$ gives the usual inner product on $\mathbb{R}^n$ with the usual distance.

Definition 4.16.4 An *orthogonal matrix* $A \in M(n, \mathbb{R})$ is an isometry of $\mathbb{R}^n$, that is, it satisfies $(Ax|Ay) = (x|y)$ for all $x, y \in \mathbb{R}^n$. The *orthogonal group* is the subgroup $O(n)$ of $GL(n, \mathbb{R})$ consisting of all such isometries.

Clearly

$$O(n) = \{A \in M(n, \mathbb{R}) \mid A^T A = I_n\}.$$

Geometrically, the orthogonal group $O(3)$ is generated by rotations and reflections in $\mathbb{R}^3$, so vectors that are pairwise orthogonal, or perpendicular, remain so after being transformed by an orthogonal matrix.

For any $A \in O(n)$, we have $1 = \det(I_n) = \det(A^T A) = \det(A)^2$, so the determinant of an orthogonal matrix is ± 1. Denoting the kernel of the epimorphism $\det \colon O(n) \to \{\pm 1\}$ by $SO(n)$, we get

$$O(n)/SO(n) \cong \mathbb{Z}_2.$$

Definition 4.16.5 The group $SO(n) = O_n(\mathbb{R}) \cap SL(n, \mathbb{R})$ is called the *special orthogonal group*.

The group $SO(3)$ consists of those orthogonal transformations in $\mathbb{R}^3$ that preserve orientation, which are the rotations.

The center of $SU(2)$ is $\pm I_2$, and it can be shown that $SU(2)/\mathbb{Z}_2 \cong SO(3)$.

Example 4.16.6 It is easily checked that the map

$$x + iy \mapsto \begin{pmatrix} x & y \\ -y & x \end{pmatrix}$$

is an isomorphism from the circle $\mathbb{T}$ to $SO(2)$.

◇

Let

$$J = \begin{pmatrix} 0 & -I_n \\ I_n & 0 \end{pmatrix}$$

and define the *symplectic bilinear form* on $\mathbb{C}^{2n}$ by $B(x, y) = x^T J y$.

4.16 Inner Products and Linear Subgroups 153

Definition 4.16.7 The group $Sp(n, \mathbb{C}) \subset M(2n, \mathbb{C})$ of isometries of $\mathbb{C}^{2n}$ with respect to the symplectic bilinear form is called the *complex symplectic group*.

Hence
$$Sp(n, \mathbb{C}) = \{A \in M(2n, \mathbb{C}) \mid A^T J A = J\}.$$

Note that $J^2 = -I_{2n}$, so $\det(A) = \pm 1$ for $A \in Sp(n, \mathbb{C})$.

One can also consider the subgroup $Sp(n, \mathbb{R})$ of $Sp(n, \mathbb{C})$ consisting of those matrices with real entries.

Definition 4.16.8 The group $Sp(n) = Sp(n, \mathbb{C}) \cap U(2n)$ is called the *symplectic group*.

Some of these groups have non-trivial centers.

Definition 4.16.9 The *projective groups* $PSU(2n)$, $PSO(2n)$ and $PSp(n)$ for $n \in \mathbb{N}$ are quotients of $SU(2n)$, $SO(2n)$ and $Sp(n)$, respectively, by their normal subgroup $\{\pm 1\}$.

Remark 4.16.10 All the groups we have introduced in this section are Lie groups. These are groups that are also differentiable manifolds and such that the map $(g, h) \mapsto gh^{-1}$ is smooth. The Lie groups we have considered here are known as the classical groups. There are many more Lie groups, including e.g. spin groups associated to other scalar products.

We include the following major result.

Theorem 4.16.11 *The number n appearing below is assumed to be an integer. The groups*

(A) $SU(2n + 1)$ and $PSU(2n)$ for $n \geq 1$,
(B) $SO(2n + 1)$ for $n \geq 2$,
(C) $PSp(n)$ for $n \geq 3$,
(D) $PSO(2n)$ for $n \geq 4$,

are non-abelian and have no proper normal subgroups. What is more important, they are mutually non-isomorphic.

Remark 4.16.12 In fact, the groups in the theorem are part of an $ABCDEFG$-classification of simple Lie groups; a classification that goes via Lie algebras, root systems and Dynkin diagrams [12, 24].

The Lie groups in the theorem are also connected and compact. The classification result says that every compact, connected non-abelian Lie group with no proper normal subgroups is isomorphic to exactly one of the groups listed in the theorem, or to exactly one of the five versions of compact and connected exceptional Lie groups E_6, E_7, E_8, F_4, G_2 with trivial center. We do not introduce these groups here.

4.17 Actions

Definition 4.17.1 An *action of a group* G on a set X is a map

$$G \times X \to X; \ (a, x) \mapsto ax$$

such that $ex = x$ and $a(bx) = (ab)x$ for $a, b \in G$ and $x \in X$. A set with an action of a group G is called a *G-space*.

We skip parentheses, and write abx for $a(bx)$.

If X is a G-space and $a \in G$, the map $f_a \colon X \to X$ given by $f_a(x) = ax$ is a permutation of X with $(f_a)^{-1} = f_{a^{-1}}$. Moreover, the map $G \to \text{Perm}(X)$ given by $a \mapsto f_a$ is a homomorphism.

Conversely, given a homomorphism $G \to \text{Perm}(X); \ a \mapsto f_a$, then the set X is a G-space under the action $(a, x) \mapsto f_a(x)$.

Examples of actions are supplied by the permutation groups we have considered so far. Hence a group acts on itself by left multiplication $(a, b) \mapsto ab$ and by conjugation $(a, b) \mapsto aba^{-1}$, or by more general automorphism groups.

We compare G-spaces according to the following definition.

Definition 4.17.2 A map $f \colon X \to Y$ of G-spaces is *equivariant*, or a *morphism of G-spaces*, if $f(ax) = af(x)$ for $a \in G$ and $x \in X$. If f is bijective, we say that it is an *isomorphism of G-spaces*.

Example 4.17.3 If we have an action $G \times X \to X$, then G acts on the power set $P(X)$ by $(a, Y) \mapsto aY$ and by conjugation $(a, Y) \mapsto aYa^{-1}$. One can also define actions on a restricted class of subsets. ◇

If $H < G$, then G acts on G/H by $(a, bH) \mapsto abH$, so G/H is a G-space. As we shall soon see, this case is quite typical for G-spaces.

Definition 4.17.4 A subset Y of a G-space is an *invariant subset of a G-space* if $aY \subset Y$ for all $a \in G$. If Y consists of a single point, we say that this point is a *fixed point*.

An invariant subset of a G-space is obviously a G-space under the restriction of the action.

Definition 4.17.5 The *orbit* of an element x of a G-space X is the subset Gx of X. The action is *transitive* if the whole space X is an orbit, and then X is said to be a *homogeneous space*.

An action $G \times X \to X$ is transitive if and only if for any $x, y \in X$, there is an element $a \in G$ such that $y = ax$.

4.17 Actions

The orbits are the smallest invariant subsets of a G-space. To see this, say Y is an invariant subspace of a G-space that is contained in an orbit Gx. Any $y \in Y$ is of the form $y = ax$, so $x = a^{-1}y \in Y$ since Y is invariant, which yet again implies that $Gx \subset Y$.

In a G-space we can define an equivalence relation $\sim$ by saying that $x \sim y$ if $x \in Gy$. This is clearly an equivalence relation, and the equivalence classes are the orbits, so they form a partition of X. We have decomposed the G-space into smallest blocks, and each one of these blocks carries an action given by restriction. We call this an *orbit decomposition*.

Definition 4.17.6 Suppose G acts on the set X. The *isotropy (stabilizer) group* of an element $x \in X$ is the subgroup of G given by

$$G_x \equiv \{a \in G \mid ax = x\}.$$

The action is a *free action* if all its isotropy groups are trivial.

Clearly a point x of a G-space is fixed if and only if $G_x = G$. Hence, if an action $G \times X \to X$ is free, then none of the permutations $x \mapsto ax$ on X for $a \neq e$ has fixed points.

The isotropy group of a coset aH in the G-space G/H is the subgroup aHa^{-1} of G.

We see that if X is a G-space, then the homomorphism $G \to \mathrm{Perm}(X)$ given by $a \mapsto f_a$ has kernel $\cap \{G_x \mid x \in X\}$. If this kernel if trivial, that is, if the unit is the only group element that fixes all members of X, we say that the action is a *faithful (or effective) action*. Obviously a free action is faithful.

Clearly, the action of G on a set X is faithful if and only if $G < \mathrm{Perm}(X)$. So we recover Cayley's theorem by considering the G-space G/H for H trivial.

Proposition 4.17.7 *Suppose X is a G-space, and let $x \in X$. Then G/G_x and Gx are isomorphic as transitive G-spaces with isomorphism*

$$G/G_x \to Gx; \quad aG_x \mapsto ax$$

for $a \in G$.

Proof Both G/G_x and Gx are obviously transitive as G-spaces. The formula $f(aG_x) = ax$ defines a map $f : G/G_x \to Gx$ because if $aG_x = bG_x$, then $a = bc$ for $c \in G_x$, so $f(aG_x) = ax = bcx = bx = f(bG_x)$, so f is well-defined. It is obviously equivariant and surjective, and it is injective because if $f(aG_x) = f(bG_x)$, then $ax = bx$, so $b^{-1}a = c \in G_x$ and $aG_x = bcG_x = bG_x$. $\square$

This shows that all homogeneous G-spaces are of the form G/H for a subgroup H of G. And if the action is also free, then the G-space is isomorphic to G with action given by left multiplication with elements of G.

Of course, for a homogeneous G-space X, the isomorphism $G/G_x \cong X$ depends on the chosen element $x \in X$ used to form the isotropy group. The following result shows that the isotropy groups for various choices of x are essentially the same.

Proposition 4.17.8 *In a homogeneous G-space X all isotropy groups are conjugate, that is, for $x, y \in X$, there is an element $a \in G$ such that $G_y = aG_x a^{-1}$.*

Proof Pick $a \in G$ such that $y = ax$. If $b \in G_y$, then $ax = y = by = bax$, so $a^{-1}ba \in G_x$ and $b \in aG_x a^{-1}$. So $G_y \subset aG_x a^{-1}$ and $G_x \subset a^{-1}G_y a$ as $x = a^{-1}y$. □

The formula below is known as the *orbit decomposition formula*.

Proposition 4.17.9 *Let X be a finite G-space. Then*

$$|X| = \sum_{x \in C}[G : G_x],$$

where C is a subset of X containing exactly one element from each orbit.

Proof This is immediate from Proposition 4.17.7 and the fact that the orbits form a partition of X. □

The formula in the corollary below is known as the *class formula*.

Corollary 4.17.10 *If G is a finite group, then*

$$|G| = \sum_{x \in C}[G : N(x)],$$

where C is a subset of G that contains exactly one element from each conjugacy class $C(a) \equiv \{bab^{-1} \mid b \in G\}$.

Proof Apply the proposition to the conjugate action of G on itself, and note that the isotropy group G_x is then the normalizer $N(x)$, whereas the orbits are the conjugacy classes. □

We can evidently write the class formula for a finite group G as

$$|G| = Z(G) + \sum_{x \in D}[G : N(x)],$$

where D is a subset of G that contains exactly one element from each non-single conjugacy class.

The following result, known as *Burnside's theorem*, is useful in combinatorics.

Theorem 4.17.11 *Let G be a finite group acting on a finite space X. Then the number n of orbits is*

$$n = \frac{1}{|G|} \sum_{a \in G} |X_a|,$$

where $X_a \equiv \{x \in X \mid ax = x\}$.

Proof We count the number of ordered pairs (a, x) with $ax = x$ in two ways, fixing either elements of G or of X in the counting process. This gives

$$\sum_{a \in G} |X_a| = \sum_{x \in X} |G_x|.$$

Let C be a subset of G that contains exactly one element from each orbit. As $|Gx| = [G : G_x] = |G|/|G_x|$, and $Gy = Gx$ when $x \in Gy$, we therefore get

$$\sum_{a \in G} |X_a| = |G| \sum_{x \in X} |Gx|^{-1} = |G| \sum_{y \in C} \sum_{x \in Gy} |Gx|^{-1} = |G| \sum_{y \in C} 1 = n|G|.$$

□

Note that $\cap_{a \in G} X_a$ are the fixed points of a G-space X.

We can of course do all this for *right actions* $X \times G \to X$ as well. They are maps $(x, a) \mapsto xa$ for which $(a, x) \mapsto xa^{-1}$ is a left action.

4.18 Spheres, Projective Spaces and Grassmannians

We will here study various actions of the classical groups.
The sphere

$$S^{n-1} \equiv \{x \in \mathbb{R}^n \mid (x|x) = 1\}$$

is obviously an invariant space for the action of $O(n)$ on $\mathbb{R}^n$. It transpires that spheres with arbitrary radii are the orbits in $\mathbb{R}^n$ considered as an $O(n)$-space. Indeed, the next result shows that they are all homogeneous space.

Proposition 4.18.1 *The action of $O(n)$ on S^{n-1} is transitive, and the isotropy group of the vector $e_n = (0, \ldots, 0, 1)^T$ is $O(n-1)$ when positioned in the upper left corner of $O(n)$, so $O(n)/O(n-1) \cong S^{n-1}$ as $O(n)$-spaces. Similarly, the action of $SO(n)$ on S^{n-1} is transitive, and the isotropy group of e_n is $SO(n-1) \subset SO(n)$, so $SO(n)/SO(n-1) \cong S^{n-1}$ as $SO(n)$-spaces.*

Proof To reach any $x \in S^{n-1}$ from e_n, complete $\{x\}$ to an orthonormal basis for $\mathbb{R}^n$, and consider the matrix A with x as the last column and the rest of the basis as the remaining columns. Then $A \in O(n)$ as $A^T A = I_n$, and $x = Ae_n$, so the action is transitive. The isotropy group of e_n will obviously consist of all matrices in $O(n)$ with entry 1 at the lower right corner with zeros above, and hence also to the left, of it. This subgroup is isomorphic to $O(n-1)$ under the prescribed embedding, so $O(n)/O(n-1) \cong S^{n-1}$ by Proposition 4.17.7.

Transitivity for the action of $SO(n)$ on S^{n-1} goes as for $O(n)$ except that one but the last of the columns of A might have to be multiplied by -1 in order to get $\det(A) = 1$. The rest is completely analogous. □

Similarly, we can view spheres as homogeneous spaces in the following way.

Proposition 4.18.2 *The group $U(n)$ acts transitively on*

$$S^{2n-1} = \{x + iy \in \mathbb{C}^n \mid (x|x) + (y|y) = 1\},$$

and the isotropy group of e_n is $U(n-1) \subset U(n)$, so $U(n)/U(n-1) \cong S^{2n-1}$ as $U(n)$-spaces. Likewise, we get $SU(n)/SU(n-1) \cong S^{2n-1}$ under the $SU(n)$-equivariant map $[A] \mapsto Ae_n$. In a similar fashion, we see that $Sp(n)/Sp(n-1) \cong S^{4n-1}$ as $Sp(n)$-spaces.

These ideas can be generalized, leading to generalizations of the sphere.

Definition 4.18.3 The *Stiefel manifold* $V_k(\mathbb{R}^n)$ consists of all k-tuples of orthonormal vectors in $\mathbb{R}^n$, often referred to as *orthonormal k-frames*.

The group $O(n)$ acts on $V_k(\mathbb{R}^n)$ by $A(x_1, \ldots, x_k) = (Ax_1, \ldots, Ax_k)$. This action is transitive because $(y_1, \ldots, y_k) \in V_k(\mathbb{R}^n)$ can be reached from $p \equiv (e_{n-k+1}, \ldots, e_n)$ by completing $\{y_i\}$ to an orthonormal basis for $\mathbb{R}^n$, and letting A be the $n \times n$-matrix with y_i as the $(n - k + i)$th column and with the rest of the basis as the $n - k$ remaining columns. The isotropy group of p is obviously $O(n-k) \subset O(n)$. We have proved the following result.

Proposition 4.18.4 *The map $[A] \mapsto Ap$ is an $O(n)$-equivariant bijection from $O(n)/O(n-k)$ to the Stiefel manifold $V_k(\mathbb{R}^n)$. Similarly, we get $U(n)/U(n-k) \cong V_k(\mathbb{C}^n)$ as $U(n)$-spaces, where $V_k(\mathbb{C}^n)$ consists of all k-tuples of orthonormal vectors in $\mathbb{C}^n$.*

Definition 4.18.5 The *Grassmannian* $G_k(\mathbb{R}^n)$ consists of all k-dimensional subspaces of $\mathbb{R}^n$.

The group $O(n)$ acts on $G_k(\mathbb{R}^n)$ by sending (A, V) to $A(V) \equiv \{Ax \mid x \in V\}$. Since we can reach any orthonormal k-frame from a given one, it is clear that this action is transitive. Considering $\mathbb{R}^k$ as a subspace of $\mathbb{R}^n$ with $\mathbb{R}^k$ sitting as the top k-coordinates and with zeros below, we see that the isotropy group of $\mathbb{R}^k \in G_k(\mathbb{R}^n)$ is the image of $O(k) \times O(n-k)$ under the monomorphism

$$(A, B) \mapsto \begin{pmatrix} A & 0 \\ 0 & B \end{pmatrix}.$$

This proves the following result.

Proposition 4.18.6 *We have*

$$G_k(\mathbb{R}^n) \cong O(n)/(O(k) \times O(n-k))$$

as $O(n)$-spaces. Similarly, we get $G_k(\mathbb{C}^n) \cong U(n)/(U(k) \times U(n-k))$ as $U(n)$-spaces, where $G_k(\mathbb{C}^n)$ consists of all k-dimensional subspaces of $\mathbb{C}^n$.

4.18 Spheres, Projective Spaces and Grassmannians

We can also think of $Sp(n)/(Sp(k) \times Sp(n-k))$ as a Grassmannian of k-planes in the nth direct product of the quaternion algebra, which we will define in the chapter on rings.

The Grassmannian $G_k(\mathbb{C}^n)$ can be though of as a set of projections. Clearly the group $U(n)$ acts via the adjoint action on the set of orthogonal projections in $M(n, \mathbb{C})$ with trace equal to k.

Proposition 4.18.7 *We have*

$$\{P \in M(n, \mathbb{C}) \mid P^2 = P = P^* \text{ and } \operatorname{Tr} P = k\} \cong G_k(\mathbb{C}^n)$$

as $U(n)$-spaces.

Proof The map $P \mapsto P(\mathbb{C}^n)$ for an orthogonal projection P with trace k identifies the two spaces because there is only one projection that projects perpendicular onto a subspace $V \in G_k(\mathbb{C}^n)$, and that is the one given by

$$P = \sum_i (\cdot | x_i) x_i,$$

where $\{x_i\}$ is any orthonormal basis for V. By completing $\{x_i\}$ to an orthonormal basis for $\mathbb{C}^n$, and calculating the trace of P with respect to this basis, we see that $\operatorname{Tr} P = k$, as required. The map also respects the two actions $(A, V) \mapsto A(V)$ and $(A, P) \mapsto APA^{-1}$. □

From Proposition 4.18.6 we see that $G_k(\mathbb{C}^n) \cong G_{n-k}(\mathbb{C}^n)$, which reflects the fact that to every k-plane with projection P, there is a unique orthogonal $(n-k)$-plane with projection $I_n - P$.

We obviously have similar identifications for $G_k(\mathbb{R}^n)$.

The Grassmannians $G_1(\mathbb{R}^n)$ and $G_1(\mathbb{C}^n)$ are known as the *real and complex projective spaces*, and denoted by $\mathbb{R}P^{n-1}$ and $\mathbb{C}P^{n-1}$, respectively. Since $O(1) \cong \mathbb{Z}_2$, we see that $\mathbb{R}P^{n-1} \cong S^{n-1}/\mathbb{Z}_2$. In particular, we get $\mathbb{R}P^2 \cong S^2/\mathbb{Z}_2$, which geometrically means that a line that goes through the origin in 3-space intersects S^2 in two antipodal points, which determines the line, and explains the presence of $\mathbb{Z}_2$.

The action of $SU(n)$ on $\mathbb{C}P^{n-1}$ is transitive since $SU(n)$ acts transitively on the orthonormal 1-frames in $\mathbb{C}^n$. The isotropy group of the line $\mathbb{C}e_n$ through the origin is the image of $U(n-1)$ under the monomorphism

$$A \mapsto \begin{pmatrix} A & 0 \\ 0 & |A|^{-1} \end{pmatrix},$$

so $\mathbb{C}P^{n-1} \cong SU(n)/U(n-1)$. Similarly, we get $\mathbb{R}P^{n-1} \cong SO(n)/O(n-1)$.

The special case $\mathbb{C}P^1 \cong SU(2)/U(1)$ is the *Hopf-fibration* of the *Riemann sphere* S^2, and is normally though of as $S^3/S^1 \cong S^2$ on account of the identifications $U(1) = S^1$ and $SU(2) \cong SU(2)/SU(1) \cong S^{2 \cdot 2 - 1} = S^3$ and $\mathbb{C}P^1 \cong S^2$.

Note also that except for trivial cases, the isotropy groups are not normal subgroups and the homogeneous spaces are not groups.

Remark 4.18.8 The objects constructed here are important to topologists, who usually consider them in the context of fibrations or smooth bundles. Generalizations of Grassmannians include e.g. flag varieties [17].

4.19 Groups of Prime Power Orders

Definition 4.19.1 Let p be a prime number. A finite group is a *p-group* if its order is some positive power of p.

We have the following structure theorem for p-groups.

Theorem 4.19.2 *The center $Z(G)$ of any p-group G is non-trivial. Even stronger, if K is a non-trivial normal subgroup of G, then $Z(G) \cap K$ is non-trivial. Any proper subgroup H of G is properly contained in its normalizer $N(H)$. In particular, if $|H| = |G|/p$, then $H \triangleleft G$.*

Proof Let $a \in D$, where D contains exactly one element from each conjugacy class in G with more than one element. Either $C(a) \cap K = \phi$. Or $C(a) \cap K \neq \phi$, and then $bab^{-1} \in K$ for some $b \in G$, so $cac^{-1} = (cb^{-1})(bab^{-1})(cb^{-1})^{-1} \in K$ for any $c \in G$ as K is normal. Thus $C(a) \subset K$ and $C(a) \cap K = C(a)$. Hence $|C(a) \cap K|$ is zero or $[G : N(a)]$.

Since $a \in D$, we know by Lagrange's theorem that $[G : N(a)]$ is a positive power of p, and the same is true for $|K|$ as K is non-trivial. Thus p divides $|K|$ and $|C(a) \cap K|$ for all $a \in D$.

Now $K = G \cap K$ is a disjoint union of $Z(G) \cap K$ and of all $C(a) \cap K$, so

$$|K| = |Z(G) \cap K| + \sum_{a \in D} |C(a) \cap K|.$$

Hence $|Z(G) \cap K|$ must be divisible by p, and $Z(G) \cap K$ is non-trivial. With $K = G$, we see that $Z(G)$ is non-trivial.

Suppose K is a maximal normal subgroup of G with $K \subset H$. The existence of such a K is a trivial application of Zorn's lemma to the family of normal subgroups of G contained in H and ordered under inclusion. Now G/K has positive prime order, and therefore has a non-trivial center, which by the correspondence theorem is of the form L/K for $L \triangleleft G$. Since L/K is non-trivial, the normal subgroup L is strictly larger than K, which is a maximal normal subgroup contained in H. Thus L is not contained in H. However, since L/K is the center of G/K, we see that $abK = baK$ for $a \in H$ and $b \in L$, so $b^{-1}ab \in H$, which shows that $L \subset N(H)$. Hence H is properly contained in $N(H)$.

If $H < G$ and $|H| = |G|/p$, then $|N(H)| = |G|$ by Lagrange's theorem, so $H \triangleleft G$. □

Corollary 4.19.3 *Every group with an order the square of a prime number is abelian.*

Proof If G is a non-abelian group with $|G| = p^2$ for a prime p, then by the theorem, its center $Z(G)$ is non-trivial, and yet $Z(G) \neq G$ as G is non-abelian. By Lagrange's theorem, we conclude that $|Z(G)| = p$, and that the normalizer $N(a)$ of any $a \in G \setminus Z(G)$ must have order p^2 as $Z(G)$ is properly contained in $N(a)$. Hence $N(a) = G$ and $a \in Z(G)$, which is a contradiction. □

Example 4.19.4 There are exactly two non-abelian groups of order eight. If G is a non-abelian group with $|G| = 8$, it cannot contain a member of order 8. If each element is of order two, then it is also abelian because $ba = a^2bab^2 = a(ab)^2b = ab$ for any $a, b \in G$. Thus G has an element a of order 4. Pick any $b \in G \setminus \langle a \rangle$. Then the cosets $\langle a \rangle$ and $b\langle a \rangle$ form a partition of G, so $b^2 \in \langle a \rangle$, otherwise $b^2\langle a \rangle = b\langle a \rangle$ and by cancelling b, we get the contradiction $b \in \langle a \rangle$. If $b^2 = a$ or $b^2 = a^3$, then b has order 8, which is impossible. So $b^2 = e$ or $b^2 = a^2$. By Theorem 4.19.2, we also know that $\langle a \rangle \triangleleft G$, so $b^{-1}ab \in \langle a \rangle$. Since the order of two conjugate elements are the same, we therefore get $b^{-1}ab = a^3$ or $b^{-1}ab = a$, and the latter option would mean that G is abelian.

Thus we have arrived at two non-isomorphic non-abelian groups; one generated by a and b with relations

$$a^4 = e = b^2, \quad ab = ba^3,$$

and the other one with generators a and b and relations

$$a^4 = e, \quad a^2 = b^2, \quad ab = ba^3.$$

The one with the first set of relations is isomorphic to D_4, and the second one is known as the quaternion group.

4.20 Cauchy's Theorems and Sylow's First Theorem

By further application of the class formula, we can extract considerable information about a finite group solely on the basis of its order.

The following result is known as *Cauchy's theorem* for abelian groups.

Lemma 4.20.1 *If a prime number p divides the order of a finite abelian group, then the group has en element of order p.*

Proof We proceed by induction on the order of the group G is question, assuming that the lemma holds for all groups of order less than $|G|$. Pick any $a \in G$ different from the unit. If p divides $\langle a \rangle$, we are done, so assume that it does not. Then by Lagrange's theorem and Euclid's lemma, it must divide $G/\langle a \rangle$. By our induction hypothesis there exists $b \in G$ such that the equivalence class $[b]$ has order p. Therefore, if n is the order of b, then $[b]^n = [b^n] = [e]$, so $n = pr$ for some integer r, and b^r has order p. □

The following result, known as *Sylow's first theorem*, provides a criteria for when a group contains subgroups that are p-groups. It also allows us to generalize Cauchy's theorem to non-abelian groups.

Theorem 4.20.2 *If p is a prime and p^n divides the order of a finite group, then it has a subgroup of order p^n.*

Proof We proceed by induction on the order m of the group G in question. The theorem obviously holds for $m = 1$. Assume that it holds for all groups of order less than m.

If $|Z(G)|$ is divisible by p, then by Cauchy's theorem for abelian groups, the center contains an element a of order p. Now $\langle a \rangle \triangleleft G$ and $G/\langle a \rangle$ has order m/p, which is divisible by p^{n-1}. Thus by the induction assumption and the correspondence theorem, the group $G/\langle a \rangle$ contains a subgroup of the form $H/\langle a \rangle$ with order p^{n-1}. By Lagrange's theorem the subgroup H of G has therefore order p^n.

If $|Z(G)|$ is not divisible by p, we invoke the class formula

$$|G| = |Z(G)| + \sum_{a \in D} [G : N(a)],$$

where D contains one element from each conjugacy class in G with more than one element. As p divides $|G|$ but not $|Z(G)|$, it cannot divide all the terms in the sum. So p does not divide $[G : N(a)]$ for some $a \notin Z(G)$. Since p^n divides $|G|$, we conclude from Lagrange's theorem and repeated us of Euclid's lemma that p^n divides $|N(a)|$, and $|N(a)| < |G|$ as $a \notin Z(G)$. By our induction assumption $N(a)$, and hence G, has a subgroup of order p^n. □

Corollary 4.20.3 *If a prime number p divides the order of a finite group, then the group has en element of order p.*

Proof By the theorem there exists a subgroup of order p, so any member of this cyclic subgroup except the unit has order p. □

The following result allows for an alternative definition of p-groups that also works for infinite groups.

Corollary 4.20.4 *A non-trivial finite group is a p-group if and only if the order of each member is some power of p.*

Proof The forward implication is obvious. Conversely, if the order of each member in a non-trivial group G is some power of p, and $|G|$ is divisible by a prime number q, then by Cauchy's theorem for non-abelian groups, the group G has an element of order q. This is only possible if $q = p$. □

The following result shows that the converse of Lagrange's theorem is true for abelian groups.

Corollary 4.20.5 *If a natural number n divides the order of a finite abelian group, then the group contains a subgroup of order n which is a direct product of p_i-groups for distinct primes p_i. In particular, any finite abelian group is of this form for unique p_i-groups. If the order of the group is square-free, then it is cyclic.*

Proof Let $n = p_1^{m_1} \cdots p_k^{m_k}$ be the prime number decomposition of n. Since n divides the order of the group, so does $p_i^{m_i}$, and by the theorem, the group has a subgroup G_i of such an order. Since we are in an abelian setting, the set $G_1 \cap (G_2 \cdots G_k)$ is a subgroup both of G_1 and $G_2 \cdots G_k$, so by Lagrange's theorem their intersection is trivial. Hence

$$G_1 \cdots G_k \cong G_1 \times (G_2 \cdots G_k) \cong \cdots \cong G_1 \times \cdots \times G_k$$

by induction. Uniqueness is clear since G_i consists of all elements of the group having order some power of p_i.

If n is the order of the group, then the group itself is a direct product of the p_i-groups. Finally observe that if all the m_i's are one, then $G_1 \times \cdots \times G_k$ is cyclic with generator $(a_1, \ldots, a_k)$, where a_i is a generator for G_i. □

4.21 Sylow's Second and Third Theorems

Definition 4.21.1 Let p be a prime number. A *Sylow p-subgroup* of a group G is a subgroup with order the greatest positive power of p that divides $|G|$.

Theorem 4.21.2 *All Sylow p-subgroups of a finite group are conjugate. Their number n divides the order of the group, and $n \equiv 1 \pmod{p}$.*

Proof Let $C(H)$ denote the family $\{aHa^{-1} \mid a \in G\}$ of subgroups of a finite group G that are conjugate to a Sylow p-subgroup H. Then p cannot divide $|C(H)| = [G : N(H)]$ as $[G : H] = [G : N(H)][N(H) : H]$ is not divisible by p.

To show that $K \in C(H)$ for any Sylow p-subgroup K, consider $C(H)$ as a K-set under conjugation. Let $L_K = \{aLa^{-1} \mid a \in K\}$ be the orbit of $L \in C(H)$, and note that these orbits form a partition of $C(H)$. The stabilizer of L under the action of K is $N(L) \cap K$, so

$$|L_K| = [K : N(L) \cap K] = p^m$$

for some $m \geq 0$ because $|K|$ is a positive power of p.

Combining these two observations gives the formula

$$|C(H)| = \sum p^m,$$

where we are summing over the orbits. As the left-hand-side is not divisible by p, one of the terms on the right-hand-side must be one, say $|L_K| = 1$ for some $L \in C(H)$. Then $K \cap N(L) = K$, so $K \subset N(L)$, which means that $KL = LK$. By the second isomorphism theorem we get $KL/L \cong K/K \cap L$ with order a power of p, so $KL = L$ by Lagrange's theorem since L is a Sylow p-subgroup. Hence $K = L$, and $K \in C(H)$. This also shows that exactly one term in the sum $\sum p^m$ is one, so $n = 1 \pmod{p}$.

The number of conjugate Sylow p-subgroups is $|C(H)| = |G|/|N(H)|$, so $|G|$ is divisible by n. □

Corollary 4.21.3 *A Sylow p-subgroup of a group is unique if and only if it is normal.*

Proof This is clear since a subgroup is normal if it coincides with all its conjugates. □

Corollary 4.21.4 *A normal subgroup H of a finite group G contains all Sylow p-subgroups of G if $[G : H]$ and p are relatively prime.*

Proof The highest powers of p in $|G|$ and $|H|$ are by assumption the same, so a Sylow p-subgroup of H will also be a Sylow p-subgroup of G, and by Sylow's first theorem, there is at least one Sylow p-subgroup of H, say K.

If L is any Sylow p-subgroup of G, then by Sylow's second theorem it must be conjugate to K. Hence there is some $a \in G$ such that

$$L = aKa^{-1} \subset aHa^{-1} \subset H$$

as $H \triangleleft G$. □

Corollary 4.21.5 *If all the subgroups of a p-group have different orders, then it is cyclic.*

Proof By Lagrange's theorem the subgroups of a p-group G are also p-groups, and by Sylow's first theorem there is a chain of p-subgroups of all powers of p less than $|G|$ contained in each other. By assumption these are all the subgroups. Let H be the largest proper subgroup. Then any $a \in G \backslash H$ is a generator for G because if $\langle a \rangle \neq G$, then $a \in \langle a \rangle \subset H$, which is absurd. □

4.22 Some Examples

Here we furnish a few examples where we apply Sylow's theorems.

Example 4.22.1 There is no simple group of order 63. If G is a group of order 63, then as $63 = 7 \cdot 3^2$, it has at least one Sylow 7-subgroup. The overall number of such subgroups is of the form $1 + 7n$ for an integer n such that $1 + 7n$ divides $|G|$. This is only possible if $n = 0$, so we have only one Sylow 7-subgroup, which then has to be normal. So G is not simple. ◇

Example 4.22.2 There is no simple group of order 56. If G is such a group, then as $56 = 7 \cdot 2^3$, it has a Sylow 7-subgroup. The total number of such subgroups is of the form $1 + 7n$ for an integer such that $1 + 7n$ divides 56. This can happen when $n = 0$, and then G has a normal subgroup of order 7, so G is not simple, or when $n = 1$. In this case the 8 conjugate cyclic subgroups of order 7 have only the unit in common, since all non-unital elements in such subgroups are generators. Thus there are $8 \cdot (7 - 1)$ elements in G of order 7. The remaining 8 elements form a Sylow 2-group, which exists since $|G| = 7 \cdot 2^3$. This subgroup is unique since otherwise there would be elements with both even order and order 7. So it is normal and G is not simple. ◇

Example 4.22.3 There exists a normal subgroup of order 9 or 27 in a group G of order 108. Observe that $|G| = 2^2 \cdot 3^3$. The number of Sylow 3-subgroups is of the form $1 + 3n$ for an integer n such that $1 + 3n$ divides $|G|$. Then either $n = 0$, and in this case we have a unique normal Sylow 3-subgroup with order $3^3 = 27$. Or $n = 1$, and then we have four Sylow 3-subgroups, and we can pick two distinct ones H and K. From the discussion after the second isomorphism theorem, we see that

$$108 = |G| \geq |HK| = |H||K|/|H \cap K| = 27 \cdot 27/|H \cap K|,$$

so $|H \cap K| \geq 27/4$, and since $H \neq K$, the subgroup $H \cap K$ of H has less elements than H. By Lagrange's theorem we therefore get $|H \cap K| = 9$. By Theorem 4.19.2 we actually know that $H \cap K$ is a normal subgroup of both H and K, so H and K, and hence HK, are contained in the normalizer $N(H \cap K)$. Since

$$|HK| = |H||K|/|H \cap K| = 27 \cdot 27/9 = 81,$$

and $HK \subset N(H \cap K) < G$, we conclude by Lagrange's theorem that $N(H \cap K) = G$, so $H \cap K$ is a normal subgroup of G with order 9. ◇

4.23 Groups with Order the Product of Two Primes

By Corollary 4.19.3 a group of order p^2, for a prime number p, is abelian, and later we shall see that it therefore is isomorphic either to $\mathbb{Z}_{p^2}$ or to $\mathbb{Z}_p \times \mathbb{Z}_p$.

More generally, if a group has an order that is the product of two distinct primes, there are also only two possibilities.

Theorem 4.23.1 *Say p and q are prime numbers and $p < q$. A group of order pq is either cyclic, or if p divides $q - 1$, then it is isomorphic to the non-abelian group generated by a and b with defining relations*

$$a^p = e = b^q, \quad ba = ab^n,$$

where n is any integer such that $n^p \equiv 1 \pmod{q}$ and $n \not\equiv 1 \pmod{q}$.

Remark 4.23.2 We will later return to what it means precisely that a group is presented by generators and relations.

What n we pick is immaterial, as long as it satisfies the stated congruence equations.

When $q = 3$ and $p = 2$, we can for instance pick $n = 2$. This way we obtain the relations for the symmetric group S_3, which together with the cyclic group $\mathbb{Z}_6$ are all groups of order 6 up to isomorphism.

Proof First notice that the relations determine a group with pq distinct elements $a^i b^j$, where $i \in \{0, 1, \ldots, p-1\}$ and $j \in \{0, 1, \ldots, q-1\}$.

Say G is a group with order pq. It has a Sylow q-subgroup, and the total number of such subgroups is $1 + mq$ for an integer m such that $1 + mq$ divides pq. This is only possible if $m = 0$, so there is exactly one Sylow q-subgroup. It is therefore normal, and since it has prime order, it is cyclic, say $\langle b \rangle$ for $b \in G$ with order q.

There also exists a Sylow p-subgroup, and the total number of such subgroups is $1 + kp$ for an integer k such that $1 + kp$ divides pq. This forces k to be either 0 or such that $1 + kp = q$. In the first case there is only one Sylow p-subgroup, which is both normal and cyclic, say $\langle a \rangle$ for $a \in G$ with order p. We claim that ab has order pq.

To see this, first note that $\langle a \rangle \cap \langle b \rangle = \{e\}$ since any element in the intersection apart from e has order larger than 1, which by Lagrange's theorem, divides both p and q, and this is absurd. Since both $\langle a \rangle$ and $\langle b \rangle$ are normal subgroups, we see that $a^{-1}bab^{-1} \in \langle a \rangle \cap \langle b \rangle$, so $a^{-1}bab^{-1} = e$, or $ba = ab$. Hence $(ab)^{pq} = (a^p)^q (b^q)^p = e$ and the order d of ab divides pq. Now $a^d b^d = (ab)^d = e$, so $a^d = b^{-d} \in \langle a \rangle \cap \langle b \rangle$, and $a^d = e = b^d$. Hence d is divisible by both p and q, whence by pq. Thus $d = pq$.

The other alternative is that the number of Sylow p-subgroups is q, which happens when $q - 1$ is divisible by p.

Let $\langle a \rangle$ be one such cyclic subgroup, so $a \in G$ has order p. Then $\langle a, b \rangle = G$ by Lagrange's theorem since $\langle a, b \rangle$ has subgroups $\langle a \rangle$ and $\langle b \rangle$ with orders p and

q, respectively. So a and b generate G. Since $\langle b \rangle \triangleleft G$, we know that $a^{-1}ba = b^n$ for some integer n. Now $n \not\equiv 1 \pmod{q}$ since otherwise $ba = ab$, and G would be abelian and $\langle a \rangle$ normal, so there would be only one (and not q) Sylow p-subgroups, yet an absurdity.

The number n does also satisfy $n^p \equiv 1 \pmod{q}$. To check this, note that

$$a^{-2}ba^2 = a^{-1}(a^{-1}ba)a = a^{-1}b^n a = (a^{-1}ba)^n = (b^n)^n = b^{n^2},$$

and by induction we get $b = a^{-p}ba^p = b^{n^p}$, as required. Hence the relations in the theorem are satisfied.

Since as we have seen $a^{-i}ba^i = b^{n^i}$, the act of replacing n by any other solution n^i for $i \in \{1, \ldots, p-1\}$ of the congruence equations, produces the same group with an isomorphism that leaves b unaltered but sends a to a^i. □

Example 4.23.3 For instance, any group of order 15 is cyclic, and groups of order 10, 14, 21 are either cyclic or non-abelian and of the type described in the theorem.

4.24 Normal Series

Definition 4.24.1 A sequence $\{G_i\}_{i=0}^n$ of subgroups of a group G is a *normal series* if

$$\{e\} = G_0 \triangleleft G_1 \triangleleft \cdots \triangleleft G_{n-1} \triangleleft G_n = G.$$

The quotient groups G_i/G_{i-1} are called *factors*. If all the factors are simple, the series is called a *composition series*.

Note that G_i need not be normal in G. Usually this is not the case.

Proposition 4.24.2 *Every finite group has a composition series.*

Proof We may assume that the finite group G in question is neither simple nor trivial. Assuming the result holds for all groups with order less than $|G|$, let H be a maximal normal subgroup of G. By the induction hypothesis H has a composition series $\{H_i\}$. But then $\{H_i\} \cup \{G\}$ is a composition series for G as G/H is simple. □

Example 4.24.3 The group $\mathbb{Z}_{18}$ has a composition series

$$\{0\} \subset \{0, 9\} \subset \{0, 3, 6, 9, 12, 15\} \subset \mathbb{Z}_{18},$$

which we can also write as $\{0\} \triangleleft \mathbb{Z}_2 \triangleleft \mathbb{Z}_6 \triangleleft \mathbb{Z}_{18}$. The inclusions

$$\{e\} \subset \{e, \sigma^2\} \subset \{e, \sigma, \sigma^2, \sigma^3\} \subset D_4$$

is a composition series for the optic group. Composition series need not be unique. In fact, the inclusions

$$\{e\} \subset \{e, \tau\} \subset \{e, \sigma^2, \tau, \sigma^2\tau\} \subset D_4$$

is another composition series for the same group. ◇

Definition 4.24.4 Two normal series $S = \{G_i\}$ and $T = \{H_i\}$ are *equivalent series*, written $S \sim T$, if all

$$H_i/H_{i-1} \cong G_{\sigma(i)}/G_{\sigma(i)-1}$$

for some permutation σ of the indices.

Evidently $\sim$ is an equivalence relation on the family of normal series of a group. Note that equivalent series must be equally long, and repetitions are not allowed for composition series.

Lemma 4.24.5 *The intersection of two distinct maximal normal subgroups of a group is a maximal normal subgroup of each of the subgroups.*

Proof Let H and K be distinct maximal normal subgroups of a group G. Recall the diamond isomorphism theorem

$$H/H \cap K \cong HK/K,$$

and observe that $K \triangleleft HK \triangleleft G$. Since K is maximal, either $HK = G$ or $HK = K$. The latter option is impossible because then $H \subset K$, so $H = K$ as H is maximal. The first option means that HK/K is simple as K is maximal, and by the isomorphism above, we see that $H \cap K$ is a maximal normal subgroup of H. By interchanging the roles of H and K, we also see that $H \cap K$ is a maximal normal subgroup of K. □

The following uniqueness result is due to Jordan-Hölder, and reduces in some sense the study of finite groups to simple ones.

Theorem 4.24.6 *Any two composition series of a finite group are equivalent.*

Proof Let G be a finite group, and assume that the theorem holds for all groups of order less than $|G|$. Consider two composition series of G, say

$$S: \quad G_0 \triangleleft G_1 \triangleleft \cdots \triangleleft G_{m-1} \triangleleft G$$

and

$$T: \quad H_0 \triangleleft H_1 \triangleleft \cdots \triangleleft H_{n-1} \triangleleft G.$$

If $G_{m-1} = H_{n-1}$, then $S \sim T$ by the induction hypothesis. If $G_{m-1} \neq H_{n-1}$, then $K \equiv G_{m-1} \cap H_{n-1}$ is by the lemma above, a maximal normal subgroup of G_{m-1} and H_{n-1}. By the proposition above, we know that K has a composition series, say $U: K_0 \triangleleft K_1 \triangleleft \cdots \triangleleft K_{k-1} \triangleleft K$, which gives two more composition series of G, namely

$$V: K_0 \triangleleft K_1 \triangleleft \cdots \triangleleft K \triangleleft G_{m-1} \triangleleft G$$

and

$$W: K_0 \triangleleft K_1 \triangleleft \cdots \triangleleft K \triangleleft H_{n-1} \triangleleft G.$$

Now $G_{m-1} H_{n-1}$ is a normal subgroup of G since G_{m-1} and H_{n-1} are normal, and it is also strictly larger than G_{m-1}, which by assumption is maximal. Hence $G_{m-1} H_{n-1} = G$. By the second isomorphism theorem, we therefore see that

$$G/G_{m-1} \cong H_{n-1}/K \quad \text{and} \quad G/H_{n-1} \cong G_{m-1}/K.$$

Thus $V \sim W$ with a permutation that interchanges the two last factors. By the induction hypothesis we also have $S \sim V$ and $T \sim W$. Hence $S \sim T$. □

The theorem tells us that the factors of a composition series for a finite group are unique up to reordering of factors, so this is a property of the group.

Example 4.24.7 Let $\{G_i\}$ be a composition series for an abelian group G. Then $|G_i/G_{i-1}|$ is a prime number, say p_i, since the composition factors are simple and abelian, and thus cyclic of prime order. Hence

$$|G| = \prod |G_i/G_{i-1}| = \prod p_i.$$

This shows that an abelian group has a composition series if and only if it is finite. Moreover, the prime factors of $|G|$ determine the composition factors up to reordering.

The uniqueness part of the fundamental theorem of arithmetic follows from the Jordan-Hölder theorem because to any two prime factor decompositions

$$p_1 \cdots p_m = q_1 \cdots q_n$$

of the same number, there are composition series of cyclic groups with cyclic factors of orders $p_1, \ldots, p_m$ and $q_1, \ldots, q_n$, respectively, and these composition factors are unique up to isomorphism and reordering.

4.25 The Theorem of Schreier

We need the following result, which is known as the *butterfly lemma*, due to the diagram of maps or arrows one might wish to draw.

Lemma 4.25.1 *Let U and V be subgroups of a group with further subgroups $u \triangleleft U$ and $v \triangleleft V$. Then $u(U \cap V)$ and $(U \cap V)v$ are subgroups with normal subgroups $u(U \cap v)$ and $(u \cap V)v$, respectively, and*

$$u(U \cap V)/u(U \cap v) \cong (U \cap V)v/(u \cap V)v.$$

Proof It is easy to check that $u(U \cap v) \triangleleft u(U \cap V)$ and $(u \cap V)v \triangleleft (U \cap V)v$, so the quotient groups make sense.

To see that they are isomorphic we shall see that both are isomorphic to

$$(U \cap V)/(u \cap V)(U \cap v).$$

By the second isomorphism theorem

$$HK/K \cong H/(H \cap K)$$

with $H = U \cap V$ and $K = u(U \cap v)$ and the easily verified facts $HK = u(U \cap V)$ and $H \cap K = (u \cap V)(U \cap v)$, we get one of the isomorphisms. The other one is obtained analogously. $\square$

Definition 4.25.2 A *refinement of a normal series* is a normal series obtained by inserting a finite number of subgroups in the original series.

The following theorem is a strengthening of the Jordan-Hölder theorem since composition series obviously do not have refinements with proper inclusions. Also, we make no assumptions on the order of the group.

Theorem 4.25.3 *All normal series of a group have equivalent refinements.*

Proof Say we have two normal series $\{G_i\}$ and $\{H_j\}$ of a group G. Let

$$G_{ij} = G_i(G_{i+1} \cap H_j) \text{ and } H_{ji} = (G_i \cap H_{j+1})H_j.$$

By the first part of the butterfly lemma, we get two normal series $\{G_{ij}\}$ and $\{H_{ji}\}$ of G under lexicographic ordering of the double indices, and these series are obviously refinements of $\{G_i\}$ and $\{H_j\}$, respectively.

By the second part of the butterfly lemma these two refinements are equivalent because

$$G_{i,j+1}/G_{ij} \cong H_{j,i+1}/H_{ji}.$$

$\square$

Example 4.25.4 The two normal series $\{0\} \subset 8\mathbb{Z} \subset 4\mathbb{Z} \subset \mathbb{Z}$ and $\{0\} \subset 9\mathbb{Z} \subset \mathbb{Z}$ of $\mathbb{Z}$ have for instance the equivalent refinements

$$\{0\} \subset 72\mathbb{Z} \subset 8\mathbb{Z} \subset 4\mathbb{Z} \subset \mathbb{Z}$$

and

$$\{0\} \subset 72\mathbb{Z} \subset 18\mathbb{Z} \subset 9\mathbb{Z} \subset \mathbb{Z},$$

respectively.

4.26 Solvable Groups

Definition 4.26.1 The *derived group* of a group G is the subgroup G' of G generated by all *commutators* $[a, b] \equiv aba^{-1}b^{-1}$ of a and b in G.

Proposition 4.26.2 *The derived group G' of a group G is a normal subgroup and G/G' is abelian. In fact, if $H \triangleleft G$, then G/H is abelian if and only if $G' \subset H$.*

Proof Note that $c[a, b]c^{-1} = [cac^{-1}, cbc^{-1}] \in G'$ for $a, b, c \in G$, and since $[a, b]^{-1} = [b, a]$, every element of G' is a product of commutators, so $G' \triangleleft G$ as $\mathrm{Ad}(c)$ is a homomorphism.

Now

$$aG'bG'(aG')^{-1}(bG')^{-1} = [a, b]G' \subset G'$$

for $a, b \in G$ shows that G/G' is abelian.

If $G' \subset H \triangleleft G$, then $G/H \subset G/G'$, so G/H is abelian. Conversely, if G/H is abelian, then $[a, b]H = aH(aH)^{-1}bH(bH)^{-1} = H$, so $[a, b] \in H$ and $G' \subset H$. □

Definition 4.26.3 The nth *derived group* of a group G is the subgroup $G^{(n)}$ of G defined inductively $G^{(n)} = (G^{(n-1)})'$ with $G^{(0)} = G$. A group G is *solvable* if $G^{(n)}$ is trivial for some natural number n.

Obviously any abelian group is solvable as its derived group is trivial.

Proposition 4.26.4 *Subgroups and homomorphic images of solvable groups are solvable. If H is a normal subgroup of a group G, and H and G/H are solvable, then so is G.*

Proof If $H < G$, then $H^{(n)} \subset G^{(n)}$, so H is solvable if G is solvable.

If $f: G \to H$ is a surjective homomorphism, then $f([a, b]) = [f(a), f(b)]$, so $H^{(n)} = f(G^{(n)})$ by induction, and H is solvable if G is solvable.

If $H \triangleleft G$ and $H^{(m)}$ and $G^{(n)}H/H = (G/H)^{(n)}$ are trivial for some m and n, then $G^{(n)} \subset H$ and $G^{(m+n)} \subset H^{(m)} = \{e\}$, so G is solvable. □

Theorem 4.26.5 *A group is solvable if and only if it has a normal series with abelian factors. A finite solvable group has a composition series with cyclic factors of prime orders. The simple solvable groups are exactly the cyclic groups of prime order.*

Proof If G is solvable with $G^{(n)}$ trivial, then

$$G^{(n)} \subset G^{(n-1)} \subset \cdots \subset G^{(1)} \subset G$$

is a composition series for G with abelian factors $G^{(i-1)}/G^{(i)}$.

Conversely, if G has a normal series $\{H_i\}_{i=0}^n$ with H_i/H_{i-1} abelian, then by Proposition 4.26.2, we see that $H_i' \subset H_{i-1}$, so $G' = H_n' \subset H_{n-1}$ and by induction $G^{(n)} = H_{n-n} = \{e\}$.

If G is a finite solvable group, then by what we have already shown, it has a normal series $\{H_i\}_{i=0}^n$ with H_i/H_{i-1} abelian. Since H_i/H_{i-1} is finite, it has a composition series, and since it is abelian, the factors must be cyclic and of prime order. Inserting the corresponding subgroups of H_i between the terms H_{i-1} and H_i in the normal series, we get a composition series with the desired properties.

Any composition series of a simple solvable group G has only two terms, so G is abelian, and thus cyclic of prime order. The converse is obvious. □

So in some sense simplicity and solvability are two opposite extremes.

Example 4.26.6 The dihedral group D_n is solvable with normal series

$$\{e\} \subset \{e, \sigma, \ldots, \sigma^{n-1}\} \subset D_n$$

consisting of abelian factors.

Example 4.26.7 Let T and D denote the subgroups of $GL(n, F)$ of matrices with zero's below and off the diagonals, respectively. Then $f \colon T \to D$ given by $f(A) = \mathrm{diag}(A)$ is an epimorphism with kernel $I + N$, where N consists of matrices with zero entries on and below the diagonal. Set $T_n = T$ and $T_i = I + N^{n-i}$ for $i \in \{0, \ldots, n-1\}$. Then $\{T_i\}_{i=0}^n$ is a normal series of T with $T_i/T_{i+1} \cong F^i$, so T is solvable.

4.27 Nilpotent Groups

Given a group G, the center of $G/Z(G)$ is a normal subgroup of the quotient group, and corresponds to a unique normal subgroup $Z_2(G)$ of G such that $Z_2(G)/Z(G) = Z(G/Z(G))$.

Definition 4.27.1 The *nth center of a group* is the normal subgroup $Z_n(G)$ of G defined inductively by $Z_n(G)/Z_{n-1}(G) = Z(G/Z_{n-1}(G))$ with $Z_0(G) = \{e\}$.

4.27 Nilpotent Groups

Clearly $Z_1(G) = Z(G)$ and

$$Z_n(G) = \{a \in G \mid [a, b] \in Z_{n-1}(G) \text{ for } b \in G\}$$

by the correspondence theorem. The ascending series

$$Z_0(G) \subset Z_1(G) \subset \cdots \subset Z_n(G) \subset \cdots$$

of normal subgroups of G is called the *upper central series* of G.

Definition 4.27.2 A group is *nilpotent* if it coincides with its nth center for some n. The smallest n for which this holds is the *class of nilpotency of the group*.

A non-trivial abelian group is nilpotent with class of nilpotency equal to one.

Proposition 4.27.3 *Nilpotent groups are solvable.*

Proof By theorem 4.26.5 a group is solvable if it has a normal series with abelian factors, and the upper central series is such a series. □

Clearly, any non-trivial group with trivial center cannot be nilpotent, so S_3 is not nilpotent. But S_3 is solvable as $S_3^{(2)}$ is trivial, which shows that the converse of the proposition fails.

Proposition 4.27.4 *Any p-group is nilpotent, and hence solvable.*

Proof The structure theorem for p-groups tells us that they have non-trivial center. If G is a p-group, then $G/Z_1(G)$ is by Lagrange's theorem either trivial, or a p-group with non-trivial center and $|Z_2(G)| > |Z_1(G)|$. We can continue this till $|Z_n(G)| = |G|$. □

Theorem 4.27.5 *A group G is nilpotent if and only if it has a normal series $\{G_i\}$ such that $G_i/G_{i-1} \subset Z(G/G_{i-1})$.*

Proof If G is nilpotent, then the upper central series will terminate at n with $Z_n(G) = G$, and it is the required normal series as $Z_i(G)/Z_{i-1}(G) = Z(G/Z_{i-1}(G))$.

Conversely, if G has a normal series $\{G_i\}_{i=0}^n$ with $G_i/G_{i-1} \subset Z(G/G_{i-1})$, then $G_1 \subset Z(G)$. As $G_2/G_1 \subset Z(G/G_1)$, then if $a \in G_2$, we have $aG_1 bG_1 = bG_1 aG_1$, and hence $[a, b] \in G_1$, for $b \in G$. Thus $[a, b] \in Z_1(G)$, so $a \in Z_2(G)$ and $G_2 \subset Z_2(G)$. Continuing this, we see that $G = G_n \subset Z_n(G)$, so G is nilpotent. □

Proposition 4.27.6 *Subgroups, quotients and finite direct products of nilpotent groups are nilpotent.*

Proof If $H < G$, obviously $H \cap Z(G) \subset Z(H)$. As $[a, b] \in Z(G)$ for $a \in Z_2(G)$ and $b \in G$, we get for $a \in H \cap Z_2(G)$ and $b \in H$, that $[a, b] \in H \cap Z(G) \subset Z_1(H)$. Hence $a \in Z_2(H)$ and $H \cap Z_2(G) \subset Z_2(H)$. By induction, $H \cap Z_n(G) \subset Z_n(H)$ for every natural number n. Thus if G is nilpotent, then so is H.

Suppose we have an epimorphism $f: G \to H$ between two groups. We know that $[a, b] \in Z(G)$ for $a \in Z_2(G)$ and $b \in G$, so

$$[f(a), f(b)] = f([a, b]) \in f(Z(G)) \subset Z(H)$$

and $f(a) \in Z_2(H)$. So $f(Z_2(G)) \subset Z_2(H)$. By induction, $f(Z_n(G)) \subset Z_n(H)$ for every natural number n. This shows that H is nilpotent if G is, and therefore quotient groups of nilpotent groups are nilpotent.

As for finite direct products it suffices to show that $G \times H$ is nilpotent if G and H are. But this is immediate from $Z_n(G \times H) = Z_n(G) \times Z_n(H)$, which we leave to the reader to check. □

Definition 4.27.7 The *lower central series* of a group G is the descending series $G \supset G_{(1)} \supset G_{(2)} \supset \cdots$ of subgroups $G_{(n)}$ defined inductively by

$$G_{(n+1)} = [G_{(n)}, G] \equiv \langle\{[a, b] \mid a \in G_{(n)}, b \in G\}\rangle.$$

Clearly $G_{(1)}$ is a normal subgroup of G, and if $G_{(n)}$ is assumed to be a normal subgroup of G, then $G_{(n+1)}$ is a normal subgroup of G. By induction we conclude that all $G_{(n)}$ are normal subgroups of G. So we can form the quotient groups $G_{(n)}/G_{(n+1)}$, and these are obviously abelian. In fact, we see that $G_{(n)}/G_{(n+1)}$ is in the center of $G/G_{(n+1)}$, hence *central series*.

By induction we also see that $G^{(n)} \subset G_{(n)}$. The following result explains the terminology upper and lower.

Proposition 4.27.8 *If m is the class of nilpotency of a nilpotent group G, then $G_{(n)} \subset Z_{m-n}(G)$. Hence $G_{(m)} = \{e\}$.*

Proof We have $G_{(0)} \equiv G = Z_m(G)$. Assuming $G_{(n)} \subset Z_{m-n}(G)$, we get $G_{(n+1)} \subset [Z_{m-n}(G), G] \subset Z_{m-(n+1)}(G)$, so the result follows by induction. □

Proposition 4.27.9 *Let G be a group with a normal subgroup H and quotient map $f: G \to G/H$. Then $(G/H)_{(n)} = f(G_{(n)})$.*

Proof We have $(G/H)_{(1)} = [f(G), f(G)] = f([G, G]) = f(G_{(1)})$, and if $(G/H)_{(n)} = f(G_{(n)})$, then

$$(G/H)_{(n+1)} = [(G/H)_{(n)}, G/H] = [f(G_{(n)}), f(G)] = f([G_{(n)}, G]) = f(G_{(n+1)}),$$

so the result follows by induction. □

4.28 Simplicity of the Alternating Group

The alternating group A_n is non-abelian for $n \geq 4$ because we can pick distinct numbers $a, b, c, d \in \{1, \ldots, n\}$ and then

$$(a\ b\ c)(a\ b\ d) = (a\ c)(b\ d) \neq (a\ d)(b\ c) = (a\ b\ d)(a\ b\ c).$$

Observe that A_1 and A_2 are trivial, and that $A_3 = \{e, (1\ 2\ 3), (1\ 3\ 2)\}$ is cyclic.

Lemma 4.28.1 *The alternating group A_n is generated by the 3-cocycles in S_n.*

Proof Every 3-cocycle is a product of two transpositions, and so is even and belongs to A_n.

On the other hand, the elements of A_n are even products of transpositions. The product of two neighboring transpositions is the identity or a 3-cocycle, or they are disjoint, say $\sigma = (a\ b)$ and $\tau = (c\ d)$, and then $\sigma\tau = (a\ b\ c)(b\ c\ d)$. □

Cycles of the same length are conjugate in S_n because

$$\sigma(1\ 2\ \cdots\ m)\sigma^{-1} = (\sigma(1)\ \sigma(2)\ \cdots\ \sigma(m))$$

for $\sigma \in S_n$.

When $n \geq m + 2$, all m-cocycles are conjugate within A_n because if $\sigma \in S_n$ is not even, replace it by $\sigma\tau$, where τ is a transposition that leaves $\{1, \ldots, m\}$ fixed. Then $\sigma\tau \in A_n$ and

$$\sigma\tau(1\ 2\ \cdots\ m)(\sigma\tau)^{-1} = (\sigma(1)\ \sigma(2)\ \cdots\ \sigma(m)).$$

Lemma 4.28.2 *The derived group of S_n is A_n.*

Proof Commutators are obviously even, so $S_n' \subset A_n$.

Both S_n' and A_n are trivial for $n = 1$ and $n = 2$. Say $n \geq 3$. Then

$$(1\ 2\ 3) = (1\ 2)(1\ 2\ 3)(2\ 1)(3\ 2\ 1) \in S_n',$$

and as S_n' is normal in S_n, and all 3-cocycles are conjugate, we see that S_n' contains all 3-cocycles, so $S_n' = A_n$ by the previous lemma. □

Theorem 4.28.3 *The group A_n is simple and S_n is not solvable for $n \geq 5$.*

Proof We claim that any non-trivial normal subgroup H of A_n with $n \geq 5$ contains a 3-cocycle. Let $\sigma \in H$ be a non-trivial permutation that moves the least number of numbers in $\{1, \ldots, n\}$. Now σ cannot be a cycle of even length as such cycles are not even. So σ is either a 3-cocycle, and we are done, or it has a disjoint decomposition of the form

$$(a\ b\ c\ \cdots)\cdots$$

or

$$(a\ b)(c\ d)\cdots$$

for distinct numbers $a, b, c, d \in \{1, \ldots, n\}$.

In the first case, since σ cannot be a 4-cocycle, it must move two more numbers, say d and f. Let $\tau = (c\ d\ f)$. As $H \triangleleft A_n$, the permutation $\sigma^{-1}\tau^{-1}\sigma\tau$ belongs to H, and fixes a but also every element that is fixed by σ since any such number cannot belong to $\{a, b, c, d, f\}$. Thus we have an element of H that moves fewer elements than σ, which is a contradiction.

In the second case, we can pick a number $f \in \{1, \ldots, n\}$ distinct from a, b, c, d, and define τ as before. Then $\sigma^{-1}\tau^{-1}\sigma\tau$ belongs to H, and fixes a and b but also every element that is fixed by σ and different from f since any such number cannot belong to $\{a, b, c, d, f\}$. Again we have an element of H that moves fewer elements than σ, which is a contradiction.

So we have a 3-cocycle $\sigma \in H$. As noted prior to this theorem, all the 3-cocycles in A_n for $n \geq 5$ are conjugate within A_n, and as $H \triangleleft A_n$, we get $H = A_n$, so A_n is simple for $n \geq 5$. □

Corollary 4.28.4 *The alternating group A_n is the only proper normal subgroup of S_n when $n \geq 5$.*

Proof Say H is a proper normal subgroup of S_n and $n \geq 5$. If $H \cap A_n$ is non-trivial, then as $H \cap A_n \triangleleft A_n$, we get $H \cap A_n = A_n$, so $A_n \subset H$. Then $|S_n/H| \leq |S_n/A_n| = 2$, so $H = A_n$ by Lagrange's theorem.

If $H \cap A_n$ is trivial, then $\frac{1}{2}|H|n! = |HA_n| \leq n!$, so $|H| = 2$, say $H = \{e, \sigma\}$. As $H \triangleleft S_n$, we must have $\tau\sigma\tau^{-1} = \sigma$ for all $\tau \in S_n$, which is impossible as $Z(S_n)$ is obviously trivial. □

It is easy to check that A_4 has exactly one proper normal subgroup. This subgroup has order four, and is therefore abelian. So A_4 is not simple, and it is solvable since it has a normal series with abelian factors. We also see that A_4 cannot contain any subgroup of order 6 as such a subgroup has index 2 and would have to be normal; if $H < G$ with $[G : H] = 2$ and $aH \neq Ha$, then either $aH = H$ or $Ha = H$, so $a \in H$ and $H \neq H$. Therefore the converse of Lagrange's theorem does not hold for non-abelian groups.

4.29 Transfer Homomorphisms

In this section we introduce a peculiar map that will be needed later.

Let G be a group with a subgroup H of finite index. A *transversal* X of H in G is a collection of representatives of the cosets of H in G, that is, a set consisting of exactly one member from each coset. So $G = \cup_{a \in X} aH$ is a disjoint union. Notice that aX for any $a \in G$ is evidently again a transversal of H in G.

4.29 Transfer Homomorphisms

For any two transversals X, Y of H in G and any homomorphism g from H to an abelian group K, define

$$\{X|Y\} = \prod g(a^{-1}b),$$

where the product is over all pairs $(a, b) \in X \times Y$ such that $a^{-1}b \in H$. Since K is abelian the order in the product above does not matter, so $\{X|Y\}$ is well-defined. Clearly $\{X|Y\}^{-1} = \{Y|X\}$, $\{X|Y\}\{Y|Z\} = \{X|Z\}$ and $\{aX|aY\} = \{X|Y\}$, so

$$\{aX|X\} = \{aX|aY\}\{aY|Y\}\{Y|X\} = \{X|Y\}\{Y|X\}\{aY|Y\} = \{aY|Y\}.$$

Hence the map $f\colon G \to K$ that sends a to $\{aX|X\}$ is well-defined, that is, independent of the chosen transversal. It is a homomorphism since

$$\{abX|X\} = \{abX|bX\}\{bX|X\} = \{aX|X\}\{bX|X\}$$

by well-definedness.

Definition 4.29.1 The map f above is called a *transfer homomorphism*.

Proposition 4.29.2 *Let G be a group with a subgroup H of finite index, and let g be a homomorphism from H to an abelian group K. For $a \in G$ consider the action of $\langle a \rangle$ on G/H by left multiplication, and let $b_i \in G$ be a representative of any coset in an orbit with n_i cosets. Then the transfer homomorphism $f\colon G \to K$ is given by*

$$f(a) = \prod_i g(b_i^{-1} a^{-n_i} b_i).$$

In particular, for $a \in Z(G)$ we get $f(a) = g(a^{-[G:H]})$.

Proof The orbits of the action partitions the set G/H. In the definition of f we accordingly use as a transversal for H in G, the set $X = \cup_i X_i$, where $X_i = \{b_i, ab_i, \ldots, a^{n_i-1}b_i\}$. The contribution from X_i to the defining product for f is $g(b_i^{-1} a^{-n_i} b_i)$ since $(ac)^{-1}d = b_i^{-1} a^{-1} b_i$ for each $c, d \in X_i$, and one must multiply such elements with each other n_i-times, using that g is a homomorphism.

The last result uses the fact that $\sum_i n_i$ equals the number $[G : H]$ of cosets since we have a partition of orbits. □

Let H' be the commutator subgroup of H. One often considers the quotient map g from H to the abelianization H/H' in the results above. For $H = Z(G)$ the commutator subgroup H' is trivial, so g is then the identity map on $Z(G)$.

Corollary 4.29.3 *Let G be a group with a center of finite index n. With respect to the identity map g on $Z(G)$ the transfer homomorphism $f\colon G \to Z(G)$ is then given by $f(a) = a^{-n}$ for $a \in G$. Hence G is n-abelian, meaning that $(ab)^n = a^n b^n$ for all a and b.*

Proof By the proposition above we know that $f(a) = g(a^{-[G:Z(G)]}) = a^{-n}$.

For the last assertion use that f is a homomorphism. □

If we had taken H to be a subgroup of $Z(G)$ instead, we would have gotten $f(a) = a^{-[G:H]}$. Working with right cosets would have given positive powers of a in the previous proposition and corollary.

Definition 4.29.4 A group is a *torsion-free group* if it has no element of finite order.

Corollary 4.29.5 *Any torsion-free group such that each element has only finitely many distinct conjugates is abelian.*

Proof Since subgroups of such groups satisfy the same hypothesis, we may assume that the group G in question is generated by the two elements a and b which one wants to show commutes. Since the center of G consists of all elements that commute with a and b, it has finite index n in G. By the corollary above $(a^{-1}b^{-1}ab)^n = e$, and since we have no torsion, we get $a^{-1}b^{-1}ab = e$. □

Given a group G, the subset $H \equiv \{a \in G \mid [G : N(a)] < \infty\}$ is a subgroup of G because if $a, b \in H$, then $c(ab)c^{-1} = (cac^{-1})(cbc^{-1})$ can take only finitely many values in G. By definition every element in H has only finitely many distinct conjugates. It is easy to see that H is a normal subgroup. In fact, it is a *characteristic subgroup* of G, meaning that it is invariant under any automorphism of G.

Characteristic subgroups include commutator subgroups and centers. They are obviously normal, but not all normal subgroups are characteristic subgroups. For instance the factors of a non-trivial direct product group $G \times G$ are normal subgroups that are not invariant under the automorphism that sends (a, b) to (b, a).

4.30 Finitely Generated Abelian Groups

In this section we provide a complete classification of finitely generated abelian groups. This comes down to finite direct sums of cyclic groups, which are obviously finitely generated and abelian.

We have seen that $\mathbb{Z}_{m_1} \times \cdots \times \mathbb{Z}_{m_n} \cong \mathbb{Z}_{m_1 \cdots m_n}$ if and only if the natural numbers m_i are relatively prime. For instance, we see that $\mathbb{Z}_3 \times \mathbb{Z}_5$ is cyclic of order 15, whereas $\mathbb{Z}_2 \times \mathbb{Z}_2$ and $\mathbb{Z}_4$ are not isomorphic. This points to the general pattern when we decompose the order of a finite abelian group into prime factors.

Lemma 4.30.1 *Suppose G is a finitely generated abelian group. Then there are cyclic groups H_i such that*
$$G \cong H_1 \times \cdots \times H_n,$$
where $|H_i|$ is finite and divides $|H_{i+1}|$ for i less than some k, and with the remaining cyclic groups having infinite order.

4.30 Finitely Generated Abelian Groups

Proof We consider G as an additive group, and prove the lemma by induction on the least number n of generators for G. It obviously holds for $n = 1$. Assume it holds for $n - 1$.

If there is a generating set $\{a_i\}_{i=1}^n$ for G such that $\sum x_i a_i = 0$ implies that all the integers x_i vanish, we can define a map $f : G \to \mathbb{Z}^n$ by

$$f\left(\sum x_i a_i\right) = (x_1, \ldots, x_n),$$

which clearly is an isomorphism.

If no such generating set exists, let m_1 be the least positive number among all integers x_i's such that $\sum x_i a_i = 0$ for all possible generating sets $\{a_i\}_{i=1}^n$ of G. The x_i's cannot all be negative as $\sum x_i a_i = 0$ if and only if $\sum (-x_i) a_i = 0$. Upon rearranging terms, we can assume that

$$m_1 a_1 + x_2 a_2 + \cdots + x_n a_n = 0$$

for a generating set $\{a_i\}$ and integers x_i.

By the division algorithm we can write $x_i = q_i m_1 + r_i$, where r_i are non-negative integers less than m_1. Then

$$m_1 b_1 + r_2 a_2 + \cdots + r_n a_n = 0,$$

where $b_1 \equiv a_1 + q_2 a_2 + \cdots + q_n a_n$. Since also $b_1, a_2, \ldots, a_n$ generate G, then by the minimal property of m_1, we must have $r_2 = \cdots = r_n = 0$, so $m_1 b_1 = 0$, and $H_1 \equiv \langle b_1 \rangle$ has order m_1, again on account of minimality of m_1. Let K be the subgroup of G generated by $a_2, \ldots, a_n$. Then $G \cong H_1 \times K$ because $x_1 b_1 \notin K$ for any non-negative integer x_1 less than m_1, on account of minimality of m_1.

By our induction hypothesis, we can write $K \cong H_2 \times \cdots \times H_n$, where $|H_i|$ is finite and divides $|H_{i+1}|$ for i larger than 1 and less than some k, and with the remaining cyclic groups having infinite order.

If H_2 is finite and of order m_2, then by the division algorithm, we may write $m_2 = q m_1 + r$, where r is a non-negative integer less than m_1. Let b_i be a generator of H_i for $i \geq 2$. Then

$$m_1(b_1 + qb_2) + rb_2 + 0b_3 + \cdots + 0b_n = m_1 b_1 + m_2 b_2 = 0,$$

and as $b_1 + qb_2, b_2, \ldots, b_n$ generate G, again by the minimality of m_1, we get $r = 0$. Hence m_1 divides m_2. □

The following result is known as the *fundamental theorem for finitely generated abelian groups*, and furnishes a complete classification of such groups.

Theorem 4.30.2 *Every finitely generated abelian group is of the form*

$$\mathbb{Z}_{m_1} \times \cdots \times \mathbb{Z}_{m_k} \times \mathbb{Z}^n$$

for unique integers n and $m_i \geq 2$ such that m_i divides m_{i+1}.

Proof That every finitely generated abelian group is isomorphic to a group of the stated form is immediate from the lemma.

The integer n is unique since if the group is isomorphic to another one of the prescribed form with n replaced by $n' \neq n$, then upon removing $\min(n, n')$ factors of $\mathbb{Z}$ from both products, one has two isomorphic groups where one is infinite whereas the other is finite, and this is absurd.

Removing the equal number of factors of $\mathbb{Z}$, we are left with

$$\mathbb{Z}_{m_1} \times \cdots \times \mathbb{Z}_{m_k} \cong \mathbb{Z}_{n_1} \times \cdots \times \mathbb{Z}_{n_l}$$

for some integers $n_i \geq 2$ such that n_i divides n_{i+1}.

Since every element of these groups have order not greater than m_k, we must have $n_l \leq m_k$. By the same argument, we get $m_k \leq n_l$, so $m_k = n_l$.

Now

$$m_{k-1}\mathbb{Z}_{m_k} \cong m_{k-1}\mathbb{Z}_{m_1} \times \cdots \times m_{k-1}\mathbb{Z}_{m_k} \cong m_{k-1}\mathbb{Z}_{n_1} \times \cdots \times m_{k-1}\mathbb{Z}_{n_l}$$

forces $|m_{k-1}\mathbb{Z}_{n_i}| = 1$ for $i \leq l-1$ as $n_l = n_k$. So in particular, the integer n_{l-1} divides m_{k-1}. By symmetry $m_{k-1} = n_{l-1}$, and $m_{k-i} = n_{l-i}$ by induction. Since $m_1 \cdots m_k = n_1 \cdots n_l$, we get $k = l$ and $m_i = n_i$ for all i. □

The number n in the theorem is called the *rank of the group*, and the product of the remaining factors is called the *torsion part of the group*. The tuple $(m_1, \ldots, m_k)$ is called the *type of the torsion part*. The theorem tells us that the rank and the type of a finitely generated abelian group is a complete invariant. Finite abelian groups are obviously finitely generated, and in this case one speaks of the *type of the group* since there is only torsion.

Example 4.30.3 The types of abelian groups of order 360 are (360), (3, 120), (2, 180), (6, 60), (2, 2, 90), (2, 6, 30), so there are 6 non-isomorphic abelian groups of order 360.

To find the types of a finite group can be somewhat cumbersome, so we will look at things slightly differently. By Corollary 4.20.5 any finite abelian group is a direct product of p-groups for distinct primes, which in general are not cyclic of course. This was a corollary of Sylow's first theorem, and can also be proved using the fundamental theorem of finitely generated abelian groups. Alternatively we could use this corollary to reduce the proof of the fundamental theorem to the special case of abelian p-groups. Anyway, being in favor of proving things only once, we might as well use this theorem to classify abelian p-groups in terms of certain partitions.

4.30 Finitely Generated Abelian Groups 181

Definition 4.30.4 A *partition of a natural number* k is a tuple $(k_1, \ldots, k_j)$ of natural numbers such that $k = \sum k_i$ and $k_i \leq k_{i+1}$. Let $P(k)$ denote the number of partitions of k.

Corollary 4.30.5 *The number of non-isomorphic abelian groups of order n is*

$$\prod_{i=1}^{m} P(n_i),$$

where $n = p_1^{n_1} \cdots p_m^{n_m}$ is the prime number decomposition of n.

Proof By Corollary 4.20.5 such groups are of the form $G_1 \times \cdots \times G_m$ with $|G_i| = p_i^{n_i}$, so it suffices to show that the number of non-isomorphic abelian groups of order p^k is $P(k)$. But the types of such groups are exactly $(p^{k_1}, \ldots, p^{k_j})$, where $(k_1, \ldots, k_j)$ is a partition of k. □

Example 4.30.6 Returning to the abelian groups of order $360 = 2^3 \cdot 3^2 \cdot 5^1$, there are $P(3)P(2)P(1) = 3 \cdot 2 \cdot 1 = 6$ non-isomorphic ones, and they are:

$$\mathbb{Z}_2 \times \mathbb{Z}_2 \times \mathbb{Z}_2 \times \mathbb{Z}_3 \times \mathbb{Z}_3 \times \mathbb{Z}_5 \cong \mathbb{Z}_2 \times \mathbb{Z}_6 \times \mathbb{Z}_{30}$$
$$\mathbb{Z}_2 \times \mathbb{Z}_2 \times \mathbb{Z}_2 \times \mathbb{Z}_9 \times \mathbb{Z}_5 \cong \mathbb{Z}_2 \times \mathbb{Z}_2 \times \mathbb{Z}_{90}$$
$$\mathbb{Z}_2 \times \mathbb{Z}_4 \times \mathbb{Z}_3 \times \mathbb{Z}_3 \times \mathbb{Z}_5 \cong \mathbb{Z}_6 \times \mathbb{Z}_{60}$$
$$\mathbb{Z}_2 \times \mathbb{Z}_4 \times \mathbb{Z}_9 \times \mathbb{Z}_5 \cong \mathbb{Z}_2 \times \mathbb{Z}_{180}$$
$$\mathbb{Z}_8 \times \mathbb{Z}_3 \times \mathbb{Z}_3 \times \mathbb{Z}_5 \cong \mathbb{Z}_3 \times \mathbb{Z}_{120}$$
$$\mathbb{Z}_8 \times \mathbb{Z}_9 \times \mathbb{Z}_5 \cong \mathbb{Z}_{360},$$

where we have also indicated the types.

Another way of looking at the fundamental theorem of a finitely generated abelian group G is to pick generators $\{a_i\}_{i=1}^{r}$ of the group and define an epimorphism $f: \mathbb{Z}^r \to G$ by

$$f(x_1, \ldots, x_r) = \sum_i x_i a_i.$$

Then $G \cong \mathbb{Z}^r / \ker f$ by the first isomorphism theorem. The non-triviality of $\ker f$ causes the torsion. In much the same way as we proved the fundamental theorem one can show that there are integers $m_i \geq 2$ such that m_i divides m_{i+1} and

$$\ker f \cong m_1 \mathbb{Z} \times \cdots \times m_k \mathbb{Z} \times \{0\}^n,$$

where $n = r - k$. The quotient will therefore be of the form prescribed in the fundamental theorem.

4.31 Free Groups

We have seen that groups often are presented in terms of generators and relations. We will be more precise what is meant by this. First we need to talk about groups with generators freed of any relations between them apart from those imposed by the axioms of a group, like $aa^{-1} = e$. At first, the following definition seems rather ad hoc.

Definition 4.31.1 The *free group generated by a set X* is a group F together with a map $f \colon X \to F$ such that for any map $g \colon X \to G$ into any group G, there is a unique homomorphism $h \colon F \to G$ satisfying $hf = g$.

For the moment we do not know whether there is anything like a free group. However, given that they do exist, we have the following uniqueness result.

Proposition 4.31.2 *Up to isomorphism there is only one free group generated by sets of the same cardinality. Conversely, any two generator sets of isomorphic free groups have the same cardinality.*

Proof Say we have two free groups F_i generated by sets X_i with accompanying maps $f_i \colon X_i \to F_i$.

If there is a bijection $j \colon X_1 \to X_2$, there are homomorphism $h \colon F_1 \to F_2$ and $h' \colon F_2 \to F_1$ such that $hf_1 = f_2 j$ and $h'f_2 = f_1 j^{-1}$, so $h'hf_1 = f_1$. By uniqueness, we see that $h'h$ is the identity map on F_1. Similarly, we conclude that hh' is the identity map on F_2. So $h' = h^{-1}$ and $F_1 \cong F_2$.

Conversely, the subgroup H_i of F_i generated by all squares of members of F_i is clearly a normal subgroup of F_i. From the discussion after Corollary 4.31.6, we see that $[F_i : H_i]$ equals $2^{|X_i|}$ if X_i is finite, and equals $|X_i|$ if X_i is infinite, which implies $X_i \times X_i \cong X_i$ and $\cup_n X_i^n \cong X_i$. Hence $|X_1| = |X_2|$ if $F_1 \cong F_2$. □

We first prove existence in the following simple setting.

Lemma 4.31.3 *There is a free group for every single generator x, namely, the additive cyclic group $\mathbb{Z}$ with $f \colon \{x\} \to \mathbb{Z}$ given by $f(x) = 1$.*

Proof Given a map $g \colon \{x\} \to G$, define $h \colon \mathbb{Z} \to G$ by $h(n) = g(x)^n$. Then $hf(x) = h(1) = g(x)$, so $hf = g$, and h is the only map with this property as h is uniquely determined on 1. □

Next we form a certain product of groups that imposes no further auxiliary relations between the group members. Again, the actual construction of such a group will be carried out after we have stated the properties that uniquely characterize it.

Definition 4.31.4 Given a family of homomorphism $f_i \colon G_i \to G$, we say that G is the *free product of the groups G_i*, if for any family of homomorphisms $g_i \colon G_i \to H$, there is a unique homomorphism $h \colon G \to H$ such that $hf_i = g_i$.

4.31 Free Groups

The free product is unique up to isomorphism. For suppose the second family in the definition above has the same property as the first. Then there is a homomorphism $h': H \to G$ such that $h'g_i = f_i$. Hence $h'hf_i = f_i$, and by uniqueness $h'h$ is the identity map on G. Similarly, we see that hh' is the identity map on H, so $H \cong G$. We denote the free group of the G_i's by $*G_i$.

Theorem 4.31.5 *The free product of any collection of groups exists.*

Proof Suppose we have a collection of groups G_i. A *word of length n* is an n-tuple $(a_1, \ldots, a_n)$ of non-unit elements from the collection of groups, and where no successive coordinates belong to the same group. Let W_n be the set of such words, and let $W = \cup W_n$, where W_0 consists of the empty word.

The groups G_i act on W in an obvious way. For $b \in G_i$ and $(a_1, \ldots, a_n) \in W_n$ define $b \cdot (a_1, \ldots, a_n) \in W$ to be:

$(b, a_1, \ldots, a_n)$ if $a_1 \notin G_i$ and $b \neq e$, or $(a_1, \ldots, a_n)$ if $a_1 \notin G_i$ and $b = e$,

or

$(ba_1, \ldots, a_n)$ if $a_1 \in G_i$ and $ba_1 \neq e$, or $(a_2, \ldots, a_n)$ if $a_1 \in G_i$ and $ba_1 = e$

with the obvious understanding concerning the empty word. It is straightforward to check that this is a faithful action of G_i on W, so we can consider $G_i \subset \text{Perm}(W)$. Let G be the subgroup of $\text{Perm}(W)$ generated by the groups G_i, and let $f_i : G_i \to G$ be the inclusion maps. We claim that G is the free product of the groups G_i.

To this end, first note that every non-unit element of G is a finite product of members from the groups G_i, and we can of course assume that two consecutive factors belong to different groups; the element is then in *reduced form*. Such a form is unique because if

$$a_1 \cdots a_n = b_1 \cdots b_m$$

for two reduced expressions, then their action on the empty word in W is

$$(a_1, \cdots, a_n) = (b_1, \cdots, b_m),$$

so $n = m$ and $a_i = b_i$.

Now given any family of homomorphisms $g_i : G_i \to H$ into any group H, define a map $h : G \to H$ by $h(e) = e$ and

$$h(a_1 \cdots a_n) = g_{i_1}(a_1) \cdots g_{i_n}(a_n)$$

for $a_1 \cdots a_n \in G$ in reduced form with $a_j \in G_{i_j}$. Clearly h is a homomorphism, and it is the unique one such that $hf_i = g_i$. □

Corollary 4.31.6 *The free group generated by any set exists.*

Proof Given any set $X = \{x_i\}$ consider, as prescribed in the lemma, the free group $F_i = \mathbb{Z}$ of each single generator x_i with maps f_i. Then $F \equiv *F_i$ is the free group generated by X because if we have a map $g: X \to G$, then g restricts to maps g_i on single elements x_i, so there is a unique homomorphism $h: F \to G$ such that $hf_i = g_i$, or in other words, such that $hf(x_i) = g(x_i)$ by definition of f. □

We see that the uniqueness requirement in the definition of a free group F is equivalent to requiring that $f(X)$ generates F as a group. Also it is clear from the construction above that f is injective.

We have also seen that every non-unit element of a free group F generated by a set X can be written uniquely as

$$x_1^{n_1} \cdots x_m^{n_m}$$

for non-zero integers n_i and distinct elements $x_i \in X$, and where we have suppressed the injection $f: X \to F$. Multiplication is by juxtaposition of such expressions with possible cancellations, and with the unit being the empty expression. It is easy to see that any group with a subset X such that all group elements can be written uniquely in the above fashion, is automatically free by the characterizing property of free groups.

Clearly, free groups on more than one generator are non-abelian with trivial center, but we can abelianize them by considering the derived subgroup generated by commutators. We define the *free abelian group* generated by a set as we did for free groups but with all groups involved in the definition replaced by abelian ones. Uniqueness up to isomorphism holds as before. The following existence result is also obvious.

Proposition 4.31.7 *Let F be a free group generated by X with $f: X \to F$. Then F/F' is a free abelian group generated by X with respect to the map qf, where $q: F \to F/F'$ is the quotient map.*

Free abelian groups are therefore direct sums of copies of $\mathbb{Z}$.

Proposition 4.31.8 *Any group is the homomorphic image of a free group.*

Proof Given any group G, form the free group F generated by $X \equiv G$ with $f: X \to F$. Then there is a unique homomorphism $h: F \to G$ such that hf is the identity map on G. Hence $G = h(F)$. □

Similarly, any abelian group is the homomorphic image of a free abelian group, and thus of some direct sum of $\mathbb{Z}$.

Clearly we can define free (abelian) monoids as we did for groups, and then get the result above with groups replaced by monoids everywhere.

4.32 An Example of a Free Product

The group $SL(2, \mathbb{R})$ acts on the complex upper half plane

$$H \equiv \{z \in \mathbb{C} \,|\, \mathrm{Im}\, z > 0\}$$

by *fractional linear transformations*

$$\begin{pmatrix} a & b \\ c & d \end{pmatrix} \cdot z = \frac{az+b}{cz+d}$$

for $z \in H$. The fraction belongs to H due to the easily verified fact that

$$\mathrm{Im}\,\frac{az+b}{cz+d} = \frac{\mathrm{Im}\, z}{|cz+d|^2}.$$

It is easily checked that we indeed have an action. Moreover, this action is transitive, and the isotropy group of $i \in H$ is

$$\begin{pmatrix} \cos\theta & \sin\theta \\ -\sin\theta & \cos\theta \end{pmatrix}$$

for $\theta \in \mathbb{R}$, which is evidently isomorphic to the circle. So one can identify H and $SL(2, \mathbb{R})/\mathbb{T}$ as $SL(2, \mathbb{R})$-spaces.

It is natural to look at the action restricted to various subgroups of $SL(2, \mathbb{R})$. One such is the *modular group* $SL(2, \mathbb{Z})$ consisting of the matrices with integer entries. The matrices

$$A = \begin{pmatrix} 0 & -1 \\ 1 & 0 \end{pmatrix} \quad \text{and} \quad C = \begin{pmatrix} 1 & 1 \\ 0 & 1 \end{pmatrix}$$

of the modular group act on H as $z \mapsto -1/z$ and $z \mapsto z+1$, respectively. It turns out that these two transformations generate all possible fractional linear transformations coming from the modular group. In fact, we have the following result.

Lemma 4.32.1 *The modular group is generated by A and $B \equiv AC$.*

Proof If A and B do not generate all of $SL(2, \mathbb{Z})$, we can pick an element

$$D \equiv \begin{pmatrix} a & b \\ c & d \end{pmatrix} \in SL(2, \mathbb{Z}) \backslash \langle A, B \rangle$$

with $|a| + |c|$ minimal. The elements

$$(AB)^n D = \begin{pmatrix} a+nc & b+nd \\ c & d \end{pmatrix} \quad \text{and} \quad (BA)^{-m} D = \begin{pmatrix} a & b \\ c+ma & d+mb \end{pmatrix}$$

obviously also belong to the complement of $\langle A, B \rangle$.

We cannot have $ac \ne 0$ because if $|a| \ge |c|$, we can pick an integer n such that $|a+nc| + |c| < |a| + |c|$, and if $|a| < |c|$, we can pick and integer m such that $|a| + |c+ma| < |a| + |c|$, so the matrices above will contradict the choice of D.

We are left with the possibilities $a = 0$ or $c = 0$. In the first case D must be

$$\begin{pmatrix} 0 & 1 \\ -1 & d \end{pmatrix} \quad \text{or} \quad \begin{pmatrix} 0 & -1 \\ 1 & d \end{pmatrix},$$

and these matrices can be written as $BA^2(AB)^{-d-1}$ and $(AB)^{d-1}$, respectively, which is impossible. In the second case D must be $(AB)^b$ or $A^2(AB)^{-b}$, which again is impossible. □

Now observe that $A^2 = -I_2 = B^3$, so considered as linear fractional transformations, or as elements $[A]$ and $[B]$ of $PSL(2, \mathbb{Z}) \equiv SL(2, \mathbb{Z})/\{\pm\}$, both $[A]^2$ and $[B]^3$ are the unit. This gives rise to two homomorphism $g_i \colon \mathbb{Z}_{i+1} \to PSL(2, \mathbb{Z})$ with $g_1([1]) = [A]$ and $g_2([1]) = [B]$, and hence a unique homomorphism

$$h \colon \mathbb{Z}_2 * \mathbb{Z}_3 \to PSL(2, \mathbb{Z})$$

such that $hf_1([1]) = [A]$ and $hf_2([1]) = [B]$. Since $[A]$ and $[B]$ generate $PSL(2, \mathbb{Z})$, we see that h is surjective. The funny things is that it is also injective.

Proposition 4.32.2 *We have an isomorphism* $PSL(2, \mathbb{Z}) \cong \mathbb{Z}_2 * \mathbb{Z}_3$.

Proof We content that the epimorphism h defined above is injective. Denote the generators $f_1([1])$ and $f_2([1])$ of $\mathbb{Z}_2 * \mathbb{Z}_3$ by a and b, respectively. Then any element $c \in \mathbb{Z}_2 * \mathbb{Z}_3$ is a product of ab's and ab^{-1}'s with a possible initial $b^{\pm 1}$ and a possible final a. Now

$$(AB)^n = \begin{pmatrix} 1 & n \\ 0 & 1 \end{pmatrix} \quad \text{and} \quad (AB^{-1})^m = (-1)^m \begin{pmatrix} 1 & 0 \\ m & 1 \end{pmatrix},$$

so a product of such elements cannot contain both positive and negative entries. However, if $c \in \ker h$, some such products would have to equal $\pm A$ or $\pm B^{\pm 1}$ or $\pm B^{\pm 1} A$ or the unit, and apart from this last case, all these products have entries with mixed signs. Thus c must be the unit. □

Remark 4.32.3 The modular group is of great importance in complex function theory and in the study of Riemann surfaces, modular forms and elliptic curves. The work of Andrew Wiles on Fermat's last theorem is related to this [4].

4.33 Generators and Relations

We have seen that if we pick a generating subset X of the group G. Then for the free group F generated by X, there is a unique epimorphism $h\colon F \to G$ such that $hf = g$, where $g\colon X \to G$ is the inclusion map. By a *relation between the generators* of X we mean a non-unital element r in the kernel of h. The subtle point to be made is that we have created a place, namely the free group, for a 'relation to live in' because in the group G itself it collapses to the unit.

Definition 4.33.1 A *presentation for a group* G is a pair $\langle X, R \rangle$ of two sets such that if F is the free group generated by X, and if N is the normal subgroup of F generated by $R \subset F$, then $G \cong F/N$. The presentation is a *finite presentation* if both X and R are finite.

By abuse of language it is customary to regard the elements of X as generators for G and the elements of R as the actual relations between the generators. One tends to write
$$\{a, b \mid a^n = e = b^2, \ bab = a^{-1}\}$$
both for the presentation $\langle \{a, b\}, \{a^n, b^2, baba\} \rangle$ and the corresponding group.

Example 4.33.2 The group F/N of the presentation $\langle X, R \rangle$ right above is actually isomorphic to the dihedral group D_n. To deduce this, first observe from the previous discussion that we have a unique epimorphism $h\colon F \to D_n$ such that $h(a) = \sigma$ and $h(b) = \tau$. Since $R \subset \ker h$ due to the relations between σ and τ, we have an epimorphism $h\colon F/N \to D_n$ such that $h(a) = \sigma$ and $h(b) = \tau$, where we have kept the symbols h and a, b. From the relations between a and b, we see that every element in F/N is of the form $a^m b^k$ for integers $m < n$ and $k < 2$, so there are maximum $2n$ elements in F/N. Since h maps onto a set D_n with $2n$ elements, it therefore has to be injective.

If we drop the relation $a^n = e$ above, we get the *infinite dihedral group*
$$D_\infty \equiv \{a, b \mid b^2 = e, \ bab = a^{-1}\}.$$

This group has also the following presentation $\{c, d \mid d^2 = e = c^2\}$. To see this we know that there are unique epimorphisms f and g between these two groups such that $f(a) = c^{-1}d$ and $f(b) = c$ and $g(c) = b$ and $g(d) = ba$. As fg and gf fix the generators, we conclude that $f^{-1} = g$.

In fact, the infinite dihedral group is isomorphic to the semidirect product $\mathbb{Z} \rtimes \mathbb{Z}_2$ formed with respect to the homomorphism $\mathbb{Z}_2 \to \mathrm{Aut}(\mathbb{Z})$, where $[k] \in \mathbb{Z}_2$ sends $m \in \mathbb{Z}$ to $(-1)^k m \in \mathbb{Z}$. To see this first note that $(m, [0])$ and $(0, [1])$ satisfy the same relations as a and b, respectively, so there is an epimorphism $D_\infty \to \mathbb{Z} \rtimes \mathbb{Z}_2$ that respects these generators. By the relations between a and b, any element of D_∞ can be written as $a^m b^k$ for $m \in \mathbb{Z}$ and $k \in \{0, 1\}$, which under the epimorphism corresponds to $(m, [k]) \in \mathbb{Z} \rtimes \mathbb{Z}_2$. But $(m, [k])$ is trivial only if $m = 0$ and $[k] = [0]$, so $a^m b^k = e$ and the kernel of the epimorphism is trivial. ◇

Given a presentation $\langle X, R \rangle$, it is by no means obvious to decide by an algorithm whether two elements of the free group F represent the same elements in F/N, in other words, whether an element of F belongs to the normal subgroup generated by R. This is known as the *word problem*. The *conjugacy problem* asks when two elements of F are conjugate in F/N, and the *isomorphism problem* is that of deciding when two finite presentations have isomorphic groups. Such problems are central in combinatorial group theory.

Example 4.33.3 Consider the transposition $\tau_i = (i\ i+1) \in S_n$. We know that every permutation is a product of transpositions. But every transposition is also a product of adjacent transposition as is seen by repeated use of the identity

$$(i\ j+1) = \tau_j (i\ j) \tau_j$$

for $i < j$. So $S_n = \langle \tau_1, \ldots, \tau_{n-1} \rangle$. It is also easy to see that the transpositions τ_i satisfy the same relations as the generators a_i in the so *Coxeter presentation*

$$G \equiv \{a_1, \ldots, a_{n-1} \mid a_i^2 = e = (a_i a_{i+1})^3 \text{ and } (a_i a_j)^2 = e \text{ if } |i-j| > 1\}$$

of S_n. Let us prove that there is an isomorphism between these two groups.

From what we have said we already know that there is a unique epimorphism $h: G \to S_n$ such that $h(a_i) = \tau_i$. As for injectivity, remember that $|S_n| = n!$, so it suffices to show that $|G| \leq n!$. This again follows by induction if we can show that $[G : H] \leq n$, where $H \equiv \langle a_1, \cdots, a_{n-2} \rangle$. To this end consider the n cosets

$$a_1 \cdots a_{n-1} H, \quad a_2 \cdots a_{n-1} H, \ldots, \quad a_{n-1} H, \quad H$$

of G. As $[G : H]$ is the number of disjoint cosets aH in G, and the elements a_j generate G, it it enough to show that left multiplication of the listed cosets by any a_j permutes these cosets, since then the list contains all the cosets, so $[G : H] \leq n$. Let us show that we just get such permutations.

If $j < i - 1$, then $a_i a_j = (a_j a_i)^{-1} = a_j^{-1} a_i^{-1} = a_j a_i$, so

$$a_j a_i \cdots a_{n-1} H = a_i \cdots a_{n-1} a_j H = a_i \cdots a_{n-1} H$$

as $a_j \in H$. The cases $j = i - 1$ and $j = i$ are evident from $a_i^2 = e$.

Finally, if $j > i$, then again as a_k and a_j commute for $|j - k| > 1$, we get

$$a_j a_i \cdots a_{n-1} H = a_i \cdots a_{j-2} (a_j a_{j-1} a_j) a_{j+1} \cdots a_{n-1} H.$$

From $(a_{j-1} a_j)^3 = e$, we get $a_j a_{j-1} a_j = a_{j-1} a_j a_{j-1}$, and inserting this gives

$$a_j a_i \cdots a_{n-1} H = a_i \cdots a_{j-2} (a_{j-1} a_j a_{j-1}) a_{j+1} \cdots a_{n-1} H = a_i \cdots a_{n-1} a_{j-1} H,$$

which equals $a_i \cdots a_{n-1} H$ as $a_{j-1} \in H$. ◇

4.34 Ordered Groups

Definition 4.34.1 A group is *(partially) ordered* if there is a (partial) order $<$ on it as a set such that $ac < bc$ and $ca < cb$ whenever $a < b$. An *order homomorphism* between two (partially) ordered groups is a group homomorphism f such that $f(a) < f(b)$ when $a < b$.

Clearly subgroups of ordered groups are ordered groups with the induced order.

Example 4.34.2 The positive real numbers $\mathbb{R}_+$ is an order group under multiplication and the usual order.

The additive groups $\mathbb{Z}$, $\mathbb{Q}$ and $\mathbb{R}$ are ordered groups with the usual ordering, and the exponential map $a \mapsto e^a$ is an order isomorphism from $\mathbb{R}$ to $\mathbb{R}_+$. ◇

If G is a partially ordered group and X is a set, then G^X is a partially ordered group under pointwise operations and partial order. The free abelian group $\mathbb{Z}^n$ of rank n is an ordered group under the *lexicographical order*, which means that $(a_1, \ldots a_n) < (b_1, \ldots b_n)$ if $a_m < b_m$ for some m such that $a_i = b_i$ for $i < m$.

If G is an ordered group, then $P = \{a \in G \mid e < a\}$ satisfies the following definition.

Definition 4.34.3 A *positive cone for a group* G is a subset P of G such that $P \cdot P \subset P$ and $P \cup \{e\} \cup P^{-1}$ is a disjoint union of G and $aPa^{-1} \subset P$ for $a \in G$.

The following trivial result shows that ordered groups can equivalently be described in terms of positive cones.

Proposition 4.34.4 *A group G with a positive cone P is an ordered group with order defined by $a < b$ if $ba^{-1} \in P$, and its positive cone $\{a \in G \mid e < a\}$ will be P.*

Notice that if P is a positive cone for a group, then so is P^{-1}.

Example 4.34.5 Consider $\mathbb{Z}^2$ with positive cone consisting of those elements in the integer lattice that belong to one side of a straight line through the origin with irrational slope. Thus $\mathbb{Z}^2$ is an ordered group for uncountable many different orderings, one for each such slope.

Proposition 4.34.6 *Ordered groups are torsion-free.*

Proof If $e < a$, then $e < a < a^2 < \cdots$ and if $e > a$, then $e > a > a^2 > \cdots$, so no power of $a \neq e$ can be e. □

Thus since -1 has order two, the multiplicative group F_* of a non-trivial field F cannot be turned into an ordered group.

Not all torsion-free groups can be turned into ordered groups.

Example 4.34.7 Let G be the group generated by a and b subject to the relation $aba^{-1} = b^{-1}$. Then every element of the group can evidently be written as $a^m b^n$ for some $m, n \in \mathbb{Z}$. If $a^m b^n = e$, then by applying $\text{Ad}(a)$ to both sides n times, we get $a^m = e$. This is impossible because

$$a \mapsto A = \frac{1}{2}\begin{pmatrix} 0 & 1 \\ 1 & 0 \end{pmatrix} \text{ and } b \mapsto I = \begin{pmatrix} 1 & 0 \\ 0 & 1 \end{pmatrix}$$

is a well-defined homomorphism from G to the multiplicative group $M_2(\mathbb{Q})$, and $A^m \neq I$ for all $m \in \mathbb{Z}$. Hence G is torsion-free. But G has no positive cone P because if $b \in P$, then $b^{-1} = aba^{-1}$ belongs to both P^{-1} and P, which is absurd. As the homomorphism $G \to M_2(\mathbb{Q})$ given by $a \mapsto I$ and $b \mapsto 2A$ shows, the case $b = e$ is no option either. So G cannot be turned into an ordered group. ◇

However, we have the following result.

Theorem 4.34.8 *A group can be turned into an ordered group if it is torsion-free and abelian.*

Proof Suppose G is torsion-free and additive. Let X be a set of representatives of the orbits of the action of $\mathbb{Z}$ on G given by $(n, a) \mapsto na$. Let H be the additive subgroup of $\mathbb{Z}^X$ under pointwise addition of functions that are non-zero for only finitely many elements of X. Pick an order on X by Zorn's lemma, see the preliminaries. Obviously H is an ordered group with order $g < h$ if there is some $x \in X$ such that $g(y) = h(y)$ for $y < x$ and $g(x) < h(x)$.

Define $G \to H$ by $a \mapsto f_a$, where $f_a(x)x = a$ for all $x \in X$. Since the orbits form a partition of G such an integer $f_a(x)$ exists, and it is uniquely determined by a and x since G has no torsion. Clearly $a \mapsto f_a$ is additive and injective, and provides an order on G inherited from H. □

4.35 Groups in Algebraic Topology

Topological spaces are sets where one can talk about convergence and continuity by declaring what should be the neighborhoods of points [23]. If you are not familiar with the definition, think of a subset of the Euclidean space $\mathbb{R}^3$ with convergence from calculus defined in terms of gradually smaller balls playing the role of neighborhoods. Bijections between topological spaces that are continuous in both directions are called homeomorphisms, and we identify topological spaces up to homeomorphisms. This means that we are not concerned about distinguishing spaces that can be continuously deformed into each other, say by stretching. Quantities that are insensitive to this are called topological invariants; they remain the same for spaces that are homeomorphic. Note that we do not allow tearing, as this is not a continuous operation. We are particularly interested in invariants that can distinguish topological spaces, and that

4.35 Groups in Algebraic Topology

can be computed. In algebraic topology [5] one looks for such notions in the realm of algebraic objects. Here we will consider topological invariants that are groups.

Let us first explain what the fundamental group of a topological space is. A coffee cup and a doughnut can be deformed into each other. The crucial topological feature they have in common is the hole in the doughnut, which is also found in the handle of the coffee cup. A loop around either of these holes cannot be shrunk to a point without leaving the subset at hand. This way such holes can be detected, and it suggests that topological insight might be gained by studying loops, which essentially constitute the elements of the fundamental group.

Recall that a *loop* in a topological space X at a distinguished point $* \in X$ is a continuous map $f: [0, 1] \to X$ such that $f(0) = * = f(1)$. A *homotopy* from such a loop f to another g is a continuous map $H: [0, 1] \times [0, 1] \to X$ such that $H(0, \cdot) = f(\cdot)$ and $H(1, \cdot) = g(\cdot)$ and $H(s, 0) = * = H(s, 1)$ for all $s \in [0, 1]$. Intuitively, this means that the loops can be deformed into each other within X, keeping the point $*$ fixed. It is not hard to see that homotopy is an equivalence relation between the loops at the point $*$. We can form the product $[f][g]$ of the equivalence classes of f and g by letting $[f][g] = [fg]$, where fg is the loop with $(fg)(t)$ given by $f(2t)$ when $t \in [0, 1/2]$, and is equal to $g(2t - 1)$ when $t \in [1/2, 1]$. It can then be checked that this is a well-defined product turning these equivalence classes into a group $\pi_1(X, *)$ known as the *fundamental group* of X at $*$. The unit element has as a representative the constant loop, while $[f]^{-1}$ has a representative given by $f^{-1}(t) = f(1 - t)$. We say that X is *arcwise connected* if any two points $x, y \in X$ can be connected by a continuous path, i.e. a continuous map $h: [0, 1] \to X$ with $h(0) = x$ and $h(1) = y$. Loops are of course special cases of such paths, and we can obviously extend the definition of homotopy to continuous paths. For arcwise connected spaces the fundamental groups corresponding to two distinguished points are clearly isomorphic, and we simply write $\pi_1(X)$. One finally shows that it is a topological invariant, and indeed a quite useful one. The space X is said to be *simply connected* if $\pi_1(X)$ is the trivial group, which equivalently means that any loop can be shrunk to a point.

Let us see how it can be computed in cases where X can be approximated homeomorphically by a polyhedron. Such a geometric object is composed of simpler building blocks, namely simplexes that are nicely fitted together. For instance, a solid tetrahedron $\langle p_0 p_1 p_2 p_3 \rangle$ is composed of four triangular faces $\langle p_0 p_1 p_2 \rangle$, $\langle p_0 p_2 p_3 \rangle$, $\langle p_0 p_1 p_3 \rangle$, $\langle p_1 p_2 p_3 \rangle$, six straight edges $\langle p_0 p_1 \rangle$, $\langle p_0 p_2 \rangle$, $\langle p_0 p_3 \rangle$, $\langle p_1 p_2 \rangle$, $\langle p_1 p_3 \rangle$, $\langle p_2 p_3 \rangle$ and four vertices $\langle p_0 \rangle$, $\langle p_1 \rangle$, $\langle p_2 \rangle$, $\langle p_3 \rangle$. These are examples of 3-,2-, 1-, 0-simplexes associated to the points $\langle p_i \rangle = p_i \in X$. Let us say what it means that these pieces are nicely fitted together. Any finite set K of simplexes in $\mathbb{R}^n$ is called a *simplicial complex* if any face of a simplex also belongs to K, and if the intersection of any two simplexes is either empty or belongs to K. Their union $|K|$ is then by definition a *generalized polyhedron*, and any homeomorpism $|K| \to X$ is called a *triangulation* of X. In this sense simplexes are building blocks for many topological spaces. For instance, any homeomorphism from a triangle to the circle is a triangulation of the circle. But you can also build a square using two triangles, and this way get another triangulation of the circle, so we do not have uniqueness.

A triangulation of X by a simplicial complex K opens up for a combinatorial approach to finding $\pi_1(X) \cong \pi_1(|K|)$, which goes as follows:

(1) find an arcwise connected and simply connected subcomplex L of K that contains all the vertices p_i of K;

(2) assign a generator g_{ij} to each 1-simplex $\langle p_i p_j \rangle$ of $K \setminus L$ with $i < j$;

(3) impose a relation $g_{ij} g_{jk} = g_{ik}$ if there is a 2-simplex $\langle p_i p_j p_k \rangle$ in K such that $i < j < k$. Moreover, if two of the vertices p_i, p_j, p_k form a 1-simplex of L, the corresponding generator should be set to be the unit;

(4) the group with generators g_{ij} satisfying the relations from (3) is isomorphic to $\pi_1(X)$.

Rather than proving this well-known result, let us illustrate how it works in concrete cases.

Example 4.35.1 Consider a solid disk $D \subset \mathbb{R}^2$. It can be triangulated by a solid triangle, which is already simply connected, so we can choose $L = K$, yielding $\pi_1(D) = \{e\}$. The same argument shows that the solid ball has trivial fundamental group. A hollow tetrahedron gives a triangulation $|K| \to S^2$ of the 2-sphere, and we obtain a subcomplex L by removing one of the 2-faces of the tetrahedron. This gives three generators, corresponding to the edges of the removed 2-simplex, and all of them are set to be the unit. So once again we get $\pi_1(S^2) = \{e\}$. ◇

Here is a more elaborate example.

Example 4.35.2 An n-bouquet is defined to be the one-point union, or the wedge product, of n circles. Taking the common point p_0 as the distinguished point, a triangulation is obtained by considering a star with center p_0 and arms, or edges, leading to vertices $p_1, p_2, \ldots, p_{2n}$ ordered counterclockwise. This gives the complex L, while K is formed by adding the edges $\langle p_1 p_2 \rangle, \langle p_3 p_4 \rangle, \ldots, \langle p_{2n-1} p_{2n} \rangle$. Clearly L is arcwise connected and simply connected. It is an example of a tree, indeed a maximal one. One associates the generators $g_{12}, g_{34}, \ldots, g_{(2n-1)2n}$ to the outer edges, and there are no relations between these. So the fundamental group of the n-bouquet is the free group on n generators. In particular, we get $\pi_1(\mathbb{T}) = \mathbb{Z}$ as the free group on one generator is $\mathbb{Z}$. The fundamental group of the figure symbolizing the number eight, or infinity, is a free group on two generators. Topologically these generators correspond to the two loops a, b running, say counterclockwise, around the upper and lower part, respectively, of the figure eight. Clearly the loop aba^{-1} cannot be continuously deformed into the loop b. ◇

Here is perhaps a topologically more important case.

Example 4.35.3 Just like the circle can be obtained from an interval by identifying its endpoints, we can produce the torus $\mathbb{T}^2$ by identifying opposite points of a solid square, working then with the quotient topology on the collection of equivalence classes thus obtained. Geometrically this means that upon identifying the opposite points on the two horizontal lines on a square sheet of paper, we first get a cylinder, and next, upon identifying the opposite points on the two vertical lines, we glue the ends of the cylinder together. This way we obviously get a torus in ordinary space $\mathbb{R}^3$. When it comes to triangulation, it is easier to work with the 2-dimensional sheet, keeping in mind that certain points are identified with each other, when we make the triangulation. We draw a chessboard on the square consisting of nine smaller squares, and we divide these up in triangles by drawing diagonals. This gives K, while L can be chosen to consist of the four squares (with their triangles) gathered in one corner of the bigger square. This gives eleven generators and ten relations, which in the end reduces to two generators a, b and one relation $aba^{-1}b^{-1} = e$, or $ab = ba$. This is just the free abelian group on two generators. Hence $\pi_1(\mathbb{T}^2) = \mathbb{Z} \times \mathbb{Z}$. Geometrically, the generators a, b correspond to one loop around the cylinder, and another around the hole in the doughnut having the torus as its surface. It is instructive to make a drawing to convince oneself that the loop $aba^{-1}b^{-1}$ can indeed be shrunk (within the torus) to a point.

We can generalize all this to a closed surface Σ_g in $\mathbb{R}^3$ having genus g, that is, a surface with g holes in it. Then the fundamental group of Σ_g is seen to have $2g$ generators a_i, b_i correspond to obvious loops, and they satisfy the single relation $\prod_{i=1}^{g} a_i b_i a_i^{-1} b_i^{-1} = e$. So for $g > 2$ we do not get an abelian group.

Note that the torus $\mathbb{T}^2$ is the direct product of two circles $\mathbb{T}$, each having fundamental group $\mathbb{Z}$. Now, forming fundamental groups respects in general products. Hence $\pi_1(\mathbb{T}^2) = \pi_1(\mathbb{T}) \times \pi_1(\mathbb{T}) = \mathbb{Z}^2$ once again. ◇

The two linked loops a, b in the torus can in a sense be seen as a topological abelianization of the 2-bouquet. Such a type of abelianization can be obtained in much greater generality. Namely, we can attach finitely generated abelian groups to simplicial complexes having an orientation (to be explained shortly). These are the homology groups of the space, and they are also topological invariants. If K is such a simplicial complex, then the first homology group $H_1(|K|)$ is actually isomorphic to the quotient group of $\pi_1(|K|)$ by its commutator subgroup, rendering $H_1(|K|)$ an abelianization of $\pi_1(|K|)$.

An *orientation* of K means that for every simplex there is a preferred order of its vertices. We identify simplexes that differ by an even permutation of the vertices, and denote the equivalence class of the n-simplex $\langle p_0 p_1 \cdots p_n \rangle$ by $(p_0 p_1 \cdots p_n)$. We consider the free abelian group $C_n(K)$ generated by the oriented n-simplexes, and regard $-(p_0 p_1 \cdots p_n)$ as the class (with opposite orientation) obtained from $\langle p_0 p_1 \cdots p_n \rangle$ by an odd permutation of its vertices. We call $C_n(K)$ the group of n-chains of K, and define a $\mathbb{Z}$-linear boundary operator $\partial_n : C_n(K) \to C_{n-1}(K)$ on such chains by $\partial_n((p_0 p_1 \cdots p_n)) = \sum_{i=0}^{n}(-1)^i (p_0 p_1 \cdots \hat{p}_i \cdots p_n)$, where p_i is understood to be omitted. For instance, we can think of $(p_0 p_1)$ as a directed line segment traversing from p_0 to p_1, and its boundary is $\partial_1((p_0 p_1)) = p_1 - p_0$. Simplexes for which the boundary operator vanishes are called *cycles*. A hollow triangle oriented counterclockwise is therefore a cycle. It is easy to check that $\partial_n \partial_{n+1} = 0$, which allows us to define the *homology group* of K as the quotient group $H_n(K) = \ker \partial_n / \operatorname{im} \partial_{n+1}$. Different triangulations of the same space X yield isomorphic homology groups, so we can speak of the homology group of X obtaining this way a topological invariant that is a finitely generated abelian group, but no longer necessarily free. We have a family of such groups indexed by $n \in \{0, 1, 2, \ldots\}$. Note that $H_n(X) = 0$ when n is larger than the dimension of the space X; the dimension then being the dimension of the largest simplex in K.

Example 4.35.4 Let $K = \{p_0, p_1, (p_0 p_1)\}$. Clearly $H_1(K) = \ker \partial_1 = 0$ as $\partial_1(m(p_0 p_1)) = mp_1 - mp_0 = 0$ forces the integer m to be zero. Now $H_n(K) = 0$ for $n > 1$, while for $n = 0$, we get the quotient group

$$\{mp_0 + m'p_1 \mid m, m' \in \mathbb{Z}\}/\{m''(p_1 - p_0) \mid m'' \in \mathbb{Z}\},$$

which is isomorphic to $\mathbb{Z}$, signifying that $|K|$ is arcwise connected. ◇

Let us consider a triangulation of the circle, and again doggedly compute the homology groups.

Example 4.35.5 Let $K = \{p_0, p_1, p_2, (p_0 p_1), (p_1 p_2), (p_2 p_0)\}$. We obviously get $H_1(K) = \ker \partial_1 = \{m((p_0 p_1) + (p_1 p_2) + (p_2 p_0)) \mid m \in \mathbb{Z}\} = \mathbb{Z}$, while $H_0(K)$ is the quotient group of $\{mp_0 + m'p_1 + m''p_2 \mid m, m', m'' \in \mathbb{Z}\}$ by the subgroup

$$\{\partial_1(m(p_0 p_1) + m'(p_1 p_2) + m''(p_2 p_0)) \mid m, m', m'' \in \mathbb{Z}\}$$

$$= \{(m'' - m)p_0 + (m - m')p_1 + (m' - m'')p_2 \mid m, m', m'' \in \mathbb{Z}\}.$$

Hence $H_0(K) = \mathbb{Z}$. The higher homology groups are all trivial. It is worth checking that the same result is gotten by considering a triangulation giving a polyhedron which is a square. ◇

Direct computations show that the homology groups of the disc are all trivial, except the first one that is isomorphic to $\mathbb{Z}$, while for S^2 we get the same, except that also the second homology group is $\mathbb{Z}$. Using more intuition, one can convince oneself that $H_1(\Sigma_g)$ is generated by the loops that are not boundaries of some area, giving $H_1(\Sigma_g) = \mathbb{Z}^{2g}$. As there are no 3-simplexes in a triangulation of Σ_g, one can say that the surface Σ_g freely generates the second homology group, so $H_2(\Sigma_g) = \mathbb{Z}$. And as Σ_g is arcwise connected, we get $H_0(\Sigma_g) = \mathbb{Z}$, whereas all the remaining homology groups are trivial.

One can also define higher homotopy groups $\pi_n(X)$ for any $n \in \mathbb{N}$ by replacing loops by n-loops $[0, 1]^n \to X$, and considering homotopy classes of such loops. These topological invariants are related to the homology groups $H_n(X)$ in a less obvious fashion. In fact, they are already abelian when $n > 1$, and we just mention here that $\pi_n(S^n) = \mathbb{Z}$.

In general one could say that homotopy groups contain more information than homology groups, but they are harder to compute, as they tend to render futile some of the effective machinery in homological algebra. Much of algebraic topology is conveniently formulated using the language of category theory [19], but we won't discuss this here.

Instead we will look at a nice family of groups appearing in knot theory [1]. Consider the plane $\mathbb{R}^2$ with n distinguished points, and take two horizontal copies

4.35 Groups in Algebraic Topology

in $\mathbb{R}^3$ of such a plane together with n non-intersecting smooth paths joining the distinguished points of the two planes in pairs. We require the tangent vectors never to be horizontal, and identify any two such stings if they can be deformed into each other without causing intersections. The *braid group* B_n on n strands is this set of equivalence classes with product given by concatenation, deleting the middle plane. Then the unit is the braid with only vertical strings, and the inverse of a 2-braid is its mirror image. Let $a_i \in B_n$ have as representative the string connecting the ith point in the upper plane to the $(i+1)$th point in the lower plane by diagrammatically crossing over the string that connects the $(i+1)$th point in the upper plane to the ith point in the lower plane, and that otherwise has only vertical strings. Then one checks that the following braid relations hold:

$$a_i a_{i+1} a_i = a_{i+1} a_i a_{i+1} \quad \text{and} \quad a_i a_j = a_j a_i \quad \text{when } |i - j| > 1.$$

A link is obtained from a braid by connecting each point at the top of the braid with the endpoint directly below it. A link is then a finite union of knots. Assigning to a braid in B_n the permutation of its endpoints defines a homomorphism from B_n to the symmetric group S_n. Its kernel is known as the *pure braid group*. Note that while S_n is finite, the braid group B_n, being identified with the group having generators a_i satisfying the braid relations, is infinite. Already $B_1 = \mathbb{Z}$ while $|S_1| = 1$. In general, we can identify the symmetric group S_n with the group given by the same generators and relations as B_n, and by adding the relations $a_i^2 = e$, which renders S_n finite.

Knot invariants can be produced from quantum groups since braid groups occur in the representation theory of the latter [21]. This profound connection between quantum groups and knot invariants is beyond the scope of this book, see [11]. We only mention that the relevant quantum groups are quantizations of simple Lie groups in the same way as quantum mechanics is a quantization of classical mechanics [2]. In this strange world one can still talk about homology, or rather cyclic cohomology, which in some sense are groups dual to homology groups, while homology groups are better seen as certain K-groups. The pairing between these two types of groups is known as the non-commutative Chern pairing, which plays a crucial role in index theory [3].

Chapter 5
Representations of Finite Groups

Groups, given their meager structure as abstract objects, are best studied in action. Of course groups act on themselves by conjugation, and we have seen that the study of such actions involves knowing the subgroups of the group. This is quite a task. For instance, by Caley's theorem every group of order less than or equal to n is a subgroup of S_n, which shows that symmetric groups have awfully many subgroups. Linear spaces and linear transformations are structures we understand well. In representation theory a strong link exists between such structures and groups because one restricts to actions on vector spaces, thus making the whole business more manageable.

A representation of a group G on a vector space V is a map π which defines to each group element $a \in G$ a linear transformation $\pi(a) \colon V \to V$ in such a way that the group product becomes a composition of maps and where the unit element corresponds to the identity map. So group elements are represented as matrices when V is finite dimensional, and a linear basis is chosen. Some groups are already defined as matrix groups, say in $M(n, F)$. They then act on F^n, which says that the identity map is a representation. Note that the vector space F^n is considerably smaller than the vector space $M(n, F)$, which the groups also act on (by left multiplication, or by conjugation). In the first case the dimension of the representation is n, whereas in the second case it is n^2.

There is an obvious notion of an intertwiner between two representations, saying when they are to be thought of as the same, or equivalent. Finding a rich class of pairwise inequivalent representations might not be so obvious, especially when the groups are not from the outset matrix groups. One-dimensional representations can occur for large groups, notably in the abelian case. Say that we have two groups, and one of them has a one-dimensional non-trivial representation, while the other one does not. We can then immediately conclude that the two groups cannot be isomorphic. Properties of a group can be deduced by studying its representations, which, as we have said, involves techniques from linear algebra.

© The Author(s), under exclusive license to Springer Nature Switzerland AG 2025
L. Tuset, *Abstract Algebra via Numbers*,
https://doi.org/10.1007/978-3-031-74623-9_5

On the other hand, since many phenomena in nature relate to symmetries and linear structures, representation theory has many applications in natural sciences, especially in quantum physics, not least in particle physics. So also from this point of view it is important to understand representation theory well [8]. However, a full bodied theory requires a lot of geometry and analysis, especially operator algebra theory, which is linear algebra taken to infinite dimensions using techniques from functional analysis. Groups can also be recast into the framework of operator algebras as generalizations of Hopf algebras. One talks then of quantum groups, which are much more general than groups [22]. All this is beyond the scope of this book. Here we restrict our study to finite groups and finite dimensional representations, and we work mainly with complex vector spaces. Although it should be said that with minor changes many results in this chapter still hold for compact groups.

We show how new representations can be constructed from old ones. In particular, we define direct sums and tensor products of representations, and we also define the contragredient representation. The finite dimensional representations of a finite group form what category theorists refers to as a tensor category. One can actually recover the group entirely from such an abstract category. This shows that the representations of a group encodes all the information about the group.

A major goal in representation theory is to decompose a representation into irreducible (or indecomposable) ones, and to classify these simpler objects. This requires that an assembly of matrices can be decomposed, sometimes even diagonalized, simultaneously, and is achieved in the complex case by a clever averaging procedure using inner products and the Haar integral, which in this case involves a finite sum over all elements in the group.

We then discuss the (left) regular representation of the group. This is defined using left multiplication of the group on itself. It is an important source for producing representations. In our context it contains all the irreducible ones, and they occur in the decomposition with a multiplicity governed by their dimensions as representations. The decomposition can also be studied by invoking the linear maps that intertwine a representation with itself. They are simply the matrices that commute with every representation matrix. These intertwiners form an algebra. Schur's lemma tells us that in the irreducible case, this algebra is just the complex numbers, which cannot be decomposed any further as an algebra. In general, it will be a direct sum of full matrix algebras. Taking the matrix elements of the irreducible representations with respect to a certain orthonormal basis on the representation space, yields an orthonormal basis with respect to the Haar integral inner product on the space of so called regular functions on the groups. In this finite dimensional setting the regular functions on the group happens to be all the complex valued functions on the group. This complete orthonormal decomposition is often referred to as Peter-Weyl theory.

Another important algebraic object is the group algebra of a group. Think of it as a linearization of the group with the group elements forming a linear basis for the vector space, whereas the product in the algebra is a linear extension of the product in the group. Then any representation of the group extends uniquely by linearity to a representation, or an algebra homomorphism, of the algebra into the algebra of endomorphisms of the representation space. This gives a one-to-one correspondence between representations of the group and ideals in the group algebra, with ideals appearing as kernels of the extended representations. This way the representation spaces become modules over the group ring (two notions that we will return to in great detail later) and can be studied as such.

We consider one-dimensional representations, so called characters of the group. Thanks to Schur's lemma, they are the irreducible representations of abelian groups. There are plenty of them in this case, equally many as there are group elements. They capture the whole structure of the finite abelian group, in that they themselves form an abelian group, and by considering the characters of this group, one recovers the original group up to a canonical isomorphism. This is known as Pontryagin duality. We carry classical Fourier analysis to finite abelian groups by considering group characters. Applying this powerful machinery to cyclic groups, we prove some number theoretical results, including another proof of quadratic reciprocity.

Taking the trace of the matrices in a representation gives us a scalar valued function on the group, known as the character of the finite dimensional representation. They are the closest we get to characters of the group, and enjoy similar nice properties. Knowing the characters of the representations tells us almost everything about the representations themselves. We study in detail the characters of irreducible representations of the symmetric group S_3.

In the final sections we study the relation between representations of a group and those of its subgroups. We study how the intertwiners relate by establishing two results known as Frobenius reciprocity. We provide a method of inducing up representations from subgroups of a group to the whole group, and we investigate how the characters of the representations come into play. Again we study how intertwiners relate by proving a geometric version of Mackey's theorem. Then we establish an algebraic version of the theorem involving double cosets.

5.1 Basic Definitions

Definition 5.1.1 A *representation* π of a group G on a vector space V over a field F is a homomorphism $\pi : G \to \text{Aut}(V)$. Or in other words, the vector space V is a G-space under linear maps $x \mapsto ax \equiv \pi(a)x$. When we talk of vector spaces as *G-spaces* we always mean actions by linear maps. A representation is finite dimensional if the associated vector space is finite dimensional. If the representation is a monomorphism, it is called a *faithful representation*. An *intertwiner* between two representations is a linear G-morphism between their G-spaces. Two representations are *equivalent representations* if there is a bijective intertwiner between them. A non-zero finite dimensional representation is *irreducible* if it has no proper invariant subspaces. The restriction of a representation to an invariant subspace is called a *subrepresentation*.

To be explicit, given two representations $\pi\colon G \to \mathrm{Aut}(V)$ and $\rho\colon G \to \mathrm{Aut}(W)$, then an intertwiner from π to ρ is a linear map $A\colon V \to W$ such that

$$A\pi(a) = \rho(a)A$$

for $a \in G$.

Definition 5.1.2 We write $\mathrm{Mor}(\pi, \rho)$ for the vector space of intertwiners from a representation π to another ρ, and $\pi \cong \rho$ means that π and ρ are equivalent.

If $A \in \mathrm{Mor}(\pi, \rho)$ and $B \in \mathrm{Mor}(\rho, \theta)$, then $AB \in \mathrm{Mor}(\pi, \theta)$, so $\mathrm{Mor}(\pi, \pi)$ is a unital algebra. If in addition A is bijective, then $A^{-1} \in \mathrm{Mor}(\rho, \pi)$, so the relation of being equivalent is an equivalence relation.

An irreducible representation has no subrepresentations except itself and the zero-representation. We have reserved the adjective 'irreducible' only to finite dimensional representations.

The *trivial representation* $\varepsilon\colon G \to \mathrm{Aut}(F)$ is given by $\varepsilon(a)r = r$ for $a \in G$ and $r \in F$.

5.2 Regular Functions

Definition 5.2.1 A *matrix representation* is a homomorphism $G \to GL(n, F)$.

We have seen several examples of matrix representations coming from subgroups of $GL(n, F)$. Any finite dimensional representation $\pi\colon G \to \mathrm{Aut}(V)$ is equivalent to a matrix representation. Indeed, pick a basis $\{v_i\}$ for V to obtain a linear isomorphism $A\colon V \to F^n$. Then the matrix representation $\rho\colon G \to GL(n, F)$ given by $\rho(a) = A\pi(a)A^{-1}$ is equivalent to π with intertwiner $A \in \mathrm{Mor}(\pi, \rho)$. The functions $\pi_{ij}\colon G \to F$ given by $\pi_{ij}(a) = \rho(a)_{ij}$ are called *matrix coefficients* of π. Obviously

$$\pi_{ij}(ab) = \sum_k \pi_{ik}(a)\pi_{kj}(b) \quad \text{and} \quad \pi_{ij}(e) = \delta_{ij}$$

for $a, b \in G$. We also have

$$\pi(a)v_j = \sum_i \pi_{ij}(a)v_i,$$

so $\pi_{ij}(a) = x_i(\pi(a)v_j)$, where $\{x_i\}$ is a dual basis of $\{v_i\}$.

Definition 5.2.2 A *regular function* of a group G with values in a field F is a function of the form $a \mapsto x(\pi(a)v)$ for a finite dimensional representation $\pi\colon G \to \mathrm{Aut}(V)$ and elements $x \in V^*$ and $v \in V$. Let $F(G) \subset F^G$ denote the set of regular functions on a group G.

5.3 New Representations from Old Ones

Definition 5.3.1 The *direct sum of representations* $\pi_i \colon G \to \mathrm{Aut}(V_i)$ is the representation $\oplus \pi_i \colon G \to \mathrm{Aut}(\oplus_i V)$ given by

$$\oplus \pi_i(a) v = \pi_i(a) v$$

for $a \in G$ and $v \in V_i$.

Each π_j is a subrepresentation of $\oplus \pi_i$ as V_j is an invariant subspace of $\oplus V_i$. To decompose a representation $\pi \colon G \to \mathrm{Aut}(V)$ into subrepresentations π_i is the same thing as decomposing V into invariant subspaces V_i, meaning that $V = \oplus V_i$ and $\pi(G) V_i \subset V_i$. Letting π_i be the restriction of π to V_i we obviously get $\pi = \oplus \pi_i$. We often write the direct sum of n equivalent representations π as $n\pi$.

Definition 5.3.2 The *tensor product of representations* $\pi \colon G \to \mathrm{Aut}(V)$ and $\rho \colon G \to \mathrm{Aut}(W)$ is the representation $\pi \otimes \rho \colon G \to \mathrm{Aut}(V \otimes W)$ given by

$$(\pi \otimes \rho)(a)(v \otimes w) = \pi(a) v \otimes \rho(a) w$$

for $a \in G$ and $v \in V$ and $w \in W$.

Proposition 5.3.3 *The set $F(G)$ of regular functions on a group G is a unital subalgebra of F^G.*

Proof Obviously every regular function is a matrix coefficient of some representation. The sum and product of two matrix coefficients is a matrix coefficient of the direct sum and the tensor product, respectively, of the corresponding representations. The matrix coefficient of the trivial representation ε is the identity. □

Definition 5.3.4 Given a representations $\pi \colon G \to \mathrm{Aut}(V)$, then its *contragredient representation* is the representation $\pi^c \colon G \to \mathrm{Aut}(V^*)$ given by

$$\pi^c(a) x = x \pi(a^{-1})$$

for $a \in G$ and $x \in V^*$.

If $\pi \colon G \to \mathrm{Aut}(V)$ is a finite dimensional representation on V with basis $\{v_i\}$ and dual basis $\{x_j\}$, then

$$\pi^c_{ij}(a) = v_i(\pi^c(a) x_j) = x_j(\pi(a^{-1}) v_i) = \pi_{ji}(a^{-1}),$$

so $\pi^c(a) = \pi(a^{-1})^T$ by abuse of language, and the function $a \mapsto \pi_{ji}(a^{-1})$ belongs to $F(G)$.

Note that $\pi^{cc} \cong \pi$, and that π is irreducible if and only if π^c is irreducible.

5.4 Decomposition Into Irreducibles

Here we introduce the powerful technique of decomposing a representation into subrepresentations on orthogonal subspaces, where orthogonality is defined with respect to a cleverly chosen scalar product.

Lemma 5.4.1 *Given a representation $\pi : G \to \mathrm{Aut}(V)$ of a finite group on a finite dimensional complex vector space with inner product $(\cdot|\cdot)$. Then*

$$\langle u|v \rangle = \sum_{a \in G} (\pi(a)u|\pi(a)v)$$

for $u, v \in V$ defines an inner product such that $\langle \pi(b)u|v \rangle = \langle u|\pi(b^{-1})v \rangle$.

Proof If $\langle v|v \rangle = 0$, then as $(\pi(a)v|\pi(a)v) \geq 0$, they must all be zero, which for $a = e$ means $(v|v) = 0$, so $v = 0$.

Now

$$\langle \pi(b)u|\pi(b)v \rangle = \sum_{a \in G}(\pi(ba)u|\pi(ba)v) = \sum_{c \in G}(\pi(c)u|\pi(c)v) = \langle u|v \rangle$$

for any $b \in G$ and $u, v \in V$. In the second step we replaced ba by c and used that $a \mapsto ba$ is a bijection in order to replace summation over a by summation over c. Substituting v by $\pi(b^{-1})v$ gives the desired result. □

Definition 5.4.2 An inner product on a finite dimensional complex vector space of a representation is an *invariant inner product* if the representation acts by unitary linear maps.

So the inner product $\langle \cdot|\cdot \rangle$ in the lemma above is invariant.

Theorem 5.4.3 *Any non-zero representation of a finite group on a finite dimensional complex vector space can be decomposed into irreducible representations.*

Proof Say we have a non-zero representation $\pi : G \to \mathrm{Aut}(V)$ of a finite group on a finite dimensional complex vector space. Pick any inner product $(\cdot|\cdot)$ on V, and consider the associated invariant inner product $\langle \cdot|\cdot \rangle$.

The theorem obviously holds when $\dim V = 1$. Assume that it holds for all complex vector spaces of dimension less than $\dim V$. By induction it suffices to show that it holds for V.

Say V has an invariant proper subspace W. Then

$$W^{\perp} = \{u \in V \mid \langle u|W \rangle = \{0\}\}$$

is also an invariant proper subspace of V because if $u \in W^{\perp}$, then

$$\langle \pi(b)u|v \rangle = \langle u|\pi(b^{-1})v \rangle = 0$$

for $v \in W$ as $\pi(b^{-1})v \in W$. Now we are done because $V = W \oplus W^\perp$, and by assumption both W and $W^\perp$ can be decomposed into invariant subspaces having no invariant proper subspaces. □

Definition 5.4.4 The number of copies, i.e. equivalent representations, of an irreducible representation π_i in the decomposition of a representation π is called the *multiplicity* of π_i in π.

5.5 Haar Integral

The invariant inner product in the previous section was obtained by averaging, or summing, the function $f: G \to \mathbb{C}$ given by $f(a) = (\pi(a)u|\pi(a)v)$ over the group. This is only possible for finite groups. We used that the average of a positive function on the group is positive, and that $f \in \mathbb{C}^G$ and the function f_b given by $f_b(a) = f(b^{-1}a)$ have the same average.

Definition 5.5.1 Consider a finite group G and a field F of characteristic zero. The linear map $\varphi: F^G \to F$ given by

$$\varphi(f) = \frac{1}{|G|} \sum_{a \in G} f(a)$$

is called the *Haar integral* of the group G.

Define $\delta_a \in F^G$ by $\delta_a(b) = 1$ if $a = b$ and $\delta_a(b) = 0$ if $a \neq b$. Then $\{\delta_a\}$ is a linear basis for F^G, so $\dim(F^G) = |G|$, and $f = \sum f(a)\delta_a$ for $f \in F^G$.

Proposition 5.5.2 *Let G be a finite group. Up to a scalar factor there is only one linear map $\psi: F^G \to F$ such that $\psi(f_b) = \psi(f)$ for $f \in F^G$ and $b \in G$.*

Proof We have

$$\psi(f) = \sum f(a)\psi(\delta_a) = \psi(\delta_e) \sum f(a)$$

because $\delta_a = (\delta_e)_a$. So $\psi = |G|\psi(\delta_e)\varphi$. □

Remark 5.5.3 To decompose representations we need Haar integrals. They exists in great generality, and are vital in representation theory. In the language of measure theory the Haar integral on a finite group is the integral of the counting measure, and for $\mathbb{R}$ it is the Lebesgue measure. Haar integrals exists even for locally compact groups, of which the subclass of compact groups are topological generalizations of finite groups.

5.6 Regular Representation

Definition 5.6.1 The (left) *regular representation* of a group G is the representation $\lambda \colon G \to \mathrm{Aut}(F^G)$ given by $\lambda(a)f = f_a$ for $f \in F^G$ and $a \in G$.

Since $\lambda(a)\delta_e = \delta_a$ for $a \in G$, the regular representation is faithful. So finite groups have a faithful finite dimensional representation. We can regard them as subgroups of the general linear group $GL(n, F)$, where n is the order of the group.

Proposition 5.6.2 *The algebra $F(G)$ of regular functions on a finite group G coincides with the algebra of all F-valued functions on G.*

Proof By Proposition 5.3.3, we only need to verify that any function on G is regular. Consider the regular representation λ, and let x be the linear functional on F^G given by $x(f) = f(e)$, where e is the unit of G. Then for $f \in F^G$, we have $f(a) = f(\lambda^c(a)x)$, so f is regular. □

We could have avoided the contragredient representation in the proof above by introducing the *right regular representation* $\rho \colon G \to \mathrm{Aut}(F^G)$; $\rho(a)f = f^a$, where $f^a(b) = f(ba)$, because then $f(a) = x(\rho(a)f)$.

Note that for a finite group G, we have $\varphi\lambda(a) = \varphi = \varphi\rho(a)$ for all $a \in G$.

Definition 5.6.3 For a finite group G, the *standard inner product* on $\mathbb{C}^G$ is given by

$$(f|g) = \frac{1}{|G|} \sum_{a \in G} f(a)\overline{g(a)}.$$

This inner product is invariant for the regular representation because $(f|g) = \varphi(h)$, with $h \in \mathbb{C}^G$ given by $h(a) = f(a)\overline{g(a)}$, so $(\lambda(b)f|\lambda(b)g) = \varphi(\lambda(b)h) = \varphi(h)$.

5.7 Schur's Lemma

The following result is known as *Schur's lemma*, and is used again and again in representation theory.

Theorem 5.7.1 *Given representations π and ρ of a group on a vector space over a field F, and consider $A \in \mathrm{Mor}(\pi, \rho)$. Then $\ker A$ and $\mathrm{im}\, A$ are invariant subspaces for π and ρ, respectively. Hence if both representations are irreducible, there are only two possibilities: Either $A = 0$ and $\mathrm{Mor}(\pi, \rho) = \{0\}$. Or A is bijective and $\pi \cong \rho$. If in addition the vector space is finite dimensional and complex, then $\mathrm{Mor}(\pi, \rho) = \mathbb{C}A$.*

Proof If $v \in \ker A$, then $A\pi(a)v = \rho(a)Av = 0$, so $\ker A$ is invariant for π. If $w = Av$ for some vector v, then $\rho(a)w = A\pi(a)v \in \mathrm{im}\, A$, so $\mathrm{im}\, A$ is invariant for ρ.

Suppose $\pi \cong \rho$ are irreducible and that the vector space is finite dimensional and complex. If $B \in \text{Mor}(\pi, \rho)$ is bijective, let λ be an eigenvalue for $A^{-1}B$, which does exist in $\mathbb{C}$ by the fundamental theorem of algebra. Then $A^{-1}B - \lambda I \in \text{Mor}(\pi, \pi)$ and because of the eigenvector, the morphism is not bijective, so $B = \lambda A$. □

Remark 5.7.2 The last assertion of this theorem is valid for finite dimensional vector spaces over algebraically closed fields.

For a complex field we can characterize irreducible representations among finite dimensional representations by their intertwiners.

Corollary 5.7.3 *Suppose π is a non-zero representation of a finite group on a finite dimensional complex vector space. Then π is irreducible if and only if* $\text{Mor}(\pi, \pi) = \mathbb{C}I$.

Proof The forward implication is part of Schur's lemma.

Conversely, say π acts on a finite dimensional complex vector space V with an invariant proper subspace W. Then $V = W \oplus W^\perp$ with respect to an invariant inner product for π. Define a linear map $P \colon V \to V$, the *orthogonal projection* onto W, to be the identity on W and zero on $W^\perp$. Then $P \notin \mathbb{C}I$, and $P \in \text{Mor}(\pi, \pi)$ because firstly
$$P\pi(a)w = \pi(a)w = \pi(a)Pw$$
for $w \in W$, as $\pi(a)w \in W$, and secondly
$$P\pi(a)v = 0 = \pi(a)Pv$$
for $v \in W^\perp$, as $\pi(a)v \in W^\perp$. □

5.8 Characters of Abelian Groups

Definition 5.8.1 An F-valued *character of a group G* is a homomorphism $G \to F_*$.

When $F = \mathbb{C}$, then by Lemma 5.4.1, we see that $\overline{\chi(a)} = \chi(a^{-1})$ for every character χ of a finite group G and $a \in G$. So in this case $\chi \colon G \to \mathbb{T}$.

When the field F is understood we talk about a character.

Upon identifying $\text{Aut}(F)$ with F_*, we see that every character is a 1-dimensional representation, and that every such representation is a character. Two 1-dimensional representations are equivalent if and only if their characters are equal.

The characters of a group G form an abelian group under pointwise operations, i.e. if χ and η are characters of G, then their product $\chi\eta$ is the character given by $(\chi\eta)(a) = \chi(a)\eta(a)$ for $a \in G$. The unit of this group is ε.

Definition 5.8.2 Let $\hat{G}$ denote the *dual group* of $\mathbb{C}$-valued characters of an abelian group G.

Characters come from irreducible representations as all 1-dimensional non-zero representations are irreducible. But the converse is not true in general. Every non-abelian finite group have higher dimensional irreducible representations. To see this, pick any faithful representation π of the group G on a finite dimensional complex vector space, and decompose it into irreducibles. If all these were characters, then $\pi(ab) = \pi(a)\pi(b) = \pi(ba)$, so $ab = ba$ for all $a, b \in G$.

Proposition 5.8.3 *Every irreducible representation of an abelian group on a complex vector space is 1-dimensional.*

Proof Say π is an irreducible representation of an abelian group G on a complex vector space V. Then $\pi(a) \in \mathrm{Mor}(\pi, \pi)$ as $\pi(a)\pi(b) = \pi(b)\pi(a)$ for all $b \in G$. By Schur's lemma, we conclude that $\pi(a) \in \mathbb{C}I$ for every $a \in G$. Thus all subspaces of V are invariant, and since π is irreducible, we must have $\dim V = 1$. □

We can obviously form the double dual $\hat{\hat{G}}$, or the *bidual of an abelian group* G. The following result is a special case of *Pontryagin's duality theorem*.

Theorem 5.8.4 *Let G be a finite abelian group. Then the map $P: G \to \hat{\hat{G}}$ given by $P(a)(\chi) = \chi(a)$ for $\chi \in \hat{G}$ and $a \in G$, is a group isomorphism.*

Proof Obviously $a \mapsto \hat{a}$ is a homomorphism. As for injectivity, say $\chi(a) = P(a)(\chi) = 1$ for all $\chi \in \hat{G}$. By Theorem 5.4.3 and Propositions 5.8.3 and 5.6.2, the characters span $\mathbb{C}^G$, so $f(a) = f(e)$ for $f \in \mathbb{C}^G$, and $a = e$.

To see that P is surjective, it is enough to show that $|\hat{G}| \leq |G|$ because any injective map from G to a set $\hat{\hat{G}}$ with no greater cardinality must be surjective. As $|G| = \dim \mathbb{C}^G$, it therefore suffices to show that $\hat{G}$ is linear independent in $\mathbb{C}^G$.

Consider the standard inner product $(\cdot|\cdot)$ on $\mathbb{C}^G$. Clearly $(\chi|\chi) = 1$ for $\chi \in \hat{G}$. Suppose $\chi, \eta \in \hat{G}$ are distinct. For $b \in G$, we have

$$\chi(b)(\chi|\eta) = \frac{1}{|G|}\sum_a \chi(ba)\eta(a^{-1}) = \frac{1}{|G|}\sum_c \chi(c)\eta(c^{-1}b) = (\chi|\eta)\eta(b),$$

and as $\chi(b) \neq \eta(b)$ for at least one $b \in G$, we see that $(\chi|\eta) = 0$. So the characters are linear independent; they are in fact orthonormal with respect to $(\cdot|\cdot)$. □

Example 5.8.5 For infinite abelian groups it is no longer the case that the dual group has cardinality not greater than that of the original group. For instance, observe that $\chi(n) = z^n$ is a character of $\mathbb{Z}$ for every $z \in \mathbb{C}_*$, so $\mathbb{C}_*$ is a subgroup of $\hat{\mathbb{Z}}$ and $|\mathbb{C}_*| > |\mathbb{Z}|$. ◇

5.9 Fourier Analysis

In the proof of Pontryagin's duality theorem we showed the following result.

Corollary 5.9.1 *The $\mathbb{C}$-valued characters of a finite abelian group G form an orthonormal basis for $\mathbb{C}^G$ with respect to the standard inner product.*

This allows us to expand any $f \in \mathbb{C}^G$ for a finite abelian group G, as a finite sum

$$f = \sum_{\chi \in \hat{G}} c_\chi \chi,$$

where $c_\chi = (f|\chi)$ are the *Fourier coefficients* of f and $(\cdot|\cdot)$ is the standard inner product on $\mathbb{C}^G$.

Definition 5.9.2 The *Fourier transform* of $f \in \mathbb{C}^G$ for a finite abelian group G is the function $\hat{f} \in \mathbb{C}^{\hat{G}}$ on the Pontryagin dual $\hat{G}$ given by

$$\hat{f}(\chi) = (f|\chi) = \frac{1}{|G|} \sum_a f(a)\overline{\chi(a)}.$$

We immediately obtain *Plancherel's formula*

$$|G|(\hat{f}|\hat{f}) = \sum_\chi |c_\chi|^2 = (f|f)$$

and the *Fourier inversion formula* $f = \sum_\chi \hat{f}(\chi)\chi$. It is also easy to see that

$$\hat{\hat{f}}(P(a)) = \frac{1}{|G|} f(-a)$$

for all $a \in G$. A straighforward application of the Cauchy-Schwarz inequality $|(f|g)|^2 \leq (f|f)(g|g)$ for any $f, g \in \mathbb{C}^G$ shows that the *uncertainty principle* in Fourier analysis holds:

$$|\text{supp}(f)| \cdot |\text{supp}(\hat{f})| \geq |G|,$$

where the support of a function is the set where it is non-zero.

Example 5.9.3 Take the finite abelian group $\mathbb{Z}_n$, and let $w = \exp(\frac{2\pi i}{n}) \in \mathbb{T}$. Define a character ψ_k on $\mathbb{Z}_n$ for $k \in \mathbb{N}$ by

$$\psi_k([m]) = w^{km}.$$

It is well-defined, and $[k] \mapsto \psi_k$ is a well-defined isomorphism from $\mathbb{Z}_n$ to $\hat{\mathbb{Z}}_n$. This is clear since the map is injective, and $|\mathbb{Z}_n| = |\hat{\mathbb{Z}}_n|$ by Pontryagin's duality theorem.

We can now expand $f \in \mathbb{C}^{\mathbb{Z}_n}$ as $f = \sum_{k=1}^n c_{\psi_k} \psi_k$, where

$$c_{\psi_k} = (f|\psi_k) = \frac{1}{n} \sum_{m=1}^n f([m])\overline{\psi_k([m])} = \frac{1}{n} \sum_m f([m]) w^{-km}.$$

Hence

$$f([r]) = \frac{1}{n} \sum_{k,m} f([m]) w^{(r-m)k}$$

for $r \in \mathbb{N}$. For $f = \delta_{[s]}$ with $s \in \mathbb{N}$, we get the identity $\delta_{rs} = \frac{1}{n} \sum_{k=1}^n w^{(r-s)k}$. $\diamond$

Let H be a subgroup of a finite abelian group G, and let $(\mathbb{C}^G)^H$ denote the vector subspace of $\mathbb{C}^G$ consisting of those functions that are constant on each equivalence class in G/H. Let $\hat{G}^H = \hat{G} \cap (\mathbb{C}^G)^H$ denote the characters of G that are one on all elements of $H \subset G$. We then have a well-defined map $q \colon \mathbb{C}^{G/H} \to (\mathbb{C}^G)^H$ given by $q(f)(a) = f(a + H)$ for $a \in G$, which is clearly a vector space isomorphism, and which obviously restricts to a group isomorphism $\widehat{G/H} \to \hat{G}^H$.

Moreover, we have the following *Poisson summation formula*.

Proposition 5.9.4 *Let notation be as in the previous paragraph. Then*

$$\sum_{b \in H} f(b) = |H| \sum_{\chi \in \hat{G}^H} \hat{f}(\chi)$$

for $f \in \mathbb{C}^G$.

Proof Define $g \in \mathbb{C}^{G/H}$ by $g(a+H) = \sum_{b \in H} f(a+b)$ and $\eta \in \widehat{G/H}$ by $\eta(a+H) = \chi(a)$ for $a \in G$ and $f \in \mathbb{C}^G$ and $\chi \in \hat{G}^H$, so $q(\eta) = \chi$. Then

$$\hat{g}(\eta) = \frac{1}{|G/H|} \sum_{a+H \in G/H} g(a+H)\overline{\eta(a+H)} = \frac{|H|}{|G|} \sum_{a+H \in G/H} \sum_{b \in H} f(a+b)\overline{\chi(a)}$$

$$= \frac{|H|}{|G|} \sum_{a+H \in G/H} \sum_{b \in H} f(a+b)\overline{\chi(a+b)} = \frac{|H|}{|G|} \sum_{a \in G} f(a)\overline{\chi(a)} = |H|\hat{f}(\chi),$$

so by the Fourier inversion formula $g = \sum_{\eta \in \widehat{G/H}} \hat{g}(\eta)\eta$, we get

$$\sum_{b \in H} f(a+b) = g(a+H) = |H| \sum_{\chi \in \hat{G}^H} \hat{f}(\chi)\chi(a)$$

and the result is obtained by setting $a = 0$. $\square$

5.10 Orthogonality Relations

We can generalize Corollary 5.9.1 to all finite groups.

Consider representations π and ρ of a finite group G on finite dimensional complex vector spaces V and W, respectively. Let $(\cdot|\cdot)$ be an inner product on V and let $v_i \in V$ and $w_i \in W$. Define a linear map $A \colon V \to W$ by

$$Av_1 = \sum_{a \in G} (\pi(a)v_1|v_2)\rho(a^{-1})w_1. \tag{5.1}$$

Then $A \in \text{Mor}(\pi, \rho)$ because

$$A\pi(b)v_1 = \sum_a (\pi(ab)v_1|v_2)\rho(a^{-1})w_1 = \sum_c (\pi(c)v_1|v_2)\rho(bc^{-1})w_1 = \rho(b)Av_1.$$

Lemma 5.10.1 *If π and ρ are irreducible and not equivalent, their matrix coefficients are mutually orthogonal.*

Proof If this was not the case, we can pick invariant inner products on V and W and elements v_i and w_i such that

$$0 \neq \sum_a (\pi(a)v_1|v_2)\overline{(\rho(a)w_2|w_1)} = \sum_a (\pi(a)v_1|v_2)(\rho(a^{-1})w_1|w_2) = (Av_1|w_2).$$

By Schur's lemma, the non-zero intertwiner A must be an isomorphism. □

What about matrix coefficients from the same representation?

Lemma 5.10.2 *Suppose $\pi \colon G \to \text{Aut}(V)$ is irreducible with an invariant inner product $(\cdot|\cdot)$ on V. Then there is a positive constant d such that*

$$\sum_a (\pi(a)v_1|v_2)\overline{(\pi(a)w_2|w_1)} = d^{-1}(v_1|w_2)(w_1|v_2)$$

for $v_i, w_i \in V$.

Proof In the identity in the proof of the previous lemma with $\pi = \rho$, the map A is by Schur's lemma, a constant times the identity, so the right-hand-side of the identity equals $(v_1|w_2)r$, where r is a constant that depends on v_2 and w_1. The next to the left-hand-side is unaltered under the substitution $a \mapsto a^{-1}$, which gives

$$(v_1|w_2)r = \sum_a (\pi(a)w_1|w_2)\overline{(\pi(a)v_2|v_1)} = (w_1|v_2)s,$$

where s is a constant that depends on v_1 and w_2, obtained by repeating the argument above. This is only possible if the lemma holds, and d is indeed positive; put $v_1 =$

$w_2 \neq 0$ and $w_1 = v_2 \neq 0$ and observe that everything is positive since vectors are cyclic under the action of G. □

In Sect. 5.12 we will show that $d = (\dim V)/|G|$. The following result is then immediate from the two lemmas above.

Theorem 5.10.3 *Let $\{\pi^i\}$ be a collection of pairwise inequivalent irreducible representations of a finite group G on complex vector spaces V_i. Assume that the collection is complete in the sense that every irreducible representation of G on a complex vector space is equivalent to one of these members. Let π^i_{mn} denote the matrix coefficients of π^i associated to an orthonormal basis of V_i with respect to an invariant inner product. Set $d_i = (\dim V_i)^{1/2}$. Then $\{d_i \pi^i_{mn}\}$ is an orthonormal basis for $\mathbb{C}^G$ with respect to the standard inner product.*

This theorem, known as the Peter-Weyl theorem, implies that the regular representation is the mother of all representations.

Corollary 5.10.4 *Retain the terminology of the theorem, and consider the regular representation λ of G. Then*

$$\lambda \cong \bigoplus (\dim V_i) \pi^i,$$

or in other words, there is a copy in the regular representation of every irreducible representation with multiplicity equalling its dimension. In particular, this means that

$$|G| = \sum (\dim V_i)^2.$$

Proof To be concrete, let $\{v_i\}$ be the orthonormal basis of V_i that the matrix coefficients π^i_{mn} of π^i are defined with respect to. Let $V_{i,n}$ be the subspace of $\mathbb{C}^G$ spanned by π^i_{mn} for all m. Then $\lambda(a)V_{i,n} \subset V_{i,n}$ because

$$\lambda(a)\pi^i_{mn} = \sum_k \pi^i_{mk}(a^{-1})\pi^i_{kn},$$

which is readily verified. Let $\lambda_{i,n}$ denote the representation gotten by restricting $\lambda(a)$ to $V_{i,n}$ for each $a \in G$. By the theorem it is a subrepresentation of λ with multiplicity $\dim V_i$, and these representations exhaust λ.

Moreover, the linear map $A: V_i \to V_{i,n}$ given by $Av_m = \pi^i_{mn}$ is an isomorphism, and $A \in \text{Mor}((\pi^i)^c, \lambda_{i,n})$ because

$$A(\pi^i)^c(a)v_m = \sum_k (\pi^i)^c_{km}(a) Av_k = \sum_k \pi^i_{mk}(a^{-1}) Av_k$$
$$= \sum_k \pi^i_{mk}(a^{-1})\pi^i_{kn} = \lambda(a)\pi^i_{mn} = \lambda_{i,n}(a) Av_m$$

for all m. Thus $(\pi^i)^c \cong \lambda_{i,n}$, so there is a copy of π^i in λ with multiplicity $\dim V_i$. □

5.11 Three Auxiliary Representations

Say π and ρ are finite dimensional representations of a group G on complex vector spaces V and W, respectively.

Definition 5.11.1 Define representations $\pi \times \rho^c \colon G \times G \to \mathrm{Aut}(V \otimes W^*)$ and $\pi\rho^c \colon G \times G \to \mathrm{Aut}(\mathrm{Hom}(W, V))$ by

$$(\pi \times \rho^c)(a, b) = \pi(a) \otimes \rho^c(b) \quad \text{and} \quad (\pi\rho^c)(a, b)A = \pi(a)A\rho(b^{-1})$$

for $a, b \in G$ and $A \in \mathrm{Hom}(W, V)$.

The following result, which shows that $\pi \times \rho^c$ and $\pi\rho^c$ are equivalent representations, is straightforward.

Proposition 5.11.2 *Define a linear isomorphism* $T \colon V \otimes W^* \to \mathrm{Hom}(W, V)$ *by*

$$T(v, x)(w) = x(w)v$$

for $v \in V$ *and* $x \in W^*$ *and* $w \in W$. *Then* $T \in \mathrm{Mor}(\pi \times \rho^c, \pi\rho^c)$.

Let M_π denote the set of matrix coefficients of π. Then M_π is a linear subspace of $\mathbb{C}(G)$ and $\dim M_\pi \leq (\dim V)^2$, with equality if π is irreducible as the matrix coefficients are then orthogonal and non-zero.

Definition 5.11.3 Define $\mathrm{Ad}_\pi \colon G \times G \to \mathrm{Aut}(M_\pi)$ by

$$\mathrm{Ad}_\pi(a, b)f(c) = f(a^{-1}cb)$$

for $f \in M_\pi$ and $a, b, c \in G$.

We are claiming that $\mathrm{Ad}_\pi(a, b)f \in M_\pi$ for $f \in \mathbb{C}^G$ of the form $f(c) = x(\pi(c)v)$ with $x \in V^*$ and $v \in V$ and $c \in G$. This holds as

$$\mathrm{Ad}_\pi(a, b)f(c) = x(\pi(a^{-1}cb)v) = x\pi(a^{-1})(\pi(c)\pi(b)v),$$

which also proves the following result.

Proposition 5.11.4 *The surjective linear map* $S \colon V \otimes V^* \to M_\pi$ *given by*

$$S(v \otimes x) = x(\pi(\cdot)v)$$

for $x \in V^*$ *and* $v \in V$ *belongs to* $\mathrm{Mor}(\pi \times \pi^c, \mathrm{Ad}_\pi)$. *In particular, if* π *is irreducible, then* S *is an isomorphism and* $\pi\pi^c \cong \pi \times \pi^c \cong \mathrm{Ad}_\pi$.

5.12 Characters of Representations

Definition 5.12.1 Let π be a representation of a group G on a finite dimensional complex vector space V. The *character of* π is the element $\chi_\pi \in \mathbb{C}(G)$ given by

$$\chi_\pi = \text{Tr}\,\pi = \sum_i \pi_{ii}.$$

By the basic property of the trace, we see that characters of equivalent representations are identical. Also note that $\chi_\pi(e) = \dim V$, and that χ_π is a character of the group when π is 1-dimensional. In particular, we see that $\chi_\varepsilon = \varepsilon$, where ε is the trivial representation.

Proposition 5.12.2 *Let π and ρ be representations of a finite group G on finite dimensional complex vector spaces V and W, respectively. Then*

$$\chi_{\pi \oplus \rho} = \chi_\pi + \chi_\rho \quad \text{and} \quad \chi_{\pi \otimes \rho} = \chi_\pi \chi_\rho \quad \text{and} \quad \chi_{\pi^c} = \overline{\chi_\pi}.$$

Proof The first two formulas are trivial. As for the last one, note that $\pi(a)$ is unitary with respect to an invariant inner product, so it is diagonalisable with eigenvalues λ_i having absolute value one. Hence

$$\chi_{\pi^c}(a) = \text{Tr}\,\pi^c(a) = \text{Tr}(\pi(a^{-1})^T) = \text{Tr}\,\pi(a^{-1}) = \sum \lambda_i^{-1} = \sum \overline{\lambda_i} = \overline{\text{Tr}\,\pi(a)}.$$

□

Definition 5.12.3 Let π be a representation of a group G on a finite dimensional complex vector space V. The set of *G-invariants* V_G is the subspace of V given by

$$V_G \equiv \{v \in V \mid \pi(a)v = v \text{ for } a \in G\}.$$

Proposition 5.12.4 *Let π be a representation of a finite group G on a finite dimensional complex vector space V. Then*

$$\frac{1}{|G|} \sum_{a \in G} \chi_\pi(a) = \dim V_G.$$

Proof The left-hand-side of the identity in the proposition can be written as $(\chi_\pi|\varepsilon)$ with respect to the standard inner product on $\mathbb{C}(G)$, so by the orthogonality relations, we get 1 if π is trivial and 0 if π is irreducible and non-trivial.

Decompose the G-space V into irreducible G-spaces V_i with characters χ_i. Since $\chi_\pi = \sum \chi_i$, the left-hand-side of the identity in the proposition counts the number of trivial V_i's, and the direct sum of these 1-dimensional subspaces is obviously V_G, which gives the result. □

5.12 Characters of Representations

Another way to see this, is to observe that the linear map $E = \frac{1}{|G|} \sum_{a \in G} \pi(a)$ is a projection of V onto V_G, in other words, it maps V onto V_G and $E^2 = E$. This follows as $\pi(b)E(v) = E(v)$ for $v \in V$ and $b \in G$, which is gotten by summing over ba rather than a, and if $v \in V_G$, then obviously $E(v) = v$. Taking the trace of E then gives the proposition above.

By the basic property of the trace, we see that $\chi_\pi(aba^{-1}) = \chi_\pi(b)$ for $a, b \in G$, so characters are constant on conjugacy classes.

Definition 5.12.5 A *class function* is a function on a group G with values in a field F that is constant on conjugacy classes.

Theorem 5.12.6 *Let G be a finite group. Every class function in $\mathbb{C}(G)$ is a linear combination of characters of irreducible representations on finite dimensional complex vector spaces.*

Proof Say $f \in \mathbb{C}(G)$ is a class function. Write $f = \sum_i f_i$, where f_i are matrix coefficients of distinct irreducible representations π_i on finite dimensional complex vector spaces V_i. Then

$$\sum_i f_i = f = f(b^{-1} \cdot b) = \sum_i \mathrm{Ad}_{\pi_i}(b, b) f_i.$$

But $\mathrm{Ad}_{\pi_i}(b, b) f_i \in M_{\pi_i}$ and the subspaces M_{π_i} are mutually orthogonal, so f_i are all class functions.

By Schur's lemma there is up to a scalar, only one intertwiner in $\mathrm{End}(V_i)$, so the character χ_{π_i} is up to a scalar, the only class function in M_{π_i}. Hence f_i is proportional to χ_{π_i}. □

Corollary 5.12.7 *Any finite dimensional representation of a finite group on a complex vector space is completely determined by its character.*

Proof If $\pi = \oplus n_i \pi_i$ with pairwise inequivalent irreducible representations π_i, then $\chi_\pi = \sum n_i \chi_{\pi_i}$. The functions χ_{π_i} are linear independent, so the multiplicities n_i of π_i in π are fixed by χ_π. Hence π is up to equivalence, uniquely determined by χ_π. □

Theorem 5.12.8 *Let π and ρ be representations of a finite group G on finite dimensional complex vector spaces V and W, respectively. Then*

$$\frac{1}{|G|} \sum_{a \in G} \chi_\pi(a) \overline{\chi_\rho(a)} = \dim \mathrm{Mor}(\rho, \pi).$$

So the characters of a complete set of finite dimensional pairwise inequivalent irreducible representations on a finite group G form an orthonormal basis of the subspace of class functions in $\mathbb{C}^G$.

Proof Note that $\text{Hom}(W, V)$ is a G-space under the action $a \mapsto \pi \rho^c(a, a)$, and that $\text{Mor}(\rho, \pi)$ is the space of G-invariants for this action, which by Proposition 5.11.2 is isomorphic to the space of G-invariants for $\pi \otimes \rho^c$. The dimension of this space again is given by Proposition 5.12.4, and results in the left-hand-side of the theorem since the character of $\pi \otimes \rho^c$ is $\chi_\pi \overline{\chi_\rho}$ in virtue of Proposition 5.12.2.

The last statement follows now by Schur's lemma combined with the previous theorem. □

Corollary 5.12.9 *Let π be a representation of a finite group on a finite dimensional complex vector space. Then π is irreducible if and only if $(\chi_\pi | \chi_\pi) = 1$.*

Corollary 5.12.10 *The constant d from Lemma 5.10.2 equals $(\dim V)/|G|$.*

Proof Let $\{v_i\}$ be an orthonormal basis of V. Then by the theorem, we get

$$|G| = \sum_a |\chi_\pi(a)|^2 = \sum_{ij} \sum_a (\pi(a)v_i | v_i)\overline{(\pi(a)v_j|v_j)} = (\dim V)/d.$$

□

Corollary 5.12.11 *The number of pairwise inequivalent irreducible representations of a finite group G on complex vector spaces equals the number of conjugacy classes in G.*

Proof According to the theorem the characters of a collection of pairwise inequivalent irreducible representations of a finite group G form a linear basis for the space of class functions on G. This space is evidently isomorphic to the space of functions on the quotient set of conjugacy classes in G for which the delta functions form a basis. Hence the result. □

5.13 Group Algebra

Given a group G and a field F. Pick any vector space over F with dimension $|G|$. Label a basis (by choosing a bijection to G) in this vector space by the group elements. So we regard G as sitting inside the vector space as a linear basis; we have linearized G. We turn this vector space into a unital algebra $F[G]$, the *group algebra* over F, by defining the algebra product as a bilinear extension of the group multiplication. The algebra product is called the *convolution product*, and is denoted by $f * g$ for $f, g \in F[G]$. It is easy to see that $(f * g)(a) = \sum_{b \in G} f(ab^{-1})g(b)$ for $a \in G$. Clearly, we have $|G| = \dim F[G]$. It is also evident that G is abelian if and only if $F[G]$ is commutative.

For $a \in G$ define a linear map $\lambda(a)$ on basis elements $b \in G$ by $\lambda(a)b = ab$. We have linearly extended the maps λ_a in Cayley's theorem to obtain a faithful representation $\lambda \colon G \to \text{Aut}(F[G])$, and have hereby strengthened that theorem: Every group is a permutation group by linear maps.

5.13 Group Algebra

As the notation suggests, the representation λ is just the regular representation acting on $F[G]$ rather than F^G. Indeed, the linear isomorphism $A\colon F[G] \to F^G$ which sends $a \in G$ to δ_a satisfies

$$\lambda(a)Ab = \lambda(a)\delta_b = \delta_{ab} = A(ab) = A\lambda_a(b) = A\lambda(a)b$$

for all $a, b \in G$, where we in the second step used $(\lambda(a)\delta_b)(c) = \delta_b(a^{-1}c) = \delta_{ab}(c)$. Note that A is in general not an algebra homomorphism.

Any representation $\pi\colon G \to \operatorname{Aut}(V) \subset \operatorname{End}(V)$ has a linear extension to a unital algebra homomorphism $\tilde{\pi}\colon F[G] \to \operatorname{End}(V)$, which we call a *representation of the algebra $F[G]$* on V. This is a one-to-one correspondence since any representation of $F[G]$ restricts to a representation of the group and then extends uniquely to the same representation of $F[G]$.

Remark 5.13.1 The kernel $\ker \tilde{\pi}$ is a two-sided ideal of $F[G]$, which gives a correspondence between representation of the group G and ideals of $F[G]$. We will later study the group algebra from the point of view of modules.

Example 5.13.2 Consider a finite abelian group $G = \{a_1, \ldots, a_n\}$. Then it is easy to see that $\widehat{f * g} = |G|\hat{f}\hat{g}$ for any $f, g \in \mathbb{C}^G$.

An *integral operator* on $\mathbb{C}^G$ is an endomorphism Ψ on $\mathbb{C}^G$ given by $\Psi(f)(a) = \sum_{b \in G} K(a, b) f(b)$ with *kernel* $K \in \mathbb{C}^{G \times G}$. Clearly $\Psi(\delta_{a_j}) = \sum_i K(a_i, a_j)\delta_{a_i}$, so the matrix of Ψ associated to the basis $\{\delta_{a_i}\}$ of $\mathbb{C}^G$ has ij-entry $K(a_i, a_j)$. Thus $\operatorname{Tr}(\Psi) = \sum K(a_i, a_i)$.

It is easy to see that $\Psi\lambda_a = \lambda_a\Psi$ for all $a \in G$ if and only if there is $g \in \mathbb{C}^G$ such that $K(b, c) = g(b - c)$ for all $b, c \in G$. So in this case $\operatorname{Tr}(\Psi) = |G|g(0)$ and $\Psi(f) = g * f$. Hence if χ is a character on G, we see that $\Psi(\chi) = \hat{g}(\chi)\chi$. So $\hat{G}$ is a basis for $\mathbb{C}^G$ of eigenvectors of Ψ with respective eigenvalues $\hat{g}(\chi)$. Thus we get the following *trace formula* $|G|g(0) = \operatorname{Tr}(\Psi) = \sum_{\chi \in \hat{G}} \hat{g}(\chi)$. ◇

Considering the regular representation λ on $\mathbb{C}[G]$ for a finite group G, we get by Corollary 5.10.4 and Proposition 5.11.4, that $\tilde{\lambda}\colon \mathbb{C}[G] \to \oplus \operatorname{End}(V_i)$ is a faithful unital representation. Here we have extended each irreducible representation in the decomposition of λ. As the dimensions of $\mathbb{C}[G]$ and $\oplus \operatorname{End}(V_i)$ coalesh, we conclude that

$$\mathbb{C}[G] \cong \bigoplus \operatorname{End}(V_i)$$

as algebras.

We can also linearise G-spaces.

Definition 5.13.3 The *permutation representation* of a G-space X is the representation

$$P\colon G \to \operatorname{Aut}(F[X])$$

given by $P(a)x = ax$ for $x \in X$, now with X viewed as a linear basis for the vector space $F[X]$ over the field F.

Proposition 5.13.4 *If X is a finite G-space, then $\chi_P(a) = |X_a|$ for $a \in G$, where $X_a \equiv \{x \in X \mid ax = x\}$.*

Proof Define a dual basis for $X \subset F[X]$ by $\hat{x}(y) = \delta_{x,y}$ for $x, y \in X$. Then

$$\chi_P(a) = \operatorname{Tr} P(a) = \sum_{x \in X} \hat{x}(P(a)x) = \sum_{x \in X} \hat{x}(ax) = \sum_{x \in X_a} 1 = |X_a|$$

for $a \in G$. □

Viewing a group G as a G-space, the following result is then immediate.

Corollary 5.13.5 *If G is a finite group, then $\chi_\lambda(a) = |G|\delta_{a,e}$ for $a \in G$.*

We recover the familiar result.

Corollary 5.13.6 *Let π_i be an irreducible representation of a finite group G on a complex vector space. Then the multiplicity of π_i in the regular representation λ is $(\chi_{\pi_i}|\chi_\lambda)$, where $(\cdot|\cdot)$ is the standard inner product on $\mathbb{C}^G$.*

Proof By the previous corollary, we have

$$(\chi_{\pi_i}|\chi_\lambda) = \frac{1}{|G|}\chi_{\pi_i}(e)|G| = \dim V_i,$$

where V_i is the vector space acted upon by π_i. □

The Peter-Weyl theorem and Corollary 5.12.11 and the class formula

$$|G| = \sum_{x \in C}[G : N(x)]$$

for a finite group G, where C consists of one element from each conjugacy class in G, suggest that $[G : N(x)] = (\dim V_{f(x)})^2$, where $f: C \to I$ is a bijection to the index set of pairwise inequivalent irreducible representation of G on complex vector spaces.

5.14 Quadratic Reciprocity from Fourier Analysis

We will be somewhat more sketchy here, leaving intermediate steps as exercises to the reader.

Let χ be a multiplicative character on the group of units on $\mathbb{Z}_n$. Sticking with the notation from Sect. 5.9 we introduce the *Gauss sum*

$$\tau(\chi, a) = n\hat{\chi}(\psi_{-a}) = \sum_{k=0}^{n-1} \chi([k])w^{ak},$$

5.14 Quadratic Reciprocity from Fourier Analysis

where we have extended χ to an element of $\mathbb{C}^{\mathbb{Z}_n}$ by setting $\chi([k]) = 0$ whenever k and n are not relatively prime. Let now n be an odd prime p, so $w = e^{2\pi i/p}$.

If $\chi \in \widehat{\mathbb{Z}_p}$ is non-trivial, then $\tau(\chi, a) = \overline{\chi([a])}\tau(\chi, 1)$ since this clearly holds when p divides a, and when it does not, we have

$$\tau(\chi, a) = \overline{\chi([a])} \sum_{k=1}^{p-1} \chi([ak])w^{ak} = \overline{\chi([a])} \sum_{k=1}^{p-1} \chi([k])w^k$$

as $\{[a1], [a2], \ldots, [a(p-1)]\} = \mathbb{Z}_p$.

Let $\chi_p = (\cdot/p)$ be the multiplicative character on the group of non-zero elements in $\mathbb{Z}_p$ given by the Legendre symbol. Then

$$\tau(\chi_p, a) = \sum_{k=1}^{p-1}(k/p)w^{ak} = (a/p)\tau(p),$$

where $\tau(p) = \tau(\chi_p, 1)$ is the *classical Gauss sum*.

Example 5.14.1 It is easily checked that $\tau(3) = i\sqrt{3}$ and $\tau(\chi_3, 2) = -i\sqrt{3}$. ◊

Lemma 5.14.2 *If p does not divide a, then*

$$\tau(\chi_p, a)^2 = (-1)^{(p-1)/2}p.$$

If $q \neq p$ is another odd prime, then

$$\tau(\chi_p, a)^{q-1} \equiv (-1)^{(p-1)(q-1)/4}(p/q) \pmod{q}.$$

Proof Observe that $\sum_{x=1}^{p-1} w^{ax(1+y)}$ is $p-1$ if $y \equiv p-1 \pmod{p}$ and is otherwise -1. Hence

$$\tau(\chi_p, a)^2 = \sum_{x=1}^{p-1}\sum_{y=1}^{p-1}(xy/p)w^{a(x+y)} = \sum_{x=1}^{p-1}\sum_{y=1}^{p-1}(xxy/p)w^{a(x+xy)}$$

$$= \sum_{y=1}^{p-1}(y/p)\sum_{x=1}^{p-1}w^{ax(1+y)} = (-1/p)(p-1) - \sum_{y=1}^{p-2}(y/p) = (-1/p)p,$$

which proves the first statement. Using this we get

$$\tau(\chi_p, a)^{q-1} = ((-1)^{(p-1)/2}p)^{(q-1)/2},$$

which proves the second statement. □

Theorem 5.14.3 *For distinct odd primes p and q, we have*

$$\tau(p)^{q-1}(q/p) = \sum (x_1 \cdots x_q / p),$$

where we sum over all $x_i \in \{1, \ldots, p-1\}$ such that $x_1 + \cdots + x_q \equiv q \pmod{p}$.

Proof We calculate $L = \widehat{\hat{\chi}_p^q}(P([-q]))$ with $[-q] \in \mathbb{Z}_p$ in two different ways. On the one hand we have

$$L = p^{-1} \sum_{x=0}^{p-1} \hat{\chi}_p^q(\psi_x) \overline{P([-q])(\psi_x)} = p^{-1-q} \sum_{x=0}^{p-1} \tau(\chi_p, -x)^q \psi_x([q])$$

$$= p^{-1-q} \tau(p)^q \sum_{x=1}^{p-1} (-x/p) w^{qx} = p^{-1-q} \tau(p)^q (-q/p) \sum_{x=1}^{p-1} (x/p) w^x$$

$$= p^{-1-q} \tau(p)^{q+1} (-q/p) = p^{-q} \tau(p)^{q-1}(q/p)$$

by the last lemma.

On the other hand we have

$$L = p^{-q+1} \widetilde{\chi_p * \cdots * \chi_p}(P([-q])) = p^{-q}(\chi_p * \cdots * \chi_p)([q])$$

and we are done. $\square$

Corollary 5.14.4 *The law of quadratic reciprocity holds.*

Proof Combining the previous lemma with the theorem and using that the sum in the theorem is one modulo q, we get

$$(p/q)(q/p) \equiv (-1)^{(p-1)(q-1)/4} \pmod{q}$$

and we evidently also get equality without taking equivalence classes. $\square$

It is worth while studying the Gauss sum further.

Proposition 5.14.5 *We have $\tau(\chi_p, a) = \sum_{x=0}^{p-1} w^{ax^2}$.*

Proof Let $R \subset \{1, \ldots, p-1\}$ be a set of representatives of congruence classes of quadratic residues modulo p, and let R^c be its complement in $\{1, \ldots, p-1\}$. Since $x^2 \equiv k \pmod{p}$ if and only if $(p-x)^2 \equiv k \pmod{p}$, and since $x \not\equiv 0 \pmod{p}$ if and only if $x \not\equiv p - x \pmod{p}$ as p is odd, we may write

$$\sum_{x=1}^{p-1} w^{ax^2} = 2 \sum_{k \in R} w^{ak}.$$

5.14 Quadratic Reciprocity from Fourier Analysis

Hence

$$\tau(\chi_p, a) = 2\sum_{k\in R} w^{ak} - \sum_{k\in R\cup R^c} w^{ap} = 1 + 2\sum_{k\in R} w^{ak} - \sum_{k=0}^{p-1} w^{ak} = \sum_{x=0}^{p-1} w^{ax^2}.$$

□

Let F be the endomorphism on $\mathbb{C}^{\mathbb{Z}_p}$, which is the Fourier transform composed with the linear map $\mathbb{C}^{\hat{\mathbb{Z}}_p} \to \mathbb{C}^{\mathbb{Z}_p}$ induced by the group isomorphism $[a] \to \psi_a$ from $\mathbb{Z}_p$ to $\hat{\mathbb{Z}}_p$. Then $F(f)([a]) = p^{-1}\sum_{x=0}^{p-1} f([x])w^{-ax}$ and $F^2(f)([a]) = p^{-1}f([-a])$ for $[a] \in \mathbb{Z}_p$ and $f \in \mathbb{C}^{\mathbb{Z}_p}$. With respect to the basis $\{\delta_0, \ldots, \delta_{p-1}\}$ of $\mathbb{C}^{\mathbb{Z}_p}$ the endomorphims F has matrix with ij-entry $p^{-1}w^{-ij}$. Hence $\tau(p) = p\,\overline{\mathrm{Tr}(F)}$.

From the lemma above we know that $\tau(p)^2$ is p if $p \equiv 1 \pmod 4$ and is $-p$ if $p \equiv 3 \pmod 4$. To find $\tau(p)$ we must therefore determine signs. We will do so by calculating $\det(F)$ in two different ways.

Lemma 5.14.6 *We have that* $\det(F)$ *equals* $(-1)^k p^{-p/2}$ *if* $p = 4k + 1$ *and equals i times the same value if* $p = 4k + 3$.

Proof The ij-entry of the matrix associated to F^2 is $p^{-2}\sum_{k=0}^{p-1} w^{-(i+j)k}$, which equals p^{-1} if $i + j \equiv 0 \pmod p$ and equals zero otherwise. Hence $\det(F^2) = i^{p-1}p^{-p}$ and $\det(F) = \pm i^{(p-1)/2}p^{-p/2}$. To determine the sign, write $p^p \det(F)$ as a Vandermonde determinant

$$p^p \det(F) = \prod_{i<j}(w^{-i} - w^{-j}) = \prod_{i<j} w^{-(i+j)/2} \prod_{i<j}(w^{-(i-j)/2} - w^{(i-j)/2})$$

$$= w^{-\sum_{i<j}(i+j)/2}(-i)^{p(p-1)/2}\prod_{i<j} 2\sin((i-j)\pi/p)$$

$$= (-i)^{p(p-1)/2}\prod_{i<j} 2\sin((i-j)\pi/p)$$

as

$$\sum_{i<j}(i+j)/2 = p(p-1)^2/4 \equiv 0 \pmod p.$$

But $\prod_{i<j} 2\sin((i-j)\pi/p) > 0$, so we must have $\det(F) = (-i)^{p(p-1)/2}p^{-p/2}$. □

Let $c \in \{2, \ldots, p-1\}$ be a primitive root of p, so $[c]$ generates the multiplicative group of units in $\mathbb{Z}_p$. For $c \in \{0, \ldots, p-1\}$ define a multiplicative character η_b modulo p by $\eta_b(c^j) = \exp(2\pi i b j/(p-1))$. These characters exhaust the dual of the abelian group of units in $\mathbb{Z}_p$.

Lemma 5.14.7 *We have*

$$p^p \det(F) = p \prod_{b=1}^{p-2} \tau(\eta_b, 1) = (-1)^{r(r-1)/2} p^{(p-1)/2} \tau(p),$$

where $r = (p-1)/2$.

Proof We will calculate the determinant of the matrix of F associated to the orthogonal basis $\{\delta_0, \eta_0, \eta_1, \ldots, \eta_{p-2}\}$. It is easy to chech that $pF(\delta_0) = \delta_0 + \eta_0$ and $pF(\eta_0) = (p-1)\delta_0 - \eta_0$ and

$$pF(\eta_b) = (-1)^b \tau(\eta_b, 1) \eta_{p-1-b}$$

when $b \not\equiv 0 \pmod{p-1}$, which gives the first equality.

To get the second equality, apply the formula above twice to get

$$p^2 F^2(\eta_b) = \tau(\eta_b, 1) \tau(\eta_{p-1-b}, 1) \eta_b$$

and compare this with the formula gotten from $F^2(\eta_b)([a]) = p^{-1} \eta_b([-a])$. This gives

$$\tau(\eta_b, 1) \tau(\eta_{p-1-b}, 1) = p(-1)^b.$$

Hence

$$p \prod_{b=1}^{p-2} \tau(\eta_b, 1) = p\tau(p) \prod_{b=1}^{r-1} \tau(\eta_b, 1) \tau(\eta_{p-1-b}, 1) = (-1)^{r(r-1)/2} p^{(p-1)/2} \tau(p).$$

$\square$

Combining the two last lemmas gives the following signs.

Theorem 5.14.8 *The classical Gauss sum* $\tau(p)$ *is* $\sqrt{p}$ *if* $p \equiv 1 \pmod{4}$ *and it is* $i\sqrt{p}$ *if* $p \equiv 3 \pmod{4}$.

5.15 The Character Table for S_3

Let us find the irreducible representations of S_3 on complex vector spaces. Now S_3 is non-abelian, so not all irreducible representations can be 1-dimensional. As $|S_3| = 6$, and since $2^2 + 1^2 + 1^2$ is the only way up to order, of writing 6 as a sum of squares of natural numbers not all 1, the Peter-Weyl theorem tells us that there are two 1-dimensional irreducible representations, or group characters, and one 2-dimensional irreducible representation up to equivalence.

We know of two group characters, the trivial one ε, and sign: $S_3 \to \{\pm 1\}$.

To get hold of the 2-dimensional representation consider the permutation representation $P: S_3 \to \text{Aut}(\mathbb{C}^3)$ of the S_3-space $X \equiv \{1,2,3\}$ given by $P(a)e_i = e_{a(i)}$ for $a \in S_3$, where $\{e_1, e_2, e_3\}$ is the standard basis for $\mathbb{C}^3$. Observe that $\mathbb{C}(e_1 + e_2 + e_3)$ is an invariant subspace, and that P restricted to this is ε. The usual inner product on $\mathbb{C}^3$ is invariant under the action of S_3 via P, so the orthogonal subspace V of $\mathbb{C}(e_1 + e_2 + e_3)$ in $\mathbb{C}^3$ is invariant, and V has a basis $\{e_1 - e_2, e_1 - e_3\}$.

Consider the usual generators $\sigma = (123)$ and $\tau = (23)$ for S_3. We see that $P(\tau)(e_1 - e_2) = e_1 - e_3$, so P restricted to V is irreducible. We call it the *standard representation* S of S_3. This is the 2-dimensional representation we were looking for. Note that $P \cong S \oplus \varepsilon$.

What about their characters? Now $S_3 = \{e, \sigma, \sigma^2, \tau, \sigma\tau, \sigma^2\tau\}$ with relations $\sigma^3 = e = \tau^2$ and $\tau\sigma = \sigma^2\tau$. The conjugacy classes $C(a) = \{b \in S^3 \mid bab^{-1}\}$ are easily found to be $C(e) = \{e\}$ and $C(\sigma) = \{\sigma, \sigma^2\}$ and $C(\tau) = \{\tau, \sigma\tau, \sigma^2\tau\}$, so they form a partition of S_3. Characters of representations are class functions and are constant on each of these conjugacy classes.

To compute their values, first note that σ is even, so sign is known. Then observe that $\chi_P(a) = |X_a|$ with $X_e = \{1,2,3\}$ and $X_\tau = \{1\}$ and $X_\sigma = \phi$ and that $\chi_S = \chi_P - \varepsilon$ as $P \cong S \oplus \varepsilon$. This gives the following character table, which we could also have calculated more directly.

S_3	$C(e)$	$C(\tau)$	$C(\sigma)$
ε	1	1	1
sign	1	-1	1
χ_S	2	0	-1

In the first column we can read off the dimension of the representations.

Representations are uniquely determined up to equivalence by their characters. We can thus use the character table to decompose tensor products of representations. For instance, the character χ of $S \otimes \text{sign}$ is $\chi_S \chi_{\text{sign}}$. Thus $\chi(e) = 2 \cdot 1 = 2$ and $\chi(\tau) = 0 \cdot (-1) = 0$ and $\chi(\sigma) = (-1) \cdot 1 = -1$, so $\chi = \chi_S$ and this means that $S \otimes \text{sign} \cong S$, which again can be checked directly by setting up an equivalence.

5.16 Induced Representations

Here we produce representations of a group from representations of subgroups.

Suppose H is a subgroup of G, and let π be a representation of H on a vector space V over a field F. With the risk of causing confusion, let V^G be the vector space under pointwise operations of all functions $f: G \to V$ such that $f(ab) = \pi(a)f(b)$ for $a \in H$ and $b \in G$.

Definition 5.16.1 The *induced representation* $\pi^G: G \to \text{Aut}(V^G)$ is given by $(\pi^G(a)f)(b) = f(ba)$ for $a, b \in G$ and $f \in V^G$.

Clearly π^G is a representation, and we some times denote it by $\text{Ind}_H^G(\pi)$.

The following result shows that induction is transitive.

Proposition 5.16.2 *Let G be a group with $H < K < G$. Then*

$$\operatorname{Ind}_K^G \operatorname{Ind}_H^K \cong \operatorname{Ind}_H^G.$$

Proof Let π be a representation of H on a vector space V. To see that $\operatorname{Ind}_K^G(\operatorname{Ind}_H^K(\pi)) \cong \operatorname{Ind}_H^G(\pi)$ we need to set up a bijective linear map $A \colon (V^K)^G \to V^G$ that intertwines the representations $(\pi^K)^G$ and π^G.

That an element f belongs to $(V^K)^G$ means that $a \mapsto f_a \in V^K$ and that $f_{ba} = \pi^K(b) f_a$ for $a \in G$ and $b \in K$. The first condition means that $f_a(cd) = \pi(c) f_a(d)$ and the second one means that $f_{ba}(d) = f_a(db)$ for $c \in H$ and $b, d \in K$ and $a \in G$.

Let $(Af)(a) = f_a(e)$ for $a \in G$. Then $Af \in V^G$ because by the conditions above, we get

$$(Af)(ca) = f_{ca}(e) = f_a(ec) = f_a(ce) = \pi(c) f_a(e) = \pi(c)(Af)(a)$$

for $c \in H$. So we have a linear map $A \colon (V^K)^G \to V^G$.

This map is injective because $Af = 0$ means that $f_a(e) = 0$ for all $a \in G$. But then by the conditions above, we get

$$f_a(b) = f_a(eb) = f_{ba}(e) = 0$$

for $b \in K$ and $a \in G$, so $f = 0$.

To see that A is surjective, take any $g \in V^G$ and let $f_a(b) = g(ba)$ for $a \in G$ and $b \in K$. Then $f_a \in V^K$ as

$$f_a(cb) = g(cba) = \pi(c) g(ba) = \pi(c) f_a(b)$$

for $b \in K$ and $c \in H$. Also $f \colon a \mapsto f_a$ for $a \in G$ satisfies $f_{ba} = \pi^K(b) f_a$ for $a \in G$ and $b \in K$ because

$$(\pi^K(b) f_a)(d) = f_a(db) = g(dba) = f_{ba}(d)$$

for $d \in K$. Thus $f \in (V^K)^G$ and $(Af)(a) = f_a(e) = g(a)$ for $a \in G$, so $Af = g$ and A is surjective.

Finally, we have $\pi^G(a) A = A(\pi^K)^G(a)$ for $a \in G$ since

$$(\pi^G(a) Af)(a') = (Af)(a'a) = f_{a'a}(e) = ((\pi^K)^G(a) f)_{a'}(e) = (A(\pi^K)^G(a) f)(a')$$

for $f \in (V^K)^G$ and $a, a' \in G$, which completes this tedious exercise. □

5.17 Reciprocity

The following result is known as the *first version of Frobenius reciprocity*.

Proposition 5.17.1 *Let H be a subgroup of G with unit e, and let π and ρ be representations of H and G on W and V, respectively. Then*

$$\mathrm{Mor}(\rho, \pi^G) \cong \mathrm{Mor}(\rho|H, \pi)$$

under an isomorphism f that sends $A \in \mathrm{Mor}(\rho, \pi^G)$ to $f(A) \in \mathrm{Mor}(\rho|H, \pi)$, where $f(A)w = (Aw)(e)$. Moreover, if $B \in \mathrm{Mor}(\rho|H, \pi)$, then $(f^{-1}(B)w)(a) = B\rho(a)w$ for $a \in G$ and $w \in W$.

Proof Now $f(A) \in \mathrm{Mor}(\rho|H, \pi)$ since

$$f(A)\rho(a)w = (A\rho(a)w)(e) = (\pi^G(a)Aw)(e) = (Aw)(ea)$$
$$= (Aw)(ae) = \pi(a)(Aw)(e) = \pi(a)f(A)w.$$

for $a \in H$.

Next $f^{-1}(B)w \in V^G$ because

$$(f^{-1}(B)w)(ab) = B\rho(ab)w = B\rho(a)\rho(b)w = \pi(a)B\rho(b)w = \pi(a)(f^{-1}(B)w)(b)$$

for $a \in H$ and $b \in G$ and $w \in W$.

Also $f^{-1}(B) \in \mathrm{Mor}(\rho, \pi^G)$ since

$$(f^{-1}(B)\rho(a)w)(b) = B\rho(ba)w = (f^{-1}(B)w)(ba) = (\pi^G(a)f^{-1}(B)w)(b)$$

for $a, b \in G$ and $w \in W$.

It is straighforward to check that $f^{-1}(f(A)) = A$ and $f(f^{-1}(B)) = B$. □

There is also a dual isomorphism, known as the *second version of Frobenius reciprocity*.

Proposition 5.17.2 *Let π be a representation on V of a subgroup H of a finite group G, and let ρ be a representation of G on W. Define $Av \colon G \to V$ for $v \in V$ by $(Av)(a) = \pi(a)v$ if $a \in H$ and otherwise 0. Then $A \colon V \to V^G$ is H-equivariant, and*

$$\mathrm{Mor}(\pi^G, \rho) \cong \mathrm{Mor}(\pi, \rho|H)$$

under an isomorphism g that sends $B \in \mathrm{Mor}(\pi^G, \rho)$ to $g(B) = BA$. Moreover, we have such an isomorphism g if and only if

$$Bf = \sum_{[a] \in G/H} \rho(a)g(B)f(a^{-1})$$

for all $f \in V^G$.

Proof Note that $(Av)(ab)$ for $a \in H$, equals $\pi(ab)v$ if $b \in H$, and is zero if $b \in G\setminus H$, so $Av \in V^G$. Similar reasoning shows that $A\pi(a)v = \pi^G(a)Av$ for $a \in H$, so $g(B) \in \text{Mor}(\pi, \rho|H)$.

Next, we claim that
$$f = \sum_{[a] \in G/H} \pi^G(a) A f(a^{-1})$$

for $f \in V^G$. By definition of V^G and from H-equivariance of A, we see that each term in this sum is independent of the representative from $[a]$, so the sum is well-defined. Applying this sum to $b \in G$, then by definition of A, we get $\sum_a (Af(a^{-1}))(ba) = (Af(b))(e)$, which indeed equals $f(b)$, so our identity holds.

Applying $B \in \text{Mor}(\pi^G, \rho)$ to this identity we obviously get the identity in the proposition, so g is injective.

To see that g is surjective, consider any $C \in \text{Mor}(\pi, \rho|H)$. Then
$$Df \equiv \sum_{[a] \in G/H} \rho(a) C f(a^{-1})$$

for $f \in V^G$ is well-defined by the same reasons as before. Also $D \in \text{Mor}(\pi^G, \rho)$ because
$$D\pi^G(b) f = \sum_{[a] \in G/H} \rho(a) C f(a^{-1}b) = \sum_{[ba] \in G/H} \rho(ba) C f(a^{-1}) = \rho(b) Df$$

for $b \in G$. Finally, by definition of A, we have $g(D) = C$ because
$$g(D)v = DAv = \sum_{[a] \in G/H} \rho(a) C(Av)(a^{-1}) = \rho(e) C \pi(e^{-1}) v = Cv$$

for $v \in V$. $\square$

5.18 Mackey Theory

The following result is known as the *geometric version of Mackey's theorem*, and gives another description of the morphisms between induced representations.

Theorem 5.18.1 *Suppose G is a finite group with subgroups H and K represented on V and W by π and ρ, respectively. Let X denote the vector space under pointwise operations of all functions $x: G \to \text{Hom}(V, W)$ such that*
$$x(abc) = \rho(a) x(b) \pi(c)$$

for $a \in K$ and $b \in G$ and $c \in H$. Let $x \in X$ and define $A_x f \in W^G$ for $f \in V^G$ by the 'convolution product' formula

$$(A_x f)(a) \equiv \sum_{[b] \in G/H} x(b) f(b^{-1} a)$$

for $a \in G$. Then $x \mapsto A_x$ is a linear isomorphism from X to $\mathrm{Mor}(\pi^G, \rho^G)$.

Proof Obviously, the sum above is well-defined; by the definition of X and V^G, each term is independent of the coset representative.

Next, we see that $A_x f \in W^G$ as

$$(\rho(a) A_x f)(c) = \sum_{[b] \in G/H} \rho(a) x(b) f(b^{-1} c) = \sum_{[b] \in G/H} x(ab) f(b^{-1} c) = (A_x f)(ac)$$

for $a \in K$ and $c \in G$, where we have substituted b by ab in the summation over the cosets.

So we have a linear map $A_x : V^G \to W^G$, which is also G-equivariant as the actions of the induced representations are by right multiplication with elements of the group G.

To see that the linear map $x \mapsto A_x$ is bijective, consider $B \in \mathrm{Mor}(\pi^G, \rho^G)$ and let $g(B) \in \mathrm{Mor}(\pi, \rho^G|H)$, where g is dictated by the second version of Frobenius reciprocity. Define $x \in X$ by $x(b)v = (g(B)v)(b)$ for $b \in G$ and $v \in V$.

Then $g(B)v \in W^G$ if and only if $x(ab) = \rho(a) x(b)$ for $a \in K$ and $b \in G$, and $g(B)$ is H-equivariant if and only if $x(bc) = x(b) \pi(c)$ for $b \in G$ and $c \in H$. Hence we have a linear isomorphism $B \mapsto x$ from $\mathrm{Mor}(\pi^G, \rho^G)$ to X.

It remains to check that $B = A_x$. Starting with the identity in the proposition for the second version of Frobenius reciprocity, we get

$$(Bf)(a) = \sum_{[b] \in G/H} (\rho^G(b) g(B) f(b^{-1}))(a) = \sum_{[b] \in G/H} (g(B) f(b^{-1}))(ab)$$

$$= \sum_{[b] \in G/H} x(ab) f(b^{-1}) = \sum_{[c] \in G/H} x(c) f(c^{-1} a) = (A_x f)(a)$$

for $f \in V^G$ and $a \in G$. □

Definition 5.18.2 Given two subgroups H and K of a group G, the collection of *double cosets* is the set

$$K \backslash G / H \equiv \{KaH \mid a \in G\}.$$

The relation of two group elements belonging to the same double coset is an equivalence relation, so the double cosets form a partition of the group.

Any function x in the result above is clearly determined by its value on a representative of a double coset KaH. Let us look at those supported on a single double coset, in which case we say that the corresponding intertwiner is supported on that double coset.

Proposition 5.18.3 *Suppose G is a finite group with subgroups H and K represented on V and W by π and ρ, respectively. Let $a \in G$ and consider the subgroup $H_a \equiv aHa^{-1} \cap K$ of G with representations π^a and ρ^a given by $\pi^a(b) = \pi(a^{-1}ba)$ and $\rho^a(b) = \rho(b)$ for $b \in H_a$. Then the linear space of all $B \in \mathrm{Mor}(\pi^G, \rho^G)$ supported on KaH is isomorphic to $\mathrm{Mor}(\pi^a, \rho^a)$.*

Proof By Mackey's theorem, there is a unique function $x \colon G \to \mathrm{Hom}(V, W)$ such that $A_x = B$, for any $B \in \mathrm{Mor}(\pi^G, \rho^G)$ supported on KaH, and

$$x(a)\pi^a(b) = x(a)\pi(a^{-1}ba) = x(ba) = \rho(b)x(a) = \rho^a(b)x(a)$$

for $b \in H_a$. So $h \colon B \mapsto x(a)$ is a linear map from the space of all $B \in \mathrm{Mor}(\pi^G, \rho^G)$ supported on KaH to $\mathrm{Mor}(\pi^a, \rho^a)$. It is injective because x is supported on KaH and is there determined by $x(a)$.

To see that h is surjective, given any $A \in \mathrm{Mor}(\pi^a, \rho^a)$, let $x \colon G \to \mathrm{Hom}(V, W)$ be zero on all double cosets different from KaH, and let $x(bac) = \rho(b)A\pi(c)$ for $b \in K$ and $c \in H$. Then x is well-defined because if $b'ac' = bac$ with $b' \in K$ and $c' \in H$, then

$$\rho(b^{-1}b')A\pi(c'c^{-1}) = A\pi^a(b^{-1}b')\pi(c'c^{-1}) = A\pi(a^{-1}b^{-1}b'ac'c^{-1}) = A\pi(e) = A,$$

so $x(b'ac') = x(bac)$. By construction x belongs to X as defined in the theorem above, and obviously $h(A_x) = x(a) = A$. □

The extension to more general functions on double cosets is easy, and comprises what is known as the *algebraic version of Mackey's theorem*.

Theorem 5.18.4 *Suppose G is a finite group with subgroups H and K represented on V and W by π and ρ, respectively. Let $a_i \in G$ be a complete set of representatives for $K \backslash G / H$, and let π^{a_i} and ρ^{a_i} be as in the proposition above. Then*

$$\dim \mathrm{Mor}(\pi^G, \rho^G) = \sum_i \dim \mathrm{Mor}(\pi^{a_i}, \rho^{a_i}).$$

Proof By the previous theorem, any element of $\mathrm{Mor}(\pi^G, \rho^G)$ is of the form A_x for a unique $x \in X$. Define $x_i \colon G \to \mathrm{Hom}(V, W)$ to be zero on the double cosets different from Ka_iH and let $x_i(b) = x(b)$ for $b \in Ka_iH$. Then $x_i \in X$ and A_{x_i} is supported on Ka_iH. By the previous proposition, the linear space of such intertwiners is isomorphic to $\mathrm{Mor}(\pi^{a_i}, \rho^{a_i})$, and since $x = \sum x_i$, the sum being direct, we get the desired result. □

As the following result shows, the procedure of first inducing and then restricting produces the same thing as gotten by first restricting and then inducing. We will here work over the field of complex numbers.

Corollary 5.18.5 *Let K and H be subgroups of a finite group G, and let π be an irreducible representation of H on a complex vector space V. Let a_i be a complete set of representatives for $K \backslash G / H$, and let $\pi^{a_i} = \pi(a_i^{-1} \cdot a_i)$ be the representation of $H_{a_i} \equiv a_i H a_i^{-1} \cap K$ on V, as described in the proposition above. Then*

$$\pi^G | K \cong \bigoplus_i \mathrm{Ind}_{H_{a_i}}^K (\pi^{a_i}).$$

Proof Since we are working over the complex field representations are completely reducible, so it suffices to show that the multiplicity of an irreducible representation ρ of K in $\pi^G | K$ is the same as the multiplicity of ρ in the direct sum representation in the corollary.

By the theorem and the first version of Frobenius reciprocity, we see that the multiplicity of ρ in $\pi^G | K$ is

$$\dim \mathrm{Mor}(\pi^G | K, \rho) = \dim \mathrm{Mor}(\pi^G, \rho^G) = \sum_i \dim \mathrm{Mor}(\pi^{a_i}, \rho^{a_i}),$$

where ρ^{a_i} is the restriction of ρ to H_{a_i}. By the second version of Frobenius reciprocity, we therefore get

$$\dim \mathrm{Mor}(\pi^G | K, \rho) = \sum_i \dim \mathrm{Mor}(\mathrm{Ind}_{H_{a_i}}^K (\pi^{a_i}), \rho),$$

as required. $\square$

5.19 Characters of Induced Representations

Let G be a finite group with a subgroup H, and let π and ρ be representations on finite dimensional complex vector spaces V and W, respectively. By Theorem 5.12.8 and Proposition 5.17.1, we see that

$$(\chi_{\pi^G} | \chi_\rho) = \dim \mathrm{Mor}(\rho, \pi^G) = \dim \mathrm{Mor}(\rho | H, \pi) = (\chi_\pi | \chi_{\rho | H}).$$

For a class function f on a subgroup H of G, define $f^G \colon G \to \mathbb{C}$ by

$$f^G(a) = \sum_{[b] \in H \backslash G} \tilde{f}(bab^{-1}),$$

where $\tilde{f}(c)$ is $f(c)$ for $c \in H$ and is zero for $c \in G \backslash H$. Since f is a class function on H, the terms in the sum above are independent of the chosen representatives, so f^G is well-defined, and we can write

$$f^G = \frac{1}{|H|} \sum_{b \in G} \tilde{f}(b \cdot b^{-1}).$$

We also see that f^G is a class function on G.

Theorem 5.19.1 *Let π be a representation on a finite dimensional complex vector space of a subgroup H of a finite group G. Then*

$$\chi_{\pi^G} = \chi_\pi^G.$$

Proof Let ρ be a representation of G on a finite dimensional complex vector space. The characters of irreducible representations on G form an orthonormal basis for the space of class functions on G. Since χ_π^G is a class function, it therefore suffices to show that $(\chi_\pi^G | \chi_\rho) = (\chi_{\pi^G} | \chi_\rho)$. But

$$\begin{aligned}
(\chi_\pi^G | \chi_\rho) &= \frac{1}{|G|} \sum_{a \in G} \frac{1}{|H|} \sum_{b \in G} \widetilde{\chi_\pi}(bab^{-1}) \overline{\chi_\rho(a)} \\
&= \frac{1}{|G|} \sum_{a \in G} \frac{1}{|H|} \sum_{c \in H} \sum_{b \in G, c = bab^{-1}} \chi_\pi(c) \overline{\chi_\rho(a)} \\
&= \frac{1}{|G|} \frac{1}{|H|} \sum_{c \in H} \sum_{b \in G} \chi_\pi(c) \overline{\chi_\rho(b^{-1}cb)} \\
&= \frac{1}{|H|} \sum_{c \in H} \chi_\pi(c) \overline{\chi_\rho(c)} = (\chi_\pi | \chi_{\rho|H}) = (\chi_{\pi^G} | \chi_\rho),
\end{aligned}$$

where we in the third step counted differently, and in the last step we used the observation made at the beginning of this section. □

We end with an easy example.

Example 5.19.2 Consider the octic group D_4 with generators σ, τ satisfying $\tau^2 = e = \sigma^4$ and $\tau \sigma \tau = \sigma^{-1}$. It has order 8 and five conjugacy classes

$$C(e) = \{e\}, \ C(\sigma) = \{\sigma, \sigma^3\}, \ C(\sigma^2) = \{\sigma^2\}, \ C(\tau) = \{\tau, \sigma^2 \tau\}, \ C(\sigma\tau) = \{\sigma\tau, \sigma^3\tau\}.$$

It has four group characters obtained by the four possible values ± 1 for σ and τ. In addition it has a 2-dimensional irreducible representation, say with character χ. These are all the irreducible representations up to equivalence, since $1^2 + 1^2 + 1^2 + 1^2 + 2^2 = 8$, and the Peter-Weyl theorem doesn't allow adding more squares. How do we get the character χ? Well, note that $\eta(\sigma) = e^{\pi i/2}$ defines a group character on the cyclic subgroup of D_4 generated by σ. Then χ is the character of the induced representation of η, so by the previous result $\chi(C(e)) = 2$, saying that the induced representation has indeed dimension two, while $\chi(C(\sigma)) = e^{\pi i/2} + e^{-\pi i/2} = 0$ and $\chi(C(\sigma^2)) = e^{\pi i} + e^{-\pi i} = -2$ and $\chi(C(\tau)) = 0 = \chi(C(\sigma\tau))$. We leave it to the reader to ponder what this means geometrically.

5.19 Characters of Induced Representations

We can generalize this to any dihedral group D_n with n even. Again we consider the generators σ, τ, and get the four group characters. But now we get $n/2 - 1$ characters χ_r of 2-dimensional representations induced up from the group characters η_r on $\langle \sigma \rangle$ given by $\eta_r(\sigma) = e^{2\pi i r/n}$ for $r \in \{1, 2, \ldots, n/2 - 1\}$. It is checked that these are all the characters of irreducible representations of D_n. ◇

Chapter 6
Rings

In this chapter we study rings more systematically. Such an approach might seem unnecessarily general and dry in the beginning, but we do eventually hone in on the main goal of the chapter, which is the decomposition of PID's, so called principal ideal domains. The techniques used to prove the fundamental theorem of arithmetic are with only minor modifications, effective in this greater generality.

A ring is an additive group with an associative multiplication which distributes over addition. It is commutative if the order of multiplication is immaterial, and it is moreover an integral domain whenever the product of two non-zero elements is again non-zero. The integers form an integral domain, whereas the ring $\mathbb{Z}_6$ does not since $[2][3] = [0]$. As soon as one has an integral domain, one can talk about the ring of fractions, which is the smallest field containing the integral domain as a subring. Not surprisingly, the ring of fractions for the integers is the rational numbers.

By an ideal I of a ring R we mean an additive subgroup closed under multiplication both from the right and left by elements of R. One can then form the quotient ring R/I consisting of cosets aI for $a \in R$. When the ideal is maximal, the quotient ring will be simple, meaning that it contains no proper ideals. The quotient ring will moreover be a field if the original ring is a commutative unital ring. For example, in $\mathbb{Z}$ the proper ideals are of the form $n\mathbb{Z}$. Up to isomorphism, they produce the quotient rings $\mathbb{Z}_n$. If m divides n, then clearly $n\mathbb{Z} \subset m\mathbb{Z}$. Hence $n\mathbb{Z}$ is maximal exactly when n is a prime number, and then $\mathbb{Z}_n$ will be a field. It turns out that all finite integral domains are fields.

Now a PID is an integral domain where every ideal is principal, i.e. is generated by a single element. A special class of PID's are Euclidean domains. These are integral domains coming with an $\mathbb{N}$-valued 'degree' function, and which furthermore have a generalized version of Euclid's division algorithm built into them. Using the 'degree' function, the well-ordering principle can be invoked to show that such rings are indeed PID's. Two important examples of Euclidean domains are the integers, where the 'degree' function is the absolute value of an integer, and the ring of polynomials $F[x]$ in one indeterminate x over a field F, where the 'degree' function is the usual degree of polynomials. Both these examples are PID's. The ring of fractions of $F[x]$ is by the way, the ring of rational functions $F(x)$.

We devote a section to explain carefully what a polynomial is with its notion of an indeterminate. Here we also construct related rings. For instance, we show that $F(x)$ is a subfield of the field of formal Laurent series $F\langle x\rangle$, which is the ring of fractions of formal power series $F[[x]]$, which in turn extends $F[x]$. These constructions can be performed over rings, and with more than one indeterminate. As promised earlier we also construct group rings, and in the same vain we introduce twisted group rings. We also include some general nonsense about rings, quotients and homomorphisms.

Turning again to PID's, we drive towards the announced decomposition result. We say that a non-zero element in an integral domain is prime if it satisfies Euclid's lemma, in that it has to divide one of the factors in a product if it divides the product itself. A more restricted class of elements are the irreducible ones. These are the elements that essentially cannot be written as a product of two other elements. Here we have to rule out the appearance of so called units, which are the invertible elements in the ring (with respect to the product, of course). This type of problem occurs already when defining the prime numbers within the ring $\mathbb{Z}$, where the units are 1 and -1. In PID's the notions of primeness and irreducibility coalesce, and obviously the irreducible elements in $F[x]$ are the usual irreducible polynomials. We study also the corresponding ideals. In a separate section we prove the desired decomposition result, including the important uniqueness part. In the last section we study UFD's, so called unique factorization domains, which are rings where the previous decomposition result holds. They clearly include PID's.

For a couple of reasonable references, see [10, 13].

6.1 Basic Definitions

We recall the following basic definition.

Definition 6.1.1 A *ring* R is an additive group together with an associative binary operation called *multiplication* which satisfies

$$a(b+c) = ab + bc, \quad (b+c)a = ba + ca$$

for all $a, b, c \in R$. It is *unital* if it has an *identity*, that is, an element 1 such that $1a = a = a1$ for $a \in R$. If $ab = ba$ for all $a, b \in R$, then R is *commutative*. An *integral domain* is a commutative unital ring such that $ab \neq 0$ when both a and b are non-zero. A *division ring* is a unital ring where every non-zero element has a multiplicative inverse. A *field* is a commutative division ring.

Since an identity is automatically unique, we have ascribed a symbol to it.

6.1 Basic Definitions

It is easy to deduce properties like

$$a0 = 0 = 0a, \quad a(-b) = -(ab) = (-a)b, \quad a(b-c) = ab - ac, \quad (a-b)c = ac - bc$$

for any elements a, b, c in a ring.

Any field is an integral domain as $a^{-1}(ab)b^{-1} = 1$ for $a \neq 0$ and $b \neq 0$. But the converse is not true. A *left (right) zero divisor* is a non-zero element a such that $ab = 0$ (or $ba = 0$) for some non-zero element b. A *zero divisor* is either a left- or a right zero divisor. Thus a (integral) domain is a (commutative) unital ring with no zero divisors. Absence of zero divisors means that the ring has the *cancellation property*, which says that $b = c$ whenever $ab = ac$ or $ba = ca$ for $a \neq 0$.

We have seen that $\mathbb{Q}, \mathbb{R}$ and $\mathbb{C}$ are fields.

Definition 6.1.2 A *homomorphism* of a ring R into another ring S is a map $f \colon R \to S$ such that $f(a+b) = f(a) + f(b)$ and $f(ab) = f(a)f(b)$ for $a, b \in R$. If it is injective we call it a *monomorphism* or an *embedding* and if it is surjective we call it an *epimorphism*, and if it is both, then it is an *isomorphism*, and in this case we say that R is isomorphic to S, and write $R \cong S$.

The relation of being isomorphic is clearly an equivalence relation.

Definition 6.1.3 A subset S of a ring R is a *subring* of R if it is a ring with respect to the binary operations induced from R. If S is neither *trivial* $\{0\}$ nor R, it is called a *proper* subring.

Note that the identity might be different in a subring. If $f \colon R \to S$ is a homomorphism of rings with R unital, then $f(1)$ is a identity for the subring $f(R)$ of S, so if f is an epimorphism, then S is unital.

Proposition 6.1.4 *A non-empty subset S of a ring is a subring if and only if $a - b \in S$ and $ab \in S$ for $a, b \in S$.*

Definition 6.1.5 The intersection of all subrings of a ring containing a non-empty subset X is a ring called the *subring generated by X*.

Clearly the subring S generated by X is the smallest subring containing X, and any homomorphism from S to another ring is uniquely determined on the generators of S, and by a *generator* of S we simply mean an element of X.

Definition 6.1.6 The *center of a ring* R is the commutative subring $Z(R)$ of R consisting of all elements $a \in R$ such that $ab = ba$ for $b \in R$.

6.2 Prime Subfields and Characteristics

Proposition 6.2.1 *The additive group $\mathbb{Z}_n$ is a ring with multiplication $[a][b] \equiv [ab]$, and it is a field if and only if n is a prime number.*

Proof Multiplication is well-defined because if $c = a + nk$ and $d = b + nl$, then $cd = ab + n(al + bk + nkl)$, so $[cd] = [ab]$. And surely the axioms for a ring will hold.

If n is prime, Corollary 2.2.7 says that for $a \in \{1, \ldots, n-1\}$, there are integers b and c such that $bn + ca = 1$, and thus $[c][a] = [1]$, so $\mathbb{Z}_n$ is a field.

When n is not a prime number, the ring $\mathbb{Z}_n$ is not even an integral domain since $n = ab$ for some $a, b \in \{1, \ldots, n-1\}$, so $[0] = [n] = [a][b]$ and neither $[a] = 0$ nor $[b] = 0$. $\square$

Whenever $\mathbb{Z}_n$ is an integral domain, it is a field, due to the following result.

Proposition 6.2.2 *Any finite integral domain is a field.*

Proof By the cancellation property, for a non-zero element a in an integral domain, the map $b \mapsto ab$ is injective, and hence surjective. So there is an element b such that $ab = 1$. $\square$

Definition 6.2.3 The *prime subfield* of a field F is the intersection of all subfields of F, so it is generated by the identity of F.

Obviously $\mathbb{Q}$ and $\mathbb{Z}_p$ for a prime number p are both fields with no proper subfields, so they are prime subfields.

Proposition 6.2.4 *Every non-trivial prime subfield is isomorphic to $\mathbb{Q}$ or $\mathbb{Z}_p$ for a unique prime number p.*

Proof Suppose P is a prime subfield of a non-trivial field. Two cases can occur.

If $n1 \neq 0$ for every non-zero integer n, then $m/n \mapsto m1(n1)^{-1}$ is a well-defined isomorphism from $\mathbb{Q}$ to P.

If $n1 = 0$ for some $n \in \mathbb{N}$, let p be the smallest such n. Its existence is guaranteed by the well-ordering principle. If $p = mk$ for smaller natural numbers m and k, then either $m1 = 0$ or $k1 = 0$, which is impossible, so p is prime. Clearly the map $[n] \mapsto n1$ is a well-defined isomorphism from $\mathbb{Z}_p$ to P. $\square$

Definition 6.2.5 The *characteristic of a ring* is the smallest natural number n such that $na = 0$ for all a. If no such number exists, we say that the ring has characteristic 0.

A unital ring has non-zero characteristic if and only if $n1 = 0$ for some natural number n, so $\mathbb{Z}_n$ has characteristic n, whereas $\mathbb{Z}$ has characteristic 0.

The characteristic of a field is zero if the prime subfield is $\mathbb{Q}$, and it is p if the prime subfield is $\mathbb{Z}_p$. Clearly any subfield of a field has the same characteristic as the field. Thus $\mathbb{R}$ and $\mathbb{C}$ have characteristic 0.

Proposition 6.2.6 *If the absolute value of a field is bounded on the prime subfield, then the absolute value is non-archimedean. In particular, any absolute value on fields with non-zero characteristic is automatically non-archimedean as the prime field is then finite.*

Proof For any elements a, b of the field and any natural number n, the binomial formula yields

$$|(a+b)^n| \leq \sum_{m=0}^{n} |\binom{n}{m} a^m b^{n-m}| \leq c(n+1) \max(|a|, |b|)^n,$$

where we have also used that the coefficients belong to the prime field and are thus bounded by a number c. Taking n-th roots and letting n go to infinity, we thus get $|a+b| \leq \max(|a|, |b|)$. $\square$

Recalling how $\mathbb{Q}$ was constructed from $\mathbb{Z}$, the following result is obvious.

Proposition 6.2.7 *Any integral domain R can be embedded into a field F such that every element of F can be expressed as a quotient a/b of $a, b \in R$ with $b \neq 0$. The field F is unique up to isomorphism, and is called the field of quotients or fractions of R.*

6.3 Examples of Non-commutative Rings

Any additive group A is a commutative ring with multiplication $ab = 0$ for all $a, b \in A$.

Example 6.3.1 The set End(A) of endomorphisms of an additive group A is a ring with $f + g$ and fg for $f, g \in$ End(A) given by $(f + g)(a) = f(a) + f(b)$ and $(fg)(a) = f(g(a))$. Endomorphisms, and not merely maps, are needed to get $h(f + g) = hf + hg$.

The following example shows that not all division rings are fields.

Example 6.3.2 The set

$$H = \{A = \begin{pmatrix} a & b \\ -\bar{b} & \bar{a} \end{pmatrix} \mid a, b \in \mathbb{C}\}.$$

is a division ring under matrix addition and multiplication because if $A \neq 0$, then $\det(A) = a\bar{a} + b\bar{b} > 0$ and $A^{-1} \in H$. This non-commutative division ring is known as the *quaternions*. $\diamond$

Example 6.3.3 Let $M_n(R)$ denote the set of $n \times n$-matrices with elements of a ring R as entries. This is obviously a ring under matrix addition and multiplication, and it is non-commutative when $n \neq 1$ and $R \neq \{0\}$. Already $M_2(R)$ contains zero-divisors when R is non-trivial, because diagonal matrices in $M_n(R)$ with only one non-zero entry $a \in R$ placed at different locations will have 0 as product. Each of these matrices provide an embedding of R into $M_n(R)$ that does not preserve possible identities when $n \geq 2$. Yet another embedding is given by $a \mapsto (a\delta_{ij})$, and this one does preserve possible identities.

An *upper (lower) triangular matrix* is a matrix $(a_{ij}) \in M_n(R)$ with $a_{ij} = 0$ for $i > j$ (or $i < j$). The set of all upper (lower) triangular matrices is a subring of $M_n(R)$. The quaternions is another subring of $M_2(\mathbb{C})$.

If R is a commutative ring, then $\{(a\delta_{ij}) \mid a \in R\}$ is a subring of $Z(M_n(R))$ for any n. ◇

6.4 Group Rings

Here we will discuss another way of producing new rings from old ones. The group algebra $F[G]$ of a group G over a field F is certainly a ring. Let us first extend this sort of ring with the coefficient field F replaced by any ring.

The direct product and direct sum of rings are obviously rings with pointwise addition and multiplication. Given a unital ring R, the direct sum $\oplus_i R$ is unital exactly when the index set is finite. Considering $\oplus_i R$ as an additive group, we can introduce another multiplication provided the index set is a group G.

To this end let $\delta_s \colon G \to R$ for $s \in G$ be given by $\delta_s(t) = 1$ if $t = s$ and otherwise set to be 0. Then any element $f \in \oplus_{s \in G} R$ can be written as a finite sum $f = \sum f(s)\delta_s$ for unique elements $f(s) \in R$.

Definition 6.4.1 The *group ring* $R[G]$ of a group G over a unital ring R is the unital ring which is $\oplus_{s \in G} R$ as a pointwise additive group, but with multiplication uniquely determined by $\delta_s \delta_t = \delta_{st}$ for $s, t \in G$. This multiplication is called the *convolution product*.

One tends to write the convolution product of $f, g \in R[G]$ as $f * g$ to distinguish it from the pointwise product, so

$$(f * g)(s) = \sum_{uv = s} f(u)g(v),$$

where the summation is over all $u, v \in G$ with $uv = s$. Note that this is a finite sum. The convolution product can also be written as

$$(f * g)(s) = \sum_u f(u)g(u^{-1}s) = \sum_v f(sv^{-1})g(v).$$

These formulas allow to define the group rings also for non-unital rings, but here we will stick to unital ones. The map $G \to R[G]$ given by $s \mapsto \delta_s$ is then a group monomorphism, so we have a copy of the group inside the group ring. The map $R \to R[G]$ given by $a \mapsto a\delta_e$ is a unital ring monomorphism, so we also have a copy of R inside the group ring, and we will often suppress this isomorphism. The unit e in the group will then be identified with the identity 1 in the ring, and we will use these two symbols interchangeably. The group ring $R[G]$ is generated by G and R. Thus it is commutative if and only if G is abelian and R is commutative.

The definition of a group ring can obviously be extended to monoid rings by replacing group with monoid. The last convolution product formula above does not make sense then.

6.5 Polynomial Rings

Let us consider the group ring $R[\mathbb{Z}^n]$ of the abelian group $\mathbb{Z}^n$ over a unital ring R. We write x_i for the function δ_s with $s \in \mathbb{Z}^n$ having coordinate 1 at the i-th place and otherwise zeroes.

Definition 6.5.1 The subring of $R[\mathbb{Z}^n]$ generated by R and the x_i's is called the *polynomial ring in n indeterminates x_i with coefficients in R*, and is denoted by $R[x_1, \ldots, x_n]$.

The m-th power x_i^m of x_i is the function δ_s, where $s \in \mathbb{Z}^n$ has m at the i-th place and otherwise zeroes.

Any element $f \in R[x]$ can be written as

$$f = a_0 + a_1 x + a_2 x^2 + \cdots + a_m x^m$$

for unique $a_i \in R$. The convolution product (now without a $*$) is the expected one, namely, if $g = \sum_n b_m x^n$, then

$$fg = \sum_k c_k x^k,$$

where

$$c_k = \sum_{m+n=k} a_m b_n.$$

This means that we expand products and collect all terms with x^k, keeping track of the order of elements in R when we round up the coefficients. We often write $f(x)$ for $f \in R[x]$ to stress that we are dealing with an indeterminate.

We will be particularly interested in the case where the coefficient ring is a field F. Then $F[x_1, \ldots, x_n]$ is clearly commutative and unital. In fact, as the following proposition shows, it is an integral domain.

Proposition 6.5.2 *If R is an integral domain, then so is $R[x_1, \ldots, x_n]$.*

Proof It is easy to see that $R[x, y] \cong (R[x])[y]$, so it is enough to show that $R[x]$ is an integral domain.

Consider non-zero polynomials $f, g \in R[x]$ and let a and b be the coefficients of the highest power of x in f and g. Then ab will be the coefficient of the highest power of x in fg, and $ab \neq 0$ as both a and b are non-zero. But then also $fg \neq 0$. $\square$

However, the integral domain $F[x_1, \ldots, x_n]$ is not a field as x_1 has no inverse. The field of fractions of $F[x_1, \ldots, x_n]$ is denoted by $F(x_1, \cdots, x_n)$ and consists of all *rational functions in n indeterminates*. For instance, the field $F(x)$ consists of all quotients $f(x)/g(x)$ of polynomials $f(x)$ and $g(x) \neq 0$.

Remark 6.5.3 It can be misleading to write

$$f(x) = a_0 + a_1 x + \cdots + a_n x^n$$

for $f(x) \in F[x]$, and to talk about a polynomial in an indeterminate or *variable* x over a field F. One gets the impression that one can plug in something for x, and get $f(x)$ out, suggesting that f is a function from F to F. This is not the case, the element $f(x)$ is a function $f : \mathbb{Z} \to F$ such that $f(m) = a_m$ for $m \in \{0, \ldots, n\}$ and $f(k) = 0$ for k not among these finite numbers. This means that a formula of the type $1 + 3x - x^2 = 0$ can never hold since $-1, 1$ and 3 are not all 0. In fact, we see that $F[x]$ is an infinite dimensional vector space with basis $\{x^n\}$ under the convention that $x^0 = 1$.

We certainly do not want to define a polynomial as a function from F to F. Suppose we did this. Consider $F = \mathbb{Z}_2$. Then $f(t) = t$ and $g(t) = t^2$ would be the same functions, and yet we would like to be able to distinguish the polynomials x and x^2.

However, we can substitute values for the variable.

Definition 6.5.4 Let

$$f(x) = a_0 + a_1 x + \cdots + a_n x^n \in R[x]$$

for a unital subring R of a commutative unital ring S with the same identity. Then the *evaluation* of $f(x)$ at $s \in S$ is the element $f(s) \in S$ given by

$$f(s) = a_0 + a_1 s + \cdots + a_n s^n.$$

The following proposition is obvious, and uses commutativity of the ring S.

Proposition 6.5.5 *Suppose we have a unital subring R of a commutative ring S with the same identity. Then for every $s \in S$, the map $R[x] \to S$ given by $f \mapsto f(s)$ is a unital homomorphism.*

6.6 Laurent Series and Power Series

Definition 6.6.1 Let R be a unital ring. The ring $R\langle x\rangle$ of *formal Laurent series* over R is the additive subgroup of $\prod_{n\in\mathbb{Z}} R$ consisting of sequences $\{a_i\}$ with $a_i = 0$ for at most finitely many negative i's, and the product of $\{a_i\}$ and $\{b_j\}$ is the sequence $\{c_k\}$ with

$$c_k = \sum_{i+j=k} a_i b_j,$$

which is still a finite sum.

We write $\sum_{n=-m}^{\infty} a_n x^n$ for the sequence $\{a_{-m}, a_{-m+1}, \ldots\}$, and again the product is consistent with the suggestive notation. Since R is unital, we see that $x \in R\langle x\rangle$. Again if R is an integral domain, then so is $R\langle x\rangle$. This can be seen by considering the coefficients a and b of the lowest powers of x in non-zero elements f and g of $R\langle x\rangle$, and then observe that $ab \neq 0$ is the coefficient of the lowest power of x in fg.

Definition 6.6.2 The *Laurent polynomials* in one indeterminant x over a unital ring R is the subring $R[x, x^{-1}]$ of $R\langle x\rangle$ generated by x and x^{-1}.

So a Laurent polynomial is a finite sum of positive and negative powers of x with coefficients in R. Clearly $R[x]$ is a subring of $R[x, x^{-1}]$.

Definition 6.6.3 The set of elements of the type $\sum_{n=0}^{\infty} a_n x^n$ is a subring of $R\langle x\rangle$ called the *formal power series* in one indeterminate over R, denoted by $R[[x]]$.

Clearly $R[x]$ is a subring of $R[[x]]$.

One defines analogously the formal Laurent ring $R\langle x_1, \ldots, x_n\rangle$ and the subring of formal power series $R[[x_1, \ldots, x_n]]$ in n indeterminates. As $R\langle x, y\rangle \cong (R\langle x\rangle)\langle y\rangle$, it is clear that $R\langle x_1, \ldots, x_n\rangle$, and hence the subring $R[[x_1, \ldots, x_n]]$, are integral domains whenever R is. The ring of Laurent polynomials $R[x_1, x_1^{-1}, \ldots, x_n, x_n^{-1}]$ in n indeterminants is also an integral domain when the unital coefficient ring R is.

Proposition 6.6.4 *Let F be a field. Then the invertible elements of $F[[x]]$ are series $\sum_{n=0}^{\infty} a_n x^n$ with $a_0 \neq 0$.*

The ring of formal Laurent series $F\langle x\rangle$ is a field, and the field of fractions of the integral domain $F[[x]]$ is isomorphic to $F\langle x\rangle$, and contains $F(x)$ as a subfield, which is also the field of fractions of $F[x, x^{-1}]$.

Proof The series $f = \sum_{n=0}^{\infty} a_n x^n$ is invertible with inverse $g = \sum_{m=0}^{\infty} b_m x^m$ if and only if $a_0 b_0 = 1$ and

$$a_0 b_n + a_1 b_{n-1} + \cdots + a_{n-1} b_1 + a_n b_0 = 0$$

for all $n \in \mathbb{N}$.

So $a_0 \neq 0$ if f is invertible. Conversely, if $a_0 \neq 0$, then define g with $b_0 = a_0^{-1}$ and with b_n constructed inductively using the relation above.

A similar argument shows that $F\langle x\rangle$ is a field. Moreover, the field of fractions of $F[[x]]$ is the smallest field containing $F[[x]]$, so it must be contained in $F\langle x\rangle$. But $x \in F[[x]]$ has an inverse x^{-1} that belongs to the field of fractions of $F[[x]]$, and $F\langle x\rangle$ is generated by $F[[x]] \cup \{x^{-1}\}$. Hence the field of fractions of $F[[x]]$ is isomorphic to $F\langle x\rangle$.

The last statements are now obvious. □

It is natural to also consider evaluation of variables in $F[[x]]$ and $F\langle x\rangle$, but here we get infinite series of members in the field, which has to be made sense of. This is a topic of analysis, which discusses conditions for convergence of such series, which of course has been done with great success in the case of the complex numbers.

6.7 Ideals

We gather some basic results relating to ideals.

Definition 6.7.1 An *ideal* I in a ring R is an additive subgroup such that $aI \subset I$ and $Ia \subset I$ for all $a \in R$. The ideal is said to be a *proper ideal* if it is not R, and it is *trivial* if $\{0\}$.

Obviously an ideal is a subring, but it is a much more restrictive notion; most subrings are not ideals. For instance, if R is unital and an ideal I in R has the same identity as R, then $I = R$ because $a = a1 \in I$ for all $a \in R$.

On the other hand, we can also talk about a *left (or right) ideal* I in a ring R to be an additive subgroup of the ring such that $aI \subset I$ (or $Ia \subset I$), and then I need not be a subring of R, except of course when e.g. the ring is commutative.

The intersection of ideals in a ring is again an ideal, so the following definition makes sense.

Definition 6.7.2 The *ideal generated by a subset* X of a ring is the smallest ideal containing X. We denote it by (X). Set $(\phi) = \{0\}$. When X consists of finitely many elements $a_1, \ldots a_n$, we write $(a_1, \ldots, a_n)$ for (X). An ideal is *finitely generated* if it has a finite generator set. A *principal ideal* I is an ideal that is generated by a single element, so $I = (a)$ for some a.

For a ring R, the left- and right ideal generated by $a \in R$ is Ra and aR, respectively.

Given ideals I and J in a ring, we denote by IJ the ideal which consists of finite sums of elements ab with $a \in I$ and $b \in J$. Then $IJ \subset I \cap J$, and in general the inclusion is proper.

Example 6.7.3 Let R and S be rings. Then $R \times 0$ is an ideal in $R \times S$. Notice that if R is unital, then the identity of the ideal is the element $(1, 0)$, which is not the identity of $R \times S$ unless S is trivial. ◇

Similarly, the functions $X \to R$ from a set to a ring is a ring, the direct product of R over X, and the set I_Y of functions vanishing on a subset Y of X is an ideal in the direct product ring.

Example 6.7.4 Let R be a unital ring. Then all matrices of the form

$$\begin{pmatrix} 0 & a \\ 0 & 0 \end{pmatrix}$$

for $a \in R$ is an ideal I in the ring of upper triangular matrices over R, but it is neither a left nor a right ideal in $M_2(R)$. The ideal generated by all such matrices is actually $M_2(R)$. Notice that I has no identity as the product of any two elements of I is zero. ◇

6.8 Quotient Rings and Homomorphisms

Suppose I is an ideal in a ring R. Write $a \sim b$ for $a, b \in R$ if $a - b \in I$. Then $\sim$ is an equivalence relation on R. Write R/I for the set of equivalence classes, so the equivalence class of $a \in R$ is $a + I$. Define addition and multiplication in R/I by

$$(a + I) + (b + I) = a + b + I \quad \text{and} \quad (a + I)(b + I) = ab + I$$

for $a, b \in R$. Both these operations are well-defined since if $c, d \in I$, then

$$(a + c + I) + (b + d + I) = a + d + c + d + I = a + b + I$$

and

$$(a + c + I)(b + d + I) = (a + c)(b + d) + I = ab + ad + cb + cd + I = ab + I.$$

They also satisfy the same arithmetic axioms as the operations of R do, so R/I is a ring for these two operations.

Definition 6.8.1 The ring R/I is called the *quotient ring* of R modulo I, and the map $f(a) = a + I$ is called the *quotient map*.

Clearly the quotient map is an epimorphism.

As for left- and right ideals, their justification is related to modules, which we will study in Chap. 9.

Example 6.8.2 We have $\mathbb{Z}/n\mathbb{Z} \cong \mathbb{Z}_n$ as rings. ◇

Here is a ring theoretic road to the complex numbers.

Example 6.8.3 Consider the quotient ring $\mathbb{R}[x]/(x^2+1)$. Denote the equivalence class of $p \in \mathbb{R}[x]$ by $[p]$. Then $[x^2+1] = [0]$, so $[x]^2 = -[1]$, and therefore

$$\mathbb{R}[x]/(x^2+1) = \{[a] + [x][b] \mid a, b \in \mathbb{R}\}.$$

Hence $a + ib \mapsto [a] + [x][b]$ is an isomorphism from $\mathbb{C}$ to $\mathbb{R}[x]/(x^2+1)$. ◇

Definition 6.8.4 The *kernel of a homomorphism* $f : R \to S$ between rings is the subset $\ker f$ of R where f is zero.

Clearly a homomorphism f of rings is injective if and only if $\ker f = \{0\}$.

It is also immediate that the kernel of the quotient map $R \to R/I$ is the ideal I. More generally, we see that the kernel of a homomorphism $f : R \to S$ between rings is an ideal in R, whereas its image $\operatorname{im} f$ is a subring of S. This discussion leads to the *fundamental homomorphism theorem for rings*.

Theorem 6.8.5 *Suppose $f : R \to S$ is a homomorphism between rings, and let $g : R \to R/\ker f$ be the quotient map. Then the map $h : R/\ker f \to \operatorname{im} f$ given by*

$$h(a + \ker f) = f(a)$$

for $a \in R$ is an isomorphism of rings, and satisfies $hg = f$. Hence $R/\ker f \cong \operatorname{im} f$.

Proof There is not much to prove. If $a + \ker f = b + \ker f$, then $f(a) = f(b)$, so h is well-defined. Clearly h is surjective onto $\operatorname{im} f$. If $h(a + \ker f) = 0$ for some $a \in R$, then $a \in \ker f$, so $a + \ker f = \ker f$, which is the zero element in the quotient ring, so h is injective. By definition of the ring operations of the quotient ring, the bijection h is a homomorphism, and again by definition $hg = f$. □

The *correspondence theorem for rings* goes as follows.

Proposition 6.8.6 *Let $f : R \to S$ be an epimorphism. Then $I \mapsto f(I)$ is a bijection from the collection of ideals in R that contain $\ker f$ to the collection of all ideals in S.*

Proof If I is an ideal in R then $f(I)$ is an ideal in S because f is surjective. For the same reason $X = f(f^{-1}(X))$ for any subset X of S, and therefore the map $I \mapsto f(I)$ is surjective as any ideal J in S is the image of the ideal $f^{-1}(J)$ in R.

As for injectivity, say I and J are two ideals in R that contain $\ker f$ and that $f(I) = f(J)$. We claim that $I = f^{-1}(f(I))$. The inclusion $\subset$ is trivial. Let $b \in f^{-1}(f(I))$. Then $f(b) \in f(I)$, so $f(b) = f(a)$ for some $a \in I$. Thus $b - a \in \ker f \subset I$ and $b \in I$, which settles the claim. □

Corollary 6.8.7 *Suppose I is an ideal in a ring R. Then every ideal in R/I is of the form J/I for some ideal J in R that contains I.*

Proof Let $f: R \to R/I$ be the quotient map. Then according to the theorem, every ideal in R/I is of the form $f(J)$ for some ideal J in R that contains I. But clearly I is an ideal in J and $f(J)$ is the ring of equivalence classes $a + I$ with $a \in J$, so $f(J) = J/I$. □

Proposition 6.8.8 *For two ideals I and J of a ring we have*

$$I/(I \cap J) \cong (I + J)/J.$$

Proof Clearly each side of the identity makes sense, and $I \to (I+J)/J$ given by $a \mapsto a + J$ is an epimorphism with kernel $I \cap J$. □

Theorem 6.8.9 *If I_i are ideals of a commutative unital ring R, and if $I_i + I_j = R$ whenever $i \neq j$, then*

$$R/(\cap_{i=1}^n I_i) \cong R/I_1 \times \cdots \times R/I_n.$$

Proof Consider two ideals I and J of R. Then the map

$$f: I + J \to (I+J)/I \times (I+J)/J$$

given by $f(a) = (a + I, a + J)$ is an epimorphism with kernel $I \cap J$, so

$$(I+J)/(I \cap J) \cong (I+J)/I \times (I+J)/J$$

by the first isomorphism theorem.

We claim that

$$R = I_i + \cap_{j \neq i}^n I_j.$$

Assume by induction that it holds for $n - 1$ ideals. Then for $i \neq n$ we have

$$R = RR = (I_i + I_n)(I_i + \cap_{j \neq i}^{n-1} I_j) \subset I_i + \cap_{j \neq i}^n I_j$$

by commutativity.

To prove the assertion in the theorem, assume that it holds for $n - 1$ ideals, use $I = I_i$ and $J = \cap_{j \neq i}^n I_j$ in the previous paragraph. □

In view of the following example we can think of the above theorem as a generalization of the Chinese remainder theorem to general commutative unital rings.

Example 6.8.10 If $a = a_1 \cdots a_n \in \mathbb{Z}$ for relatively prime integers a_i, then

$$\mathbb{Z}_a \cong \mathbb{Z}_{a_1} \times \cdots \times \mathbb{Z}_{a_n}$$

as unital rings. To see this, consider $a = bc$ for integers b and c that are relatively prime. Then $\mathbb{Z} = b\mathbb{Z} + c\mathbb{Z}$ because there are integers k, l such that $kb + lc = 1$. We have $a\mathbb{Z} = b\mathbb{Z} \cap c\mathbb{Z}$ since if both b and c divide an integer d, then a divides d. Thus $\mathbb{Z}/a\mathbb{Z} \cong \mathbb{Z}/b\mathbb{Z} \times \mathbb{Z}/c\mathbb{Z}$ by the previous theorem. The general result follows by induction. ◇

6.9 Rings with Generators and Relations

Definition 6.9.1 The *free ring* $R\{X\}$ over a unital ring R with indeterminants in a non-empty set X is the monoid ring $R[F]$ of the free monoid F generated by X. If we replace F by the free abelian monoid generated by X, we talk about the *polynomial ring* $R[X]$ over R with indeterminants in X.

The elements of the free ring $R\{X\}$ are finite sums of finite products of indeterminants from X with coefficients in R. The same is true for the polynomial ring $R[X]$, and now the indeterminants in addition commute with each other, and when $|X| = n$ we recover the usual polynomial ring in n variables. The free ring $R\{X\}$ (the polynomial ring $R[X]$) is obviously a (integral) domain when R is a (integral) domain.

Clearly any homomorphism between monoids extends to a homomorphism between the corresponding monoid rings over the same ring R.

Example 6.9.2 Free rings behave rather differently from polynomial rings. Consider the free ring $F\{x, y\}$ in two variables over a field F. Let $z_i = xy^i$. The subring of $F\{x, y\}$ generated by $z_1, \ldots, z_n$ is isomorphic to the free ring $F\{z_1, \ldots, z_n\}$ since different monomials in $z_1, \ldots, z_n$ convert to different monomials in x and y. ◇

Obviously every unital ring is the homomorphic image of a free ring.

Definition 6.9.3 Let $R\{X\}$ be the free ring over a unital ring R with indeterminants in a non-empty set X. Write (Y) for the ideal of $R\{X\}$ generated by a subset Y of $R\{X\}$. The quotient ring $R\{X\}/(Y)$ is the *ring over R generated by the elements of X subject to the relations Y*.

To explain the terminology in the definition above, if $f \in Y$ is expressed in terms of the variables $x_1, \ldots, x_n$ with images in $R\{X\}/(Y)$ carrying the same names, then $f(x_1, \ldots, x_n) = 0$ in $R\{X\}/(Y)$. Hence we think of the elements of Y as relations between the variables from X.

Example 6.9.4 Consider the free ring $R\{X\}$ with $Y = \{xy - yx \mid x, y \in X\}$. Then $R\{X\}/(Y) \cong R[X]$. ◇

There are of course endless many examples since every unital ring is by the first ring isomorphism theorem generated over some coefficient ring by elements subject to relations.

Example 6.9.5 The *Weyl algebra* is the ring over a field generated by x and y subject to the relation $xy - yx = 1$. ◇

6.10 Twisted Group Rings

Proposition 6.9.6 *Let R be a unital ring and X a non-empty set. Suppose we have a map from X to a unital ring S, and a homomorphism $R \to S$ such that the elements of the images in S of X and R commute. Then there is a unique homomorphism $f \colon R\{X\} \to S$ that extends $X \to S$ and $R \to S$. If in addition the elements in the image of X in S commute with each other, there is a unique homomorphism $g \colon R[X] \to S$ such that $gq = f$, where $q \colon R\{X\} \to R\{X\}/(Y)$ with Y as in the example above.*

Proof Let F be the free monoid generated by X. Since S is a monoid under multiplication, the map $X \to S$ extends uniquely to a monoid homomorphism $F \to S$ by the definition of freeness of F. We extend this map further, and uniquely so, to a unital additive map $R[F] \to S$, letting $R \to S$ carry the coefficients over. Since the elements of F and R in S commute with each other, this map will also be multiplicative.

If in addition the elements in the image of X in S commute with each other, we can well-define a homomorphism $g \colon R[X] \to S$ uniquely by $gq = f$. $\square$

6.10 Twisted Group Rings

The notion of a group ring can be generalized.

Definition 6.10.1 Consider a group G acting as automorphisms of a ring R. Write $\sum a_s s$ for the function $G \to R$ that is $a_s \in R$ when evaluated at $s \in G$. The *twisted group ring* $R * G$ consists of all functions $G \to R$ that are non-zero for only finitely many group elements. Addition is pointwise and the product is a biadditive extension of $(as)(bt) = as(b)st$, where $s, t \in G$ and $a, b \in R$ and $s(b) \in R$ is the action of s on b.

When G acts trivially on R, we recover the group ring $R[G]$, but for less trivial actions the twisted group ring $R * G$ can be non-commutative even when G is abelian and R is commutative.

The following result shows that group rings of groups can sometimes be studied as twisted group rings of less complicated groups over more complicated coefficient rings.

Proposition 6.10.2 *Consider the semidirect product $K \rtimes H$ of two groups with respect to a homomorphism $f \colon H \to \mathrm{Aut}(K)$. For any ring R we have*

$$R[K \rtimes H] \cong R[K] * H,$$

where the product in the twisted group ring is with respect to $H \to \mathrm{Aut}(R[K])$ gotten by extending each $f(s)$ to an R-linear map.

Proof By Proposition 4.14.2 we know that $K \rtimes H = KH$ with $H \cap K$ trivial, and that the subgroup H acts on the normal subgroup K by $t(s) = tst^{-1}$ for $s \in K$ and $t \in H$. Then the R-linear bijection $R[K \rtimes H] \to R[K] * H$ given by

$$\sum_{(s,t) \in K \times H} a_{st} st \mapsto \sum_{t \in H} (\sum_{s \in K} a_{st} s) t$$

is easily seen to be multiplicative. □

Example 6.10.3 Suppose we have an automorphism x of a unital ring R and that $\langle x \rangle$ is an infinite cyclic subgroup of $\mathrm{Aut}(R)$. Then we can think of $R * \langle x \rangle$ as the Laurent polynomials $R[x, x^{-1}]$ under addition but with the product twisted according to the rule $xa = x(a)x$, where $x(a) \in R$ denotes the action of x on $a \in R$. Thus in $R * \langle x \rangle$ the action of x on a is by conjugation $a \mapsto xax^{-1}$.

Analogously we can define twisted versions of the polynomial ring $R[x]$ and the ring $R[[x]]$ of formal power series by twisting the products according to $xa = f(a)x$ for any endomorphism f of R. We denote these rings by $R[x; f]$ and $R[[x; f]]$, respectively. When f is an automorphism the latter ring can be extended to a twisted version $R\langle x; f\rangle$ of the ring of formal Laurent series $R\langle x\rangle$, so $R\langle x; x\rangle$ contains the subring $R * \langle x \rangle$ from above. ◇

6.11 Simple Rings and Maximal Ideals

Definition 6.11.1 A *simple ring* is non-trivial and has no proper non-trivial ideals.

Any non-trivial division ring, and hence any non-trivial field, is simple because any non-zero element a of an ideal has an inverse, so $1 = a^{-1}a$ belongs to the ideal.

In the other direction, any simple commutative unital ring R is a field. To see this, assume that we have a non-zero element $a \in R$. Then aR is a non-trivial ideal in R, so $aR = R$ and $ab = 1$ for some b.

Definition 6.11.2 A *maximal ideal in a ring R* is a proper ideal I that is not properly contained in any other proper ideal.

Proposition 6.11.3 *Suppose I is an ideal in a ring R. Then I is maximal if and only if R/I is simple.*

Proof If I is maximal, then Corollary 6.8.7 tells us that R/I is simple, for otherwise there would be a proper ideal J in R that properly contains I.

Conversely, suppose R/I is simple, and say J is an ideal that is strictly larger than I. Pick $a \in J \setminus I$. Then $(I + (a))/I$ is a non-trivial ideal in R/I, which therefore must be R/I. Hence $I + (a) = R$, and obviously $I + (a) \subset J$, so $J = R$. □

Corollary 6.11.4 *Suppose R is a commutative ring with identity. Then an ideal I in R is maximal if and only if R/I is a non-trivial field.*

6.11 Simple Rings and Maximal Ideals

Proof Clearly R/I is a commutative unital ring for any ideal I in R. If I is maximal, then by the proposition, the ring R/I is simple, and hence a field. Conversely, if R/I is a non-trivial field, then it is simple, so I is maximal by the same proposition. □

Example 6.11.5 The ideal in $\mathbb{R}[x]$ generated by $x^2 + 1$ is maximal because $\mathbb{R}[x]/(x^2+1)$ is isomorphic to the field of complex numbers. ◇

Example 6.11.6 A non-trivial ideal I in $\mathbb{Z}$ is maximal if and only if $I = (p)$ for a prime number p. ◇

The following result shows that non-trivial non-simple rings do have non-trivial maximal ideals.

Proposition 6.11.7 *Suppose R is a non-trivial unital ring. Then every proper ideal in R is contained in a maximal ideal.*

Proof Say I is an ideal in R with $I \neq R$. Let $\mathcal{F}$ be the family of proper ideals in R that contain I. Partially order $\mathcal{F}$ by inclusion. Observe that the union of all members in a chain is an upper bound for them, and belongs to $\mathcal{F}$ since the identity cannot belong to this union as one of the members would then be R. By Zorn's lemma, the family $\mathcal{F}$ contains a maximal element J, and this is a maximal ideal containing I. For suppose it was properly contained in a proper ideal, then that ideal would contradict the maximality of J in $\mathcal{F}$. □

This means that unital commutative rings either are fields or have a non-trivial field as a quotient.

Proposition 6.11.8 *If I_i are ideals of a commutative unital ring R such that $I_i + I_j = R$ for $i \neq j$, then*
$$I_1 \cdots I_n = I_1 \cap \cdots \cap I_n.$$

The condition $I + J = R$ obviously holds for any distinct ideals I and J of R with one of them maximal.

Proof Suppose I and J are ideals of R with $I + J = R$. Then
$$I \cap J = (I \cap J)R = (I \cap J)(I + J) \subset (I \cap J)I + (I \cap J)J \subset JI + IJ \subset IJ,$$
so $I \cap J = IJ$. Invoking $R = I_i + \cap_{j \neq i}^n I_j$ from the second half of the proof of Theorem 6.8.9, the statement for n ideals follows now easily by induction. □

6.12 Euclidean Domains and Principal Ideal Domains

Euclid's division algorithm is easy to generalize to the ring $F[x]$ of polynomials over a field F.

Definition 6.12.1 The *degree* deg p *of a non-zero polynomial* p *over a ring* is the highest power of the indeterminate x occurring with non-zero coefficient, called the *leading coefficient* of p. If the ring is unital and the leading coefficient of p is the identity, then p is *monic*.

We set $\deg 0 = -\infty$. Note that $\deg(fg) \leq \deg f + \deg g$ for polynomials over a ring, and that equality holds if the ring is a field.

Proposition 6.12.2 *Let F be a field, and let $f, g \in F[x]$ with $g \neq 0$. Then there exist polynomials $q, r \in F[x]$ such that $f = qg + r$, where $\deg r < \deg g$.*

Proof If g is not a factor of f, let $r = f - qg$ be the polynomial with q chosen such that r has minimal degree. If this degree is not less than $\deg g$, consider the leading terms ax^n and bx^m of r and g, respectively. Then

$$r - ab^{-1}x^{n-m}g = f - (q + ab^{-1}x^{n-m})g$$

has degree less than r, which is a contradiction. $\square$

Definition 6.12.3 A *Euclidean domain* is an integral domain R with a function λ from the non-zero elements of R to $\mathbb{N} \cup \{0\}$ such that for $a, b \in R$ with $b \neq 0$, there exist $c, d \in R$ satisfying $a = cb + d$, where either $d = 0$ or $\lambda(d) < \lambda(b)$.

Euclidean domains have a generalized version of Euclid's division algorithm built into them. The integers $\mathbb{Z}$ is a Euclidean domain with λ given by $\lambda(n) = |n|$ for $n \neq 0$. By the proposition above, we see that $F[x]$ is a Euclidean domain with $\lambda = \deg$.

Definition 6.12.4 A *principal ideal domain* (PID) is an integral domain where every ideal is principal.

So every ideal in a PID is of the form (a) for an element a of the ring.

Proposition 6.12.5 *Every Euclidean domain is a PID.*

Proof Consider a non-trivial ideal I in a Euclidean domain R. By the well-ordering principle we can pick an element a among the non-zero elements in I with 'lowest possible degree' $\lambda(a)$, meaning that $\lambda(a) \leq \lambda(b)$ for all non-zero $b \in I$. We claim that $I = aR$. Obviously $aR \subset I$. If $b \in I$, then there are elements $c, d \in R$ such that $b = ca + d$, where either $d = 0$ or $\lambda(d) < \lambda(a)$. The latter option must be ruled out since $d = b - ca \in I$. So $b = ac \in aR$, and $I = (a)$. $\square$

So the ring $F[x]$ of polynomials over field is a PID.
Here are two more examples of Euclidean domains.

6.12 Euclidean Domains and Principal Ideal Domains

Example 6.12.6 The *Gaussian integers* is the subring $\mathbb{Z}[i]$ of $\mathbb{C}$ consisting of the complex numbers $a + ib$ with $a, b \in \mathbb{Z}$. Being a subring of a field, it is an integral domain. To see that it is a Euclidean domain, consider the function $\lambda \colon \mathbb{C} \to \mathbb{R}$ given by $\lambda(z) = |z|^2$, which sends non-zero elements of $\mathbb{Z}[i]$ to $\mathbb{N}$. If $z, w \in \mathbb{Z}[i]$ with $w \neq 0$, pick $a, b \in \mathbb{Z}$ such that $|\text{Re}(z/w) - a| \leq 1/2$ and $|\text{Im}(z/w) - b| \leq 1/2$. Then with $v = a + ib \in \mathbb{Z}[i]$ and $u = z - wv$, we have $z = wv + u$, and either $u = 0$ or

$$\lambda(u) = \lambda(w(z/w - v)) = \lambda(w)\lambda(z/w - v) < \lambda(w)$$

because

$$\lambda(z/w - v) = (\text{Re}(z/w) - a)^2 + (\text{Im}(z/w) - b)^2 \leq 1/4 + 1/4.$$

◇

The second example is quite similar.

Example 6.12.7 Let $\omega = (-1 + i\sqrt{3})/2$. Then $\omega^2 = -1 - \omega$. Hence

$$\mathbb{Z}(\omega) = \{a + b\omega \mid a, b \in \mathbb{Z}\}$$

is a subring of $\mathbb{C}$. Note that $\bar{\omega} = \omega^2 = -1 - \omega$, so $\mathbb{Z}(\omega)$ is closed under complex conjugation. The map $\lambda \colon \mathbb{C} \to \mathbb{R}$ given by $\lambda(z) = |z|^2 = z\bar{z}$ sends non-zero elements of $\mathbb{Z}(\omega)$ to $\mathbb{N}$ as

$$\lambda(a + b\omega) = (a + b\omega)(a + b\bar{\omega}) = a^2 - ab + b^2.$$

Moreover, if $z, w \in \mathbb{Z}(\omega)$ with $w \neq 0$, there are real numbers x and y such that $z/w = z\bar{w}/w\bar{w} = x + y\omega$. Pick $a, b \in \mathbb{Z}$ such that $|x - a| \leq 1/2$ and $|y - b| \leq 1/2$. Then with $v = a + b\omega \in \mathbb{Z}[i]$ and $u = z - wv$, we have $z = wv + u$, and either $u = 0$ or

$$\lambda(u) = \lambda(w(z/w - v)) = \lambda(w)\lambda(z/w - v) < \lambda(w)$$

because

$$\lambda(z/w - v) = (x - a)^2 - (x - a)(y - b) + (y - b)^2 \leq 3/4.$$

◇

Hence both $\mathbb{Z}[i]$ and $\mathbb{Z}[\omega]$ are PID's, and as we shall see, elements of such rings decompose uniquely in a certain sense.

6.13 Prime Ideals and Irreducible Ideals

Definition 6.13.1 Let R be an integral domain. We say that a non-zero element $a \in R$ *divides* $b \in R$, and write $a|b$, if $b = ac$ for some $c \in R$. A *unit* is an element $u \in R$ that divides 1. An element $a \in R$ is an *associate* of $b \in R$ if $a = ub$ for a unit $u \in R$. An element $a \in R$ is *irreducible* if is not a unit and if the only elements that divide a are units or associates of a, otherwise a is *reducible*. A non-zero element $p \in R$ is *prime* if it is not a unit and if $p|ab$ only when $p|a$ or $p|b$.

Note that the units are the elements that have a multiplicative inverse, and that the set of units is an abelian group under multiplication. The relation of being associates is an equivalence relation.

The units of $\mathbb{Z}$ are 1 and -1, so the set of associates of $a \in \mathbb{Z}$ is $\{-a, a\}$. In the ring of integers the irreducible elements are $\pm p$, where p is a prime number. The definition of irreducible elements mimics that of prime numbers up to signs.

Prime elements are automatically irreducible because if p is prime and $a|p$, then $p = ab$ for some b, so $p|ab$. Then either $p|a$, in which case $a = pc$ for some c, so $p = pcb$ and $1 = bc$, showing that a is an associate of p. Or $p|b$ and then $b = pd$, so $p = apd$ and a is a unit with inverse d.

The converse is not true in general; irreducible elements need not be prime.

Some of these notions can be translated into the language of ideals, as the following easily proved result shows.

Proposition 6.13.2 *Suppose R is an integral domain. Then:*

1. $a|b$ *if and only if* $(b) \subset (a)$;
2. $u \in R$ *is a unit if and only if* $(u) = R$;
3. a *and* b *are associates if and only if* $(a) = (b)$;
4. p *is irreducible if and only if* (p) *is proper and non-trivial and not properly contained in a proper principal ideal.*
5. p *is prime if and only if* (p) *is proper and non-trivial and if* $ab \in (p)$ *implies that* $a \in (p)$ *or* $b \in (p)$.
6. p *is prime if and only if* (p) *is proper and non-trivial and if* $(a)(b) \subset (p)$ *implies that* $(a) \subset (p)$ *or* $(b) \subset (p)$.

We see that the irreducible elements in a PID are precisely the generators of non-trivial maximal ideals.

The last equivalence suggests the following definition.

Definition 6.13.3 An ideal P in a ring is *prime* whenever $I \subset P$ or $J \subset P$ for ideals I and J with $IJ \subset P$.

It is easy to check that an ideal P in a commutative ring is prime if and only if $ab \in P$ implies that $a \in P$ or $b \in P$.

Proposition 6.13.4 *Suppose R is a commutative unital ring, and let P be an ideal in R. Then P is prime if and only if R/P is an integral domain.*

Proof Note that R/P is an integral domain if and only if $(a+P)(b+P) = P$ implies that either
$$a + P = P \text{ or } b + P = P.$$
This amounts to saying that $ab \in P$ implies that either $a \in P$ or $b \in P$, so the result follows from the proposition. □

The ideal (0) in $\mathbb{Z}$ is prime, but not maximal.

Proposition 6.13.5 *Maximal ideals in a unital ring are prime.*

Proof Suppose M is a maximal ideal in a unital ring R, and that $IJ \subset M$ for ideals I and J. If I is not contained in M, then $I + M$ is an ideal that properly contains M, so $I + M = R$. In particular, there are elements $a \in I$ and $b \in M$ such that $1 = a + b$, and then $J = 1J = aJ + bJ \in IJ + M \subset M$. □

Corollary 6.13.6 *An irreducible element in a PID is prime.*

Proposition 6.13.7 *In a PID proper prime ideals are maximal.*

Proof Say R is a PID with a proper prime ideal P. If P is not maximal, there is a proper ideal M that properly contains P. Since R is a PID, there are elements $a, b \in R$ such that $P = aR$ and $M = bR$. Hence there exists $c \in R$ such that $bc = a \in aR$. As $b \notin aR$, and $aR = P$ is prime, we must have $c \in aR$. Say $c = ad$ for some $d \in R$. Then $a = bc = bad$, and by cancellation, we see that $1 = bd$, so bR is unital and $M = R$, which is a contradiction. □

We have seen that in an integral domain, non-trivial maximal ideals are prime, and that prime elements are irreducible. In a PID we have implications the other way; irreducible elements are prime, and proper prime ideals are maximal, so there is no distinction between prime elements and irreducible elements. Moreover, they are single generators of the non-trivial maximal ideals.

6.14 Unique Factorization in a PID

We will see that PID's allow for a 'fundamental theorem of arithmetic'.

Definition 6.14.1 Two elements a and b of an integral domain are *relatively prime* if their only common divisors are units. A *greatest common divisor* of two elements a and b of an integral domain is an element d that divides both a and b, and if d' is another element with this property, then $d'|d$.

Greatest common divisors need not exist, but if they do, they are unique up to unit factors. Note that a and b are relatively prime if they have a greatest common divisor which is a unit.

Note that d is a common divisor of a and b if and only if $(a, b) \subset (d)$, and d is greater than any other common divisor d' of a and b if and only if $(d) \subset (d')$.

Proposition 6.14.2 *If $(a, b) = (d)$ for elements a, b, d of an integral domain, then d is a greatest common divisor of a and b. Hence any two elements in a PID have a greatest common divisor.*

Proof As $(a, b) \subset (d)$, the element d divides both a and b. If d' also divides a and b, then $(d) = (a, b) \subset (d')$, so d is a greatest common divisor of a and b. □

Corollary 6.14.3 *If a and b are relatively prime in a PID, then (a, b) is the whole ring.*

Here is another argument for why irreducible elements in a PID are prime. Suppose p is irreducible and that $p|ab$ but that a is not divisible by p. Then a and p are relatively prime, so $R = (a, p) = aR + pR$. Thus

$$(b) = bR = b(aR + pR) = (ab, pb) \subset (p)$$

and $p|b$, so p is prime.

The following lemma uses the *ascending chain condition* for PID's; a property which we will return to in the context of modules.

Lemma 6.14.4 *Any chain $(a_1) \subset (a_2) \subset \cdots$ in a PID breaks off after finitely many steps, that is, there is an n such that $(a_n) = (a_{n+1}) = \cdots$.*

Proof Obviously the union of all ideals in the chain is an ideal, and hence of the form (a). So $a \in (a_n)$ for some n, and therefore $(a) = (a_n) = (a_{n+1}) = \cdots$. □

Lemma 6.14.5 *Every non-zero non-unit element of a PID is a product of irreducible elements.*

Proof Let a be a non-zero element that is not a unit.

We claim that a has an irreducible factor. If a is irreducible, we are done. Otherwise we can write $a = a_1 b_1$ for elements a_1 and b_1 that are not units. If a_1 is irreducible, we are done. Otherwise $a_1 = a_2 b_2$ for elements a_2 and b_2 that are not units. If a_2 is irreducible, we are done. Otherwise $a_2 = a_3 b_3$ for elements a_3 and b_3 that are not units. If this never stops, we get a chain $(a) \subset (a_1) \subset (a_2) \subset \cdots$ of proper inclusions, which contradicts the previous lemma. So a_n has to be irreducible for some n, and the claim is valid.

In fact, we assert that the element a is a product of irreducible elements. If a is irreducible, we are done. Otherwise there is an irreducible element p_1 such that $p_1|a$, so $a = p_1 b_1$ for some element b_1. If b_1 is a unit, we are done since $p_1 b_1$ is irreducible. Otherwise there is an irreducible element p_2 such that $p_2|b_1$, so $a = p_1 p_2 b_2$ for some element b_2. If b_2 is a unit, we are done. Otherwise there is an irreducible element p_3 such that $p_3|b_2$, so $a = p_1 p_2 p_3 b_3$ for some element b_3. If this never stops, we get a chain $(a) \subset (b_1) \subset (b_2) \subset \cdots$ of proper inclusions, which contradicts the previous lemma. So b_n has to be a unit for some n, and we are done. □

6.14 Unique Factorization in a PID

Theorem 6.14.6 *Suppose R is a PID. Then every non-zero element that is not a unit can be factorized into a product of finitely many irreducible elements. Furthermore, this factorization is unique in the following sense: Any two factorizations have the same number of irreducible elements, and these are modulo rearrangements, unique up to associates.*

Proof Existence was established in the previous lemma.

As for uniqueness, first notice that if an irreducible element p in a PID divides a finite product of elements, then p has to divide one of the factors. This follows by induction and the fact that irreducible elements in a PID are prime.

Now say

$$p_1 \cdots p_m = q_1 \cdots q_n$$

are two decomposition into irreducible elements of a non-zero element that is not a unit. Then p_1 has to divide the right hand side, so it must divide one of the q's, say q_1 upon reordering of factors. Hence $q_1 = p_1 u_1$, and u_1 is a unit as both p_1 and q_1 are irreducible. Cancelling p_1 in an integral domain, we therefore get

$$p_2 \cdots p_m = u_1 q_2 \cdots q_n.$$

Continuing this we finally get

$$1 = u_1 \cdots u_m q_{m+1} \cdots q_n,$$

which says that g_n is a unit, an absurdity, unless $n = m$.

In the process we have also shown that all the p's are associates of the q's. □

As we see, there is some arbitrariness in choosing the irreducible elements in the decomposition, they are only determined up to units. In $\mathbb{Z}$ there was a natural choice, we could work with the prime numbers. And in the polynomial ring $F[x]$ over a field, we could work with monic polynomials since the units are the non-zero scalars. But in general there is no preferred choice.

The same factors up to units, can of course appear several times in the decomposition.

Proposition 6.14.7 *Let p be a prime element in a PID. For every non-zero element a, there is a unique non-negative integer n such that p^n divides a and p^{n+1} does not divide a.*

Proof Otherwise there would be an element b_n such that $a = p^n b_n$ for every non-negative integer n. Then $p b_{n+1} = b_n$ by cancelling p^n. But then

$$(b_1) \subset (b_2) \subset (b_3) \subset \cdots$$

is an ascending chain of proper inclusions, since equality at any step would force p to be a unit. And such a chain contradicts Lemma 6.14.4. □

Definition 6.14.8 The *order* ord$_p$ *a of the prime element* p in a non-zero element a of a PID is the unique number n described in the proposition above.

Let p be a prime element in a PID. The function ord$_p$ enjoys the following three properties:

(i) ord$_p u = 0$ for a unit u;

(ii) ord$_p q = 1$ if q is an associate of p, and equals 0 if q is a prime that is not an associate of p;

(iii) ord$_p (ab) =$ ord$_p a +$ ord$_p b$ if a and b are non-zero.

The two first properties are obvious. As for the third, let $m =$ ord$_p a$ and $n =$ ord$_p b$. Then $a = p^m c$ and $b = p^n d$, and p will not divide c and d. Thus $ab = p^{m+n} cd$, and p will not divide cd as p is prime. Therefore ord$_p (ab) = m + n$.

Now suppose we have a decomposition

$$a = u \prod_q q^{n(q)}$$

of a non-zero element a that is not a unit, with factors consisting of a unit u and pairwise non-associate prime elements q. Then we can apply the function ord$_p$ to both sides, and we immediately get $n(p) =$ ord$_p a$. So the exponents in a decomposition are uniquely determined. But then the unit u is also determined. This is another proof of the uniqueness statement of the theorem above, and the exponent of any irreducible factor occurring in the decomposition is the maximal one for this factor.

6.15 Unique Factorization Domains

It is fruitful to consider integral domains that from the outset offer a 'fundamental theorem of arithmetic'.

Definition 6.15.1 A *unique factorization domain* (UFD) is an integral domain that satisfies the conclusion of the theorem above.

So by the same theorem, a PID is a UFD.

Any two, and thus any finite number of, elements in a UFD have obviously a greatest common divisor.

Definition 6.15.2 Suppose R is a UFD. The *content* of $f \in R[x]$ is a greatest common divisor cont$(f) \in R$ of the coefficients of f, so the content of f is uniquely determined up to a unit factor. If cont$(f) = 1$ and deg$(f) > 0$, we say that f is *primitive*.

6.15 Unique Factorization Domains

Given $f \in R[x]$ where R is a UFD, we see that $\text{cont}(af) = a \, \text{cont}(f)$ for $a \in R$. Therefore for a non-constant $f \in R[x]$, we can write $f = \text{cont}(f) f_1$, where f_1 is primitive and uniquely determined up to a unit factor.

Proposition 6.15.3 *Let R be a UFD. The product of finitely many primitive polynomials of $R[x]$ is primitive.*

Proof Let f and g be primitive. We have to show that any irreducible $p \in R$ does not divide all the coefficients of fg.

Define a homomorphism $h \to [h]$ from $R[x]$ to $(R/(p))[x]$ by applying the quotient map $R \to R/(p)$ to the coefficients of h.

By assumption $[f]$ and $[g]$ are non-zero, and then also $[fg] = [f][g] \neq 0$ because p is prime (being irreducible in a UFD), and then $R/(p)$ is an integral domain by Proposition 6.13.4. □

Corollary 6.15.4 *Let R be a UFD. Then up to unit factors, we have*

$$\text{cont}(fg) = \text{cont}(f) \, \text{cont}(g) \quad \text{and} \quad (fg)_1 = f_1 g_1$$

for non-constant $f, g \in R[x]$.

Lemma 6.15.5 *Let R be a UFD, and let F be the field of quotients of the integral domain R. Consider $f \in R[x]$. Then f factorizes non-trivially in $F[x]$ if and only if f factorizes non-trivially in $R[x]$. Moreover, the non-constant factors of $f \in R[x]$ can be chosen to be primitive polynomials, and the remaining factor in R is then unique up to multiplication by units.*

Proof If f factors non-trivially in $R[x]$, then because $R[x]$ is a subring of $F[x]$, it clearly also factors non-trivially in $F[x]$.

Conversely, if f is a product of lower degree polynomials in $F[x]$, then clearing denominators, we can write $df = gh$ for $g, h \in R[x]$ and $d \in R$. Then

$$f_1 = (df)_1 = (gh)_1 = g_1 h_1,$$

so $f = \text{cont}(f) g_1 h_1$ up to a unit factor. □

Corollary 6.15.6 *Let R be a UFD, and let F be the field of quotients of the integral domain R. The irreducible elements of $R[x]$ are the irreducible elements of R and the primitive polynomials in $R[x]$ that are irreducible as elements of $F[x]$.*

Theorem 6.15.7 *If R is a UFD, then $R[x]$ is a UFD.*

Proof Existence of decomposition of an element in $R[x]$ into irreducible elements follows immediately from the lemma and its corollary and the fact that $F[x]$ is a UFD.

Uniqueness of the decomposition is again clear from the lemma and its corollary because as $F[x]$ is a UFD, the factors are unique up to associates in $F[x]$. □

Corollary 6.15.8 *If R is a UFD, then $R[x_1, \ldots, x_n]$ is a UFD.*

Proof Follows by induction as $R[x, y] = (R[x])[y]$. □

There are UFD's that are not PID's.

Example 6.15.9 Let F be a field. Then the ideal I generated by the elements x and y of the integral domain $F[x, y]$ cannot be principal, so $F[x, y]$ is not a PID. However, it is a UFD, as the corollary shows. ◇

Clearly $\mathbb{Z}[x]$ is a UFD as $\mathbb{Z}$ is a PID and hence a UFD. But $\mathbb{Z}[x]$ is not a PID because elements of the form $a + xf$, where $f \in \mathbb{Z}[x]$ and $a \in 2\mathbb{Z}$, is an ideal in $\mathbb{Z}[x]$ that is not principal, as can be checked by trivial calculations with coefficients of polynomials.

Example 6.15.10 The set of all $a + b\sqrt{5}$ for $a, b \in \mathbb{Z}$ is a subring R of $\mathbb{C}$, and hence an integral domain. The elements $3, 2 \pm \sqrt{5}$ are irreducible in R, and

$$3^2 = (2 + \sqrt{5})(2 - \sqrt{5}),$$

so R is not a UFD. ◇

Example 6.15.11 Consider the quotient ring $R \equiv \mathbb{C}[x, y, z]/I$, where I is the ideal in $\mathbb{C}[x, y, z]$ generated by $x^2 + y^2 + z^2 - 1$. Identifying x, y and z with their images in R under the quotient map, we get $x^2 + y^2 + z^2 = 1$. The ring R is not a UFD because

$$(x + iy)(x - iy) = (1 + z)(1 - z).$$

◇

Chapter 7
Field Extensions

In this chapter we investigate the relationship between roots and extensions of fields. Let F be a field with an extension E, meaning that E is a field containing F as a subfield. We can regard E as a vector space over F, and we say that the extension is finite if the dimension $[E, F]$ of E is finite. A root of $p(x) \in F[x]$ in E is an element $a \in E$ such that $p(a) = 0$. The element a is then said to be algebraic over F, and if E consists solely of such elements, we say it is an algebraic extension of F. If the root belongs to F, then by the division algorithm for polynomials, we can write $p(x) = (x-a)q(x)$ for $q(x) \in F[x]$. Continuing in this fashion, we see that if we have $\deg(p(x))$ distinct roots in F, then we obtain a complete factorization of $p(x)$ into first order polynomials. This stresses the relationship between reducibility of a polynomial in a field and the roots of the polynomial in the field. Be aware though, that if $p(x)$ has no roots in F, it doesn't necessarily mean that it is irreducible over F; it can split into higher order irreducible polynomials thanks to roots lying in some extension of F.

It is true that any irreducible polynomial has a root in some extension. Indeed, assume that $p(x)$ is irreducible. Then the ideal (p) in $F[x]$ generated by p is a maximal ideal in a PID, so its quotient will be an extension of F, and the equivalence class $[x]$ will be a root of p, seen now as a polynomial over this extension. If on the other hand, an element $a \in E$ is algebraic over F, then the kernel of the homomorphism $f \in F[x] \mapsto f(a) \in E$ is a proper ideal in a PID, so it must be generated by a unique irreducible monic polynomial, called the minimal polynomial of $a \in E$ over F. This means, by a simple application of the division algorithm, that the subfield $F(a)$ in E generated by F and a, called a simple extension, has the same dimension as the degree of the minimal polynomial. But any finite extension is algebraic since the powers of an element cannot form a linear independent set. Hence $F(a)$ is algebraic, and so is any extension of F generated by F and finitely many algebraic elements.

Recall that a field is algebraically closed if it has no proper algebraic extensions, and an algebraic extension of a field F is called its algebraic closure, denoted by

$\overline{F}$, if it is algebraically closed. The notation is justified as we furnish the non-trivial result saying that F always has an algebraic closure, and such a closure is unique up to isomorphism. The algebraic closure is a convenient recipient for the roots of F, and of course every polynomial over such a closure decomposes completely into first order ones. In general, the algebraic closure is big, but still countable if the original field is countable. Typically, the algebraic numbers $\overline{\mathbb{Q}}$ is a countable set, and the further extension $\mathbb{C}$ consists of uncountable many transcendental elements in addition to the elements in $\overline{\mathbb{Q}}$.

At the opposite extreme is the splitting field of a polynomial $p(x)$ over F. This is the subfield in $\overline{F}$ generated by F and all the roots of $p(x)$. It is up to isomorphism, the unique smallest field where $p(x)$ splits into linear factors. A more general extension is that of a normal extension of F. It has by definition the property that if an irreducible polynomial over F has one root there, then all the roots will be there. One can show that an algebraic extension over F is normal if and only if it is a splitting field for a family of polynomials over F.

We study also multiple roots, and the notion of a separable extension of F consisting solely of elements having minimal polynomials over F with only simple roots. We say F is perfect if all its algebraic extensions are separable. Examples of perfect fields are fields of characteristic zero, and all finite fields. We characterize the simple fields among the finite extensions of a field F as those having only finitely many intermediate fields between.

We devote a section to finite fields. Their order is always a power of a prime number. We also devote a section to the antique problems of the impossibility of constructing certain geometric objects by rulers and compasses. We associate to such a construction a chain of subfields of $\mathbb{R}$, and deduce a property that such a chain leads to. For instance, to divide an angle by three will violate this property, thus making the construction impossible.

For suggested reading for this and the next chapter, see [15] and references therein.

7.1 Roots and Reducible Polynomials

Definition 7.1.1 Suppose F is a subfield of a field E. A *root* of $f \in F[x]$ in E is an element $a \in E$ such that $f(a) = 0$.

We should specify what field we are referring to when we talk about roots and reducibility. When we say that a polynomial is *reducible over a field* F we mean that it is not irreducible regarded as an element of the integral domain $F[x]$.

Example 7.1.2 The number $\sqrt{3}$ is a root of $x^2 - 3$ in $\mathbb{R}$, but not in $\mathbb{Q}$. Similarly, the numbers $\pm i$ are roots of $x^2 + 1$ in $\mathbb{C}$, but not in $\mathbb{R}$. ◇

Proposition 7.1.3 *Any polynomial over a field with degree greater than one that has a root in this field is reducible over the field.*

7.1 Roots and Reducible Polynomials

Proof Say F is a field and $a \in F$ is a root of $f \in F[x]$ with $\deg(f) \geq 2$. By the division algorithm for polynomials, we can write $f(x) = q(x)(x - a) + r$ for $r \in F$ and $q \in F[x]$ with $\deg(q) \geq 1$. But then $r = q(a)(a - a) + r = f(a) = 0$, so f is reducible. □

The converse is true for polynomials $f \in F[x]$ of degree 2 and 3, because then we can write $f(x) = g(x)(ax + b)$ for $a \neq 0$, and f has root $-ba^{-1}$.

Corollary 7.1.4 *A polynomial of degree n over a field has at most n distinct roots in the field.*

Proof Say f is a polynomial over a field with $\deg(f) = n$. If a is a root of f in the given field, then f is divisible by $x - a$, resulting in a polynomial of degree $n - 1$. This opens for an obvious induction argument. □

The corollary immediately implies that any polynomial over a field vanishes if it has infinitely many distinct roots in the field.

Proposition 7.1.5 *Any polynomial over $\mathbb{Z}$ can be non-trivially factorized over $\mathbb{Z}$ if and only if it is reducible over $\mathbb{Q}$.*

Proof The forward implication is trivial. Conversely, say $f \in \mathbb{Z}[x]$ is reducible over $\mathbb{Q}$. We can assume that f is primitive and $f = agh$ for non-constant primitive polynomials $g, h \in \mathbb{Z}[x]$ and $a \in \mathbb{Q}$. But then, since also gh is primitive, we get $a = \pm 1$. □

The following *integral root test* pulls in the same direction.

Proposition 7.1.6 *Any rational root of a monic polynomial over $\mathbb{Z}$ is an integer and divides the constant term of the polynomial.*

Proof Say $f(x) = a_0 + \cdots + a_{n-1}x^{n-1} + x^n \in \mathbb{Z}[x]$ has a root r/s with integers r and s that are relatively prime. Then

$$a_0 s^{n-1} + a_1 r s^{n-2} + \cdots + a_{n-1} r^{n-1} = -r^n/s$$

shows that $s = \pm 1$ and that r divides a_0. □

A more subtle criterion for irreducibility is the following one by Eisenstein.

Proposition 7.1.7 *If a prime number p divides all the coefficients of a polynomial $f \in \mathbb{Z}[x]$ except the coefficient of the highest order of x, and if the constant term of f is not divisible by p^2, then f cannot be non-trivially factorized.*

Proof Suppose to the contrary that

$$f(x) = a_0 + \cdots + a_n x^n = (b_0 + \cdots + b_m x^m)(c_0 + \cdots + c_k x^k)$$

with $a_i, b_i, c_i \in \mathbb{Z}$ and $b_m c_k \neq 0$ and $k < n$. Since p divides $a_0 = b_0 c_0$, whereas p^2 does not, then either p divides b_0 and not c_0, or p divides c_0 and not b_0. By symmetry it is enough to consider one of these cases, say the second one.

Since p does not divide $a_n = b_m c_k$, it cannot divide c_k. Let c_r be the first coefficient in $c_0 + \cdots + c_k x^k$ that is not divisible by p. Then

$$a_r = b_0 c_r + b_1 c_{r-1} + \cdots + b_r c_0$$

cannot be divisible by p because otherwise $b_0 c_r$, and hence c_r, would be. But a_n is the only coefficient of f with this property, so $r = n$, which is absurd. □

Using Eisenstein's criterion and Proposition 7.1.5, we conclude that $x^2 - 2$ is irreducible over $\mathbb{Q}$. The same is true for $x^n - m$ for any natural number n and any square free integer $m \neq \pm 1$.

Here is a standard example, where the criterion is applied to a translation of the polynomial of interest.

Example 7.1.8 The polynomial

$$f(x) = 1 + x + \cdots + x^{p-1}$$

with p prime, is irreducible over $\mathbb{Q}$. To see this, observe that

$$g(x) \equiv f(x+1) = \frac{(x+1)^p - 1}{(x+1) - 1} = x^{p-1} + \binom{p}{1} x^{p-2} + \cdots + \binom{p}{p-1}$$

obviously satisfies Eisenstein's criterion. Thus g is irreducible over $\mathbb{Q}$, and so is f, otherwise g would be reducible. ◇

7.2 Algebraic Extensions

Definition 7.2.1 If F is a subfield of a field E, then E is an *extension field* of F.

We also talk about extensions when we have a ring monomorphism $F \to E$, and we often suppress this map.

Note that an extension field E_i of E_{i-1} is a vector space over E_{i-1}. Denote the dimension of E_i over E_{i-1} by $[E_i, E_{i-1}]$. If all these dimensions are finite, that is, if the extension field E_i of E_{i-1} is finite for each i, then

$$[E_n : E_0] = [E_n : E_{n-1}] \cdots [E_1 : E_0]$$

because $\{x_{j_1}^1 \cdots x_{j_n}^n\}$ is by an easy successive argument, a linear basis for E_n over E_0 whenever $\{x_{j_i}^i\}$ is a linear basis for E_i over E_{i-1}.

Here is one of the reasons why we introduce extension fields.

7.2 Algebraic Extensions

Proposition 7.2.2 *Any non-constant polynomial over a field has a root in an extension field.*

Proof Suppose f is an irreducible polynomial over a field F. Then (f) is a maximal ideal in the principal ideal domain $F[x]$, so $E \equiv F[x]/(f)$ is an extension field of F. The equivalence class $[x]$ of the indeterminate x is a root in E of f because $f([x]) = [f(x)] = 0$.

Given any non-constant polynomial, we can decompose it into factors of irreducible ones, and a root in an extension field of any of these, will also be a root in the same extension field of the original polynomial. $\square$

Corollary 7.2.3 *Any finite collection of non-constant polynomials over a field all have roots in a common extension field.*

Proof Say $f_1, \ldots, f_n$ are non-constant polynomials over a field F. Let E_1 be an extension field of F such that f_1 has a root in E_1. View f_2 as a polynomial over E_1 and let E_2 be an extension field of E_1 such that f_2 has a root in E_2. Proceeding inductively this way completes the proof. $\square$

Definition 7.2.4 Let E be an extension field of F with $a_i \in E$. Write $F(a_1, \ldots, a_n)$ for the subfield of E generated by F and the a_i's. If E itself is of this form for some a_i's, then it is *finitely generated*. When $n = 1$ we talk about *simple extension fields*.

Example 7.2.5 The subfield $\mathbb{Q}(\sqrt{2})$ of $\mathbb{R}$ consists of all elements of the form $a + b\sqrt{2}$ for $a, b \in \mathbb{Q}$ because such elements form a subfield of $\mathbb{R}$ with

$$(a + b\sqrt{2})^{-1} = \frac{a - b\sqrt{2}}{a^2 + 2b^2},$$

and it is the smallest one that contains both $\mathbb{Q}$ and $\sqrt{2}$. Since $\sqrt{2}$ is irrational, we see that $\{1, \sqrt{2}\}$ is a basis for $\mathbb{Q}(\sqrt{2})$ over $\mathbb{Q}$, so $[\mathbb{Q}(\sqrt{2}) : \mathbb{Q}] = 2$. $\diamond$

Definition 7.2.6 Suppose E is an extension field of F. An element of E is *algebraic* over F if it is a root of a non-zero polynomial over F. If all elements of E are algebraic over F, then E is an *algebraic extension* of F.

Example 7.2.7 In the example above we actually have an algebraic extension because $a + b\sqrt{2}$ is a root of $(x - a)^2 - 2b^2 \in \mathbb{Q}[x]$. $\diamond$

Proposition 7.2.8 *A finite extension of a field F is an algebraic extension over F.*

Proof All integer powers of any non-zero element in the extension field cannot be linear independent over F, and any non-trivial linear combination between these says that the element is algebraic over F. $\square$

The field of fractions $F(x)$ of polynomials over a field F is clearly an extension field of F generated by F and x, but it is not an algebraic extension as the indeterminant $x \in F(x)$ by definition is not a root of any polynomial over F.

The following example shows that there are infinite algebraic extension fields.

Example 7.2.9 The field embeddings

$$\mathbb{Q} \subset \mathbb{Q}(\sqrt{2}) \subset \mathbb{Q}(\sqrt{2}, \sqrt{3}) \subset \mathbb{Q}(\sqrt{2}, \sqrt{3}, \sqrt{5}) \subset \cdots$$

within $\mathbb{R}$ are proper. This is certainly true for the first inclusion. Suppose the inclusion $F \equiv \mathbb{Q}(\sqrt{p_1}, \ldots, \sqrt{p_{n-1}}) \subset F(\sqrt{p_n})$ is proper, where $p_1, \ldots, p_n$ are n consecutive primes. For any prime q different from these, we have $\sqrt{q} \notin F(\sqrt{p_n})$ because otherwise we can write $\sqrt{q} = a + b\sqrt{p_n}$ for $a, b \in F$, so

$$q = a^2 + b^2 p_n + 2ab\sqrt{p_n},$$

which shows that $\sqrt{p_n} \in F$, and this contradicts our induction hypothesis.

Let E be the subfield of $\mathbb{R}$ generated by square roots of all primes. Clearly any $c \in E$ will belong to one of the fields in the ascending chain of inclusions above. Each of these fields is a finite extension field, and hence an algebraic extension, of $\mathbb{Q}$. So c is algebraic over $\mathbb{Q}$. Thus E is an algebraic extension of $\mathbb{Q}$ of infinite dimension. ◇

Suppose E is an extension field of a field F, and that $a \in E$ is algebraic over F. Then the kernel of the evaluation homomorphism $f \in F[x] \mapsto f(a) \in E$ is a nontrivial ideal in the principal ideal domain $F[x]$, so it is generated by a single monic polynomial. Since this polynomial divides any other polynomial over F with root a, it must have least degree among all polynomials over F with root a, and clearly it is uniquely determined by this property.

Definition 7.2.10 Let F be a field with an extension field E. The *minimal polynomial* Irr(a, F) of an algebraic element $a \in E$ over F is the monic polynomial of least degree with root a.

Minimal polynomials are irreducible because if they could be factorized nontrivially, then one of the factors would have the same root, and this factor would have lower degree than the minimal polynomial. By the division algorithm for polynomials the minimal polynomial of an element is clearly also the unique irreducible monic polynomial having that element as a root.

Proposition 7.2.11 *Suppose F is a field with an extension field E. Let n be the degree of the minimal polynomial of $a \in E$ over F. Then $\{a^m\}_{m=0}^{n-1}$ is a linear basis of the vector space $F(a)$ over F. So $[F(a) : F] = n$ and $F(a)$ is an algebraic extension of F.*

Proof Since Irr(a, F) is irreducible in the principle ideal domain $F[x]$, and since it generates the kernel of the evaluation homomorphism $f \mapsto f(a)$, the quotient ring of $F[x]$ by this kernel is a field. By the fundamental ring isomorphism theorem the image $\{f(a) \mid f \in F[x]\}$ of the evaluation homomorphism is therefore also a field, and this field is clearly the smallest subfield $F(a)$ of E that contains both a and F. By the division algorithm we can write any $f \in F[x]$ as $f = q \text{Irr}(a, F) + r$,

7.2 Algebraic Extensions

where $q, r \in F[x]$ and $\deg(r) < \deg(\mathrm{Irr}(a, F))$. So $f(a) = r(a)$, which shows that the vectors $\{a^m\}_{m=0}^{n-1}$ span $F(a)$ over F. They are also linear independent since any non-trivial F-linear combination of them shows that a is a root of a polynomial with lower degree than the minimal polynomial. $\square$

We have seen that $F(a) = \{f(a) \mid f \in F[x]\}$. To check more directly that $f(a)^{-1}$ belongs to this latter set when $f(a) \neq 0$, observe that by Corollary 6.14.3, there are polynomials $u, v \in F[x]$ such that $u\mathrm{Irr}(a, F) + vf = 1$ as $\mathrm{Irr}(a, F)$ cannot divide f because $f(a) \neq 0$. So $u(a)0 + v(a)f(a) = 1$ and $f(a)^{-1} = v(a)$.

Corollary 7.2.12 *Let F be a field with an extension field E. The field $F(a_1, \cdots a_n)$, where $a_i \in E$ are algebraic over F, is a finite extension field of F. In particular, it is an algebraic extension.*

Proof Let $E_i \equiv F(a_1, \ldots, a_i)$. Then a_1 is algebraic over F, and a_2 is algebraic over F and hence over E_1, and a_3 is algebraic over F and hence over E_2, and so on. Thus a_i is algebraic over E_{i-1} and $E_i = E_{i-1}(a_i)$. By the proposition we know that $[E_i : E_{i-1}]$ is finite, and hence $[E_n : F] = [E_n : E_{n-1}] \cdots [E_1, F] < \infty$. $\square$

Corollary 7.2.13 *Suppose F is a field with an extension field E. Then the subset of all elements of E that are algebraic over F is an algebraic extension of F.*

Proof Let K be the subset of all elements of E that are algebraic over F. We must show that K is a subfield of E. Take any $a, b \in K$. Then the subfield $F(a, b)$ of E is an algebraic extension of F, so $F(a, b) \subset K$, and as a, b belong to the field $F(a, b)$, so will $a \pm b$, ab and a^{-1}. $\square$

The previous result is less trivial than it seems. Of course the product x of $\sqrt{2}$ and $\sqrt{3}$ is algebraic over $\mathbb{Q}$ since $x^2 - 6 = 0$. And so is their sum y, since upon squaring two times, we get $y^4 - 10y^2 + 1 = 0$. And there are no problems with divisions either, say the reciprocal z of $\sqrt{3}$, which satisfies $3z^2 - 1 = 0$. The problem arises when one picks roots of general polynomials of arbitrary degree and wants to check that sums and products and inverses of such roots are algebraic.

We can nevertheless prove the previous result more directly using linear algebra. Consider two non-zero elements $a, b \in E$ that are algebraic over F. The crucial observation is that the vector space V over F spanned by the monomials $a^m b^n$ as m, n range over the integers, is finite dimensional. This is so because as soon as for instance m reaches the degree of the polynomial that a is a root of, then a^m is a F-linear combination of lower degree monomials in a. A similar reduction happens for negative m when it hits minus the degree of the polynomial. Now V is clearly invariant under multiplication by $a + b$, so with respect to any basis $\{x_i\}$ of V, we have $(a + b) \cdot x_i = \sum_j a_{ij} x_j$ for $a_{ij} \in F$. Thus $a + b$ is an eigenvalue with eigenvector $\sum_i x_i$ of the matrix (a_{ij}), and will be a root of its characteristic equation, so $a + b$ is algebraic over F. The same argument works for ab and a^{-1}, so we do get a field.

This method also helps to find polynomials having as roots combinations of other roots.

Example 7.2.14 We know that the monic polynomial $x^3 - 2$ is irreducible over $\mathbb{Q}$, so it is the minimal polynomial of $a \equiv \sqrt[3]{2} \in \mathbb{R}$ over $\mathbb{Q}$. Therefore the three vectors $x_1 = 1, x_2 = a$ and $x_3 = a^2$ form a basis for $\mathbb{Q}(\sqrt[3]{2})$ over $\mathbb{Q}$. Let $b \equiv a + a^2$, and let $A = (a_{ij}) \in M(3, F)$ be given by $bx_i = \sum_j a_{ij} x_j$. Then as $a^3 = 2$, we get $bx_1 = a + a^2 = x_2 + x_3$ and $bx_2 = a^2 + a^3 = 2x_1 + x_3$ and $bx_3 = a^3 + a^4 = 2x_1 + 2x_2$, so

$$A = \begin{pmatrix} 0 & 2 & 2 \\ 1 & 0 & 2 \\ 1 & 1 & 0 \end{pmatrix}$$

with characteristic equation

$$0 = \det(A - \lambda I_3) = -\lambda^3 + 6\lambda + 6,$$

which is consistent with what we get by hand:

$$b^3 = a^3(1+a)^3 = 2(1 + 3a + 3a^2 + a^3) = 6(a + a^2 + 1) = 6(b+1).$$

◇

7.3 Algebraic Closures

Definition 7.3.1 A field is algebraically closed if it possesses no proper algebraic extension.

Proposition 7.3.2 *For any field F the following conditions are equivalent:*

(i) *The field F is algebraically closed;*

(ii) *Every irreducible polynomial over F has degree one;*

(iii) *Every non-constant polynomial over F factors completely into linear factors;*

(iv) *Every non-constant polynomial over F has a root in F.*

Proof Assume that (i) holds and that f is an irreducible polynomial over F of degree n. By the previous section we know that F has an algebraic extension E in which f has a root, and that $[E : F] = n$. In order for this not to be a proper extension, we must have $n = 1$, so (ii) holds.

From the previous chapter any non-constant polynomial in the principal ideal domain $F[x]$ factors completely into irreducible polynomials, so if we assume (ii), we get (iii). Trivially (iii) implies (iv).

7.3 Algebraic Closures

Assuming (iv) holds and that E is some algebraic extension of F. The minimal polynomial of $a \in E$ over F is irreducible, and since it has a root in F, its degree must be one, so a is this root and $E = F$. □

Definition 7.3.3 An algebraic extension E of a field F is an *algebraic closure* of F if it is algebraically closed.

We will show that algebraic closures exist and are unique in a natural sense. For this fundamental result we need a lemma.

Lemma 7.3.4 *Every field has an extension field that is algebraically closed.*

Proof First we show that there exists an extension field E_1 of a field F in which every non-constant polynomial over F has a root.

Let I denote the subset of $F[x]$ consisting of non-constant polynomials. Consider the group ring $F[\mathbb{Z}^I]$ of the additive group $\mathbb{Z}^I$. Let $F[I]$ denote the subring generated by the unit 1 and the members x_i of $\mathbb{Z}^I \subset F[\mathbb{Z}^I]$, where $x_i(j) = \delta_{ij}$ for $i, j \in I$. Thus $F[I]$ consists of 1 and all elements that are finite sums of the form

$$\sum a_i x_{i_1} \cdots x_{i_n},$$

where the coefficients $a_i \in F$ are uniquely determined by the element. We think of $F[I]$ as a polynomial ring in an infinite number of commuting variables x_i.

Let J be the ideal in $F[I]$ generated by the singled variable polynomials $f(x_f)$ as f ranges over $I \subset F[x]$. We claim that $J \neq F[I]$. If on the contrary $1 \in J$, then

$$g_1 f_1(x_{f_1}) + \cdots + g_n f_n(x_{f_n}) = 1$$

for some $f_m \in F[x]$ and $g_m \in F[I]$. By Corollary 7.2.3 the polynomials f_m have roots b_m in a common extension field. Replacing x_{f_m} by b_m in the identity above, and with zeroes for the variables of the polynomials g_m, we get $0 = 1$, which is absurd. So J is a proper ideal in $F[I]$.

Therefore, by Zorn's lemma, the ideal J is contained in a maximal ideal, and the quotient of $F[I]$ by this maximal ideal is a field E_1 that extends F. Moreover, every non-constant polynomial f over F has a root in E_1 since the maximal ideal contains $f(x_f)$.

Inductively we can thus form an ascending chain $E_1 \subset E_2 \subset \cdots$ of fields E_k such that every non-constant polynomial over E_{k-1} has a root in E_k. Obviously $E \equiv \cup_k E_k$ is a field that extends F. It is also algebraically closed because any non-constant polynomial over E will be a polynomial over E_k for some k, and will therefore have a root in $E_{k+1} \subset E$. □

The algebraically closed field in the lemma above is not an algebraic extension. We had no reservations making it large since it just serves as a recipient for roots of the original field.

Theorem 7.3.5 *Every field has an algebraic closure.*

Proof Let F be a field, and let E be any extension field of F that is algebraically closed; the existence of such a field is guaranteed by the lemma above.

Let K be the set of all elements of E that are algebraic over F. By Corollary 7.2.13 we know that K is an algebraic extension of F. It is also algebraically closed. For suppose f is a non-constant polynomial over K. Viewed as an element of $E[x]$, it has a root a in E since E is algebraically closed. Let $b_1, \ldots, b_m \in K$ be the coefficients of f, and consider the finite extension $E_0 \equiv F(b_1, \ldots, b_m)$ of F. As $E_0(a)$ is a finite extension of E_0, it is therefore also a finite extension of F, and hence an algebraic extension. Thus $a \in E_0(a)$ is algebraic over F, so by definition it belongs to K. □

In the proof above we showed that if we have a chain of fields $F \subset K \subset E$ such that K is an algebraic extension of F and $a \in E$ is algebraic over K, then a is algebraic over F. Let us record the obvious generalization of this result.

Corollary 7.3.6 *If we have an ascending chain $E_0 \subset \cdots \subset E_n$ of fields such that E_i is algebraic over E_{i-1} for all i, then E_n is algebraic over E_0.*

Using linear algebra we can give a more direct proof of the theorem above provided we know that the field F has some extension field E that is algebraically closed. Using linear algebra we have already seen that the set K of elements of E that are algebraic over F is a field, and hence an algebraic extension of F. To see that it is algebraically closed, first note that any non-constant polynomial f over K obviously has a root a in E. To verify that $a \in K$, let $V \subset E$ be the vector space over F spanned by the coefficients of f and all integer powers of a. Arguing as before we conclude that V is finite dimensional, and the element a will be a root of the characteristic equation with coefficients in F of the matrix corresponding via a chosen linear basis of V to multiplication by a. So by definition $a \in K$.

The fundamental theorem of algebra says that $\mathbb{C}$ is algebraically closed, see Theorem 8.4.2. Using this we see that all subfields of $\mathbb{C}$ have algebraic closures without referring to the lemma above. However, the enveloping field that the lemma provides us with is by construction countable if the field we start with is countable, and this is not the case for $\mathbb{C}$.

Let us now turn to uniqueness, to the task of proving that every field has an algebraic closure that is unique up to isomorphisms that fix elements of the original field.

Lemma 7.3.7 *Suppose α is a monomorphism from a field F to an algebraically closed field L. Let $F(a)$ be a simple algebraic extension of F. Then α can be extended to a homomorphism $F(a) \to L$.*

Proof Say $p(x) = a_0 + \cdots + a_n x^n \in F[x]$ is the minimal polynomial of a. Let $b \in L$ be a root of the polynomial $q(x) = \alpha(a_0) + \cdots + \alpha(a_n) x^n \in L[x]$. Such a root

7.3 Algebraic Closures

exists because L is algebraically closed. Since $\{a^k\}_{k=0}^{n-1}$ is a linear basis of $F(a)$ over F, we can extend α to a linear map $F(a) \to L$ by

$$c_0 + c_1 a + \cdots + c_{n-1} a^{n-1} \mapsto \alpha(c_0) + \alpha(c_1)b + \cdots + \alpha(c_{n-1})b^{n-1}$$

for $c_i \in F$, and this map is clearly a homomorphism. □

Next we prove the following general extension result.

Theorem 7.3.8 *Let K be an algebraic extension of a field F, and let $\alpha \colon F \to L$ be a monomorphism into an algebraically closed field L. Then α can be extended to a monomorphism $K \to L$.*

Proof Let S be the set of pairs (E, γ), where E is a subfield of K that contains F, and $\gamma \colon E \to L$ is an extension of α. The set S is non-empty as $(F, \alpha) \in S$. It has an obvious partial order with $(E, \gamma) \leq (E', \gamma')$ if E is a subfield of E' and if γ' restricted to E equals γ. Also, if (E_i, γ_i) is a chain, then $(\cup_i E_i, \beta)$ with $\beta(a) = \gamma_i(a)$ if $a \in E_i$, is an upper bound. It is straightforward to check that $\cup_i E_i$ is a field, and that $\beta \colon \cup_i E_i \to L$ is a well-defined homomorphism that extends α.

Therefore S has a maximal element by Zorn's lemma, which we claim is the required extension. Otherwise, there exists $a \in K$ that does not belong to the maximal extension, and this contradicts the lemma above since a then provides a simple extension that is strictly larger.

Since the extension $K \to L$ takes the identity to the identity, and is a homomorphism between fields, it has to be injective because non-zero elements of K are invertible. □

Corollary 7.3.9 *If F is a field with algebraic closures K and L, then there exist an isomorphism $\alpha \colon K \to L$ such that $\alpha(a) = a$ for $a \in F$.*

Proof By the theorem above the identity map $F \to F$ extends to an monomorphism $\alpha \colon K \to L$. Then $\alpha(K)$ is an algebraically closed field and an algebraic extension of F, and since L is also an algebraic extension of F, it cannot be larger than $\alpha(K)$. So α is an isomorphism from K onto L. □

Definition 7.3.10 We denote the algebraic closure of a field F by $\overline{F}$. The *algebraic numbers* are the members of $\overline{\mathbb{Q}}$.

So the algebraic numbers are the complex roots of polynomials with integer coefficients not all zero. Clearly nothing is gained by considering rational coefficients, and of course, all complex numbers with rational real- and imaginary parts are in particular algebraic.

As we have seen, the set of algebraic numbers is countable. We can also convince us of this by observing that an n-th degree equation has maximally n distinct roots, and that there are only countable many n-th degree equations with integer coefficients.

Definition 7.3.11 A *transcendental number* is a complex number that is not algebraic. We say that an element in an extension field of a field F is *transcendental over a field* F if it is not algebraic over F, and the extension field is then a *transcendental field*.

So the transcendental numbers are uncountable, and it is easy to see that they have cardinality $|\mathbb{C}| = |\mathbb{R}|$. In any case they certainly do exist, and this countability argument was actually Cantor's proof of their existence.

The algebraic numbers meet the quest for numbers solving algebraic equations, provided one can manage without the transcendental numbers necessary for Cauchy completeness; in this respect the complex numbers is an overkill.

We have proper field inclusions $\mathbb{Q} \subset \overline{\mathbb{Q}} \subset \mathbb{C}$, but the field of complex numbers is not an algebraic extension of $\mathbb{Q}$. Neither is it an algebraic extension of $\overline{\mathbb{Q}}$ because the latter is already algebraically closed. However, it is an algebraic extension of $\mathbb{R}$ since any complex number can be written in normal form and $i^2 + 1 = 0$.

Let us record the following useful result.

Corollary 7.3.12 *Let a be an algebraic element of a field F, and let L be an algebraically closed field. Then the number of monomorphic extensions $F(a) \to L$ of a monomorphism $F \to L$ equals the number of distinct roots in $\overline{F}$ of the minimal polynomial over F of a.*

Proof Recall that if $p(x) = a_0 + \cdots + a_n x^n$ is the minimal polynomial of a, then an extension of a monomorphism $\alpha \colon F \to L$ is given by

$$c_0 + \cdots + c_{n-1} a^{n-1} \mapsto \alpha(c_0) + \cdots + \alpha(c_{n-1}) b^{n-1}$$

for $c_i \in F$, where $b \in L$ is a root of $q(x) = \alpha(a_0) + \cdots + \alpha(a_n) x^n$.

Clearly, this gives a bijection between the set of distinct roots of q in L and the monomorphic extensions $F(a) \to L$ of α. We claim that these roots are in one-to-one correspondence with the distinct roots of p in $\overline{F}$.

By the theorem above extend $\alpha \colon F \to L$ to a monomorphism $\beta \colon \overline{F} \to L$, and define a ring monomorphism $\eta \colon \overline{F}[x] \to L[x]$ by

$$\eta(d_0 + \cdots + d_m x^m) = \beta(d_0) + \cdots + \beta(d_m) x^m$$

for $d_i \in \overline{F}$. Then $q = \eta(p)$ and uniqueness of decompositions of p over $\overline{F}$ and q over L shows that p and q have the same number of distinct roots. $\square$

7.4 Ruler and Compass

According to the Greeks circles and straight lines are perfect figures, so everything ought to be constructed by compasses and rulers, and by rulers was meant unmarked straight edges. However, using only these tools, they ran into problems. In particular,

7.4 Ruler and Compass

they could not duplicate the cube, trisect the angle or square the circle. This was not due to lack of skill. As it turns out, these challenges were not merely difficult, they were simply impossible. In this section we will explain why.

Let us first be clear about what we here mean by constructing.

Definition 7.4.1 Given a set X of points in the Euclidean plane $\mathbb{R}^2$. A point $x \in \mathbb{R}^2$ is constructed from X by ruler and compass if it is obtained after finitely many steps starting from X and at each step adding new points gotten as intersections of two distinct straight lines or circles drawn respectively with a ruler between old points or with a compass centered at some old point and adjusted to pass through another old point.

This way one can for instance construct the midpoint between two given points; first draw the straight line between them, then draw two distinct circles with center at these points and with radius the distance between the two points, and finally draw the straight line between the intersection of these two circles, which then cuts the former straight line at the midpoint. From the original two points we constructed in the process three new intersection points p_1, p_2 and p_3, of which p_3 was the midpoint obtained at the third step.

How does field theory enter the picture?

Definition 7.4.2 Suppose a point is constructed in n steps from $X \subset \mathbb{R}^2$ by ruler and compass, producing as intersection points $p_1, \ldots, p_n$. The chain of fields associated to this construction is the ascending chain

$$E_0 \subset E_1 \subset \cdots \subset E_n$$

of subfields of $\mathbb{R}$, where E_0 is the subfield of $\mathbb{R}$ generated by the coordinates of X, and E_1 is the subfield generated by E_0 and the coordinates of p_1, and so on. Thus, if $p_i = (x_i, y_i)$, then $E_i = E_{i-1}(x_i, y_i)$.

Proposition 7.4.3 *Suppose $E_i = E_{i-1}(x_i, y_i)$ is a chain of fields associated to a construction of a point from $X \subset \mathbb{R}^2$ by ruler and compass producing intersection points $p_i = (x_i, y_i)$ with $i \in \{1, \ldots, n\}$. Then x_i and y_i are roots in E_i of second degree polynomials over E_{i-1}.*

Proof The point p_i is an intersection of either two straight lines, a straight line and a circle, or two circles. The straight lines considered here are assumed to pass through at least two points with coordinates in E_{i-1}, and the circles have centers with coordinates in E_{i-1} and with circumferences that pass through points with coordinates in E_{i-1}.

Let us look at the case where a straight line through points a and b meets a circle with center c and having radius r equaling the distance between c and d, where the coordinates a_j, b_j, c_j, d_j of a, b, c, d, respectively, all belong to E_{i-1}. Note that $r^2 = (c_1 - d_1)^2 + (c_2 - d_2)^2 \in E_{i-1}$. Since p_i lies on the straight line that goes through a and b, we have

$$\frac{y_i - b_2}{x_i - a_1} = \frac{b_2 - a_2}{b_1 - a_1}.$$

The same point p_i also lies on the circumference of the circle with center c and radius r, so

$$(x_i - c_1)^2 + (y_i - c_2)^2 = r^2.$$

This gives a second degree equation

$$(x_i - c_1)^2 + (\frac{b_2 - a_2}{b_1 - a_1}(x_i - a_1) + b_2 - c_2)^2 = r^2$$

in x_i of required form. One gets a similar quadratic equation for y_i. The two other intersection cases are verified in a similar fashion. □

Corollary 7.4.4 *If (x, y) is constructed by ruler and compass from $X \subset \mathbb{R}^2$, and if E is the subfield of $\mathbb{R}$ generated by the coordinates of X, then $[E(x) : E]$ and $[E(y) : E]$ are powers of 2.*

Proof Let $\{E_i\}$ be a chain of fields associated to a construction of (x, y) from $X \subset \mathbb{R}^2$ by ruler and compass with intersection points $p_i = (x_i, y_i)$, where $E_0 = E$ and $(x_n, y_n) = (x, y)$. By the proposition we see that

$$[E_i : E_{i-1}] = [E_{i-1}(x_i, y_i) : E_{i-1}(x_i)][E_{i-1}(x_i) : E_{i-1}]$$

is either 1, 2 or 4. Therefore $[E_n : E]$ is a power of 2, and so are $[E(x) : E]$ and $[E(y) : E]$ because e.g.

$$[E_n : E(x)][E(x) : E] = [E_n : E].$$

□

Having now translated the inherent limitations of constructions by ruler and compass to the language of algebra, we can return to the earlier problems that caused such headache for the old Greeks.

Proposition 7.4.5 *One cannot by ruler and compass construct a cube with volume twice the volume of a given cube, nor an angle one-third of a given angle, nor a square with area equal to that of a given circle.*

Proof Say we have a cube with one corner at the origin $(0, 0)$ and another one at $(1, 0)$. If we are to construct a cube with volume 2, we must be able to construct the point $(a, 0)$, where $a^3 = 2$. Now the subfield of $\mathbb{R}$ containing 0 and 1 is $\mathbb{Q}$, whereas the subfield of $\mathbb{R}$ containing 0, 1 and a is $\mathbb{Q}(a)$. But $[\mathbb{Q}(a), \mathbb{Q}] = 3$ as $x^3 - 2$ is the minimal polynomial of a over $\mathbb{Q}$, and 3 is certainly not a power of 2. This contradicts the corollary above, so we cannot duplicate the cube by ruler and compass.

7.5 Splitting Fields and Normal Extensions

To construct an angle trisecting $\pi/3$ is equivalent to constructing the point $(a, 0)$ given $(0, 0)$ and $(1, 0)$, where $a = \cos(\pi/9)$. From this we could construct $(b, 0)$, where $b = 2\cos(\pi/9)$. Plugging $u = \pi/9$ into the trigonometric identity

$$\cos(3u) = 4\cos^3 u - 3\cos u,$$

we get $b^3 - 3b - 1 = 0$. But $f(x) = x^3 - 3x - 1$ is the minimal polynomial of b over $\mathbb{Q}$ because $f(x + 1) = x^3 + 3x^2 - 3$ is irreducible by Eisenstein's criterion. Thus $[\mathbb{Q}(b) : \mathbb{Q}] = 3$, which is not a power of 2.

To square the circle by ruler and compass is equivalent to constructing the point $(\sqrt{\pi}, 0)$ from $(0, 0)$ and $(1, 0)$. But then by the proposition below, we could also construct $(\pi, 0)$. This is impossible because $[\mathbb{Q}(\pi) : \mathbb{Q}]$ is not a power of 2. The extension $\mathbb{Q}(\pi)$ is not even algebraic over $\mathbb{Q}$ due to Lindeman's famous theorem, which says that π is transcendental. □

Definition 7.4.6 A real number a is *constructable* if $(a, 0)$ is constructable by ruler and compass from $(0, 0)$ and $(1, 0)$.

Proposition 7.4.7 *The subset of $\mathbb{R}$ consisting of constructable real numbers is a subfield of $\mathbb{R}$, and hence an algebraic extension of $\mathbb{Q}$.*

Proof If a, b belong to this subset E, then we can easily construct the points $(a + b, 0)$ and $(a - b, 0)$ from $(a, 0)$ and $(b, 0)$ by ruler and compass.

Let us construct ab for positive $a, b \in E$. Form any ray from the origin not parallel with the x-axis, and set off the points P and B at distance 1 and b, respectively, from the origin O. Let $A = (a, 0)$ and $C = (c, 0)$, where C is constructed as the intersection of the x-axis and the line parallel to AP. Comparing the similar triangles OAP and OCB, we get $1/a = b/c$, so $ab = c$ is constructable. The cases with other signs for a and b are easy enough.

One constructs a/b from $a, b \in E$ with $b \neq 0$, in a similar fashion. □

7.5 Splitting Fields and Normal Extensions

Definition 7.5.1 The *splitting field* of a polynomial f over a field F is the subfield of $\overline{F}$ generated by F and the roots of f in $\overline{F}$.

So the splitting field E of a non-constant polynomial $f \in F[x]$ is the smallest field where f decomposes into linear factors over E. More concretely, if $a_i \in \overline{F}$ are the roots of f, then $E = F(a_1, \ldots, a_n)$ and

$$f(x) = b(x - a_1) \cdots (x - a_n)$$

for some $b \in F$. In particular, the splitting field E is a finite extension, and hence an algebraic extension, of F.

The field $\mathbb{Q}(\sqrt{2})$ is a splitting field of $x^2 - 2$ over $\mathbb{Q}$.

Splitting fields are unique, as the following result shows.

Theorem 7.5.2 *If E and K are splitting fields of a polynomial over a field F, then there is an isomorphism $\alpha \colon E \to K$ that is the identity on F.*

Proof Since $\overline{K}$ is an algebraic extension of K, then by Corollary 7.3.6, it is also an algebraic extension of F, so $\overline{K} = \overline{F}$. By Theorem 7.3.8 there exists a monomorphism $\alpha \colon E \to \overline{K}$ that is the identity on F. It remains to check that $\alpha(E) = K$.

Say $f \in F[x]$ is the polynomial for which both E and K are splitting fields, and let $a_i \in E$ and $b_i \in K$ be the roots of f in these two fields. Since the coefficients of f are in F, and are thus fixed by α, which also is a homomorphism, it is clear that $\alpha(a_i)$ are roots of f in K, decomposing f into linear factors. So $\{b_i\} = \{\alpha(a_i)\}$ and

$$\alpha(E) = \alpha(F(a_1, \ldots, a_n)) = F(\alpha(a_1), \ldots, \alpha(a_n)) = F(b_1, \ldots, b_n) = K.$$

$\square$

Example 7.5.3 The splitting field of $x^4 - 2$ over $\mathbb{Q}$ has degree 8 despite the fact that $x^4 - 2$ is the minimal polynomial of the positive root $2^{1/4}$ over $\mathbb{Q}$. This is due to the fact that $x^4 - 2$ has the factor $x^2 + \sqrt{2}$, which is irreducible over $\mathbb{Q}(2^{1/4})$ with roots $\pm 2^{1/4} i$, so the splitting field is $\mathbb{Q}(2^{1/4}, i)$, which has degree $2 \cdot 4$ over $\mathbb{Q}$. $\diamond$

Example 7.5.4 The splitting field of $f(x) = x^3 + x^2 + 1$ over $\mathbb{Z}_2$ consists of 8 elements. To see this, first observe that neither 0 nor 1 are roots of f, which has degree 3, so f is irreducible over $\mathbb{Z}_2$. If a is a root of f, then

$$f(x) = (x + a)(x + a^2)(x + 1 + a + a^2).$$

Therefore the splitting field of f is $\mathbb{Z}_2(a)$ and a has f as the minimal polynomial over $\mathbb{Z}_2$. Thus $\{1, a, a^2\}$ is a basis for $\mathbb{Z}_2(a)$ over $\mathbb{Z}_2$, and

$$\mathbb{Z}_2(a) = \{0, 1, a, a^2, a+1, a^2+1, a^2+a, a^2+a+1\}.$$

Definition 7.5.5 A *normal extension* of a field F is a field E such that every irreducible polynomial over F that has at least one root in E decomposes into linear factors over E.

So if an irreducible polynomial over a field F has a root in a normal extension E of F, then all its roots belong there.

To say that a field E is the *splitting field of a family* $\{f_i\}$ of polynomials over a field F means that E is the subfield of $\overline{F}$ generated by all the roots of each member of this family. Every polynomial f_i decomposes then into linear factors over E. When the family is finite, then E is an algebraic extension of F, and can obviously be regarded as the splitting field of the polynomial that is the product of all the members of the family. It is therefore unique up to isomorphisms that fix elements of F, and by essentially the same proof, this remains true when the family is infinite.

7.5 Splitting Fields and Normal Extensions

Theorem 7.5.6 *Suppose F is a field with an algebraic extension $E \subset \overline{F}$. Then the following conditions are equivalent:*

(i) *E is a normal extension of F;*
(ii) *E is the splitting field of some family of polynomials over F;*
(iii) *Every monomorphism $\alpha: E \to \overline{F}$ that fixes elements of F satisfies $\alpha(E) = E$.*

Condition (iii) means that α can be regarded as an automorphism of E.

Proof To see that (i) implies (ii), observe that E is the splitting field of the family of minimal polynomials of all elements of E.

Assume that (ii) holds, so E is the splitting field of some family $\{f_i\} \subset F[x]$. Suppose $\alpha: E \to \overline{F}$ is a monomorphism that is the identity on F. If $a \in E$ is a root of f_i, then so is $\alpha(a)$, and since E is generated by the roots of f_i, we see that $\alpha(E) \subset E$. But then $\alpha(E) = E$ by the lemma below. So (iii) holds.

Suppose (iii) is true, and let f be an irreducible polynomial over F with a root $a \in E$. To get (i), we must show that if $b \in \overline{F}$ is another root of f, then $b \in E$.

Since f is irreducible, there is an isomorphism

$$\alpha: F(a) \to F[x]/(f) \to F(b)$$

that fixes elements of F and satisfies $\alpha(a) = b$. By Theorem 7.3.8 there is a monomorphic extension $E \to \overline{F}$ of α, which by assumption must map E onto E, so $b = \alpha(a) \in E$. □

Lemma 7.5.7 *If E is an algebraic extension of a field F, and if $\alpha: E \to E$ is a monomorphism that fixes elements of F, then $\alpha(E) = E$.*

Proof Let $a \in E$, and let $f \in F[x]$ be the minimal polynomial of a. Let K be the subfield of E generated by F and the roots of f in E. Since α maps roots of f to other roots of f, it maps K into K, and since K is a finite extension of F and α is an injective F-linear map, we get $\alpha(K) = K$. So there is an element $b \in K \subset E$ such that $\alpha(b) = a$. □

Since $x^3 - 2$ is irreducible over $\mathbb{Q}$, and has only one real root, the field $\mathbb{Q}(2^{1/3})$ is not a normal extension of $\mathbb{Q}$, nor is any other subfield of $\mathbb{R}$.

Example 7.5.8 Any extension E of a field F with $[E:F] = 2$ is normal because any $a \in E \setminus F$ has a minimal polynomial f over F with degree greater than one, so

$$2 = [E:F] = [E:F(a)][F(a):F]$$

shows that $\deg(f) = 2$ and $[E:F(a)] = 1$. Therefore $E = F(a)$ is a splitting field of f. ◇

7.6 Multiple Roots

Define the *derivative* of a polynomial $f(x) = a_0 + a_1 x + \cdots + a_n x^n$ over a field F to be the polynomial $f'(x) \in F[x]$ given by

$$f'(x) = a_1 + 2a_2 x + \cdots + n a_n x^{n-1}.$$

It is easy to check that $(f+g)' = f' + g'$ and $(fg)' = f'g + fg'$ for $g \in F[x]$.

This notion of a formal derivative of polynomials over a field requires some caution. For instance, we see that $(x^7)' = 7x^6 = 0$ over $\mathbb{Z}_7$ despite the fact that x^7 is non-constant.

When we talk about roots without any explicit reference to the field they belong, we usually mean that they belong to the algebraic closure of the field, or to a splitting field of the field.

Definition 7.6.1 A root a of a polynomial f over a field F has *multiplicity* m if $f(x) = g(x)(x-a)^m$, where $g \in \overline{F}[x]$ and $g(a) \neq 0$. It is a *simple root* if $m = 1$.

Counted with multiplicities we see that any polynomial of degree n over a field has at most n roots.

Proposition 7.6.2 *Any root a of a polynomial f over a field F has multiplicity greater than one if and only if $f'(a) = 0$.*

Proof Write f as $f(x) = g(x)(x-a)^m$ with $g \in \overline{F}[x]$ and $g(a) \neq 0$. Then

$$f'(x) = g'(x)(x-a)^m + mg(x)(x-a)^{m-1}.$$

So if $m \geq 2$, then $f'(a) = 0$, and conversely, if $m = 1$, then $f'(a) = g(a) \neq 0$. □

Corollary 7.6.3 *Suppose f is an irreducible polynomial over a field F. Then f has a non-simple root if and only if $f' = 0$.*

Proof Clearly f has some root a in $\overline{F}$, and if $f' = 0$, then in particular $f'(a) = 0$, so by the proposition this root is not simple.

Conversely, if f has a root $b \in \overline{F}$ with multiplicity greater than one, then since f is irreducible, the minimal polynomial of b over F is proportional to f. By the proposition we know that $f'(a) = 0$, so $f' = 0$ since f' has one degree less than f, and hence lower degree than the minimal polynomial of a. □

Say f is a polynomial over a field F with root $a \in \overline{F}$, and that $f(x) = g(x^n)$ for some $g \in F[x]$ and $n \in \mathbb{N}$. Then clearly the multiplicity of a is at least n. We will see that every irreducible polynomial over a field of non-zero characteristic is of this form whenever it has a non-simple root.

Corollary 7.6.4 *Every irreducible polynomial f over a field F has simple roots if F has characteristic zero. When F has prime characteristic p, then f has non-simple roots if and only if there is a $g \in F[x]$ such that $f(x) = g(x^p)$.*

7.6 Multiple Roots

Proof Suppose $f(x) = a_0 + a_1 x + \cdots + a_n x^n$. By the corollary above we know that f has non-simple roots if and only if $f' = 0$. This happens precisely when $m a_m = 0$ for $m \in \{1, \ldots, n\}$.

If F has characteristic zero and if f has a non-simple root, then these a_m's vanish, and $f(x) = a_0$, which is absurd.

If F has characteristic p, then for $m \geq 1$, either $a_m = 0$ or p divides m. Therefore we can write $f(x) = g(x^p)$ for some $g \in F[x]$. □

Theorem 7.6.5 *All roots of an irreducible polynomial over a field have the same multiplicity.*

Proof Let f be an irreducible polynomial over a field F having roots a and b in $\overline{F}$ with multiplicity k and l, respectively. As f is irreducible, there is an isomorphism

$$\alpha \colon F(a) \to F[x]/(f) \to F(b)$$

that fixes elements of F and satisfies $\alpha(a) = b$.

By Theorem 7.3.8 we can extend α to an automorphism β of $\overline{F}$. This automorphism can be extended further to a ring endomorphism η on $\overline{F}[x]$ such that

$$\eta(a_0 + \cdots + a_n x^n) = \beta(a_0) + \cdots + \beta(a_n)x^n$$

for $a_i \in \overline{F}$. Since $\eta(f(x)) = f(x)$ and $\eta((x-a)^k) = (x-b)^k$, we see that $(x-b)^k$ is a factor of $f(x)$, so $k \leq l$. By symmetry we also get $l \leq k$. □

We can therefore write any irreducible polynomial f over a field F as a finite product

$$f(x) = c \prod_i (x - a_i)^k$$

in its splitting field E over F, where $c \in F$ and the roots $a_i \in E$ of f have multiplicity k.

The theorem above combined with the previous corollary tell us that all the roots of an irreducible polynomial over a field of characteristic zero are simple. In Proposition 7.7.10 we will see that all the roots of an irreducible polynomial over a finite field are also simple. So non-simple roots can only occur for irreducible polynomials over infinite fields of prime characteristic. Here is an example of this.

Example 7.6.6 Consider the field $F(x)$ of fractions of polynomials over a field F of characteristic 3. Then $f(y) = y^3 - x \in F(x)[y]$ is irreducible and has non-simple roots. To see that this polynomial of degree 3 is indeed irreducible, it clearly suffices to show that there are no polynomials $g, h \in F[x]$ with $h \neq 0$ such that g/h is a root of f. Suppose it was, then $g(x)^3 = xh(x)^3$, which is impossible as the polynomial on the left-hand-side has degree a multiple of 3, whereas the one on the right-hand-side has degree a multiple of 3 plus 1.

If a and b are two roots of f in its splitting field, then $a^3 = x = b^3$. Calculating in characteristic 3, we get $(a-b)^3 = a^3 - b^3 = 0$, which shows that $a = b$. So f has one root with multiplicity 3, and this is consistent with the fact that $F(x)$ is an infinite field of characteristic 3. ◇

7.7 Finite Fields

We have seen that a finite field F has characteristic p for some prime number, and will then contain the prime subfield $\mathbb{Z}_p$. Say $[F : \mathbb{Z}_p] = n$. Using a basis of F over $\mathbb{Z}_p$ we get a ring isomorphism $F \cong \mathbb{Z}_p^n$, so $|F| = p^n$. Conversely, if F is any field having order a power of a prime number p, then its characteristic is p because if it was a prime number q, then $q^m = p^n$, which is impossible unless $p = q$. We record this.

Proposition 7.7.1 *Every finite field has order a power of a prime number, and this prime number is the characteristic of the field.*

Proposition 7.7.2 *A finite field of order p^n for p prime, is the splitting field of the polynomial $x^{p^n} - x$ over $\mathbb{Z}_p$. Consequently, a finite field is determined up to isomorphism by its order.*

Proof The set F_* of non-zero elements of a field F of order p^n is a multiplicative group of order $p^n - 1$, so if $a \in F_*$, then $a^{p^n - 1} = 1$. Therefore the elements of F are the roots of the polynomial $x^{p^n} - x$.

Two fields with the same order have the same characteristic and are splitting fields of the same prime fields. The result is then immediate from Theorem 7.5.2. □

As the following result shows there is a field with order p^n for every prime number p and natural number n.

Proposition 7.7.3 *For every prime number p and natural number n, the roots of $f(x) \equiv x^{p^n} - x \in \mathbb{Z}_p[x]$ are distinct and form a field with p^n elements, and this is indeed the splitting field of f.*

Proof Since $f'(x) = p^n x^{p^n - 1} - 1 = -1 \neq 0$, then by Proposition 7.6.2 we see that f has p^n distinct roots. It remains to show that they form a field. If a and $b \neq 0$ are two roots of f, then so are ab^{-1} and $a \pm b$ since

$$(ab^{-1})^{p^n} = a^{p^n}(b^{p^n})^{-1} = ab^{-1} \quad \text{and} \quad (a \pm b)^{p^n} = a^{p^n} \pm b^{p^n} = a \pm b,$$

where we in the second last step used the binomial formula and calculated in characteristic p. □

We now show that any finite field has an extension field of any finite degree. Since these are all algebraic extensions, no finite field is algebraically closed. So the algebraic closure of a finite field is an infinite countable field.

7.7 Finite Fields

Proposition 7.7.4 *Let F be a field with p^n elements for p prime, and let $m \in \mathbb{N}$. Then up to isomorphism there is a unique extension field E of F with $[E : F] = m$.*

Proof Since the multiplicative group F_* has order $p^n - 1$, we have $a^{p^n-1} = 1$ for any non-zero $a \in F$. The formula

$$(p^n)^m - 1 = (p^n - 1)(1 + p^n + \cdots + (p^n)^{m-1})$$

shows that $p^n - 1$ divides $p^{mn} - 1$, so we also get $a^{p^{mn}-1} = 1$. Thus every element of F is a root of $f(x) = x^{p^{mn}} - x \in F[x]$.

According to the proposition above the roots of f in $\overline{F}$ form a field E with p^{mn} elements. We have just seen that F is a subfield of E, so E is an extension of F of degree m as $|F|^m = p^{mn}$. This extension is unique up to isomorphism as any finite field is determined up to isomorphism by its order. □

Proposition 7.7.5 *Any finite subgroup of the multiplicative group of a field is cyclic.*

Proof A finite subgroup G of the multiplicative group of a field is by Corollary 4.20.5 isomorphic to a direct product $G_1 \times \cdots \times G_n$ of cyclic groups G_i with prime power order m_i. If m is the least common multiplier of all the m_i's, then the $m_1 \cdots m_n$ elements of G are roots of the polynomial $x^m - 1$, so $m = m_1 \cdots m_n$, which shows that the numbers m_i are relatively prime. But then G is cyclic. □

Corollary 7.7.6 *The multiplicative group of a field is cyclic if and only if the field is finite.*

Proof Suppose the multiplicative group F_* of a field F is cyclic with generator a.

If F has characteristic p, then obviously $F = \mathbb{Z}_p(a)$. If $a + 1 \in F_*$ and $a \neq 0$, then $a + 1 = a^n$ for some $n \in \mathbb{N}$, so the minimal polynomial of a over $\mathbb{Z}_p$ has degree not greater than n, and $|F| \leq p^n$.

If F has characteristic zero, then $-1 \in F_*$ is not the unit, so $-1 = a^n$ for some $n \in \mathbb{N}$, and $a^{2n} = 1$. Thus F_* and F are finite.

The opposite direction is immediate from the proposition above. □

Corollary 7.7.7 *Any finite extension of a finite field is simple.*

Proof Suppose E is a finite extension of a finite field F. As E is a finite field, then by the corollary above, its multiplicative group is generated by an element a, and then $E = F(a)$. □

Corollary 7.7.8 *There is an irreducible polynomial of any given degree over a finite field.*

Proof Given a finite field F, then by Proposition 7.7.4, there is an extension field E of F having any degree. By the corollary above $E = F(a)$ for some algebraic element a over F. Then the minimal polynomial of a over F will do. □

Proposition 7.7.9 *A finite field F has exactly one subfield of order that divides $|F|$.*

Proof Say F has p^n elements for a prime number p, and that $m \in \mathbb{N}$ divides n, which is the only way the order p^m of a subfield of F can divide p^n. Regard F as the splitting field of $x^{p^n} - x$ over $\mathbb{Z}_p$. Then as $p^m - 1$ divides $p^n - 1$, the required subfield is the splitting field of $x^{p^m} - x$ over $\mathbb{Z}_p$. □

Proposition 7.7.10 *The roots of an irreducible polynomial over a finite field are distinct.*

Proof Suppose F is a field with p^n elements for a prime number p, and that f is an irreducible polynomial over F. By Corollary 7.6.4 the polynomial f has non-simple roots if and only if $f(x) = \sum_k a_k (x^p)^k$ for some $a_k \in F$. Set $b_k = a_k^{p^{n-1}}$, so $b_k^p = a_k^{p^n} = a_k$. Calculating in characteristic p we therefore get

$$f(x) = \sum_k (b_k x^k)^p = (\sum_k b_k x^k)^p$$

by repeated use of the binomial formula. Hence f has distinct roots if and only if it is irreducible. □

Theorem 7.7.11 *Let F be a field with p^n elements for p prime. The group of automorphisms of F is cyclic of order n, and is generated by the automorphism ϕ of F given by $\phi(a) = a^p$, known as the* Frobenius *endomorphism.*

Proof Using the binomial formula and calculating in characteristic p, we see that the map ϕ is a homomorphism. Clearly it has trivial kernel, so it is also surjective on the finite set F. Thus ϕ is an automorphism of F, and ϕ^n is the identity because $\phi(a)^n = a^{p^n} = a$ for $a \in F$. Let m be the order of ϕ. Then every element of F is a root of $x^{p^m} - x$, so $p^m \geq p^n$, or $m \geq n$, which means that $m = n$. So ϕ has order n.

Let a be a generator of the cyclic multiplicative group F_*, so $F = \mathbb{Z}_p(a)$. Let f be the minimal polynomial of a over $\mathbb{Z}_p$, so $\deg(f) = n$. Since automorphisms of F are unital and therefore fix the elements of $\mathbb{Z}_p$, they correspond to all possible extension of the identity map $\mathbb{Z}_p \to \overline{F}$ to monomorphisms $F \to \overline{F}$. These will automatically have range F as F is a splitting field and therefore a normal extension of $\mathbb{Z}_p$, so Theorem 7.5.6 kicks in. By Corollary 7.3.12 there are as many such monomorphisms as there are distinct roots of f. By the proposition above all the roots of f are indeed distinct, and since the degree of f is n, there are n automorphism of F. These are the n ones generated by ϕ. □

7.8 Separable Extensions

Definition 7.8.1 A polynomial over a field F is *separable* if its irreducible factors are separable, that is, if all their roots are simple. An algebraic element in an extension field E of F is a *separable element* if its minimal polynomial is separable. If E is an algebraic extension of F consisting only of separable elements, then E is a *separable extension* of F. The field F is *perfect* if all its algebraic extensions are separable.

7.8 Separable Extensions

So finite fields and fields with characteristic zero are perfect. We have seen an example of an infinite field of characteristic 3 that has an inseparable extension.

Proposition 7.8.2 *Finite separable extensions are simple.*

Proof Let F be a field with a finite separable extension E. By Corollary 7.7.7 we may assume that F is infinite. Since E is a finite extension field of F, it is generated over F by finitely many elements of E that are algebraic over F. By induction if therefore suffices to show that if $E = F(a, b)$ for algebraic elements $a, b \in E$ over F, then there exists $c \in E$ such that $E = F(c)$.

Let f and g be the minimal polynomials over F of a and b, respectively. Since E is separable, the roots a_i of f are distinct, and so are the roots b_j of g. Say $a_1 = a$ and $b_1 = b$. As F is infinite it has an element d different from $(a - a_i)/(b_j - b)$ for all i and $j \neq 1$. Let $c = a + db$ and define $h(x) = f(c - dx) \in F(c)[x]$. Then $h(b) = f(a) = 0$ and $h(b_j) \neq 0$ for $j \neq 1$ as $c - db_j = a - d(b_j - b) \neq a_i$ for all i. So b is the only common root of h and g, and therefore also the only root of its minimal polynomial p over $F(c)$ since p divides both h and g. Thus $p(x) = x - b$ and $b \in F(c)$. But then also $a = c - db \in F(c)$, so $F(c) = F(a, b)$. □

Theorem 7.8.3 *Suppose E is a finite extension of a field F. Then E is a simple extension of F if and only if there are only finitely many intermediate fields between F and E.*

Proof Assume $E = F(a)$ and that a has minimal polynomial $f \in F[x]$. We define a map η from the intermediate fields between F and E to divisors of f by letting $\eta(K) = g$, where g is the minimal polynomial of a over the intermediate field K. The polynomial g will indeed be a divisor of f because f can be regarded as an element of $K[x]$ and $f(a) = 0$. Since there are only finitely many divisors of f we know that there are only finitely many intermediate fields between F and E provided we can show that η is injective. But this follows because if L is the subfield of K generated by F and the coefficients of g, so L is uniquely determined by g, then $L = K$. To verify this equality observe that g is evidently also irreducible over L, and $K(a) = E = L(a)$ as $E = F(a)$, so $[E : K] = \deg(g) = [E : L]$, which shows that K and L have the same degree over F.

To prove the opposite direction we may by Corollary 7.7.7 assume that F is infinite, and of course with only finitely many fields between F and E.

We claim that the field $F(a, b)$ for $a, b \in E$ is generated over F by a single element of E. By assumption there are only finitely many fields between F and E, and hence finitely many fields of the form $F(a + cb)$ for $c \in F$. Since F is infinite there is a non-zero element c of F such that $F(a + cb) = F(a)$. Thus $cb = (a + cb) - a \in F(a + cb)$, so $b \in F(a + cb)$ and $F(a, b) = F(a + cb)$, proving the claim.

Choose $d \in E$ such that $[F(d) : F]$ is as large as possible, and this is by assumption a finite number. If there exists an element u of E that does not belong to $F(d)$, then by the previous paragraph we can find an element $v \in E$ such that $F(v)$ contains both d and u, so $F(v)$ is strictly larger than $F(d)$, and this is a contradiction. Hence $E = F(d)$. □

Proposition 7.8.4 *Let a be an algebraic element over a field F. Then a is separable over F if and only if $F(a)$ is a separable extension of F.*

Proof We only need to show that the roots of the minimal polynomial f over F of an element $b \in F(a)$ are simple. So if f has n distinct roots, we must show that $n = \deg(f)$.

By Corollary 7.3.12 there are correspondingly n distinct extensions α_i to $F(b)$ of any monomorphism $\alpha \colon F \to L$ into an algebraically closed field. If the minimal polynomial g of a over $F(b)$ has m distinct roots, then again by the same corollary each $\alpha_i \colon F(b) \to L$ has m extensions α_{ij} to $F(a)$. These are the mn possible extensions of $\alpha \colon F \to L$ to $F(a)$.

Let h be the minimal polynomial of a over F. Since a is separable over F, the number of distinct roots of h coincides with its degree. By the same corollary we conclude that $[F(a) : F] = mn$. But a is evidently also separable over $F(b)$, so by the same corollary $[F(a) : F(b)] = \deg(g) = m$. Hence

$$mn = [F(a) : F] = [F(a) : F(b)][F(b) : F] = m \deg(f).$$

□

Proposition 7.8.5 *If E is a finite separable extension of a field F, and K is a finite separable extension of E, then K is a finite separable extension of F.*

Proof By Proposition 7.8.2 we know that $E = F(a)$ for some $a \in E$. We need to show that any element $b \in K \setminus F(a)$ is separable over F. Now b is separable over $F(a)$, and $F(a)$ is a separable extension of F. So by the same reasoning as in the previous proof, the number of extensions to $F(a, b)$ of any monomorphism $\alpha \colon F \to L$ into an algebraically closed field equals $[F(a, b) : F(a)][F(a) : F]$. This also equals $[F(a, b) : F(b)][F(b) : F]$, so to obtain these extensions via $F(b)$ requires by Corollary 7.3.12 that a is separable over $F(b)$ and that b is separable over F. □

Proposition 7.8.6 *Let F be a field of prime number characteristic p. Then F is perfect if and only if every element of F has a p-th root in F.*

Proof Suppose F is perfect and let $a \in F$. Let b be a root of $x^p - a \in F[x]$. Since F is perfect the minimal polynomial f of b over F has only simple roots, and it must divide $x^p - a = x^p - b^p = (x - b)^p$, so $f(x) = x - b$ and $b \in F$.

Conversely, assume that every element of F has a p-th root. To show that F is perfect, it is certainly enough to show that every irreducible polynomial f over F has only simple roots. If f has non-simple roots, then by Corollary 7.6.4 it is of the form $f(x) = a_0 + a_1 x^p + \cdots + a_n x^{np}$, where $a_i \in F$. Let $b_i \in F$ be the p-th root of a_i. Then by repeated use of the binomial formula in characteristic p, we get the contradiction

$$f(x) = b_0^p + b_1^p x^p + \cdots + b_n^p x^{np} = (b_0 + b_1 x + \cdots + b_n x^n)^p.$$

□

Chapter 8
Galois Theory

Having an extension E of a field F, consider the group $G(E/F)$ of automorphisms of E that leave F fixed. Such automorphisms permute the roots of any polynomial over F. We are interested in the situation when all the roots belong to E and are distinct, so we require E to be a finite separable normal extension of F; henceforth called a Galois extension. Given this setup, one can define a map from the subgroups of $G(E/F)$ to the intermediate extensions of F, which sends a subgroup H to its fixed field $E_H = \{a \in E \mid \alpha(a) = a \text{ for } \alpha \in H\}$. It has an inverse map sending an intermediate field K to $G(E/K)$. Moreover, one has $[E:K] = |G(E/K)|$ and $[K:F] = [G(E/F):G(E/K)]$. Finally, this correspondence restricts to a correspondence between normal extensions and normal subgroups, and then $G(K/F)$ is isomorphic to $G(E/F)/G(E/K)$. We spend the first three sections proving this beautiful result, known as the fundamental theorem in Galois theory.

Using this result, we then prove the fundamental theorem of algebra. Crucial in these investigations is the Galois group of a polynomial over F. This is just $G(E/F)$ where E now is the splitting field of the polynomial. This will always be a subgroup of S_n, where n is the number of distinct roots of the polynomial.

Returning to rulers and compasses, we characterize exactly what n-gons can be constructed by these means. This is achieved by studying the Galois group of the cyclotomic nth polynomial Φ_n, which by definition is the monic polynomial over $\mathbb{C}$ having the primitive nth roots of unity in $\mathbb{C}$ as its roots. It turns out that Φ_n has only integer coefficients, and is irreducible over $\mathbb{Q}$. Also, if $a \in \mathbb{C}$ is any of these roots, then $\mathbb{Q}(a)$ is the splitting field both of Φ_n and of $x^n - 1$. In addition, one has $|G(\mathbb{Q}(a)/\mathbb{Q})| = \phi(n)$, and $G(\mathbb{Q}(a)/\mathbb{Q})$ is isomorphic to the group $U(\mathbb{Z}_n)$ of units in the ring $\mathbb{Z}_n$. We know that $U(\mathbb{Z}_n)$ is cyclic exactly when n is either 2, 4, p^m or $2p^m$ for odd primes p. Any Galois extension E of a field F for which $G(E/F)$ is cyclic is called cyclic. We characterize such extensions of degree n in terms of splitting fields of irreducible polynomials over F of the form $x^n - b$.

In the last three sections of the chapter we focus on the problem of solving polynomial equations by radicals. This loosely speaking, means finding some generalized *abc*-formula involving extracting recursively square roots, cubic roots, etc. Mathematically this means that the splitting field of the polynomial should ultimately belong to some extension of the original field by radicals. We prove the spectacular result that in a field of characteristic zero, this can be done precisely when the Galois group of the polynomial is solvable. We know that this group is a subgroup of S_n when the polynomial has n distinct roots. But we have also seen that S_n is solvable if and only if $n \leq 4$. Indeed, the complex cubic and quartic equations were solved by radicals already in the Renaissance. We refresh this work in the final section. However, no general such formulas exists for $n \geq 5$. In fact, there exist polynomials over $\mathbb{C}$ in any degree having symmetric groups as Galois groups. We prove this in two different ways. One of these methods involves symmetric functions, to be defined and studied in a separate section.

8.1 Automorphisms and Fixed Fields

Let E be an extension field of a field F. We denote by $G(E/F)$ the group of automorphisms of E that leave the elements of F fixed. Trivially we get a group under composition this way. Now the crucial observation is that any element of $G(E/F)$ permutes the roots in E of any polynomial over F. The picture is complete when all the roots belong to E, and when they are distinct. This sets the focus on finite separable normal extensions.

Proposition 8.1.1 *If E is a finite separable extension of a field F, then*

$$|G(E/F)| \leq [E : F].$$

Proof By Proposition 7.8.2 we have $E = F(a)$ for some algebraic element a. By Corollary 7.3.12 the number of possible extensions $F(a) \to \overline{F}$ of the identity map $F \to F$ equals the number of distinct roots of the minimal polynomial of a over F. The number of these roots is obviously less than the degree $[F(a) : F]$ of the minimal polynomial. $\square$

Example 8.1.2 By the proposition above we have $|G(\mathbb{C}/\mathbb{R})| \leq [\mathbb{C} : \mathbb{R}] = 2$. We actually get equality here because if $\alpha \in G(\mathbb{C}/\mathbb{R})$, then $\alpha(a + ib) = a + \alpha(i)b$ for $a, b \in \mathbb{R}$, and as $\alpha(i)^2 = -1$, there are two possibilities $\alpha(i) = \pm i$ corresponding to the identity map and complex conjugation, which are indeed $\mathbb{R}$-automorphisms of $\mathbb{C}$. $\diamond$

Example 8.1.3 Any $\alpha \in G(\mathbb{Q}(5^{1/3})/\mathbb{Q})$ is determined by its value $\alpha(5^{1/3})$ on the generator $5^{1/3}$. This value must also be a root of the irreducible polynomial $x^3 - 5$, and can only be real if α is the identity map. So $G(\mathbb{Q}(5^{1/3})/\mathbb{Q})$ is trivial, whereas $[\mathbb{Q}(5^{1/3}) : \mathbb{Q}] = 3$. $\diamond$

8.1 Automorphisms and Fixed Fields

Definition 8.1.4 Let G be a subgroup of the group of automorphisms of a field E. Then the *fixed field of* G is the subfield

$$E_G \equiv \{a \in E \mid \alpha(a) = a \text{ for } \alpha \in G\}$$

of E.

Lemma 8.1.5 *Any finite collection of distinct monomorphisms from a non-trivial field F into an extension field E of F is linear independent over E.*

Proof Suppose this is not true. Then from a finite collection of monomorphisms from F to E we may pick a least number of members $\alpha_1, \ldots, \alpha_n$ such that

$$a_1\alpha_1 + \cdots + a_n\alpha_n = 0$$

for some non-zero elements $a_i \in E$. Clearly $n > 1$. Pick $b \in E$ with $\alpha_1(b) \neq \alpha_n(b)$. Since

$$a_2(\alpha_1(b) - \alpha_2(b))\alpha_2 + \cdots + a_n(\alpha_1(b) - \alpha_n(b))\alpha_n = 0,$$

we get a contradiction as a less number of members with non-zero coefficients can obviously be picked from $\{\alpha_2, \ldots, \alpha_n\}$. $\square$

From the proof we see that the lemma remains valid when the monomorphisms are replaced by e.g. multiplicative maps from a semigroup to the multiplicative semigroup of an integral domain with the same definition of linear independence. In particular, it holds for characters of abelian groups.

Theorem 8.1.6 *If G is a finite subgroup of the group of automorphisms of a field E, then*

$$[E : E_G] = |G|$$

provided $[E : E_G] < \infty$.

Proof Say $G = \{\alpha_1, \ldots, \alpha_n\}$ with no repetitions, and let $\{a_1, \ldots, a_m\}$ be a basis for E over E_G.

If $m < n$, then the system

$$\sum_j \alpha_j(a_i)x_j = 0$$

of m linear equations in the variables x_j has a non-trivial solution $\{b_j\} \subset E$.

Writing $a \in E$ as $a = \sum c_i a_i$ for $c_i \in E_G$, we therefore get

$$\sum_j \alpha_j(a)b_j = 0$$

which contradicts the lemma.

If $m > n$, then by arguing as above, we can find a set $\{b_1, \ldots, b_k\} \subset E$ of non-zero members such that

$$\sum_{i=1}^{k} \alpha_j(a_i) b_i = 0,$$

and we may assume that the number $k \leq n+1$ is the least such possible. Applying $\alpha \in G$ to these equations and observing that $\{\alpha\alpha_1, \ldots, \alpha\alpha_n\} = \{\alpha_1, \ldots, \alpha_n\}$, gives

$$\sum_{i=1}^{k} \alpha_j(a_i)\alpha(b_i) = 0.$$

Combining these two systems we see that

$$\sum_{i=2}^{k} \alpha_j(a_i)(b_i\alpha(b_1) - \alpha(b_i)b_1) = 0.$$

By assumption all the coefficients $b_i\alpha(b_1) - \alpha(b_i)b_1$ must vanish. So $c_i \equiv b_i b_1^{-1} \in E_G$ for $i \geq 1$.

From $\sum_{i=1}^{k} \alpha_j(a_i)b_i = 0$, we get $\sum_{i=1}^{k} \alpha_1(a_i)c_i = 0$. Thus $\sum_{i=1}^{k} a_i c_i = 0$ and $c_i = 0$ as the a_i's are linear independent over E_G. But then $b_i = 0$, which is a contradiction. So $m = n$. □

Theorem 8.1.7 *Let E be a finite separable extension of a field F. If H is a subgroup of $G(E/F)$, then $G(E/E_H) = H$ and $[E : E_H] = |G(E/E_H)|$.*

Proof Obviously H is a subgroup of $G(E/E_H)$. By Theorem 8.1.6 and Proposition 8.1.1 we have

$$|H| = [E : E_H] \geq |G(E/E_H)| \geq |H|,$$

which gives the desired result. □

Theorem 8.1.8 *Suppose E is a finite separable extension of a field F. Then the following conditions are equivalent:*

(i) *E is a normal extension of F;*

(ii) *F is the fixed field of $G(E/F)$;*

(iii) *$[E : F] = |G(E/F)|$.*

Proof By Theorem 8.1.6 we have $[E : K] = |G(E/F)|$ for the fixed field K of $G(E/F)$.

By Proposition 7.8.2 we have $E = F(a)$ for an algebraic element a. By Corollary 7.3.12 the number of possible extensions $F(a) \to \overline{F}$ of the identity map $F \to F$

8.1 Automorphisms and Fixed Fields

equals the number of distinct roots n of the minimal polynomial of a over F, which is $[E : F]$ as a is separable.

If E is a normal extension of F, these extensions map onto E by Theorem 7.5.6, so $|G(E/F)| = n$. But then $[E : K] = [E : F]$, so $[K : F] = 1$ and $K = F$. So (i) implies (ii).

Assume (ii) holds. Consider $f \in E[x]$ given by

$$f(x) = (x - \alpha_1(a)) \cdots (x - \alpha_n(a)),$$

where α_i are the elements of $G(E/F)$.

Let $\eta_i : E[x] \to E[x]$ be the ring homomorphism obtained by letting α_i act on the coefficients. Then as $G(E/F) = \{\alpha_i\alpha_1, \ldots, \alpha_i\alpha_n\}$, we get

$$\eta_i(f)(x) = (x - \alpha_i\alpha_1(a)) \cdots (x - \alpha_i\alpha_n(a)) = f(x),$$

so the coefficients of f are in the fixed field of $G(E/F)$. By assumption these are therefore in F, and f is actually a polynomial over F. By construction all the roots of f belong to E, and a is one of them since $G(E/F)$ contains the identity map. Therefore E is the splitting field of f, and E is a normal extension by Theorem 7.5.6. So (i) holds.

The implication $(ii) \Rightarrow (iii)$ is immediate from the first paragraph in this proof. Conversely, if (iii) holds, then $[E : K] = [E : F]$, so $K = F$. □

Example 8.1.9 Let $a \in \overline{\mathbb{Q}}$ be a non-unital root of $x^5 - 1$. Clearly $\mathbb{Q}(a)$ is the splitting field of $x^5 - 1$ with distinct roots $1, a, a^2, a^3, a^4$. By Theorem 7.5.6 it is a finite separable normal extension of $\mathbb{Q}$. Thus $|G(\mathbb{Q}(a)/\mathbb{Q})| = [\mathbb{Q}(a) : \mathbb{Q}] = 4$ as a is a root of the polynomial

$$f(x) = (x^5 - 1)/(x - 1) = 1 + x + x^2 + x^3 + x^4,$$

which is irreducible by Example 7.1.8. The element $\alpha \in G(\mathbb{Q}(a)/\mathbb{Q})$ which sends a to a^2 is clearly a generator, so $G(\mathbb{Q}(a)/\mathbb{Q}) \cong \mathbb{Z}_4$. ◇

Example 8.1.10 Let $a \in \overline{\mathbb{Q}}$ be a non-unital root of $x^3 - 1$, and let α be the $\mathbb{Q}$-automorphism of $E \equiv \mathbb{Q}(2^{1/3}, a)$ given by $\alpha(a) = a^2$ and $\alpha(2^{1/3}) = a2^{1/3}$. Let $G = \{\iota, \alpha\}$. We claim that $E_G = \mathbb{Q}(a^2 2^{1/3})$. To see this write any $b \in E$ as a linear combination of the basis $\{1, 2^{1/3}, 2^{2/3}, a, a2^{1/3}, a2^{2/3}\}$ for E over $\mathbb{Q}$. Then the requirement $\alpha(b) = b$ forces b to be of the form

$$b = c_1 + c_2(1 + a)2^{1/3} + c_3 a 2^{2/3}$$

for $c_i \in \mathbb{Q}$. The result now follows from $1 + a = -a^2$ and $a = a^4$. ◇

8.2 The Galois Group of a Polynomial

Definition 8.2.1 The *Galois group of a polynomial f* over a field F is the group $G(E/F)$, where E is the splitting field of f.

Example 8.2.2 Let $f(x) = x^2 - a$ be an irreducible polynomial over a field F of characteristic different from 2. If b is a root of f, then $-b$ is also a root, and $b \neq -b$ in characteristic 2. So f is separable over F, and by Theorem 7.5.6 its splitting field $F(b)$ is a finite separable normal extension of F. Thus the Galois group of f has order two as $|G(F(b)/F)| = [F(b) : F] = 2$. ◇

Proposition 8.2.3 *The Galois group $G(E/F)$ of a polynomial over a field F with n distinct roots in E is a subgroup of S_n.*

Proof Say $a_1, \ldots, a_n$ are the distinct roots. Every $\alpha \in G(E/F)$ produces a permutation $f_\alpha \in S_n$ of these roots, where $f_\alpha(a_i) = \alpha(a_i)$. The map $G(E/F) \to S_n$ which sends α to f_α is a monomorphism as $E = F(a_1, \ldots, a_n)$. □

Example 8.2.4 The Galois group of $x^4 - 2 \in \mathbb{Q}[x]$ is the dihedral group D_4, known as the octic group. As for the details, first note that

$$x^4 - 2 = (x - 2^{1/4})(x + 2^{1/4})(x - i2^{1/4})(x + i2^{1/4}).$$

Thus $E \equiv \mathbb{Q}(2^{1/4}, i)$ is the splitting field of $x^4 - 2$ over $\mathbb{Q}$. Since $\mathbb{Q}$ has characteristic zero, we know that E is a finite separable normal extension of $\mathbb{Q}$. Therefore $|G(E/\mathbb{Q})| = [E : \mathbb{Q}] = 8$.

Any $\mathbb{Q}$-automorphism of E is determined by its value on i and $2^{1/4}$, and must send these elements to $\pm i$ and to $\pm 2^{1/4}$ or $\pm i 2^{1/4}$. The generators σ and τ of D_4 correspond to permutations of the roots $2^{1/4}, i2^{1/4}, -2^{1/4}, -i2^{1/4}$. The first sends $2^{1/4}$ to $i2^{1/4}$ and fixes i, and the second sends i to $-i$ and fixes $2^{1/4}$. In the complex plane the roots are vertices of a square with center at the origin, and σ and τ correspond to the 90° counterclockwise rotation and the reflection about the real axis, respectively. ◇

Proposition 8.2.5 *Let E be the splitting field of a polynomial $x^n - a$ over a field F that contains all the nth roots of unity. Then $E = F(b)$, where b is a root of $x^n - a$, and the Galois group of the polynomial is abelian.*

Proof Let $c \in F$ be a generator for the nth roots of unity. Then all the roots of $x^n - a$ are of the form bc^i for a non-negative integer i. Hence $E = F(b)$. Moreover, if $\alpha, \beta \in G(E/F)$, then $\alpha(b) = bc^i$ and $\beta(b) = bc^j$. Thus $\alpha\beta(b) = bc^{i+j} = \beta\alpha(b)$. □

Example 8.2.6 Let a be a root of the irreducible polynomial $x^2 + x + 1$ over $\mathbb{Q}(2^{1/3})$. Since

$$x^3 - 2 = (x - 2^{1/3})(x - a2^{1/3})(x - a^2 2^{1/3}),$$

the splitting field of $x^3 - 2 \in \mathbb{Q}[x]$ is $E \equiv \mathbb{Q}(2^{1/3}, a)$, so $|G(E/\mathbb{Q})| = [E : \mathbb{Q}] = 6$.

The six $\mathbb{Q}$-automorphisms of E are those that send $2^{1/3}$ to itself, or to $a2^{1/3}$ or $a^2 2^{1/3}$, and a to itself or a^2. It is easy to see that $G(E/\mathbb{Q})$ is isomorphic to the dihedral group D_3. While D_3 is not abelian, the Galois group of $x^3 - 2$ over $\mathbb{Q}(a)$ is, as $E = \mathbb{Q}(a)(2^{1/3})$ is the splitting field over $\mathbb{Q}(a)$, and $G(E/\mathbb{Q}(a)) \cong \mathbb{Z}_2$. ◇

Example 8.2.7 Let $E \equiv \mathbb{Q}(a)$, where $a = e^{\pi i/4}$. Then the roots of $x^4 + 1$ are a, a^3, a^5, a^7, so E is the splitting field of this irreducible polynomial over $\mathbb{Q}$. Therefore $|G(E/\mathbb{Q})| = [E : \mathbb{Q}] = 4$. Any $\mathbb{Q}$-automorphism of E is determined by its value on a, which must also be a root of $x^4 + 1$. Apart from the identity map, we get the three automorphisms $a \mapsto a^3$, $a \mapsto a^5$ and $a \mapsto a^7$, which all have degree two. Therefore $G(E/\mathbb{Q}) \cong \mathbb{Z}_2 \times \mathbb{Z}_2$. ◇

8.3 The Fundamental Theorem in Galois Theory

Definition 8.3.1 A *Galois extension* E of a field F is a finite separable normal extension of F.

For example, the splitting field of a polynomial over a field F of characteristic zero is a Galois extension of F.

Proposition 8.3.2 *Suppose E is a Galois extension of a field F with an intermediate field K. Then E is a Galois extension of K, and K is a finite separable extension of F.*

Proof By Theorem 7.5.6 the field E has a splitting family of polynomials over F. Regarding these as polynomials over K, we conclude by the same theorem that E is a normal extension of K. Clearly it is also a finite extension. To see that it is a separable one, consider any $a \in E$ and its minimal polynomials f and g over F and K, respectively. By assumption all the roots of f are distinct. To see that all the roots of g are also distinct, by the division algorithm for polynomials, write $f = gq + r$ for $q, r \in K[x]$ with $\deg(r) < \deg(g)$. But $f(a) = g(a) = 0$ implies that $r(a) = 0$, which contradicts the minimality of g unless $r = 0$. Hence all the roots of g are also roots of f, and are therefore distinct.

Clearly K is a finite separable extension of F, but in general it cannot be expected to be a normal one. □

The following result, fundamental in Galois theory, provides a correspondence between fixed fields and automorphisms.

Theorem 8.3.3 *Suppose E is a Galois extension of a field F. Let S be the set of subfields of E containing F, and let $\mathcal{G}$ be the set of subgroups of $G(E/F)$. Then the maps $\mathcal{G} \to \mathcal{S}$ and $\mathcal{S} \to \mathcal{G}$ given by*

$$H \mapsto E_H \quad \text{and} \quad K \mapsto G(E/K),$$

respectively, are mutual inverses, i.e. $K = E_{G(E/K)}$ and $H = G(E/E_H)$. Also $[E : K] = |G(E/K)|$ and $[K : F] = [G(E/F) : G(E/K)]$. Furthermore, the maps restrict to bijective maps between normal extensions and normal subgroups, so K is a normal extension of F if and only if $G(E/K)$ is a normal subgroup of $G(E/F)$, and in this case $G(K/F) \cong G(E/F)/G(E/K)$.

Proof By the proposition E is a normal extension of K. By Theorem 8.1.8, we therefore get $K = E_{G(E/K)}$. By the same theorem

$$|G(E/F)| = [E : F] = [E : K][K : F] = |G(E/K)|[K : F].$$

Now $H = G(E/E_H)$ holds by Theorem 8.1.7, so we have mutual inverse maps.

We claim that K is a normal extension of F if and only if $\alpha(K) = K$ for $\alpha \in G(E/F)$. The forward implication is immediate from Theorem 7.5.6. For the backward implication, let $\alpha \colon K \to \overline{F}$ be a monomorphism that fixes the elements of F. By Theorem 7.3.8 we can extend α to the monomorphism $\beta \colon E \to \overline{F}$, and Theorem 7.5.6 tells us that $\beta \in G(E/F)$. By assumption $\beta(K) = K$, so $\alpha(K) = K$ and Theorem 7.5.6 tells us that K is a normal extension of F. So our claim is true.

Thus if K is a normal extension of F, then for $a \in K$ and $\alpha \in G(E/F)$, we have $\alpha(a) \in K$, so $\gamma(\alpha(a)) = \alpha(a)$ for $\gamma \in G(E/K)$. Hence $\alpha^{-1}\gamma\alpha \in G(E/K)$ and $G(E/K)$ is a normal subgroup of $G(E/F)$.

Conversely, if $G(E/K)$ is a normal subgroup of $G(E/F)$, then $\alpha^{-1}\gamma\alpha \in G(E/K)$ for $\alpha \in G(E/F)$ and $\gamma \in G(E/K)$, so $\gamma(\alpha(a)) = \alpha(a)$ for $a \in K$. Since K is the fixed field of $G(E/K)$, we see that $\alpha(a) \in K$. Similarly, we get $\alpha^{-1}(K) \subset K$. So $\alpha(K) = K$, and K is a normal extension of F.

Again by our claim, restricting $\alpha \in G(E/F)$ to a normal extension K of F gives a homomorphism $G(E/F) \to G(K/F)$ with kernel $G(E/K)$. The last assertion in the theorem now follows by the fundamental theorem for homomorphisms because

$$[G(E/F) : G(E/K)] = [K : F] = |G(K/F)|.$$

□

The maps $\mathcal{G} \to \mathcal{S}$ and $\mathcal{S} \to \mathcal{G}$ in the theorem above clearly reverse the inclusions of subgroups and subfields.

Example 8.3.4 Continuing Example 8.2.4, the Galois group G of $x^4 - 2$ over $\mathbb{Q}$ was identified with D_4, where σ sends $2^{1/4}$ to $i2^{1/4}$ and fixes i, and τ sends i to $-i$ and fixes $2^{1/4}$.

The group G has four normal subgroups, namely

$$G_1 = \{\iota, \sigma\tau, \sigma^2, \sigma^3\tau\}, \quad G_2 = \langle\sigma\rangle, \quad G_3 = \{\iota, \tau, \sigma^2, \sigma^2\tau\}, \quad G_4 = \langle\sigma^2\rangle.$$

In addition, it has four non-normal subgroups of order two, namely

$$H_1 = \langle\sigma^3\tau\rangle, \quad H_2 = \langle\sigma\tau\rangle, \quad H_3 = \langle\sigma^2\tau\rangle, \quad H_4 = \langle\tau\rangle.$$

These are all the proper subgroups of G. Note that H_1 and H_2 are included in G_1, that H_3 and H_4 are included in G_3, whereas G_4 is included in G_1, G_2, G_3, and that these are all the non-trivial inclusions.

The corresponding fixed fields with inclusions reversed are

$$E_{G_1} = \mathbb{Q}(i2^{1/2}), \quad E_{G_2} = \mathbb{Q}(i), \quad E_{G_3} = \mathbb{Q}(2^{1/2}), \quad E_{G_4} = \mathbb{Q}(i, 2^{1/2}),$$

which are normal extensions and indeed splitting fields of $x^2 + 2, x^2 + 1, x^2 - 2$ and $(x^2 + 1)(x^2 - 2)$, respectively, and

$$E_{H_1} = \mathbb{Q}((1-i)2^{1/4}), \quad E_{H_2} = \mathbb{Q}((1+i)2^{1/4}), \quad E_{H_3} = \mathbb{Q}(i2^{1/4}), \quad E_{H_4} = \mathbb{Q}(2^{1/4}).$$

Note that E_{G_3} is included in both E_{H_3} and E_{H_4}, as it should be.

To calculate for instance E_{G_4}, let $a \equiv 2^{1/4}$ and write any $b \in \mathbb{Q}(a, i)$ as

$$b = b_1 + b_2 a + b_3 a^2 + b_4 a^3 + b_5 i + b_6 i a + b_7 i a^2 + b_8 i a^3.$$

Since $b \in E_{G_4}$ if and only if $\sigma^2(b) = b$, and as $\sigma^2(a) = -a$ and $\sigma^2(i) = i$, the b_j's with even j will vanish, whereas there are no constrains on those with odd j. Hence $E_{G_4} = \mathbb{Q}(i, 2^{1/2})$, as claimed. Similarly, we see that $b \in E_{G_3}$ if and only if $\sigma^2(b) = b = \tau(b)$. This will again force the b_j's with even j to vanish, but now, in addition $b_5 = 0 = b_7$, with no further constraints, so $E_{G_3} = \mathbb{Q}(2^{1/2})$. The other fixed fields are calculated similarly. ◇

8.4 Proof of the Fundamental Theorem of Algebra

There are many proofs of the fundamental theorem of algebra. We choose a rather elaborate one to illustrate the algebraic theory developed. It uses analytic properties of real numbers from calculus, which we for the sake of completeness, prove rigorously. The second assertion of the lemma below is known as the intermediate value theorem, based on the principle that in the plane, always pressing the pencil down, you need to cross an infinite line to get from one side to another. All this can be put in the conceptual context of continuity and connectedness, but this belongs to analysis, and ought perhaps to be minimized here.

Lemma 8.4.1 *The square root of a complex number exists as a complex number. Any polynomial over $\mathbb{R}$ of odd degree has a real root.*

Proof We claim that the square root of a positive real number r exists as a real number. The reader is encouraged to work out a recursive procedure using the decimal system and the division algorithm to find a Cauchy sequence $\{a_n\}$ of positive rational numbers such that $a_n^2 \to r$. Then $\sqrt{r} \equiv \lim a_n$ belongs to $\mathbb{R}$ by completeness, and as the notation suggests, we have $\sqrt{r}^2 = (\lim a_n)^2 = \lim a_n^2 = r$. The square root of the complex number $re^{i\theta}$ is then $\sqrt{r} e^{i\theta/2}$.

Say $f \in \mathbb{R}[x]$ has odd degree, and consider the evaluation at numbers as a function f from $\mathbb{R}$ to $\mathbb{R}$. The highest power of x, when replaced by a real number with large absolute value, will dominate the value of f at that number, and since this power is odd, we can find $a, b \in \mathbb{R}$ with $a < b$ and such that $f(a) < 0$ and $f(b) > 0$. Let $s \in \mathbb{R}$ be the supremum of the non-empty set $X \equiv \{c \in [a, b] \mid f(c) \leq 0\}$. Clearly $s \in [a, b)$. By definition of s there are numbers $c_n \in X$ such that $c_n \to s$. Then $f(s) = \lim f(c_n)$ since we are dealing with a polynomial, and can use the triangle inequality. Thus $f(s) \leq 0$. If $f(s) < 0$, again using that f is a polynomial, we can find $\varepsilon > 0$ small enough such that $s + \varepsilon \in X$, which is absurd. So $f(s) = 0$. □

Theorem 8.4.2 *The field of complex numbers is the algebraic closure of the real numbers.*

Proof We show that any polynomial $f(x) = a_0 + \cdots + a_n x^n \in \mathbb{C}[x]$ factors into first degree polynomials over $\mathbb{C}$. Consider

$$g(x) = (x^2 + 1)f(x)\overline{f(x)} \in \mathbb{R}[x],$$

where bar means complex conjugating the coefficients of f. Let E be the splitting field of g over $\mathbb{R}$. Thus E is a Galois extension of $\mathbb{R}$ that contains $\mathbb{C}$ as $x^2 + 1$ has roots $\pm i$. It suffices to prove $E = \mathbb{C}$.

Let H be a Sylow 2-subgroup of $G \equiv G(E/\mathbb{R})$. Then by the fundamental result of Galois theory, we have $[E : E_H] = |H| = 2^m$ and $[E_H : \mathbb{R}] = [G : H] = k$ for a non-negative integer m and an odd natural number k. By Proposition 7.8.2 the field E_H is a simple extension of $\mathbb{R}$ with a root of a minimal polynomial of degree k. By the lemma this irreducible polynomial over $\mathbb{R}$ must have a real root, which means that $k = 1$. Thus

$$2[E : \mathbb{C}] = [E : \mathbb{C}][\mathbb{C} : \mathbb{R}] = [E : \mathbb{R}] = [E : E_H][E_H : \mathbb{R}] = 2^m,$$

so $|G(E/\mathbb{C})| = 2^{m-1}$. If $m \geq 2$, pick a subgroup K of $G(E/\mathbb{C})$ of order 2^{m-2}. The existence of such a subgroup is again guaranteed by Theorem 4.20.2. Then

$$2^{m-2}[E_K : \mathbb{C}] = [E : E_K][E_K : \mathbb{C}] = [E : \mathbb{C}] = 2^{m-1},$$

so $[E_K : \mathbb{C}] = 2$. By Proposition 7.8.2 the field E_K is a simple extension of $\mathbb{C}$ with a root of a second degree minimal polynomial h, say $h(x) = x^2 + 2ax + b$ for $a, b \in \mathbb{C}$. By the lemma we can form the complex number $\sqrt{a^2 - b}$. But then we get a contradiction because h is irreducible over $\mathbb{C}$, and yet

$$h(x) = (x + a + \sqrt{a^2 - b})(x + a - \sqrt{a^2 - b}).$$

To avoid this, we must have $m = 1$, and then $[E : \mathbb{R}] = 2$, so $E = \mathbb{C}$. □

8.5 Primitive Roots and Cyclotomic Polynomials

Definition 8.5.1 Let $n \in \mathbb{N}$. A *primitive nth root of unity in a field F* is an element $a \in F$ such that $a^n = 1$, but $a^m \neq 1$ for every natural number $m < n$.

By Proposition 7.7.5 we know that the nth roots of unity in a non-trivial field F form a cyclic group. A primitive nth root of unity in F will obviously be a generator of this group, but the converse is not true; there might not even be any primitive nth roots of unity in F.

For instance, the number -1 is a generator for the cyclic group $\{\pm 1\} \cong \mathbb{Z}_2$ of roots of $x^4 - 1$ in $\mathbb{R}$, but it is not a primitive 4th root of unity. Obviously, the number i is a primitive 4th root of unity in the extension field $\mathbb{C}$ of $\mathbb{R}$. There are two such primitive roots in $\mathbb{C}$, namely $\pm i$. In general there are $\phi(n)$ primitive nth roots of unity in $\mathbb{C}$, namely the numbers $e^{2\pi i m/n}$, where m is coprime to n.

Proposition 8.5.2 *There exists a primitive nth root of unity in some extension field E of a field F if and only if the characteristic of F is either zero or does not divide n.*

Proof If a is a primitive nth root of unity in some extension field E of F, then $1, a, \ldots, a^{n-1}$ are n distinct roots of $f(x) = x^n - 1$. Thus $0 \neq f'(x) = nx^{n-1}$, which means that the characteristic of F is zero or cannot divide n.

If the characteristic of F is zero or does not divide n, and $f(x) = x^n - 1$, then $f'(x) = nx^{n-1} \neq 0$ and f has n distinct roots in its splitting field E over F. Any generator of the cyclic group of these roots will then be a primitive n-root of unity in E. □

Definition 8.5.3 The nth *cyclotomic polynomial* is the monic polynomial

$$\Phi_n(x) = \prod_a (x - a)$$

over $\mathbb{C}$, where the product runs over all primitive nth roots of unity in $\mathbb{C}$.

Now if we factorize the polynomial $x^n - 1$ into linear factors and gather those factors where the roots of unity have the same periode d, we get the formula

$$x^n - 1 = \prod \Phi_d(x),$$

where we take the product over all natural numbers d that divide n.

By what has been said, the polynomial $\Phi_n(x)$ has degree $\phi(n)$, and if n is a prime number, then

$$\Phi_n(x) = 1 + x + \cdots + x^{n-1}.$$

It is also easy to see that $\Phi_4(x) = x^2 + 1$ and $\Phi_6(x) = x^2 - x + 1$. This suggests that all the coefficients of $\Phi_n(x)$ are ± 1, but this is not true; the first counter example occur for $n = 105$. However, we have the following fundamental result.

Theorem 8.5.4 *The cyclotomic polynomials belong to $\mathbb{Z}[x]$ and are irreducible over $\mathbb{Q}$.*

Proof The splitting field E of $x^n + 1 \in \mathbb{Q}[x]$ is a Galois extension of the field $\mathbb{Q}$ of characteristic zero. By the fundamental result of Galois theory, the rational numbers is the fixed field of $G(E/\mathbb{Q})$. Any $\mathbb{Q}$-automorphism of E takes a primitive nth root of unity to another primitive nth root of unity, so the corresponding ring endomorphism of $E[x]$ fixes Φ_n. Thus the coefficients of $\Phi_n(x)$ belong to $\mathbb{Q}$. But Φ_n is a monic factor of $x^n - 1$, and from the proof of Proposition 7.1.5, we see that this is only possible if $\Phi_n \in \mathbb{Z}[x]$.

The same proposition tells us that Φ_n is irreducible over $\mathbb{Q}$ if it is irreducible over $\mathbb{Z}$. Say $f \in \mathbb{Z}[x]$ is an irreducible factor of Φ_n. We aim to show that $\Phi_n = f$ by proving that all the nth primitive roots of unity are roots of f. Say a is a root of f. Then it is clearly a primitive nth root of unity. If m is prime number that does not divide n, then a^m, being a generator of the cyclic group of all nth roots of unity, is again a primitive nth root of unity, and clearly all such roots are obtained by repeating this process finitely many times. Hence it suffices to show that a^m is a root of f.

If this is not the case, we can write $\Phi_n = fg$, where $g \in \mathbb{Z}[x]$ and $g(a^m) = 0$. By the division algorithm for polynomials over $\mathbb{Q}$ we can write $g(x^m) = f(x)q(x) + r(x)$ for $r, q \in \mathbb{Q}[x]$ with $\deg(r) < \deg(f)$. But $r(a) = 0$, so $r = 0$ since f is irreducible over $\mathbb{Q}$ and is therefore the minimal polynomial of a over $\mathbb{Q}$. As above we see that $q \in \mathbb{Z}[x]$. By $\bar{h}$ we mean the polynomial over $\mathbb{Z}_m$ obtained by applying the quotient map $\mathbb{Z} \to \mathbb{Z}_m$ to the coefficients of $h \in \mathbb{Z}[x]$. As $a^m \equiv a \pmod{m}$, we see that $\bar{g}$ and $\bar{f}$ have a common root $b = [a]$, which must also be a root of $x^n - [1]$. Thus the derivative of this polynomial must vanish at b, so $[n]b^{n-1} = [0]$. Since m and n are relatively prime, we get $b^{n-1} = 0$ and $b = 0$, which obviously cannot be a root of $x^n - [1]$. This is a contradiction. □

Definition 8.5.5 We denote the group of units in a unital ring R by $U(R)$.

Note that $U(F) = F_*$ for a field F.

The invertible elements of the ring $\mathbb{Z}_n$ are the elements $[m]$ with $\gcd(m, n) = 1$, that is, those m having integers k and l such that $mk + nl = 1$, or $[m][k] = 1$ for some integer k. So the order of $U(\mathbb{Z}_n)$ is $\phi(n)$.

Theorem 8.5.6 *Suppose a is a primitive nth root of unity in $\mathbb{C}$. Then $\mathbb{Q}(a)$ is the splitting field both of Φ_n and of $x^n - 1$ regarded as polynomials over $\mathbb{Q}$. Moreover, we have $|G(\mathbb{Q}(a)/\mathbb{Q})| = [\mathbb{Q}(a) : \mathbb{Q}] = \phi(n)$ and $G(\mathbb{Q}(a)/\mathbb{Q}) \cong U(\mathbb{Z}_n)$.*

Proof Since a generates all roots of unity in $\mathbb{C}$, the first statement is immediate. Clearly $\mathbb{Q}(a)$ is a Galois extension of $\mathbb{Q}$. Since Φ_n has degree $\phi(n)$ and is by the theorem above, the minimal polynomial of a over $\mathbb{Q}$, we therefore get $|G(\mathbb{Q}(a)/\mathbb{Q})| = [\mathbb{Q}(a) : \mathbb{Q}] = \phi(n)$.

Define a map $f \colon U(\mathbb{Z}_n) \to G(\mathbb{Q}(a)/\mathbb{Q})$ by $[m] \mapsto \sigma_m$, where σ_m is uniquely determined by $\sigma_m(a) = a^m$. This is clearly a well-defined monomorphism. It is surjective because if $\sigma \in G(\mathbb{Q}(a)/\mathbb{Q})$, then $\sigma(a)$ is again a primitive nth root of unity, that is, of the form a^m with m is coprime to n, so $\sigma = \sigma_m$ and $[m] \in U(\mathbb{Z}_n)$. □

8.6 Constructable Polygons

By Theorem 1.12.13 we know that $U(\mathbb{Z}_n)$ is cyclic exactly when n is 2, 4, p^m or $2p^m$, where p is an odd prime number.

Example 8.5.7 The Galois groups of Φ_8 and $x^8 - 1$ are both isomorphic to $U(\mathbb{Z}_8)$, which is the group with four elements such that all non-unital elements have square one. ◇

Example 8.5.8 The Galois groups of both $x^4 + x^2 + 1$ and $x^6 - 1$ are $\mathbb{Z}_2$. Indeed, since $y^2 + y + 1 = \Phi_3(y)$, the splitting field E of $x^4 + x^2 + 1$ will contain the square root of $e^{2\pi i/3}$, and thus the primitive 6th root of unity $a = e^{\pi i/3}$, so $E = \mathbb{Q}(a)$, which is also the splitting field of $x^6 - 1$ over $\mathbb{Q}$. In both cases the Galois group is $G(\mathbb{Q}(a)/\mathbb{Q}) = U(\mathbb{Z}_6) = \{[1], [5]\} \cong \mathbb{Z}_2$, which is cyclic, as expected. ◇

8.6 Constructable Polygons

We will here investigate what regular n-gons are constructable by ruler and compass.

Lemma 8.6.1 *We can construct the square root of any positive real number.*

Proof Say OA has length $a \in \mathbb{R}$. Prolong this line to the left with a unit length arriving at the point P. Going via the midpoint of AP, form an upper semicircle that intersects A and P. Erect a perpendicular line from O that intersects the semicircle at Q. The triangles OQP and OQA are similar, and comparing two and two sides, we see that the length of OQ is the square root of a. □

Lemma 8.6.2 *The number $\phi(n)$ is a power of two if and only if the odd primes dividing n are Fermat primes with squares that do not divide n.*

Proof Say $n = 2^{n_0} p_1^{n_1} \cdots p_m^{n_m}$ for odd primes p_i and $n_i, m \in \mathbb{N}$. Now

$$\phi(n) = 2^{n_0-1} p_1^{n_1-1} \cdots p_m^{n_m-1}(p_1 - 1) \cdots (p_m - 1)$$

is a power of two if and only if for all $i \geq 1$, we have $n_i = 1$ and $p_i = 2^{k_i} + 1$ for $k_i \in \mathbb{N}$. For $2^{k_i} - 1$ to be a prime number, the number k_i must be a power of two because if $k_i = rp$ for an odd prime p, then $2^{k_i} - 1$ would be divisible by $2^r + 1$. □

Theorem 8.6.3 *The regular n-gon is constructable by ruler and compass if and only if the odd primes dividing n are Fermat primes with squares that do not divide n.*

Proof Clearly a regular n-gon is constructable by ruler and compass if and only if the angle $2\pi/n$ is constructable if and only if $\cos(2\pi/n)$ is constructable if and only if $2\cos(2\pi/n) = a + 1/a$, with $a = e^{2\pi i/n}$, is constructable.

Let E be the splitting field of $x^n - 1$ over $\mathbb{Q}$. By Theorem 8.5.6 we have $[E : \mathbb{Q}] = \phi(n)$. Any non-trivial $\alpha \in G(E/\mathbb{Q})$ obviously satisfies $\alpha(a) = a^m$ for some integer m between 1 and n, and then

$$\alpha(a+1/a)) = a^m + 1/a^m = 2\cos(2m\pi/n),$$

which equals $2\cos(2\pi/n)$ if and only if $m = n - 1$. Thus the subgroup of $G(E/\mathbb{Q})$ with fixed field $\mathbb{Q}(a+1/a)$ is $\{\iota, \beta\}$, where $\beta(a) = 1/a$. By the fundamental result in Galois theory we therefore get $[\mathbb{Q}(a+1/a) : \mathbb{Q}] = \phi(n)/2$.

If a regular n-gon is constructable, by Corollary 7.4.4 we know that $[\mathbb{Q}(a+1/a) : \mathbb{Q}]$ is a power of two, and the same must therefore be true for $\phi(n)$. By the lemma above we have proved the forward implication.

For the converse, we may assume by the lemma above that $\phi(n)$ is a power of two. Then by Theorem 4.20.2 there is an ascending chain of subgroups G_i of $G(\mathbb{Q}(a)/\mathbb{Q})$ of order 2^i with $G_1 = \{\iota, \beta\}$ and $G_k = G(\mathbb{Q}(a)/\mathbb{Q})$. By the fundamental result in Galois theory we have for the corresponding fixed fields that

$$[E_{G_{i-1}} : E_{G_i}] = 2$$

for $i \geq 2$, and $E_{G_1} = \mathbb{Q}(a+1/a)$. As $a + 1/a$ is real they are all subfields of $\mathbb{R}$, so $E_{G_{i-1}} = E_{G_i}(a_i)$, where the a_i's are real roots of quadratic polynomials over E_{G_i} that can be solved using the high school formula for second degree equations. Therefore $E_{G_{i-1}} = E_{G_i}(b_i)$ for roots b_i of positive real numbers. By Lemma 8.6.1 and Proposition 7.4.7 we conclude that all numbers in $\mathbb{Q}(a+1/a)$ are constructable. In particular, the number $a + 1/a$ is constructable, and we can construct the regular n-gon. □

Example 8.6.4 The numbers $2^{2^k} + 1$ are prime for $k = 0, 1, 2, 3, 4$. These are the first 5 Fermat primes. Now $k = 5$ gives a number that is divisible by 641, disproving Fermat's conjecture that all numbers of such a form are primes. Anyway, we conclude that the first five regular p-gons for p a prime that are constructable by ruler and compass correspond to the numbers 3, 5, 17, 257 and 65537. Already the construction of the regular 17-gon is quite intricate.

As for non-primes, the regular 60-gon is constructable because $60 = 2^2 \cdot 3 \cdot 5$ and both 3 and 5 are Fermat primes. However, the regular 18-gon is not constructable, for while 3 is a Fermat prime, its square divides 18. ◇

8.7 Cyclic Extensions

Definition 8.7.1 A Galois extension E of a field F is a *cyclic extension of F* if $G(E/F)$ is cyclic.

In Sect. 8.5 we saw that the splitting fields of $x^n - 1 \in \mathbb{Q}[x]$ are cyclic extensions of $\mathbb{Q}$ precisely when n is 2, 4, p^m or $2p^m$, where p is an odd prime number.

Splitting fields of polynomials over finite fields are separable extensions, and are therefore cyclic extensions by Theorem 7.7.11.

8.7 Cyclic Extensions

Lemma 8.7.2 *Suppose E is a field, and that we have a map $f : G \to E_*$, where G is a subgroup of the group of all automorphism of E. Then there exists an element $a \in E_*$ such that $f(\alpha) = \alpha(a^{-1})a$ if and only if $f(\alpha\beta) = \alpha(f(\beta))f(\alpha)$.*

Proof The forward implication is immediate. As for the backward implication, by Lemma 8.1.5 there exists $b \in E_*$ such that

$$a \equiv \sum_{\beta \in G} f(\beta)\beta(b) \neq 0$$

as all $f(\beta) \neq 0$. Thus

$$\alpha(a) = \sum_\beta \alpha(f(\beta))\alpha\beta(b) = f(\alpha)^{-1} \sum_\beta f(\alpha\beta)\alpha\beta(b) = f(\alpha)^{-1}a.$$

□

Lemma 8.7.3 *Suppose E is a finite extension of a field F, and that $G(E/F)$ is cyclic of order n with generator α. If E contains a primitive nth root b of unity, then $b = \alpha(a)a^{-1}$ for some $a \in E_*$.*

We are going to apply the lemma for $b \in F$, which simplifies matters. Nevertheless we supply a proof in the general case.

Proof By the lemma above it suffices to define a map $f : G(E/F) \to E_*$ such that $f(\alpha) = b$ and with the property

$$\alpha^i(f(\alpha^j))f(\alpha^i) = f(\alpha^i\alpha^j)$$

for all $i, j \in \mathbb{N}$. Set $f(\iota) = 1$ and $f(\alpha^m) = \alpha^{m-1}(a)\cdots\alpha(a)a$ for $m \in \mathbb{N}$.

For consistency we must show that $\alpha^{n-1}(a)\cdots\alpha(a)a = 1$. As $\alpha(a)$ is another primitive nth root of unity, we can write $\alpha(a) = a^k$ for k coprime to n. Thus $\alpha^{n-1}(a)\cdots\alpha(a)a = a^r$, where

$$r(k-1) = (1 + k + \cdots + k^{n-1})(k-1) = k^n - 1.$$

But $k^n \equiv 1 \pmod{n}$ as $\alpha^n = \iota$, so $a^r = 1$.

Now

$$f(\alpha^i\alpha^j) = f(\alpha^{i+j}) = \alpha^{i+j-1}(a)\cdots\alpha(a)a,$$

whereas

$$\alpha^i(f(\alpha^j))f(\alpha^i) = \alpha^i(\alpha^{j-1}(a)\cdots\alpha(a)a)\alpha^{i-1}(a)\cdots\alpha(a)a = \alpha^{i+j-1}(a)\cdots\alpha(a)a.$$

□

Theorem 8.7.4 *Let F be a field that contains a primitive nth root of unity. Then E is a finite cyclic extension of F of degree n if and only if E is the splitting field of an irreducible polynomial over F of the form $x^n - b$. Moreover, in this case $E = F(a)$ for a root a of $x^n - b$.*

Proof For the forward implication, let α be a generator of $G(E/F)$. By the previous lemma there is $a \in E_*$ such that $\alpha(a) = ca$, where $c \in F$ is a primitive nth root of unity. Then $\alpha^m(a) = c^m a$ for all $m \in \mathbb{N}$, so $\alpha^m(a^n) = c^{mn} a^n = a^n$ and $a^n \in F$. Set $b = a^n$. Then all the roots of $x^n - b \in F[x]$ are of the form $c^m a$, and one can obviously get from one to another by applying integer powers of α. So any factor of $x^n - b$ over F will have n distinct roots, showing that $x^n - b$ is irreducible over F. This shows that $E = F(a)$ and that E is the splitting field of $x^n - b$.

Conversely, if E is the splitting field of an irreducible polynomial $x^n - b \in F[x]$ with root $a \in E$, then all its roots are of the form $c^i a$ for an integer i. Clearly there will be n distinct ones, so $E = F(a)$ is a Galois extension of F. Define a map $\mathbb{Z}_n \to G(E/F)$ by $[i] \mapsto \alpha_i$, where $\alpha_i(a) = c^i a$. This is clearly a well-defined monomorphism which is surjective because $|G(E/F)| = [E:F] = n = |\mathbb{Z}_n|$. □

8.8 Polynomials Solvable by Radicals

Definition 8.8.1 We say that a field E is an *extension of a field F by radicals* if there are elements $a_i \in E$ and $n_i \in \mathbb{N}$ such that $E = F(a_1, \ldots, a_m)$ and $a_i^{n_i} \in F(a_1, \ldots, a_{i-1})$ with $a_1^{n_1} \in F$.

Thus $\mathbb{Q}(2^{1/3}, (5 + 2^{1/3})^{1/2})$ is an extension of $\mathbb{Q}$ by radicals as $2 \in \mathbb{Q}$ and $5 + 2^{1/3} \in \mathbb{Q}(2^{1/3})$.

Definition 8.8.2 A polynomial over a field F is *solvable by radicals* if its splitting field is contained in some radical extension of F.

Hence a polynomial over a field F is solvable by radicals if every root of it can be obtained using field operations and taking nth roots in any finite combination.

Lemma 8.8.3 *If E is a splitting field of $x^n - a$ over a field F of characteristic zero, then $G(E/F)$ is a solvable group.*

Proof By Proposition 8.5.2 we may pick a primitive nth root b of unity in $\overline{F}$. If c is a root of $x^n - a$, then cb is also a root, so $b = c^{-1}(cb) \in E$ and $F(b) \subset E$. Now $F(b)$ is a Galois extension of F being the splitting field of $x^n - 1$ over a field of characteristic zero. By the fundamental result in Galois theory, we deduce that $G(E/F(b))$ is a normal subgroup of $G(E/F)$. So we have a normal series

$$\{\iota\} \triangleleft G(E/F(b)) \triangleleft G(E/F).$$

By Proposition 8.2.5 the group $G(E/F(b))$ is abelian, and again by the fundamental result of Galois theory, we see that $G(E/F)/G(E/F(b)) \cong G(F(b)/F)$, which also

8.8 Polynomials Solvable by Radicals   297

is abelian because any F-automorphism α of $F(b)$ is of the form $\alpha(b) = b^m$ for an integer m and such automorphisms will commute. □

Lemma 8.8.4 *Say F_r is an extension by radicals of a field F of characteristic zero with intermediate fields F_i. Then there exists a normal extension E_s by radicals of F containing F_r and with intermediate fields E_j such that E_j is a splitting field of a polynomial of the form $x^{m_j} - a_j \in E_{j-1}[x]$.*

Proof Say F_i is a simple extension of a root of $x^{n_i} - b_i \in F_{i-1}[x]$. Let $n = n_1 \cdots n_r$ and let $c \in \overline{F}$ be a primitive nth root of unity. Let K be the splitting field of $g_1(x) \equiv (x^n - 1)(x^{n_1} - b_1) \in F[x]$. Clearly $F_1 \subset K$ and $c \in K$. We obviously also get an ascending chain of intermediate fields E_j between F and K such that each one is a splitting field of a polynomial of the form $x^m - a$ over the former one.

Next we construct a normal extension L of F that contains K and F_2. Set

$$f_2(x) = \prod_{\alpha \in G(K/F)} (x^{n_2} - \alpha(b_2)) \in K[x].$$

Since we are working in characteristic zero, we know that K is a Galois extension of F, and since f_2 is clearly fixed under the action of the elements of $G(K/F)$ on its coefficients, we see that $f_2 \in F[x]$. Then the splitting field L of $g_2 \equiv g_1 f_2 \in F[x]$ clearly has the desired properties. We get further intermediate fields E_j between K and L such that each one is a splitting field of a polynomial of the form $x^m - a$ over the former one. Continuing this way gives the desired normal extension by radicals. □

Theorem 8.8.5 *A polynomial is solvable by radicals over a field of characteristic zero if and only if its Galois group is solvable.*

Proof Say we have a polynomial f over a field F of characteristic zero, and let E be its splitting field over F, so E is a Galois extension of F since we are working in characteristic zero.

Suppose that f is solvable by radicals. By the previous lemma we can assume that E is contained in a normal extension by radicals E_s of F which contains F_r and has intermediate fields E_j such that E_j is a splitting field of a polynomial of the form $x^{m_j} - a_j \in E_{j-1}[x]$.

By the fundamental result of Galois theory, we get a normal series

$$\{\iota\} \lhd G(E_s/E_{s-1}) \lhd G(E_s/E_{s-2}) \lhd \cdots \lhd G(E_s/F)$$

which by the first lemma above has solvable quotients

$$G(E_s/E_{s-i})/G(E_s/E_{s-i+1}) \cong G(E_{s-i+1}/E_{s-i}).$$

Thus $G(E_s/F)$ is solvable, and so is its homomorphic image

$$G(E/F) \cong G(E_s/F)/G(E_s/E).$$

Conversely, suppose $G(E/F)$ is solvable. Say $[E:F] = n$ and assume that F has a primitive nth root of unity.

Since the group is finite, it has a normal series of groups G_i with cyclic quotients G_{i-1}/G_i. By the fundamental result in Galois theory, the fixed field $E_i \equiv E_{G_i}$ is a normal extension of F, and thus also normal over E_{i-1}, and therefore is a cyclic extension of E_{i-1}. By Theorem 8.7.4 we know that E_i is the splitting field of a polynomial $x^{n_i} - b_i \in E_{i-1}[x]$, so $E_i = E_{i-1}(a_i)$, where $a_i^{n_i} = b_i$. Thus f is solvable by radicals.

Suppose F does not contain a primitive nth root of unity. Since we are working in characteristic zero there is such a root a in $\overline{E}$. Now $E(a)$ is the splitting field of f regarded as a polynomial over $F(a)$. Moreover, we have a monomorphism $G(E(a)/F(a)) \to G(E/F)$ given by restriction of automorphisms. They will map into E under restriction because E is a normal extension of F. A subgroup of a solvable group is solvable, so $G(E(a)/F(a))$ is solvable. By the paragraph above we know that $E(a)$ is an extension by radicals of $F(a)$, and hence also of F. Since $E \subset E(a)$ we conclude that f is solvable by radicals also in this case. □

We know that the Galois group of a polynomial over a field F with n distinct roots can be embedded in the symmetric group S_n. Since S_n is not solvable if and only if $n \geq 5$, we see that polynomials of degree less than 5 are solvable by radicals when F has characteristic zero. Explicit formulas to this effect can be found for such polynomials over $\mathbb{C}$, and we will return to that in Sect. 8.10. For $n \geq 5$ no such general formulas exist because in this case we can find polynomials over subfields of $\mathbb{C}$, and sometimes over $\mathbb{Q}$, such that their Galois groups are all of S_n. We look at criteria that will guarantee this.

Definition 8.8.6 A subgroup G of S_n is *transitive* if it acts transitively on the set $\{1, \ldots, n\}$.

Proposition 8.8.7 *If p is a prime number and G is a transitive subgroup of S_p that contains a transposition, then $G = S_p$.*

Proof Since G acts transitively on $X \equiv \{1, \ldots, p\}$, we have $[G : G_x] = p$, where G_x is the isotropy group of $x \in X$. Thus p divides the order of G, and by Corollary 4.20.3 it has an element of order p, or in other words, a p-cocycle σ. We may assume that $\sigma = \begin{pmatrix} 1 & \cdots & p \end{pmatrix}$ and $\tau \equiv \begin{pmatrix} 1 & n \end{pmatrix} \in G$ for some $n \leq p$.

Now $\sigma^m \tau \sigma^{-m} = \begin{pmatrix} m & m+n \end{pmatrix}$, so $\begin{pmatrix} p-1 & p \end{pmatrix} \in G$ if $n = p$. If $n < p$, then

$$\begin{pmatrix} 1 & 2n \end{pmatrix} = \begin{pmatrix} n & 2n \end{pmatrix}\begin{pmatrix} 1 & n \end{pmatrix}\begin{pmatrix} n & 2n \end{pmatrix}^{-1} \in G.$$

Continuing this shows that $\begin{pmatrix} 1 & mn \end{pmatrix} \in G$. Pick the largest integer m such that $mn \leq p$. Since p is prime, we know that $1 \leq k \equiv p - mn < n$, so we get an element $\begin{pmatrix} 1 & k \end{pmatrix} \in G$ after conjugating $\begin{pmatrix} 1 & mn \end{pmatrix}$ by σ^k. By repeating the whole process finitely many times we get $k = 2$, so $\begin{pmatrix} 1 & 2 \end{pmatrix} \in G$, and again $\begin{pmatrix} p-1 & p \end{pmatrix} \in G$. The result then follows from Proposition 4.11.3. □

8.8 Polynomials Solvable by Radicals

Proposition 8.8.8 *A polynomial over a field is irreducible if and only if its Galois group is a transitive subgroup of the symmetric group on its distinct roots.*

Proof Say f is a polynomial over a field F with n distinct roots a_i in its splitting field E. Consider its Galois group $G(E/F)$ as a subgroup of S_n.

If f is irreducible over F, then $F(a_i) \cong F[x]/(f) \cong F(a_j)$ under an isomorphism that sends a_i to a_j and fixes each element of F. Since E is a normal extension of F we can extend this map to an automorphism $\alpha \in G(E/F)$, showing that the Galois group is a transitive subgroup of S_n.

Conversely, if $G(E/F)$ is transitive, then if a_i is a root of an irreducible factor g of f, then $\alpha(a_i)$, and thus every root of f will also be a root of g. Thus f is proportional to g. □

Theorem 8.8.9 *Say $f \in \mathbb{Q}[x]$ is an irreducible polynomial of prime number degree p that has exactly two non-real roots in $\mathbb{C}$. Then the Galois group of f is isomorphic to S_p. Thus f is not solvable by radicals when $p \geq 5$.*

Proof Let $E \subset \mathbb{C}$ be the splitting field of f over $\mathbb{Q}$. Since $\mathbb{Q}$ has characteristic zero, the irreducible polynomial f has p distinct roots. By the proposition above $G(E/\mathbb{Q})$ is isomorphic to a transitive subgroup of S_p. Since the coefficients of f are real, the operation of complex conjugation sends a root to a root and fixes each element of $\mathbb{Q}$. Since E is a normal extension of $\mathbb{Q}$ we can extend this operation to an automorphism $\alpha \in G(E/\mathbb{Q})$. Clearly α fixes each real root and permutes the two non-real ones, so it is a transposition. By Proposition 8.8.7 we therefore get $G(E/\mathbb{Q}) \cong S_p$. □

Example 8.8.10 The polynomial $f(x) = x^5 - 6x + 3 \in \mathbb{Q}[x]$ is not solvable by radicals. To see this, first observe that by Eisenstein's criterion it is irreducible over $\mathbb{Q}$. Next, drawing its graph, we see that f has exactly three real roots, and thus two non-real ones by the fundamental theorem of algebra. The theorem above then kicks in.

Being slightly more rigorous about the roots, we notice that f changes signs at real values sufficiently many times for it to have at least three real roots according to the proof of Lemma 8.4.1. Also we see that $5x^4 - 6 = f'(x) = 0$ has precisely two real roots $\pm(6/5)^{1/4}$, which accounts for no more changes of signs for f.

Similarly, we see that $x^5 - 4x + 2 \in \mathbb{Q}[x]$ is not solvable by radicals. Using Descartes' rule of signs in combination with the proof of Lemma 8.4.1, one also sees that the irreducible polynomial $x^7 - 10x^5 + 15x + 5 \in \mathbb{Q}[x]$ has exactly two non-real roots, and is therefore not solvable by radicals.

Of course, there are also polynomials over $\mathbb{Q}$ of degree higher than five that are solvable by radicals. The polynomial $x^{13} - 1$ is one such example. In the next section we will however construct polynomials of any degree greater than four over subfields of $\mathbb{C}$ that are not solvable by radicals. ◇

Proposition 8.8.11 *Any irreducible polynomial over a field F with a root in an extension by radicals of F is solvable by radicals.*

Proof By the lemma above any extension by radicals of F with a root of the polynomial can be contained in a normal extension by radicals of F. Since the polynomial is irreducible, this normal extension must also contain a splitting field of the polynomial, so it is solvable by radicals. $\square$

8.9 Symmetric Functions

Consider the field $F(x_1, \ldots, x_n)$ of rational functions over a field F. Any $\sigma \in S_n$ induces an F-automorphism η_σ on $F(x_1, \ldots, x_n)$ uniquely determined by $\eta_\sigma(x_i) = x_{\sigma(i)}$ for all i. Clearly the map $\sigma \mapsto \eta_\sigma$ is a monomorphism

$$S_n \to G(F(x_1, \ldots, x_n)/F).$$

Definition 8.9.1 A *symmetric function* in $x_1, \ldots, x_n$ over a field F is an element $f \in F(x_1, \ldots, x_n)$ such that $\eta_\sigma(f) = f$ for all $\sigma \in S_n$.

Let $E \subset F(x_1, \ldots, x_n)$ be the fixed field of S_n. Consider $f \in F(x_1, \ldots, x_n)[y]$ given by

$$f(y) = \prod_{i=1}^{n}(y - x_i).$$

Clearly f is fixed by the mapping induced by η_σ, so its coefficients belong to E.

Definition 8.9.2 The number $(-1)^i$ times the coefficient of y^{n-i} in $\prod(y - x_i)$ is called the *elementary symmetric function* s_i in $x_1, \ldots, x_n$.

Note that

$$s_1 = x_1 + x_2 + \cdots + x_n$$
$$s_2 = x_1 x_2 + x_1 x_3 + \cdots + x_{n-1} x_n$$
$$\cdot$$
$$s_n = x_1 x_2 \cdots x_n$$

Theorem 8.9.3 *Every symmetric function in $x_1, \ldots, x_n$ is a rational function of the elementary symmetric functions s_i, and $F(x_1, \ldots, x_n)$ is a normal extension of $F(s_1, \ldots, s_n)$ of degree $n!$ with*

$$G(F(x_1, \ldots, x_n)/F(s_1, \ldots, s_n)) \cong S_n.$$

Proof Clearly $K \equiv F(s_1, \ldots, s_n) \subset E$. Note that $F(x_1, \ldots, x_n)$ is a splitting field of a separable polynomial $\prod(y - x_i)$ which is of degree n over K. By transitivity of the degree of extensions, we have

$$[F(x_1,\ldots,x_n) : K] \leq n!$$

which combined with

$$[F(x_1,\ldots,x_n) : E] = |S_n| = n!$$

yields $E = K$. For the last claim in the theorem observe that for our Galois extension we have

$$[F(x_1,\ldots,x_n) : K] = |G(F(x_1,\ldots,x_n)/K)|,$$

so S_n is embedded into a group of the same finite order. □

Example 8.9.4 The symmetric polynomial $x_1^2 + x_2^2 + x_3^2$ can obviously be written as $s_1^2 - 2s_2$. ◇

By a countability argument we can pick real numbers $a_1,\ldots,a_n$ such that a_i is transcendental over $\mathbb{Q}(a_1,\ldots,a_{i-1})$ with a_1 transcendental over $\mathbb{Q}$. Then $a_1,\ldots,a_n$ are said to be *independent transcendental elements* over $\mathbb{Q}$.

The $\mathbb{Q}$-homomorphism $F(x_1,\ldots,x_n) \to F(a_1,\ldots,a_n)$ that sends x_i to a_i is clearly a $\mathbb{Q}$-automorphism. On the level of polynomials in y over these fields it sends $\prod(y-x_i)$ to $\prod(y-a_i)$. Evaluating the elementary polynomial s_i at the numbers $a_1,\ldots,a_n$ produce $t_i \in F(a_1,\ldots,a_n)$. As $G(F(x_1,\ldots,x_n)/F(s_1,\ldots,s_n)) \cong G(F(a_1,\ldots,a_n)/F(t_1,\ldots,t_n))$ we immediately get the following result from the theorem above.

Theorem 8.9.5 *Let $a_1,\ldots,a_n$ be independent transcendental elements over $\mathbb{Q}$. Then the Galois group of the polynomial $\prod(y-a_i)$ over the subfield of $\mathbb{R}$ generated by the elementary symmetric functions s_i evaluated at the a_i's is isomorphic to S_n, rendering the polynomial not solvable by radicals for $n \geq 5$.*

8.10 Cubic and Quartic Equations

As we have seen, cubic and quartic polynomials over fields of characteristic zero are solvable by radicals.

The renaissance Italians solved the complex cubic equation

$$x^3 + ax^2 + bx + c = 0$$

by the following trickery. Translate away the quadratic term by $y = x + a/3$. Then

$$y^3 + py = q$$

for constants p and q that depend on a, b and c. The ingenious step is to make the ansats $y = u + v$, where $uv = r$ should only depend on p and q and should be adjusted correctly to simplify matters. Plugging this in gives

$$q = u^3 + 3u^2v + 3uv^2 + v^3 + p(u+v) = u^3 + v^3 + (3r+p)(u+v).$$

As $v^3 = r^3/u^3$, we get a quadratic equation in u^3 if we set $r = -p/3$, namely the equation

$$(u^3)^2 - qu^3 - p^3/27 = 0.$$

This gives

$$u^3 = q/2 \pm \sqrt{q^2/4 + p^3/27}$$

and

$$v^3 = r^3/u^3 = q/2 \mp \sqrt{q^2/4 + p^3/27},$$

where the square root is any complex number with square $q^2/4 + p^3/27$. Hence we get three solutions

$$y_n = \omega^n \sqrt[3]{q/2 + \sqrt{q^2/4 + p^3/27}} + \omega^{3-n} \sqrt[3]{q/2 - \sqrt{q^2/4 + p^3/27}},$$

where $\omega \in \mathbb{C}$ is a 3-rd primitive root of unity and the two cube roots are any chosen ones. We have not missed any complex number in the substitution $y = u + r/u$ as $u^2 - yu + r = 0$ has complex solutions u for any y and r.

How do we approach this from the point of view of Galois theory? Let a_i be n complex independent transcendental elements over $\mathbb{Q}$, and let t_i be the elementary symmetric polynomial s_i evaluated at these a_i's. Then the Galois group of $\prod(x - a_i)$ over the subfield of $\mathbb{C}$ generated by the t_i's is S_n. Let us first consider $n = 2$.

The Galois group of the quadratic polynomial

$$x^2 - t_1 x + t_2$$

is $S_2 \cong \{\iota, \tau\}$, where τ transposes a_1 and a_2. Clearly $(a_1 - a_2)^2$ is fixed by S_2, so by the previous section, it belongs to the rational field $\mathbb{Q}(t_1, t_2)$ generated by the coefficients t_1 and t_2. Explicitly, we get

$$(a_1 - a_2)^2 = t_1^2 - 4t_2,$$

which combined with $a_1 + a_2 = t_1$ gives the familiar formula

$$a_i = t_1/2 \pm \sqrt{t_1^2/4 - t_2}$$

for the roots in terms of the coefficients.

The Galois group of the cubic polynomial

8.10 Cubic and Quartic Equations

$$x^3 - t_1 x^2 + t_2 x - t_3$$

has the normal series

$$\{\iota\} \triangleleft A_3 \triangleleft S_3$$

with $A_3 \cong \mathbb{Z}_3$ and $S_3/A_3 \cong \mathbb{Z}_2$. Let

$$U = a_1 + \omega a_2 + \omega^2 a_3 \quad \text{and} \quad V = a_1 + \omega^2 a_2 + \omega a_3.$$

Since the generator of A_3 permutes a_1, a_2 and a_3 cyclically, and thus multiplies U by ω and V by ω^2, we see that both U^3 and V^3 are fixed by A_3. This generator and τ generate the whole group S_3. As $\tau(U) = \omega V$, we see that $U^3 + V^3$ and $U^3 V^3$ are fixed by S_3, and thus belong to $\mathbb{Q}(t_1, t_2, t_3)$. One then finds explicit expressions for $U^3 + V^3$ and $U^3 V^3$ in terms of the t_i's, solve the quadratic equations in U^3 and in V^3, and extract U and V as cube roots. Finally, using $t_1 = a_1 + a_2 + a_3$ and remembering that $\omega^2 + \omega + 1 = 0$, we arrive at

$$a_1 = \frac{1}{3}(t_1 + U + V), \quad a_2 = \frac{1}{3}(t_1 + \omega^2 U + \omega V), \quad a_3 = \frac{1}{3}(t_1 + \omega U + \omega^2 V).$$

The Galois group of the quartic polynomial

$$x^4 - t_1 x^3 + t_2 x^2 - t_3 x + t_4$$

has the normal series

$$\{\iota\} \triangleleft G \triangleleft A_4 \triangleleft S_4,$$

where

$$G = \{\iota, (1\ 2)(3\ 4), (1\ 3)(2\ 4), (1\ 4)(2\ 3)\}$$

and the quotients are abelian. We therefore consider the elements

$$b_1 = (a_1 + a_2)(a_3 + a_4), \quad b_2 = (a_1 + a_3)(a_2 + a_4), \quad b_3 = (a_1 + a_4)(a_2 + a_3),$$

which permute among themselves under the action of S_4. Hence the symmetric expressions

$$b_1 + b_2 + b_3, \quad b_1 b_2 + b_1 b_3 + b_2 b_3, \quad b_1 b_2 b_3$$

belong to $\mathbb{Q}(t_1, t_2, t_3, t_4)$. Finding explicit expressions of these, we see that the b_i's are roots of a cubic polynomial over this coefficient field. Solving this *resolvent cubic* we find the b_i's. Next, invoking

$$a_1 + a_2 + a_3 + a_4 = t_1$$

we get three quadratic polynomials with roots $a_1 + a_2$ and $a_3 + a_4$, $a_1 + a_3$ and $a_2 + a_4$, $a_1 + a_4$ and $a_2 + a_3$. From these we then find a_1, a_2, a_3 and a_4.

Alternatively, we can follow Ferrari's method for solving the complex quartic equation
$$x^4 + ax^3 + bx^2 + cx + d = 0$$
by first translating $y = x + a/4$, obtaining the equation in depressed form
$$y^4 + py^2 + qy + r = 0$$
for constants p, q and r that depend on a, b, c and d, and then basically insisting on factorizing the left-hand-side into two second order polynomials with appropriate coefficients. A straightforward calculation shows that this can be done if the square of the first order coefficients in either case satisfy a cubic equation, the resolvent cubic. Solving this cubic, extracting square roots, and determining the remaining coefficients, one is left with two quadratic equations which are easily solved.

Definition 8.10.1 The *discriminant* of a monic polynomial over a field F with roots a_i is $\Delta = \delta^2 \in \overline{F}$, where
$$\delta = \prod_{i<j}(a_i - a_j).$$

Note that the discriminant of a polynomial is zero precisely when it has multiple roots.

Proposition 8.10.2 *Let f be a monic separable polynomial of degree n over a field F of characteristic zero. Then its discriminant Δ belongs to F. Moreover, the discriminant is a perfect square in F if and only if the Galois group of f is contained in the alternating group A_n.*

Proof Note that the splitting field of f is a Galois extension of F, and that we may consider the Galois group of f as a subgroup of S_n. Next note that δ is unaltered up to a sign under the action of elements from S_n, so $\Delta \in F$. This sign is actually what defines the sign of a permutation on n elements.

Thus if Δ is a perfect square, then $\delta \in F$, so it is fixed by the Galois group G of f, which means that $G \subset A_n$.

Conversely, if $G \subset A_n$, then δ is fixed by G, so it belongs to the fixed field F. Thus Δ is a perfect square in F. □

Example 8.10.3 Consider the cubic polynomial $x^3 + ax + b$ over a field F of characteristic zero. If all its roots are in F, then its Galois group is trivial, and if this is not the case but that it is still reducible, then its Galois group is S_2.

Consider now the case when the cubic polynomial is irreducible over F, or in other words, when none of its roots belong to F. In characteristic zero it is then automatically separable, and on order grounds, its Galois group is either S_3 or A_3.

8.10 Cubic and Quartic Equations

By the proposition above, the Galois group is A_3 precisely when the discriminant Δ is a perfect square in F. This in turn can be easily checked because

$$\Delta = -4a^3 - 27b^2.$$

One can get this formula by brute force, or by considering the roots to be independent transcendental elements, and then argue that as Δ is homogeneous of degree 6 in t_1 and t_2, while a and b are homogeneous of degree 2 and 3, respectively, it must be of the form $\mu a^3 + \nu b^2$ for universal integers μ and ν. These integers are then determined by picking two simple cubic polynomials, like $x^3 - x$ and $x^3 - 1$.

To find the discriminant for higher degree polynomials, observe that δ is given by the Vandermonde determinant of the roots, so that Δ is the determinant of the product of the Vandermonde matrix and its transpose, whose entries can be found recursively from the polynomial.

The polynomial $x^3 - x + 1$ is irreducible over $\mathbb{Q}$ since any rational root would have to be ± 1, and its discriminant is -23, which is not the square of a rational number, so its Galois group is S_3.

The polynomial $x^3 - 3x + 1$ is also irreducible over $\mathbb{Q}$ since it has no integer roots, and its discriminant is $81 = 9^2$, so its Galois group is A_3, which is isomorphic to $\mathbb{Z}_3$. ◇

Chapter 9
Modules

In this monster chapter we consider modules. A module is a generalization of a vector space over a field, where the field is replaced by a general ring R. One speaks then of an R-module, or a module over R. However, many facts about vector spaces do not carry over as expected; caution is required.

An abelian group G is an example of a module over $\mathbb{Z}$, where $\pm na$ means adding $\pm a \in G$ to itself n-times. When $a \in \mathbb{Z}_n$, we get $na = 0$. So a single non-zero element is not linear independent in the $\mathbb{Z}$-module $\mathbb{Z}_n$. Also note that $\{2, 3\}$ generate the $\mathbb{Z}$-module $\mathbb{Z}$, but $3 \cdot 2 + (-2) \cdot 3 = 0$ shows that it isn't (and cannot be reduced to) a linear basis. Yet, we see that $\{1\}$ is a linear basis for the $\mathbb{Z}$-module $\mathbb{Z}$. Modules that admit a linear basis are called free. If a ring R is unital, the direct sum $\oplus_i R$ is a free R-module with R-action given by ring multiplication from the left.

More interesting types of modules are the projective ones. Over unital rings they are the direct summands of free modules. They can alternatively be described in terms of module maps between three modules, represented by fulfilling a certain completion of a diagram of map-arrows. Reversing the arrows in the diagram leads to the notion of injectivity. All modules over unital rings are submodules of injective modules. A crucial notion in dealing with diagrams is exactness, meaning that the image of any incoming arrow equals the kernel of the outgoing arrow. Finitely generated projective modules are particularly relevant in algebraic topology, as sections of bundles over manifolds form such modules over rings of continuous (or smooth) functions on the manifolds. In homological algebra and K-theory, which are essential parts of algebraic topology, diagrammatic representations of notions for modules are very convenient. Basically, all mathematics can be expressed using category theory, where the emphasis is on morphisms and composition of these, or in other words, on arrows and the formation of diagrams. In this book we won't dwell excessively with such an abstract approach since in particular number theory doesn't exactly scream after this machinery. Nevertheless, we include a section about diagram chase, and use this to study flatness of modules, which for unital rings generalizes projectivity. Along the way we introduce and interrelate basic concepts for modules, like homomorphisms, bimodules, tensor products, duals. We discuss how some of these notions are related for modules over specific rings. For instance, free modules over PID's behave a bit more like vector spaces, and a notion of dimension makes sense. Projective modules over PID's are automatically free, and so are finitely generated flat modules.

In the middle sections we focus on decomposition of modules, starting with a generalization of the fundamental theorem for finitely generated abelian groups to modules over PID's. This echoes the fundamental theorem of arithmetic. The study of Smith normal forms is more in the spirit of linear algebra. We even apply this to linear algebra, talk about generalized Jordan blocks, and of the Jordan–Chevalley decomposition of a matrix into a diagonalizable and nilpotent part. This brings us to simple and semisimple modules. The latter modules are direct sums of simple ones, and they in turn are modules having non-trivial ring actions and with no non-trivial proper submodules. A left ideal I in a ring R is an important example of an R-module, and so is the quotient module R/I. The simple modules over a unital ring are precisely the quotient modules with respect to maximal left ideals. Semisimplicity passes to submodules and to quotients. It is a notion more related to representation theory. An analog of Schur's lemma says that non-zero homomorphisms between simple modules must be isomorphisms, which implies that the endomorphism ring of a simple module is a division ring. Jacobson's density theorem, which says something about the density of the image of the natural map in a certain bicommutant of a unital ring provided with a semisimple module, yields Burnside's theorem as a corollary, which again implies that an irreducible representation of a finite group on a vector space V over an algebraically closed field has a full image in the endomorphism ring of V. We consider also balanced modules over unital rings, those where the natural map into the bicommutant is an isomorphism. Meanwhile we look more carefully into (simple) semisimple rings. These are the non-trivial unital rings which are (simple) semisimple as left-modules over themselves. Modules over such rings are automatically semisimple. Division rings are all semisimple. Conversely, every semisimple ring is a direct product of full matrix rings over division rings, and this decomposition is unique up to trivial alterations.

We then look at noetherian and artinian modules. They are the modules where every ascending, and respectively descending, chain of submodules eventually stabilizes. A module is noetherian if and only if all its submodules are finitely generated if and only if any non-empty family of submodules has a maximal element. A reverse statement holds for artinian modules. Completely reducible modules are examples of noetherian and artinian modules, and any such module is completely reducible if it is semisimple. One can push the theory in a similar way to the case of groups and series of subgroups. The Hilbert basis theorem says that polynomial rings over noetherian unital rings are noetherian, and this is also the case for rings of formal power series. Another major result is the characterization of finite direct products of matrix rings over division rings as the unital artinian rings having no non-trivial nilpotent ideals, where by nilpotent, we mean that some power of the ideal is trivial.

We study in length various radicals, the Jacobsen radical $\text{rad}(A)$ of a module A being the most versatile. This is just the intersection of all the maximal submodules.

It is itself a submodule, which is trivial when A is semisimple. We prove various results related to the Jacobson radical, including Nakayama's lemma. For artinian unital commutative rings the Jacobson radical coincides with the more historical Wedderburn radical, which by definition consists of the nilpotent elements of the ring. We also study the relation between the Jacobson radicals of a ring and its subrings before we look in detail at radicals of polynomial rings and group rings, which are of course quite different in nature. We consider in particular group rings over fields and of their extensions. On the way we consider the Jacobson radical of an algebra over a field, and show that the elements of the radical has vanishing spectrum; the spectrum of an element a are the scalars λ such that $a - \lambda I$ is not invertible. Finally, we devote a section to units in group rings, and one to division rings, due to their importance in decomposition results.

Going further, we suggest the references [7, 14].

9.1 Basics

Definition 9.1.1 A *module over a ring R*, or an *R-module*, or simply a *module*, is an additive group A with a map $R \times A \to A$; $(a, x) \mapsto ax$ such that

$$(ab)x = a(bx), \quad (a+b)x = ax + bx, \quad a(x+y) = ax + ay,$$

and $1x = x$ if R is unital. An *R-submodule B* of A is an additive subgroup of A such that $RB \subset B$. We say B is a *proper submodule* if $B \neq A$.

A *right R-module* is a (left) module over the ring R^{op} with the same addition and *opposite product* $a \cdot b \equiv ba$. If we have a homomorphism $f \colon R^{op} \to R$, or an *antiendomorphism* of R, then we can turn R-modules into right R-modules by $x \cdot a = f(a)x$. For commutative rings the identity map is an antiendomorphism, and for the group ring the unique extension of the inverse map of the group is such a map, so in both these cases modules can be turned into right modules and vice versa.

Note that a submodule is automatically a module with restricted operations. Left ideals in a ring are by definition the submodules of the ring regarded as a module over itself under multiplication.

Definition 9.1.2 If A, B are modules over a ring R, then $f \colon A \to B$ is a *module homomorphism*, or an *R-module map*, or simply a *module map*, if it is a group homomorphism and $f(ax) = af(x)$. Endomorphisms, monomorphisms, epimorphisms, isomorphisms and automorphisms have the obvious meaning, and so do $\mathrm{Hom}(A, B)$ and $\mathrm{End}(A)$ etc., where we often write $\mathrm{Hom}_R(A, B)$ etc. to stress the ring R. Similarly, we use $A \cong B$, or $A \cong_R B$, to say that A and B are isomorphic R-modules.

Note that the kernel $\ker f$ and image $\mathrm{im} f$ of a module map $f \colon A \to B$ are submodules of A and B, respectively. Also note that $\mathrm{Hom}(A, B)$ is an R-module under pointwise operations when R is commutative, otherwise this statement is not true. For a unital commutative ring R, the map $\text{Hom}(R, B) \to B$ which evaluates homomorphisms at 1, is obviously an isomorphism.

Clearly vector spaces over a field F are modules over F, subspaces are submodules, and linear maps are module maps.

We have seen that a vector spaces over a field F acted upon by a group G via a representation of G are precisely the $F[G]$-modules. This important observation casts the theory of representations into the language of rings and modules, so subrepresentations correspond to submodules, and intertwiners to module maps.

We can regard an additive group G as a $\mathbb{Z}$-module under the operation na. Then subgroups and homomorphisms will be submodules and module maps.

Example 9.1.3 The additive group of $(m \times n)$-matrices $M_{m,n}(R)$ with entries in a ring R is a module over $M_m(R)$ under multiplication from the left, and it is a right $M_n(R)$-module under multiplication from the right. In particular, the additive group R^n is an R-module under the pointwise operation $a\{b_i\} = \{ab_i\}$, an observation that extends to R^X for any set X. ◇

Trivially the direct product $\prod A_i$ and direct sum $\oplus A_i$ of R-modules A_i as additive groups and with R acting by pointwise operations are again R-modules, and $\oplus A_i$ is the submodule of $\prod A_i$ generated (in the sense explained just below) by the submodules A_i.

Notice that intersections of submodules are submodules.

Definition 9.1.4 If A is an R-module with a subset X, then the *submodule* $\langle X \rangle$ *generated by* X is the smallest submodule of A containing X. In other words, it is the intersection of all submodules that contain X, so $\langle X \rangle$ consists of all finite sums $\sum_x a_x x$ with $x \in X$ and $a_x \in R$. If there exists a subset X of A such that $A = \langle X \rangle$, then A is generated by X, and if X is finite we say that A is a *finitely generated module*, and if it consists of a single element x, or *generator*, we say that $\langle x \rangle$ is a *cyclic module*. The least $|X|$ among all finite generator sets X of a finitely generated module A is the *rank of the module* A. We denote by $\sum B_i$ the submodule of A generated by the submodules B_i, or more presicely, by the subset $\cup B_i$.

Note that an abelian group is cyclic if it is cyclic as a $\mathbb{Z}$-module. Any submodule of a PID has rank one. A finite dimensional vector space over a field F acted upon by a group G via an irreducible representation is a cyclic $F[G]$-module.

Definition 9.1.5 The *quotient module* of an R-module A by a submodule B is the additive group A/B turned into an R-module with operation $a(x + B) = ax + B$.

Clearly the quotient map $A \to A/B$ of the additive groups will then be an epimorphism of modules.

The homomorphic image of a finitely generated module is a finitely generated module since the image of the generators will be a generator set, so quotients of finitely generated modules are finitely generated. We leave the following converse result as an exercise to the reader.

9.1 Basics

Proposition 9.1.6 *Suppose A is a module with a submodule B. If both B and A/B are finitely generated, then so is A with rank not greater than the sum of the ranks of B and A/B.*

In order not to repeat ourselves excessively, we state in this context the first-, second-, third isomorphism theorem, and the correspondence theorem without the by now, standard proofs.

Theorem 9.1.7 *If $f: A \to B$ is a module map, then $A/\ker f \to \operatorname{im} f$ given by $x + \ker f \mapsto f(x)$ is an isomorphism.*

Theorem 9.1.8 *If A is a module with submodules B and C, then the quotient map $A \to A/C$ restricted to B has kernel $B \cap C$ and image $(B+C)/C$ and induces an isomorphism*

$$B/(B \cap C) \cong (B+C)/C.$$

Theorem 9.1.9 *If A is a module with submodules B, C such that $C \subset B$, then the module map $A/C \to A/B$ given by $x + C \mapsto x + B$ induces an isomorphism*

$$(A/B)/(B/C) \cong A/B.$$

Theorem 9.1.10 *If A is a module with a submodule B, then the quotient map $A \to A/B$ provides a bijection from the intermediate submodules of A and B to the submodules of A/B.*

We also state the following proposition without a proof.

Proposition 9.1.11 *Suppose A is a module generated by submodules A_i such that each A_i intersects $\sum_{j \neq i} A_j$ trivially. Then $\oplus A_i \to A$ given by $\{x_i\} \mapsto \sum x_i$ is an isomorphism.*

Definition 9.1.12 A submodul B of a module A is a *direct summand of the module A*, or is *complemented in the module A*, if there is a submodule C of A such that $A \cong B \oplus C$.

Example 9.1.13 Let p be a prime number. Then $\mathbb{Z}_p$ is not complemented in the $\mathbb{Z}$-module $\mathbb{Z}_{p^2}$ since it is the only subgroup of order p. ◇

Definition 9.1.14 The *annihilator of a subset X of a module* over a ring R is the left ideal $\operatorname{Ann}(X)$ of R consisting of all $a \in R$ such that $ax = 0$ for all $x \in X$.

Clearly the annihilator of a submodule B is an ideal. If B is a cyclic submodule over a commutative ring, then $\operatorname{Ann}(B) = \operatorname{Ann}(\{x\}) \equiv \operatorname{Ann}(x)$, where x is a generator for B, and one then talks about the *order ideal* of x. If a belongs to an additive group and has finite order p, then $\operatorname{Ann}(a) = \langle p \rangle \subset \mathbb{Z}$, which explains the terminology.

Proposition 9.1.15 *If $\langle x \rangle$ is a cyclic module over a ring R, then $\langle x \rangle \cong R/\operatorname{Ann}(x)$. In particular, any non-zero cyclic module over a field F is isomorphic to F.*

Proof For the first claim apply the first isomorphism theorem to the module map $R \to \langle x \rangle; a \mapsto ax$.

If $\langle x \rangle$ is a non-zero cyclic module over a field F, then $\mathrm{Ann}(x) = \{0\}$. □

Definition 9.1.16 An element x in a module A over an integral domain is a *torsion element* if $\mathrm{Ann}(x)$ is non-trivial. The set A_τ of torsion elements in A is called the *torsion submodule* of A. We say that A is a *torsion-free module* if A_τ is trivial, and that A is a *torsion module* if $A_\tau = A$.

Proposition 9.1.17 *Suppose A is a module over an integral domain R. Then A_τ is a submodule of A, and A/A_τ is torsion-free.*

Proof Consider $x, y \in A_\tau$ and $a, b \in R$, so there are non-zero elements $c, d \in R$ such that $cx = 0 = dy$. Then $ax + by \in A_\tau$ because $(cd)(ax + by) = ad(cx) + bc(dy) = 0$ and $cd \neq 0$ in an integral domain.

For the second assertion, assume we have a non-zero element $a \in R$ such that $a(x + A_\tau) = 0$ for $x \in A$. Then $ax \in A_\tau$, so there is a non-zero $b \in R$ such that $(ba)x = b(ax) = 0$. Thus $x \in A_\tau$ as $ba \neq 0$. □

A non-trivial vector space over a non-trivial field is torsion-free. An additive group with no non-trivial elements of finite order, like $\mathbb{Z}^n$, is torsion-free as a $\mathbb{Z}$-module. An additive group is a torsion module over $\mathbb{Z}$ if and only if every element is of finite order. So finite additive groups are torsion modules over $\mathbb{Z}$. Every element in the infinite additive group $\mathbb{Q}/\mathbb{Z}$ has finite order because $n(m/n + \mathbb{Z})$ is zero in $\mathbb{Q}/\mathbb{Z}$.

9.2 Exactness

Definition 9.2.1 Let R be a ring. A sequence

$$\cdots \longrightarrow A_{i-1} \xrightarrow{f_{i-1}} A_i \xrightarrow{f_i} A_{i+1} \longrightarrow \cdots$$

of R-modules is *exact* if it is *exact at A_i*, meaning $\mathrm{im}\, f_{i-1} = \ker f_i$, for each i.

Note that $0 \longrightarrow A \xrightarrow{f} B$ is exact if and only if f is injective, and that $B \xrightarrow{g} C \longrightarrow 0$ is exact if and only if g is surjective.

Definition 9.2.2 An exact sequence of the form

$$0 \longrightarrow A \xrightarrow{f} B \xrightarrow{g} C \longrightarrow 0$$

is called *short extact*, and if addition $\mathrm{im}\, f$ is a direct summand of B, then it is called *split exact*.

9.2 Exactness

By the first isomorphism theorem we see that $C \cong B/A$ whenever the sequence of modules in the definition above is exact, and that $B \cong A \oplus C$ whenever it is split exact.

Example 9.2.3 Given two primes numbers p and q, we get a short exact sequence

$$0 \longrightarrow \mathbb{Z}_p \xrightarrow{f} \mathbb{Z}_{pq} \xrightarrow{g} \mathbb{Z}_q \longrightarrow 0$$

of $\mathbb{Z}$-modules, where g is the quotient map and f is given by $f([n]) = [qn]$. We also see that it is split exact exactly when $p \neq q$. ◇

We have the following characterization of split exactness.

Proposition 9.2.4 *For a short exact sequence*

$$0 \longrightarrow A \xrightarrow{f} B \xrightarrow{g} C \longrightarrow 0$$

the following assertions are equivalent:

(i) *The sequence of modules is split exact;*

(ii) *There is a homomorphism $r \colon B \to A$ such that $rf = \iota$;*

(iii) *There is a homomorphism $s \colon C \to B$ such that $gs = \iota$.*

We say the maps r and s *split the sequence*, and will see that $C \cong \ker r \cong \operatorname{im} s$.

Proof If the sequence of modules is split extact, then $B \cong A \oplus C$, where f is the inclusion map and g is the projection map. Thus the projection map $r \colon B \to A$ and the inclusion map $s \colon C \to B$ will satisfy $rf = \iota$ and $gs = \iota$, so (i) implies both (ii) and (iii).

Suppose that (ii) holds for a homomorphism $r \colon B \to A$. Writing the identity map on B as $fr + (\iota - fr)$ and using $rf = \iota$, we see that $B = \operatorname{im} f + \ker r$ and $\operatorname{im} f \cap \ker r = \{0\}$, so $B \cong \operatorname{im} f \oplus \ker r$ and (i) holds.

Similarly one shows that (iii) implies $B \cong \ker g \oplus \operatorname{im} s$. □

Definition 9.2.5 An *idempotent in a ring* is any element a that satisfies $a^2 = a$. A family of idempotents is *orthogonal* if $ab = 0$ for distinct elements a and b of the family.

Note that with notation as in the proposition above we see that fr and sg form an orthogonal family of idempotents in the ring $\operatorname{End}(B)$ such that $fr + sg = \iota$. Thus $A \cong frB$ and $C \cong sgB$ in the decomposition $B \cong A \oplus C$.

If $\{e_i\}$ is an orthogonal family of idempotents in a ring, then $\sum_i Re_i$ is a direct sum of left ideals Re_i because if $be_i = \sum_{j \neq i} c_j e_j$, then $be_i = c_j e_j e_i = 0$. If on the other hand a unital ring R is a direct sum of left ideals A_i, then

$$1 = e_1 + \cdots + e_n$$

for elements $e_i \in A_i$. Upon multiplying with e_i we see that $\{e_i\}$ is an orthogonal family of idempotents such that $A_i = Re_i$ and $R = \oplus_{i=1}^n Re_i$ as modules.

Proposition 9.2.6 *If a unital ring R is a direct sum of simple left ideals, then every left ideal of R is of the form Re for an idempotent $e \in R$.*

Proof Since R is unital the direct sum must be finite, say $R = A_1 \oplus \cdots \oplus A_n$ for simple left ideals A_i. If A is a proper left ideal of R, then after renumbering we may assume that A_1 is not contained in A. Then $A_1 \cap A = \{0\}$ since $A_1 \cap A$ is a proper left ideal of the simple left ideal A_1. So $A + A_1$ is a direct sum. If $A \oplus A_1 \neq R$, then by renumbering we may assume $A_2 \cap (A \cap A_1) = \{0\}$. Continuing this way, we get $R = A \oplus (\oplus_{i=1}^m A_i)$ for some $m \leq n$. By the discussion before the proposition we conclude that $A = Re$ for an idempotent e. □

Given a module map $f \colon A \to B$, define an additive map

$$f_* \colon \mathrm{Hom}(C, A) \to \mathrm{Hom}(C, B) \quad \text{by} \quad f_*(u) = fu.$$

Proposition 9.2.7 *The sequence*

$$0 \longrightarrow A \xrightarrow{f} B \xrightarrow{g} C$$

of modules is exact if and only if the sequence

$$0 \longrightarrow \mathrm{Hom}(D, A) \xrightarrow{f_*} \mathrm{Hom}(D, B) \xrightarrow{g_*} \mathrm{Hom}(D, C)$$

of $\mathbb{Z}$-modules is exact for all modules D. Moreover, if the first sequence is split, then the second sequence is short exact.

Proof Suppose the first sequence of modules is exact. If $u \in \ker f_*$, then $fu = 0$, so $u = 0$ and f_* is injective.

Also $\mathrm{im} f_* \subset \ker g_*$ as $g_*(f_*(u)) = gfu = 0u = 0$ for any u is the domain of f_*.

If $v \in \ker g_*$, then $g(v(x)) = 0$ for $x \in D$. As $\ker g = \mathrm{im} f$ and f is injective, there is a unique $y \in A$ such that $v(x) = f(y)$. Thus we can define a module map $w \colon D \to A$ by $w(x) = y$, and clearly $v = f_*(w) \in \mathrm{im} f_*$.

Conversely, assume that the second sequence of modules is exact for all D. Let $D = \ker f$ and consider the inclusion map $u \colon D \to A$. Then $f_*(u) = fu = 0$ and as f_* is injective, we get $u = 0$, so f is injective.

Next let $D = A$. Then $0 = g_* f_*(\iota) = gf\iota = gf$. Letting $D = \ker g$ and $v \colon D \to B$ be the inclusion map, then $g_*(v) = gv = 0$, so $v \in \ker g_*$. By exactness there exists $w \in \mathrm{Hom}(D, A)$ with $v = f_*(w) = fw$. So $\ker g = \mathrm{im} v \subset \mathrm{im} f$ and the first sequence of modules is exact.

If $s \colon C \to B$ splits the first sequence, then s_* splits the second sequence because $g_* s_* = (gs)_* = \iota$. □

9.2 Exactness

When R is commutative the second sequence in the proposition above is a sequence of R-modules.

Given a module map $f: A \to B$, define an additive map

$$f^*: \mathrm{Hom}(B, C) \to \mathrm{Hom}(A, C) \quad \text{by} \quad f^*(u) = uf.$$

Just as above one can prove the following result.

Proposition 9.2.8 *The sequence*

$$A \xrightarrow{f} B \xrightarrow{g} C \longrightarrow 0$$

of modules is exact if and only if the sequence

$$0 \longrightarrow \mathrm{Hom}(C, D) \xrightarrow{g^*} \mathrm{Hom}(B, D) \xrightarrow{f^*} \mathrm{Hom}(A, D)$$

of $\mathbb{Z}$-modules is exact for all modules D. Moreover, if the first sequence is split, then the second sequence is short exact.

In this case the arrows are reversed. When R is commutative the second sequence in the proposition above is a sequence of R-modules.

The following example shows that one cannot hope to preserve short exact sequences in full generality.

Example 9.2.9 We claim that $\mathrm{Hom}_{\mathbb{Z}}(\mathbb{Z}_m, \mathbb{Z}_n) \cong \mathbb{Z}_d$, where d is the greatest common divisor of m and n. To see this, consider the short exact sequence

$$0 \longrightarrow \mathbb{Z} \xrightarrow{f} \mathbb{Z} \xrightarrow{g} \mathbb{Z}_m \longrightarrow 0$$

with quotient map g and with f given by $f(k) = mk$. By the proposition above, we get an exact sequence

$$0 \longrightarrow \mathrm{Hom}(\mathbb{Z}_m, \mathbb{Z}_n) \xrightarrow{g^*} \mathrm{Hom}(\mathbb{Z}, \mathbb{Z}_n) \xrightarrow{f^*} \mathrm{Hom}(\mathbb{Z}, \mathbb{Z}_n)$$

and $\mathrm{Hom}(\mathbb{Z}_m, \mathbb{Z}_n) \cong \ker f^*$. Define a monomorphism $\ker f^* \to \mathbb{Z}_n$ by $u \mapsto u(1)$. As $f^*(v)(1) = vf(1) = v(m1) = mv(1)$ for any $v \in \mathrm{Hom}(\mathbb{Z}, \mathbb{Z}_n)$, we see that the image of the monomorphism is $\mathbb{Z}_d$, which proves the claim.

Picking $m = n$ and using $\mathrm{Hom}(\mathbb{Z}, \mathbb{Z}_n) \cong \mathbb{Z}_n$ we thus get the exact sequence

$$0 \longrightarrow \mathbb{Z}_n \xrightarrow{g^*} \mathbb{Z}_n \xrightarrow{f^*} \mathbb{Z}_n$$

with $f^* = 0$, which is not surjective, so we do not get a short exact sequence.

Similarly, applying Proposition 9.2.7 with $D = \mathbb{Z}_m$ to the initial sequence, and using $\mathrm{Hom}(\mathbb{Z}_m, \mathbb{Z}) = 0$, we get

$$0 \longrightarrow 0 \xrightarrow{f_*} 0 \xrightarrow{g_*} \mathbb{Z}_m$$

that is also not short exact. ◇

Corollary 9.2.10 *If A, B, C are modules over a ring R, then*

$$\mathrm{Hom}(C, A \oplus B) \cong \mathrm{Hom}(C, A) \oplus \mathrm{Hom}(C, B)$$

and

$$\mathrm{Hom}(A \oplus B, C) \cong \mathrm{Hom}(A, C) \oplus \mathrm{Hom}(B, C)$$

as $\mathbb{Z}$-modules, and as R-modules if R is commutative.

Proof Apply the two propositions above to the split exact sequence

$$0 \longrightarrow A \xrightarrow{\iota_1} A \oplus B \xrightarrow{\pi_2} B \longrightarrow 0$$

where ι_1 is the inclusion map and π_2 is the projection map. □

The first and second isomorphism in the corollary are $f \mapsto (\pi_1 f, \pi_2 f)$ for $f \in \mathrm{Hom}(C, A \oplus B)$ and $g \mapsto (g\iota_1, g\iota_2)$ for $g \in \mathrm{Hom}(A \oplus B, C)$, respectively. Extending this we get the following result.

Proposition 9.2.11 *We have as $\mathbb{Z}$-modules that*

$$\mathrm{Hom}(\oplus A_i, \oplus B_j) = \prod_i \oplus_j \mathrm{Hom}(A_i, B_j)$$

for any modules A_i and B_j over a ring.

9.3 Projectivity

Definition 9.3.1 A non-empty subset X of a module A over a ring R is *linear independent*, as opposed to *linear dependent*, if the elements $a_x \in R$ are zero whenever $\sum a_x x = 0$ for a finite collection of $x \in X$. If in addition X generates A, it is a *basis* of A, and then A is a *free module* over R.

In an R-module A with a basis X any element $a \in A$ can be written as a finite R-linear combination $a = \sum a_x x$ for unique $a_x \in R$. Thus $A \cong \oplus_{x \in X} Rx$. Conversely, if R viewed as an R-module contains an element with trivial annihilator, then $\oplus_i R$ over any index set is a free R-module. This holds for instance when R is unital.

Given any module A with basis X, we can uniquely extend any map $f \colon X \to B$ into a module over the same ring to a module map $A \to B$, and this map will be an isomorphism if $f(X)$ is a basis of B.

9.3 Projectivity

Vector spaces are free modules over fields as they have a basis by Zorn's lemma.

If A_i are free modules with bases X_i, then the module $\oplus A_i$ is also free with basis consisting of all sequences with $x \in X_i$ in place i, and zeroes in all other components. For instance, the set $\{(1, 0), (0, 1)\}$ is a basis for the $\mathbb{Z}$-module $\mathbb{Z}^2$.

Proposition 9.3.2 *Any basis of a finitely generated module is finite.*

Proof Writing each member of a finite generator set of a free module as a linear combination of elements in a basis, we see that only finitely many basis elements generate the module, so the basis must be finite. □

Corollary 9.3.3 *Suppose A and B are finitely generated free modules over a commutative unital ring R. Then*

$$\mathrm{Hom}(A, B) \cong \oplus_{ij} R,$$

where i and j range over finite bases for A and B, respectively.

Proof From the proposition above we know that A and B have indeed finite bases. By Corollary 9.2.10 we therefore have

$$\mathrm{Hom}(A, B) \cong \mathrm{Hom}(\oplus_i R, \oplus_j R) \cong \oplus_{ij} \mathrm{Hom}(R, R) \cong \oplus_{ij} R.$$

□

The following result shows that no finite additive group is a free $\mathbb{Z}$-module.

Proposition 9.3.4 *Any free module over an integral domain is torsion-free.*

Proof If A is a module with basis X over an integral domain R, and $x \in A_\tau$, then $ax = 0$ for a non-zero $a \in R$. Writing $x = \sum a_i x_i$ for $a_i \in R$ and $x_i \in X$, we get $0 = \sum aa_i x_i$, so $aa_i = 0$ for all i. In an integral domain all $a_i = 0$, so $x = 0$. □

For any ideal I of a commutative ring R to be free, it must be principal because $ab + (-b)a = 0$ for $a, b \in R$.

Definition 9.3.5 A short exact sequence

$$0 \longrightarrow A \longrightarrow B \longrightarrow C \longrightarrow 0$$

of modules with B free is called a *free presentation* of C.

Proposition 9.3.6 *Every module over a unital ring has a free presentation, and can moreover be chosen to be a quotient of a free module with the same rank.*

Proof Suppose X is a generating set for a module C over a unital ring R, we can always pick $X = C$. Consider the free module $B = \oplus_{x \in X} R$. Then $f: B \to C$ given by $f(\{a_x\}) = \sum_x a_x x$ is an epimorphism, so $C \cong B/\ker f$. Since the image of a generator set under the quotient map will be a generator set for the quotient module, the latter cannot have larger rank. Picking X to be of minimal cardinality, we also see that the rank of C cannot be larger than the rank of B. □

Definition 9.3.7 A module A is a *projective module* if for any modules B, C and any module maps $f: A \to C$ and $g: B \to C$ with g surjective, there exists a module map $h: A \to B$ such that the diagram

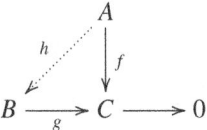

commutes, that is, we have $f = gh$.

Proposition 9.3.8 *A module over a unital ring is projective if and only if it is a direct summand of a free module. Moreover, the statement also holds if we add 'finitely generated' on both sides of the equivalence.*

Proof Suppose A is a projective module over a unital ring. By the proposition above there is a free module B of the same rank and an epimorphism $g: B \to A$. In the diagram above pick $f: A \to C \equiv A$ to be the identity map. Then the existence of $h: A \to B$ shows that we get a splitting, so $B \cong A \oplus \ker g$, and B is finitely generated if A is.

Conversely, suppose there exists a module D such that $A \oplus D$ is free, and that we are given B, C and f and g as in the diagram above. To get h, pick a basis X for $A \oplus D$, and by the Axiom of choice define a module map $r: A \oplus D \to B$ such that $r(x) \in g^{-1}(\{f\pi_1(x)\})$ for all $x \in X$, so $gr = f\pi_1$. Then $h \equiv r\iota_1$ satisfies $f = gh$ as $\pi_1 \iota_1 = \iota$. And if $A \oplus D$ is finitely generated, then so is $A \cong (A \oplus D)/D$. □

Corollary 9.3.9 *A module A over a unital ring is projective if and only if every sequence*

$$0 \longrightarrow C \longrightarrow B \longrightarrow A \longrightarrow 0$$

of modules is split exact.

Proof If A is projective, then picking the identity map on A in the sequence above, we get a module map $A \to B$ that splits the sequence.

If every such sequence is split exact, we may split a free presentation of A, to get A as a direct summand of a free module, so A is projective. □

Corollary 9.3.10 *A module A over a unital ring is projective if and only if*

$$f_*: \mathrm{Hom}(D, B) \to \mathrm{Hom}(D, A)$$

is surjective for any epimorphism $f: B \to A$ and all modules D.

Proof The forward implication is immediate from Proposition 9.2.7 since

$$0 \longrightarrow \ker f \longrightarrow B \xrightarrow{f} A \longrightarrow 0$$

is split exact. As for the backward implication, if $f_*\colon \text{Hom}(A, B) \to \text{Hom}(A, A)$ is surjective for an epimorphism $f\colon B \to A$, then there is a module map $s\colon A \to B$ such that $fs = f_*(s) = \iota$, so A is projective by the corollary above. □

Cleary free modules are projective, and projective modules over integral domains are torsion-free by Propositions 9.3.4 and 9.3.8. So no finite additive group is projective as a $\mathbb{Z}$-module. In fact, by Corollary 9.9.6 any projective module over a *PID* is free.

Over rings that are not integral domains, torsion does not forbid projectivity. For instance, the module $\mathbb{Z}_2$ over $\mathbb{Z}_6$ is projective because $\mathbb{Z}_6 \cong \mathbb{Z}_2 \times \mathbb{Z}_3$, but $\mathbb{Z}_2$ is not free as a $\mathbb{Z}_6$-module. Neither is the module $\mathbb{R} \times \{0\}$ over $\mathbb{R} \times \mathbb{R}$ with pointwise operations, yet it is evidently projective.

Remark 9.3.11 The continuous sections of a vector bundle form a finitely generated projective module over the ring of continuous functions on a compact Hausdorff manifold, and every such module can be regarded as the sections of some underlying vector bundle. If the manifold is contractible, then all these modules are free, but already for the circle this is no longer true. Thus the importance of finitely generated projective modules in algebraic topology, in formation of K-groups, and in homological algebra, which measures the failor of sequences being exact.

We include three result without their easy proofs.

Proposition 9.3.12 *Each member A_i of a family of modules over a unital ring is projective if and only if $\oplus A_i$ is projective.*

Proposition 9.3.13 *If A and B are finitely generated projective modules over a commutative unital ring R, then so is the R-module $\text{Hom}(A, B)$.*

Proposition 9.3.14 *A module A is finitely generated projective over a unital ring R if and only if there is a non-negative integer n such that $A \cong eR^n$ for some idempotent $e \in \text{End}(R^n)$.*

Idempotents are also referred to as *projections*, hence the name projective module.

Definition 9.3.15 Two elements a and b in a unital ring are *conjugate* if there exists an invertible element c in the ring such that $a = cbc^{-1}$.

Proposition 9.3.16 *Let R be a unital ring, and let n be a non-negative integer. Two idempotents f and g in the ring $\text{End}(R^n)$ are conjugate if and only if*

$$fR^n \cong gR^n \quad \text{and} \quad (\iota - f)R^n \cong (\iota - g)R^n$$

as modules.

Proof If $gh = hf$ for invertible $h \in \text{End}(R^n)$, then $h_1 \colon fR^n \to gR^n$ given by restriction of h has an inverse given by restriction of h^{-1}. Similarly h restricts to an isomorphism h_2 from $(\iota - f)R^n$ to $(\iota - g)R^n$.

Conversely, if $h_1: fR^n \to gR^n$ and $h_2: (\iota - f)R^n \to (\iota - g)R^n$ are isomorphisms, then $h = h_1 f + h_2(\iota - f) \in \mathrm{End}(R^n)$ is invertible with $h^{-1} = h_1^{-1} g + h_2^{-1}(\iota - g)$ and
$$gh = gh_1 f = h_1 f = h_1 f^2 = hf.$$

□

One needs both isomorphisms in the proposition above to assure that the idempotents are conjugate; it is easy to give examples showing that this is required.

9.4 Injectivity

We included here the less common notion of an injective module.

Definition 9.4.1 A module A is an *injective module* if for any modules B, C and any module maps $f: C \to A$ and $g: C \to B$ with g injective, there exists a module map $h: B \to A$ such that the diagram

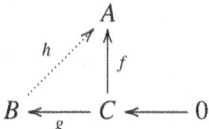

commutes, that is, we have $f = hg$.

Notice that the diagramatic definition of injectivity is the same as that of projectivity, except that we have changed the direction of all the arrows, so g is injective rather than surjective. Clearly direct products of injective modules are injective.

It is immediate from Proposition 9.2.8 and the definition of injectivity that a module A is injective if and only if the sequence

$$0 \longrightarrow \mathrm{Hom}(D, A) \longrightarrow \mathrm{Hom}(B, A) \xrightarrow{g^*} \mathrm{Hom}(C, A) \longrightarrow 0$$

of $\mathbb{Z}$-modules is exact whenever

$$0 \longrightarrow C \xrightarrow{g} B \longrightarrow D \longrightarrow 0$$

is exact.

Definition 9.4.2 The *pushout of two module maps* $f: A \to B$ and $g: A \to C$ is a triple (E, r, s) with module maps $r: B \to E$ and $s: C \to E$ such that $rf = sg$, and whenever we have another triple (F, u, v) with this property, there exists a unique module map $h: E \to F$ such that $hr = u$ and $hs = v$.

9.4 Injectivity

Lemma 9.4.3 *The pushout exists and is unique up to isomorphism.*

Proof As for uniqueness, if we have two pushouts (E, r, s) and (F, u, v), then there are module maps $h \colon E \to F$ and $h' \colon F \to E$ such that $hr = u$ and $hs = v$, and $h'u = r$ and $h'v = s$. Then $h'hr = r$ and $h'hs = s$, so $h'h = \iota$ by uniqueness in the case when the triples are chosen to be the same. Similarly we get $hh' = \iota$, so $h \colon E \to F$ is an isomorphism that carry one triple to the other.

As for existence, given module maps $f \colon A \to B$ and $g \colon A \to C$, consider the module $(B \oplus C)/D$, where D is the submodule of $B \oplus C$ of all $(f(a), -g(a))$ with $a \in A$. Define module maps $r \colon B \to E$ and $s \colon C \to E$ by $r(b) = (b, 0) + D$ and $s(c) = (0, c) + D$. Then $rf = sg$ because $rf(a) - sg(a) = (f(a), -g(a)) + D = D$.

Given another triple (F, u, v), then $h \colon E \to F$ given by $h((b, c) + D) = u(b) + v(c)$ is clearly the unique module map such that $hr = u$ and $hs = v$. □

We denote the pushout of two module maps $f \colon A \to B$ and $g \colon A \to C$ by $(B \oplus_A C, r, s)$. Notice that s is injective if f is injective. To see this consider the construction in the proof, and note that if $s(c) = 0$, then $(0, c) \in D$, so $c = -g(a)$ for some $a \in A$ with $f(a) = 0$. As f is injective, we get $a = 0$, so $c = -g(0) = 0$.

Proposition 9.4.4 *A module A is injective if and only if every short exact sequence*

$$0 \longrightarrow A \longrightarrow B \longrightarrow C \longrightarrow 0$$

of modules is split exact.

Proof If A is injective, then using the definition of injectivity to the identity map on A of any sequence of the above type, will produce a split at A.

Conversely, suppose any such short exact sequence is split exact. We need to produce a module map h making the following diagram

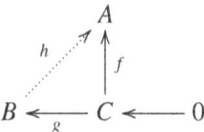

commute. Since g is injective the module map $s \colon A \to B \oplus_C A$ in the pushout $(B \oplus_C A, r, s)$ of g and f is also injective. By assumption s therefore splits, so we have a module map $q \colon B \oplus_C A \to A$ such that $qs = \iota$. Then the module map $h = qr \colon B \to A$ satisfies $hg = qrg = qsf = \iota f = f$. □

The following result is known as *Baer's criterion*.

Proposition 9.4.5 *Let R be a unital ring. An R-module A is injective if and only if any module map $I \to A$ from any left ideal I of R extends to a module map $R \to A$.*

Proof The forward implication is immediate from the definition.

Conversely, suppose we are given a submodule C of a module B and a module map $f: C \to A$, which we need to extend to a module map $B \to A$. Partially order the set of all possible module map extensions $D \to A$ of f, where D is any submodule of A with $C \subset D$. By Zorn's lemma there is a maximal one $h: E \to A$. If there is $a \in B$ with $a \notin E$, consider the left ideal $I = \{b \in R \mid ba \in E\}$ of R. By assumption the module map $I \to A$ which sends b to $h(ba)$ extends to a module map $\tilde{h}: R \to A$. Define $E + Ra \to A$ by $c + ba \mapsto h(c) + \tilde{h}(b)$ for $c \in E$ and $b \in R$. This is well-defined because $h(ba) = \tilde{h}(b)$ for $b \in R$ with $ba \in E$. But then we have a module map that is a strictly larger extension of f than h, which contradicts the maximality of h, so $E = B$ and A is injective. $\square$

Definition 9.4.6 A module A over an integral domain R is *divisible* if for any $y \in A$ and non-zero $a \in R$, there is $x \in A$ with $ax = y$.

Direct sums, direct products and quotients of divisible modules are clearly divisible. Obviously any field of characteristic zero is divisible as a module over any subring of the field.

Corollary 9.4.7 *Any module over a PID is injective if and only if it is divisible. This is in particular true for additive groups as $\mathbb{Z}$ is a PID.*

Proof Suppose we have an injective module A over an integral domain R. To see that A is divisible, consider $y \in A$ and a non-zero $a \in R$. Define a module map $f: Ra \to A$ by $f(ba) = by$ for $b \in R$. This is well-defined because if $ba = ca$ for $c \in R$, then $b = c$ as R is an integral domain, so $f(ba) = f(ca)$. By injectivity we can extend f to a module map $g: R \to A$. Then with $x = g(1) \in A$, we get $y = f(1a) = g(a1) = ag(1) = ax$.

Conversely, if A is divisible over a PID R, then using Baer's criterion it suffices to extend any module map $f: I \to A$ for any module $I \subset R$ to a module map $R \to A$. In a PID any left ideal is a singly generated ideal, so say $I = (a)$ for some $a \in I$. Since A is divisible there exists $x \in A$ such that $ax = f(a)$. Define $g: R \to A$ by $g(b) = bx$ for $b \in R$. Clearly this is a module map, and g extends f because $g(ba) = bax = bf(a) = f(ba)$ for $b \in R$. $\square$

Note that $\mathbb{Z}$ is not divisible since an odd number can never be even. Thus submodules of injective modules need not be injective.

We have seen that any module has a free presentation, and so is certainly the homomorphic image of a projective module. One says there are *enough projectives*. Similarly, we would like to see that there are *enough injectives*, in the sense that every module sits inside an injective one. To prove this we need some preliminary results.

Definition 9.4.8 The *Pontryagin dual* $\hat{A}$ of an additive group A is the additive group $\text{Hom}_{\mathbb{Z}}(A, \mathbb{Q}/\mathbb{Z})$ with pointwise operations and addition in $\mathbb{Q}/\mathbb{Z}$.

9.4 Injectivity

The following result says that the Pontryagin dual is rich enough. We proved a similar result when we established Pontryagin's duality theorem for finite abelian groups, then with $\mathbb{R}/\mathbb{Z}$ instead of $\mathbb{Q}/\mathbb{Z}$.

Proposition 9.4.9 *The additive map* $\eta \colon A \to \hat{\hat{A}}$ *given by* $(\eta(x))(f) = f(x)$ *is injective.*

Proof Say we have a non-zero element $x \in A$. We need to construct an element of $\hat{A}$ that is not zero at x. To this end let n be the least natural number such that $nx = 0$, and set $f(mx) = [m/n]$ for $m \in \mathbb{Z}$. If no such n exists, set $f(mx) = [m/2]$. In both cases we get $f \in \mathrm{Hom}_{\mathbb{Z}}(\mathbb{Z}x, \mathbb{Q}/\mathbb{Z})$ with $f(x) \neq 0$. To see that f is well-defined, say $mx = 0$. In the second case $m = 0$, so $f(mx) = 0$. In the first case, there must be an integer k such that $m = kn$, so $f(mx) = [k] = 0$.

We extend f to an element of $\hat{A}$ by partially ordering all subgroups of A that contain $\mathbb{Z}x$ and carry an additive extension of f. By Zorn's lemma there is a maximal extension $g \colon B \to \mathbb{Q}/\mathbb{Z}$. If $B \neq A$, pick $y \in A \backslash B$. Define $h \colon B + \mathbb{Z}y \to \mathbb{Q}/\mathbb{Z}$ on $\mathbb{Z}y$ by the same procedure as we defined f, and let h be g on B. Then h is a well-defined additive map because $z + y = 0$ is impossible with $z \in B$. This contradicts the maximality of g. □

An immediately consequence of the proposition above is that a module A is trivial if and only if $\hat{A}$ is trivial.

Lemma 9.4.10 *Every additive group is a submodule of an injective $\mathbb{Z}$-module.*

Proof Let A be an additive group. Every additive group is the homomorphic image of a free abelian group, and thus of some direct sum of $\mathbb{Z}$. So there exists a $\mathbb{Z}$-module map $f \colon \oplus_i \mathbb{Z} \to \hat{A}$ that is surjective. By Proposition 9.2.8 the $\mathbb{Z}$-module map $f^* \colon \hat{\hat{A}} \to \mathrm{Hom}_{\mathbb{Z}}(\oplus_i \mathbb{Z}, \mathbb{Q}/\mathbb{Z})$ is injective. But by Proposition 9.2.11 we have

$$\mathrm{Hom}(\oplus_i \mathbb{Z}, \mathbb{Q}/\mathbb{Z}) \cong \prod_i \mathrm{Hom}_{\mathbb{Z}}(\mathbb{Z}, \mathbb{Q}/\mathbb{Z}) \cong \prod_i \mathbb{Q}/\mathbb{Z},$$

which is divisible and thus injective by the corollary above. By the proposition above we can regard A as a submodule of $\hat{\hat{A}}$. Then f^* restricted to A is an injective $\mathbb{Z}$-module map into an injective $\mathbb{Z}$-module. □

We would like to dualize in a similar fashion in the case of a general module. If R is a ring and G is an additive group, we may regard the additive group $\mathrm{Hom}_{\mathbb{Z}}(R, G)$ as an R-module under the operation $(af)(b) = f(ba)$ for $f \in \mathrm{Hom}_{\mathbb{Z}}(R, G)$ and $a, b \in R$.

Lemma 9.4.11 *Suppose R is a unital ring and G is an additive group. Then for any R-module B, the map*

$$\mathrm{Hom}_{\mathbb{Z}}(B, G) \to \mathrm{Hom}_R(B, \mathrm{Hom}_{\mathbb{Z}}(R, G))$$

which sends $f\colon B \to G$ to $g\colon B \to \mathrm{Hom}_{\mathbb{Z}}(R,G)$, where $(g(x))(a) = f(ax)$, is an isomorphism of additive groups.

Proof The map g is plainly an R-module map. The map in the lemma is indeed an isomorphism with an inverse map

$$\mathrm{Hom}_R(B, \mathrm{Hom}_{\mathbb{Z}}(R,G)) \to \mathrm{Hom}_{\mathbb{Z}}(B,G)$$

that sends $g\colon B \to \mathrm{Hom}_{\mathbb{Z}}(R,G)$ to $f\colon B \to G$, where $f(x) = (g(x))(1)$. □

Lemma 9.4.12 *Suppose R is a unital ring and G is an additive group. If G is injective as a $\mathbb{Z}$-module, then $\mathrm{Hom}_{\mathbb{Z}}(R,G)$ is an injective R-module.*

Proof By the characterization of injectivity given just below the definition of injectivity, we have an additive surjective map $h\colon \mathrm{Hom}_{\mathbb{Z}}(B,G) \to \mathrm{Hom}_{\mathbb{Z}}(C,G)$ for any exact sequence $0 \longrightarrow C \longrightarrow B$ of R-modules. Composing h on both sides with the isomorphisms from the previous lemma applied to B and C, we get an additive surjective map

$$\mathrm{Hom}_R(B, \mathrm{Hom}_{\mathbb{Z}}(R,G)) \to \mathrm{Hom}_R(C, \mathrm{Hom}_{\mathbb{Z}}(R,G)),$$

which by the same characterization shows that $\mathrm{Hom}_{\mathbb{Z}}(R,G)$ is injective. □

Theorem 9.4.13 *Any module over a unital ring is a submodule of an injective module.*

Proof Suppose A is a module over a unital ring R. Regard A as an additive group. By the first lemma there is an injective additive map $f\colon A \to G$ into an additive group that is injective as a $\mathbb{Z}$-module. The R-module map $A \to \mathrm{Hom}_{\mathbb{Z}}(R,G)$ given by $x \mapsto f_x$, where $f_x(a) = f(ax)$, is clearly injective, and $\mathrm{Hom}_{\mathbb{Z}}(R,G)$ is an injective R-module by the last lemma above. □

9.5 Tensor Products and Bimodules

Definition 9.5.1 Let A be a right module over a ring R, let B be a left module over R, and let C be an additive group. A map $f\colon A \times B \to C$ is *balanced* if it is biadditive and if $f(xa, y) = f(x, ay)$ for $x \in A$, $y \in B$ and $a \in R$. The *tensor product* of A and B is an additive group $A \otimes_R B$ together with a balanced map $u\colon A \times B \to A \otimes_R B$ such that whenever $f\colon A \times B \to C$ is a balanced map, there is a unique additive map $g\colon A \otimes_R B \to C$ such that $f = gu$. We write $x \otimes y$ for $u(x, y)$.

The following result shows that the tensor product and its accompanying map are defined only up to isomorphism.

Proposition 9.5.2 *The tensor product of a right- and left module over a ring exists and is unique up to isomorphism.*

9.5 Tensor Products and Bimodules

Proof The tensor product of a right module A over a ring R with a left module B over R is formed by considering the free additive group F generated by $A \times B$, so F consists of all functions $A \times B \to \mathbb{Z}$ that are zero everywhere except on finite subsets, and is an additive group under pointwise addition. Identify a point in $A \times B$ with the delta function at that point. Let I be the subgroup of F generated by all elements of the form $(x+y, z) - (x, z) - (y, z)$, $(x, z+w) - (x, z) - (x, w)$ and $(xa, z) - (x, az)$. Then F/I and the quotient map $u \colon A \times B \to F/I$ restricted to $A \times B \subset F$ have the desired properties. By definition of I the map u is clearly a balanced map. And if $f \colon A \times B \to C$ is a balanced map, then as $A \times B$ is a $\mathbb{Z}$-basis for F, the map f extends to an additive map $F \to C$, which factors through F/I because it is balanced. In other words, the additive map $g \colon F/I \to C$ given by $g((x, y) + I) = f(x, y)$ is well-defined as $f(I) = \{0\}$. Clearly $f = gu$, and as the image of u generates F/I as an additive group, the requirement $f = gu$ uniquely determines g.

Say D is an additive group and $v \colon A \times B \to D$ is a balanced map with the same properties as $A \otimes_R B$ and u. As both u and v are balanced, there are additive maps $r \colon D \to A \otimes_R B$ and $s \colon A \otimes_R B \to D$ such that $rv = u$ and $su = v$. Thus $rsu = u$ and $srv = v$, which by uniqueness of such maps means that $rs = \iota$ and $sr = \iota$, so r is an isomorphism that carry D and v to $A \otimes_R B$ and u. □

The characterizing property of the tensor product exploited in the last paragraph of the proof above is called its *universal property*.

Note that $A \otimes_R B$ is generated by $x \otimes y$ as an additive group. Balancedness of u means that

$$(x+y) \otimes z = x \otimes z + y \otimes z, \quad x \otimes (z+w) = x \otimes z + x \otimes w, \quad xa \otimes z = x \otimes az.$$

We would also like to stress that $x \otimes z = y \otimes w$ does not imply $x = y$ and $z = w$ as u has non-trivial kernel I.

Definition 9.5.3 Given two rings R and S, then an RS-*bimodule* A is a left R-module and a right S-module with compatibel operations in the sense that $a(xb) = (ax)b$ for $a \in R$, $b \in S$ and $x \in A$. A *bimodule map* is a map that respects both module actions.

If A is an RS-bimodule and B is a ST-bimodule, then $A \times B$ is clearly an RT-bimodule under $a(x, y)b \equiv (ax, yb)$. Extending this additively to F in the proof above, and observing that I is an RT-subbimodule of F, we see that $A \otimes_S B = F/I$ is an RT-bimodule under $a(x \otimes y)b = ax \otimes yb$, and will be regarded as such when the context suggests so.

Corollary 9.5.4 *Suppose A is an RS-bimodule and B is a ST-bimodule. Then $A \otimes_S B$ is an RT-bimodule under $a(x \otimes y)b = ax \otimes yb$, and if $f \colon A \times B \to C$ is a balanced RT-bimodule map into an RT-bimodule C, there is a unique RT-bimodule map $g \colon A \otimes_S B \to C$ such that $f = gu$. Moreover, the tensor product is uniquely determined by this requirement.*

Proof The unique additive map $g\colon A \otimes_S B \to C$ such that $f = gu$, is an RT-bimodule map. To see this fix $a \in R$ and $b \in T$. Then both $g(a \cdot b)$ and $ag(\cdot)b$ are additive maps for the balanced map $af(\cdot)b$ when f is an RT-bimodule map, and by uniqueness they coincide.

As before the tensor product is uniquely determined by this up to an RT-bimodule isomorphism because the identity maps are always bimodule maps. □

We can be in a situation where for instance A is an RS-bimodule and B is only a left S-module. Then $A \otimes_S B$ is only a left R-module, or what amounts to the same thing, it is an $R\mathbb{Z}$-bimodule, where B from the outset is considered an $S\mathbb{Z}$-bimodule.

A special case occurs when both A and B are R-modules over a commutative ring R. Then A and B can be considered as RR-bimodules by letting the right actions of R be the same as the left actions. Then $A \otimes_R B$ is an R-module such that

$$a(x \otimes y) = ax \otimes y = xa \otimes y = x \otimes ay = x \otimes ya = (x \otimes y)a,$$

and balanced RR-bimodule maps will simply be R-bilinear maps, so the universal property of the tensor product holds then for such maps. In the case of vector spaces over a field we therefore recover the usual vector space tensor product.

Be aware that there can be a great deal of collapse in the formation of tensor products of modules. For instance, if m and n are relatively prime integers, then $\mathbb{Z}_m \otimes_\mathbb{Z} \mathbb{Z}_n$ is trivial because for any $a \otimes b$ in this tensor product, we have

$$m(a \otimes b) = ma \otimes b = 0 = a \otimes nb = n(a \otimes b),$$

so $a \otimes b = 0$ as there are $k, l \in \mathbb{Z}$ such that $km + ln = 1$.

Proposition 9.5.5 *Say $f\colon A \to C$ is an RS-bimodule map and $g\colon B \to D$ is a ST-bimodule map. Then there is a unique RT-bimodule map*

$$f \otimes g \colon A \otimes_S B \to C \otimes_S D$$

such that $(f \otimes g)(x \otimes y) = f(x) \otimes g(y)$.

Proof The map $(x, y) \mapsto f(x) \otimes g(y)$ is clearly a balanced RT-bimodule map from $A \otimes B$ into $C \otimes_S D$, so there is a unique RT-bimodule map $f \otimes g$ with the required property. □

The composition of such maps evidently satisfies $(f' \otimes g')(f \otimes g) = f'f \otimes g'g$.

Proposition 9.5.6 *If A and B are RR-bimodules with R commutative, then $A \otimes_R B \cong B \otimes_R A$ as RR-bimodules under the flip map $x \otimes y \mapsto y \otimes x$.*

Proof The map $(x, y) \mapsto y \otimes x$ is obviously a balanced RR-bimodule map, so we have a flip map with a flip map as its inverse map. □

We can also form repeated tensor products. The following result says that the tensor product is associative up to isomorphism.

Proposition 9.5.7 *Suppose A_i is an (R_i, R_{i+1}) module. Then there exists a unique $R_1 R_3$-bimodule isomorphism*

$$(A_1 \otimes_{R_1} A_2) \otimes_{R_2} A_3 \to A_1 \otimes_{R_1} (A_2 \otimes_{R_2} A_3)$$

that sends $(x \otimes y) \otimes z$ to $x \otimes (y \otimes z)$.

Proof The map $(x \otimes y, z) \mapsto x \otimes (y \otimes z)$ is a balanced $R_1 R_3$-bimodule map, and we get the required map with an obvious inverse map. □

The tensor product behaves well under direct sums, and we skip the easy proof of the following result.

Proposition 9.5.8 *If A_i are RS-bimodules and B_j are ST-bimodules, then*

$$(\oplus_i A_i) \otimes_S (\oplus_j B_j) \cong \oplus_{ij} (A_i \otimes_S B_j)$$

as RT-bimodules.

Proposition 9.5.9 *Suppose R is a unital ring. Let A be an RS-bimodule and consider R as an RR-bimodule under multiplication from left and right. Then there is a unique RS-bimodule isomorphism $R \otimes_R A \to A$ such that $a \otimes x \mapsto ax$.*

Proof The map $(a, x) \mapsto ax$ is a balanced RS-bimodule map with inverse map $x \mapsto 1 \otimes x$. □

Similarly, we have $A \otimes_S S \cong A$ as RS-bimodules when S is a unital ring.

Definition 9.5.10 If R is a subring of a ring S and A is an R-module, then the S-module $S \otimes_R A$ is called *extension of scalars*, where S is considered as an SR-bimodule under multiplication from left and right.

When the ring R is unital, then $A \cong R \otimes_R A$ is an R-submodule of $S \otimes_R A$ considered as an R-module. We say that the complex vector space $\mathbb{C} \otimes_\mathbb{R} V$ is a *complexification of a real vector space V*.

Proposition 9.5.11 *Suppose R is a unital subring of a ring S, and that A is an R-module. Consider the R-module monomorphism $r : A \mapsto S \otimes_R A$ given by $r(x) = 1 \otimes x$. If B is an S-module and $f : A \to B$ is an R-module map, then there is a unique S-module map $g : S \otimes_R A \to B$ such that $f = gr$.*

Proof The map $(a, x) \mapsto af(x)$ is a balanced $S\mathbb{Z}$-bimodule map from $S \times A$ to B, and provides the unique map g such that $f = gr$. □

The following result shows that induced representation can be seen as extension of scalars with the role of scalars being played by group rings.

Theorem 9.5.12 *Suppose π is a representation on a vector space V of a subgroup H of a finite group G. Denote by π^G the induced representation of π on the vector space V^G of V-valued functions f on G that satisfy $f(ab) = \pi(a)f(b)$ for $a \in H$ and $b \in G$, so $(\pi^G(b)f)(c) = f(cb)$ for $c \in G$. Regard the group ring $F[G]$ as an $F[G]F[H]$-bimodule and V as an $F[H]$-module and V^G as an $F[G]$-module. Then there is a unique $F[G]$-module isomorphism*

$$g\colon F[G] \otimes_{F[H]} V \to V^G$$

such that $g(b \otimes x) = \pi^G(b)Ax$, where $b \in G \subset F[G]$ and $Ax \in V^G$ for $x \in V$ is given by $(Ax)(c) = \pi(c)x$ if $c \in H$ and is otherwise zero.

Proof We claim that the map from $F[G] \times V \to V^G$ given by $(b, x) \mapsto \pi^G(b)Ax$ is a balanced $F[G]$-module map. To see that it is balanced take any $a \in H$. Then it sends (ba, x) to $\pi^G(ba)Ax$ and $(b, \pi(a)x)$ to $\pi^G(b)A\pi(a)x$, so we must show that $\pi^G(a)Ax = A\pi(a)x$. For $c \in G$ we get $(\pi^G(a)Ax)(c) = \pi(ca)x$ if $ca \in H$, or $c \in H$ as $a \in H$, and is otherwise zero, whereas $(A\pi(a)x)(c) = \pi(c)\pi(a)x$ if $c \in H$ and is otherwise zero. So we have equality and therefore balancedness. Clearly the map has range in V^G and is an $F[G]$-module map. Hence we get the $F[G]$-module map g. We leave the construction of the inverse map to the reader. $\square$

The proposition above can be phrased as

$$\mathrm{Hom}_R(A, B) \cong \mathrm{Hom}_S(S \otimes_R A, B)$$

under $f \mapsto g$, where $f = gr$. Thus it captures the second version of Frobenius reciprocity. This can be generalized, see Proposition 9.5.15.

Definition 9.5.13 *If A is an RS-bimodule and B is an RT-bimodule, we can consider $\mathrm{Hom}_R(A, B)$ as an ST-bimodule under*

$$afb = f(\cdot a)b$$

for $a \in S$ and $b \in T$ and $f \in \mathrm{Hom}_R(A, B)$.

Of course, one should check that $afb \in \mathrm{Hom}_R(A, B)$ and that the axioms for a bimodule hold, but all this is straightforward. When we discussed injective modules we tacitly used this bimodule structure.

We have the following generalization of Proposition 9.2.7 that we state without an almost identical proof.

Proposition 9.5.14 *The sequence*

$$0 \longrightarrow A \xrightarrow{f} B \xrightarrow{g} C$$

of RS-bimodules and bimodule maps is exact if and only if the sequence

9.5 Tensor Products and Bimodules

$$0 \longrightarrow \mathrm{Hom}(D, A) \xrightarrow{f_*} \mathrm{Hom}(D, B) \xrightarrow{g_*} \mathrm{Hom}(D, C)$$

of TS-bimodules and bimodule maps is exact for all RT-bimodules D. Moreover, if the first sequence splits by an RS-bimodule map, then the second sequence is short exact and splits by a TS-bimodule map.

We obviously have a similar generalized version of Proposition 9.2.8. The following fundamental result is also easily verified.

Proposition 9.5.15 *If A is a PQ-bimodule and B is an RP-bimodule and C is an RS-bimodule, then the map*

$$\Psi \colon \mathrm{Hom}_P(A, \mathrm{Hom}_R(B, C)) \to \mathrm{Hom}_R(B \otimes_P A, C)$$

given by $(\Psi(f))(a \otimes b) = (f(b))(a)$ *is a well-defined QS-bimodule isomorphism with inverse given by* $((\Psi^{-1}(g))(d))(c) = g(c \otimes d)$. *If* $f \colon A_1 \to A_2$ *is a PQ-bimodule map, then*

$$\Psi_1 f^* = (\iota \otimes f)^* \Psi_2,$$

where Ψ_i is to A_i as Ψ is to A.

We get the following nice result.

Corollary 9.5.16 *If A is a PQ-bimodule and B is an RP-bimodule and both are projective as left modules, then $B \otimes_P A$ is projective as a left R-module. A similar statement holds for right modules.*

Proof We must show that

$$f_* \colon \mathrm{Hom}_R(B \otimes_P A, C) \to \mathrm{Hom}_R(B \otimes_P A, D)$$

is surjective for any epimorphism $f \colon C \to D$ between R-modules. It is easily verified that with the isomorphism in the proposition, this amounts to requiring that the map

$$(f_*)_* \colon \mathrm{Hom}_P(A, \mathrm{Hom}_R(B, C)) \to \mathrm{Hom}_P(A, \mathrm{Hom}_R(B, D))$$

is surjective, but this is obvious since the first induced map is surjective as B is projective, and then the induced map of this new epimorphism is surjective as A is projective. □

Proposition 9.5.17 *If*

$$A \xrightarrow{f} B \xrightarrow{g} C \longrightarrow 0$$

is an exact sequence of RS-bimodules, then for any TR-bimodule D the sequence

$$D \otimes_R A \xrightarrow{\iota \otimes f} D \otimes_R B \xrightarrow{\iota \otimes g} D \otimes_R C \longrightarrow 0$$

is an exact sequence of TS-bimodules. Moreover, if the first sequence is split exact, then so is the second sequence.

Proof The first part follows immediately from the proposition above and the alternative version of the proposition before that.

To get the last part, tensor the split map from the initial sequence with the identity map to get a split map for the second sequence. □

We get a similar result by tensoring with D from the right.

Example 9.5.18 Consider the short exact sequence

$$0 \longrightarrow \mathbb{Z} \xrightarrow{f} \mathbb{Z} \xrightarrow{g} \mathbb{Z}_m \longrightarrow 0$$

of $\mathbb{Z}$-modules, where g is the quotient map and $f(k) = mk$ with $m \in \mathbb{N}$. Then we get the exact sequence

$$\mathbb{Z} \otimes_\mathbb{Z} \mathbb{Z}_n \xrightarrow{f \otimes \iota} \mathbb{Z} \otimes_\mathbb{Z} \mathbb{Z}_n \longrightarrow \mathbb{Z}_m \otimes_\mathbb{Z} \mathbb{Z}_n \longrightarrow 0$$

for any $n \in \mathbb{N}$. Upon identifying $\mathbb{Z} \otimes_\mathbb{Z} \mathbb{Z}_n$ with $\mathbb{Z}_n$, we see that $\mathbb{Z}_m \otimes_\mathbb{Z} \mathbb{Z}_n$ is isomorphic as a $\mathbb{Z}$-module to $\mathbb{Z}_n/(f \otimes \iota)(\mathbb{Z}_n)$. But $(f \otimes \iota)(\mathbb{Z}_n) = d\mathbb{Z}_n$, where d is the greatest common divisor of m and n. Thus

$$\mathbb{Z}_m \otimes_\mathbb{Z} \mathbb{Z}_n \cong \mathbb{Z}_n/d\mathbb{Z}_n \cong \mathbb{Z}_d$$

as $\mathbb{Z}$-modules, which again shows that we get a complete collapse when n and m are relatively prime.

We also note that the tensored sequence above is not short exact when $m = n = 2$ as $f \otimes \iota$ is then the zero map. ◇

Evidently we also get a complete collapse of $G \otimes_\mathbb{Z} \mathbb{Q}$ for any finite additive group G.

The following device reduces the study of bimodules to modules, albeit over more complicated rings. Given two rings R and S, we turn $R \otimes_\mathbb{Z} S$ into a ring with multiplication

$$(a \otimes b)(c \otimes d) = ac \otimes bd$$

which exists thanks to the universal property of the tensor product. Then an additive group A is an RS-bimodule if and only if it is an $R \otimes_\mathbb{Z} S^{op}$-module under the operation $(a \otimes b)x = axb$ for $a \in R$ and $b \in S^{op}$ and $x \in A$.

9.6 Diagram Chase

Definition 9.6.1 The *cokernel* $\operatorname{coker} f$ of a module map $f: A \to B$ is the module $B/\operatorname{im} f$.

9.6 Diagram Chase

Consider a commutative diagram

$$\begin{array}{ccc} A_1 & \xrightarrow{f} & A_2 \\ {\scriptstyle d_1}\downarrow & & \downarrow{\scriptstyle d_2} \\ B_1 & \xrightarrow{f} & B_2 \end{array}$$

of module maps, meaning that $d_2 f = f d_1$, and where we have sloppily named both horizontal maps by f. Then the upper horizontal arrow restricts to a module map $\ker d_1 \to \ker d_2$ since if $d_1(x) = 0$, then $d_2 f(x) = f d_1(x) = 0$. While the lower horizontal arrow induces a module map $\operatorname{coker} d_1 \to \operatorname{coker} d_2$ by

$$x + d_1(A_1) \mapsto f(x) + d_2(A_2)$$

for $x \in B_1$. This is well-defined because if we replace x by $x + d_1(y)$, then we get $f(x + d_1(y)) = f(x) + d_2 f(y)$.

Definition 9.6.2 A *snake diagram* is a commutative diagram

$$\begin{array}{ccccccc} & A_1 & \xrightarrow{f} & A_2 & \xrightarrow{g} & A_3 & \longrightarrow 0 \\ & {\scriptstyle d_1}\downarrow & & \downarrow{\scriptstyle d_2} & & \downarrow{\scriptstyle d_3} & \\ 0 \longrightarrow & B_1 & \xrightarrow{f} & B_2 & \xrightarrow{g} & B_3 & \end{array}$$

of modules with exact rows.

The *Snake lemma* says the following.

Lemma 9.6.3 *Suppose we are given a snake diagram as above with quotient map $q: B_1 \to \operatorname{coker} d_1$. Then the module map*

$$h: \ker d_3 \to \operatorname{coker} d_1$$

given by $h = q f^{-1} d_2 g^{-1}$ is well-defined, and gives an exact sequence

$$\ker d_1 \to \ker d_2 \to \ker d_3 \xrightarrow{h} \operatorname{coker} d_1 \to \operatorname{coker} d_2 \to \operatorname{coker} d_3$$

together with the induced maps described above.

Proof The proof is by diagram chasing. Starting with $x \in \ker d_3$, then as g is surjective, we can pick an element $y \in A_2$ such that $g(y) = x$. By commutativity of the diagram we get $g d_2(y) = d_3 g(y) = d_3(x) = 0$, and by exactness of the lower row, there is an element $z \in B_1$ such that $f(z) = d_2(y)$. Thus $z = f^{-1} d_2 g^{-1}(x)$. It is easy to check that $q(z)$ does not depend on the choice of elements picked in the inverse images. The map h one then gets is evidently a module map.

Exactness of the long sequence is gotten by numerous diagram chases left to the reader. To illustrate how these are carried out, let us show that $\ker h \subset g(\ker d_2)$. Say $h(x) = 0$, so $q(z) = 0$ with z as above. Then $z = d_1(w)$ for some $w \in A_1$. Now

$$d_2(y - f(w)) = d_2(y) - fd_1(w) = f(z) - f(z) = 0$$

shows that $y - f(w) \in \ker d_2$, and $g(y - f(w)) = g(y) - gf(w) = x - 0 = x$, so $x \in g(\ker d_2)$. □

Definition 9.6.4 A module A is a *finitely presented module* if there is an exact sequence

$$C \longrightarrow B \longrightarrow A \longrightarrow 0$$

of modules with B and C both finitely generated and free.

Equivalently, a module A is finitely presented if there is a short exact sequence

$$0 \longrightarrow D \longrightarrow B \longrightarrow A \longrightarrow 0$$

of modules with both B and D finitely generated and with only B free. If the first statement holds, one gets the second statement by letting D be the quotient module of C by the kernel of the map into B. If the second statement holds, let C be a finitely generated free presentation of D.

If the underlying ring R is unital, in the definition above we can replace C by R^m and B by R^n and the map $C \to B$ by its matrix $(a_{ij}) \in M_{nm}(R)$. Let $\{x_i\}$ be the image of the standard basis in R^n under the quotient map $R^n \to B/C \cong A$. Then the elements x_i generate A and are subject to finitely many relations

$$\sum_j a_{ij} x_j = 0.$$

Hence the term finitely presented. The image of C is to be thought of as the submodule of relations among the free generators in B.

The following result shows that if we pick any finite collection of generators of a finitely presented module over a unital ring, then there will always be finitely many relations between them.

Proposition 9.6.5 *If A is a finitely presented module with an epimorphism $g \colon E \to A$ of a finitely generated (free) module E, then $\ker g$ is finitely generated.*

Proof Since A is finitely presented, there are finitely generated free modules B and D making the first row in the following diagram

9.6 Diagram Chase

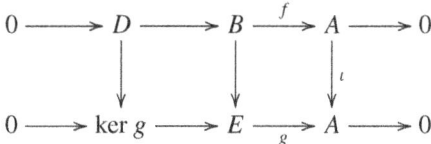

exact. The exactness of the second row is provided by g. Using that B is projective we get the middle vertical arrow which f factors through making the square diagram to the right commutative. The leftmost vertical arrow is just the restriction of the middle vertical arrow to D, which makes the leftmost square diagram commutative. Thus we have a snake diagram. The long exact sequence of the snake lemma gives the isomorphism

$$\ker g/\operatorname{im} D \cong E/\operatorname{im} B.$$

As both $E/\operatorname{im} B$ and D are finitely generated, we conclude by Proposition 9.1.6 that $\ker g$ is finitely generated. We did not use freeness of E in this argument. □

The following result is known as the *five-lemma*.

Lemma 9.6.6 *Suppose we have the following commutative diagram*

$$\begin{array}{ccccccccc} A_1 & \longrightarrow & A_2 & \longrightarrow & A_3 & \longrightarrow & A_4 & \longrightarrow & A_5 \\ d_1 \downarrow & & d_2 \downarrow & & d_3 \downarrow & & d_4 \downarrow & & d_5 \downarrow \\ B_1 & \longrightarrow & B_2 & \longrightarrow & B_3 & \longrightarrow & B_4 & \longrightarrow & B_5 \end{array}$$

of modules. If d_1 is an epimorphism and d_2 and d_4 are monomorphisms, then d_3 is a monomorphism. Dually, if d_5 is a monomorphism and d_2 and d_4 are epimorphisms, then d_3 is a epimorphism. Thus d_3 is an isomorphisms if d_1, d_2, d_4, d_5 are isomorphisms.

Proof Let f_i and g_i be the horizontal arrows starting at A_i and B_i, respectively. To prove the first statement, again by diagram chasing, if $x \in A_3$ with $d_3(x) = 0$, then $d_4 f_3(x) = g_3 d_3(x) = 0$, so $f_3(x) = 0$ as d_4 is injective. Thus by exactness, there is $y \in A_2$ such that $f_2(y) = x$. As $g_2 d_2(y) = d_3 f_2(y) = 0$, by exactness there is $z \in B_1$ with $g_1(z) = d_2(y)$. Since d_1 is surjective there is $w \in A_1$ with $d_1(w) = z$. Then $d_2 f_1(w) = g_1 d_1(w) = d_2(y)$, so $f_1(w) = y$ as d_2 is injective. Hence $x = f_2(y) = f_2 f_1(w) = 0$, so d_3 is injective.

The proof of the second statement is similar, and the last assertion is a combination of these two dual statements. Alternatively one can prove the lemma using the snake lemma. □

The need the following useful result relating to exactness under the formation of Pontryagin duals.

Proposition 9.6.7 *A module map $f: A \to B$ is injective if and only if the dual map $f^*: \hat{B} \to \hat{A}$ is surjective. Thus the sequence*

$$0 \longrightarrow A \xrightarrow{f} B \xrightarrow{g} C \longrightarrow 0$$

of modules is exact if and only if the sequence

$$0 \longrightarrow \hat{C} \xrightarrow{g^*} \hat{B} \xrightarrow{f^*} \hat{A} \longrightarrow 0$$

of $\mathbb{Z}$-modules is exact. If we cut out the last arrow in the first sequence and the first arrow of the second sequence, we still have a bijective correspondence between exact sequences.

Proof If f^* is surjective, then $f^{**}: \hat{\hat{A}} \to \hat{\hat{B}}$ is injective by Proposition 9.2.8. But A and B are contained in these double duals by Proposition 9.4.9, and f^{**} restricts to f on A, so f is injective.

If $\ker f$ is trivial, then by Proposition 9.2.8 the exact sequence $0 \to A \to \operatorname{im} f \to 0$ yields and exact sequence $0 \to \widehat{\operatorname{im} f} \to \hat{A} \to 0$, so any element in $\hat{A}$ is of the form uf for an additive map $u: \operatorname{im} f \to \mathbb{Q}/\mathbb{Z}$. By the proof of Proposition 9.4.9 we can extend u to $v \in \hat{B}$, and then $f^*(v) = vf = uf$, so $f^*: \hat{B} \to \hat{A}$ is surjective.

By Proposition 9.2.8 any sequence

$$A \xrightarrow{f} B \xrightarrow{g} C \longrightarrow 0$$

of modules is exact if and only if the sequence

$$0 \longrightarrow \hat{C} \xrightarrow{g^*} \hat{B} \xrightarrow{f^*} \hat{A}$$

of $\mathbb{Z}$-modules is exact, so the first part of the proof implies that the first of these sequences is short exact if and only the second one is short exact. □

The proposition above immediately implies that $\operatorname{coker} f^* \cong \widehat{\ker f}$ for any module map $f: A \to B$. This is consistent with the fact that f is injective if and only if f^* is surjective.

Note that the Pontryagin dual $\hat{A}$ of an R-module A is a right R-module under the operation $(fa)(b) = f(ab)$. If we have an R-module map $f: A \to B$, then $f^*: \hat{B} \to \hat{A}$ will obviously be a right R-module map. So the second sequence in the proposition above will be a short exact sequence of right R-module maps.

Proposition 9.6.8 *Consider modules A and B over a unital ring R. Then the additive map $\sigma: \hat{A} \otimes_R B \to \operatorname{Hom}_R(B, A)^\wedge$ given by $(\sigma(f \otimes x))(g) = fg(x)$ is bijective if B is a finitely presented module.*

Proof Exploiting the isomorphisms $\hat{A} \otimes_R R \cong R$ and $\operatorname{Hom}_R(R, A) \cong R$, we see that σ is bijective when B is just R. Since $B \mapsto \hat{A} \otimes_R B$ and $B \mapsto \operatorname{Hom}_R(B, A)^\wedge$ both preserve finite direct sums, we conclude that σ is also bijective if B is a finite direct sum of R, or in other words, if B is a finitely generated free module.

Since B is finitely presented, we have an exact sequence
$$D \longrightarrow C \longrightarrow B \longrightarrow 0$$
of modules with C and D both finitely generated and free. Since σ is natural in the B-variable in that it respects module maps, we therefore get a commutative diagram

$$\begin{array}{ccccccc}
\hat{A} \otimes_R D & \longrightarrow & \hat{A} \otimes_R C & \longrightarrow & \hat{A} \otimes_R B & \longrightarrow & 0 \\
{\scriptstyle \sigma}\downarrow & & {\scriptstyle \sigma}\downarrow & & {\scriptstyle \sigma}\downarrow & & \\
\mathrm{Hom}_R(D,A)^\wedge & \longrightarrow & \mathrm{Hom}_R(C,A)^\wedge & \longrightarrow & \mathrm{Hom}_R(B,A)^\wedge & \longrightarrow & 0
\end{array}$$

that has exact rows by Propositions 9.5.17, 9.2.8 and 9.6.7. The proof is completed by prolonging the diagram above trivially to the right and invoking the five-lemma remembering that the two leftmost vertical arrows are isomorphisms by the first part of this proof. $\square$

9.7 Flatness

Let us study exactness of the tensor product.

Definition 9.7.1 An R-module A is *(faithfully) flat* when the sequence of additive groups
$$0 \longrightarrow B \otimes_R A \xrightarrow{f \otimes \iota} C \otimes_R A \xrightarrow{g \otimes \iota} D \otimes_R A \longrightarrow 0$$
is exact if (and only if)
$$0 \longrightarrow B \xrightarrow{f} C \xrightarrow{g} D \longrightarrow 0$$
is an exact sequence of right R-modules. Flatness and faithful flatness for right modules are defined analogously.

Clearly a unital ring is flat as a module over itself. It is also easy to see that a direct sum of modules is flat if and only if each component is flat. So free modules over unital rings are flat.

In view of the alternative version of Proposition 9.5.17, to show flatness of an R-module A, it is enough to check that
$$f \otimes \iota \colon B \otimes_R A \to C \otimes_R A$$
is injective whenever f is a monomorphism between right R-modules B and C.

Proposition 9.7.2 *Projective modules over unital rings are flat.*

Proof Suppose an R-module A is a direct summand of a free module B, so $B = A \oplus C$, and consider a monomorphism $f \colon D \to E$ of right R-modules. As B is flat, the map
$$f \otimes \iota \colon D \otimes_R B \to E \otimes_R B$$
is injective. If $f \otimes \iota_A$ kills some element $x \in D \otimes_R A$, then the injective composition
$$D \otimes_R A \to (D \otimes_R A) \oplus (D \otimes_R C) \cong D \otimes_R B \to E \otimes_R B \cong (E \otimes_R A) \oplus (E \otimes_R C)$$
will also vanish at x, so $x = 0$. □

Later in this section we shall see that the converse is false.

Proposition 9.7.3 *A flat module over an integral domain is torsion-free.*

Proof If A is a flat module over an integral domain R, then for any non-zero element $a \in R$, the map $f \colon R \to R$ which sends b to ba is an injective module map. Thus $f \otimes \iota \colon R \otimes_R A \to R \otimes_R A$ is injective, and this map sends $x \in A \cong R \otimes_R A$ to ax, so A has no non-zero torsion elements. □

In Example 9.5.18 we saw more directly that $\mathbb{Z}_n$ is not flat as a module over $\mathbb{Z}$ for $n > 1$. So the quotient of a flat module by a flat module need not be flat. The second result below shows that if the quotient of a module by a flat submodule is flat, then the module itself is flat.

Lemma 9.7.4 *Given a short exact sequence*
$$0 \longrightarrow A \longrightarrow B \longrightarrow C \longrightarrow 0$$
of modules over a unital ring R with C flat, then the sequence
$$0 \longrightarrow D \otimes_R A \longrightarrow D \otimes_R B \longrightarrow D \otimes_R C \longrightarrow 0$$
of additive groups is exact for any right R-module D.

Proof Pick any right R-module free presentation
$$0 \longrightarrow F \longrightarrow E \longrightarrow D \longrightarrow 0$$
of D. Then we get the following commutative diagram

9.7 Flatness

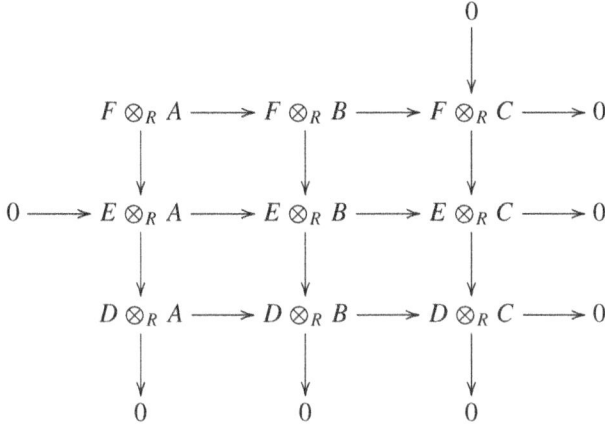

of additive groups with exact rows. The commutativity of the diagram and exactness of the first and third row is just Proposition 9.5.17 and naturality of tensor products. The upper 0 is flatness of C, whereas the left 0 is right freeness, and thus right flatness, of E. But then the kernel-cokernel map in the snake lemma shows that the third row is short exact. □

Proposition 9.7.5 *Given a short exact sequence*

$$0 \longrightarrow A \longrightarrow B \longrightarrow C \longrightarrow 0$$

of modules over a unital ring R with C flat, then B is flat if and only if A is flat.

Proof Let $D \to E$ be a monomorphism of right R-modules. Then we have a commutative diagram

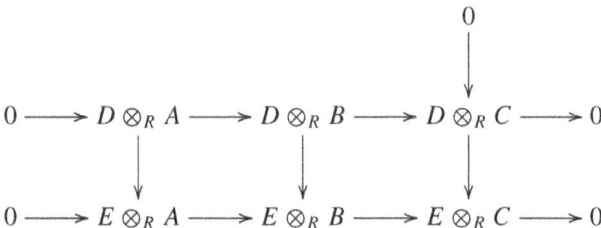

of additive groups with exact rows. The exactness of the two rows is due to the lemma above, and the upper 0 is flatness of C. If A is flat, then the leftmost vertical map is injective, and the snake lemma shows that the middle vertical arrow has trivial kernel, so B is flat. On the other hand, if B is flat, then the middle vertical arrow is injective, and then the commutativity of the leftmost square shows that the leftmost vertical arrow is injective. Thus A is flat. □

Theorem 9.7.6 *Suppose A is a module over a unital ring R. Then A is flat if and only if its Pontryagin dual $\hat{A}$ is injective as a right R-module if and only if the surjective map $f \colon I \otimes_R A \to IA$ given by $f(a \otimes x) = ax$ is an isomorphism of additive groups for every right ideal I of R.*

Proof Consider an inclusion $B \to C$ of right R-modules. By the right module version of Proposition 9.5.15 we have a commutative diagram

$$\begin{array}{ccc} \mathrm{Hom}(C, \hat{A}) & \longrightarrow & \mathrm{Hom}(B, \hat{A}) \\ \downarrow & & \downarrow \\ (C \otimes_R A)^\wedge & \longrightarrow & (B \otimes_R A)^\wedge \end{array}$$

of additive groups, where Hom here stands for homomorphisms of right R-modules, and where the vertical arrows are bijections.

By Propositions 9.6.7 and 9.4.9 we thus see that $\hat{A}$ is injective if and only if $(C \otimes_R A)^\wedge \to (B \otimes_R A)^\wedge$ is surjective for all monomorphisms $B \to C$ if and only if $B \otimes_R A \to C \otimes_R A$ is injective for all monomorphisms $B \to C$ if and only if A is flat.

By invoking Baer's criterion we similarly see that $\hat{A}$ is injective if and only if $(R \otimes_R A)^\wedge \to (I \otimes_R A)^\wedge$ is surjective for all right ideals I of R if and only if $I \otimes_R A \to R \otimes_R A$ is injective for all right ideals I of R if and only if f is an isomorphism for every right ideal I of R. To verify the last equivalence it suffices to consider the commutative diagram

$$\begin{array}{ccc} I \otimes_R A & \longrightarrow & R \otimes_R A \\ f \downarrow & & \downarrow \\ IA & \longrightarrow & A \end{array}$$

of additive groups, where the second vertical arrow is the usual isomorphism, whereas the first row is induced by $I \subset R$, and the second row is the inclusion map. □

The isomorphism $I \otimes_R A \cong IA$ for a flat module A over a unital ring R with a right ideal I did involve only the diagram consideration at the end of the proof. As for the opposite direction, there is another approach that does not use Baer's criterion and Pontryagin duals. It is based on the following trivial result, and a more flexible notion of flatness.

Proposition 9.7.7 *Let I be a right ideal of a unital ring R, and let A be an R-module. Then the map $(R/I) \otimes_R A \to A/IA$ that sends $(a + I) \otimes x$ to $ax + IA$ is a well-defined additive map with an inverse map that sends $x + IA$ to $(1 + I) \otimes x$. When I is an ideal, then $(R/I) \otimes_R A \to A/IA$ is an isomorphism of R-modules and of R/I-modules.*

9.7 Flatness

The *reduction map* $A \mapsto A/IA \cong (R/I) \otimes_R A$ in the proposition above is regarded as a *change of the base ring* from R to R/I, normally then in the setting of commutative rings.

At the risk of confusion with musical terms we now introduce the following relative notion of flatness.

Definition 9.7.8 An R-module A is *B-flat*, or *flat for B*, if for every monomorphism $C \to B$ of right modules, the tensored sequence $0 \to C \otimes_R A \to B \otimes_R A$ is exact.

Lemma 9.7.9 *A B-flat module is flat for right submodules and right quotient modules of B.*

Proof The part on submodules is true because if A is an R-module that is B-flat and $D \subset C$ are right submodules of B, then $D \otimes_R A \to C \otimes_R A$ is injective because its composition with $C \otimes_R A \to B \otimes_R A$ is the monomorphism $D \otimes_R A \to B \otimes_R A$.

As for the quotient issue, suppose we have a short exact sequence

$$0 \longrightarrow D \longrightarrow B \longrightarrow C \longrightarrow 0$$

of right R-modules, and let E be a right submodule of C. Say that the kernel of $E \otimes_R A \to C \otimes_R A$ is the additive group F. We must show that $F = 0$. Consider the commutative diagram

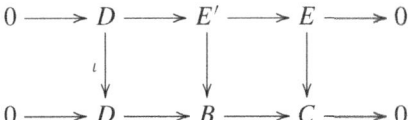

of right modules with exact rows, where $E' \subset B$ is the inverse image of E under the quotient map. This gives the commutative diagram

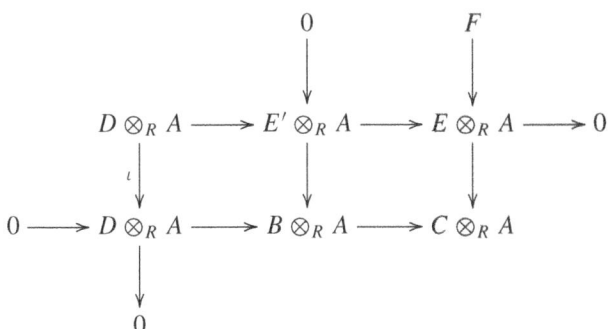

of additive groups with exact rows, where we have used B-flatness of A to the right submodules D and E' of B. The snake lemma provides $0 \to F \to 0$, so $F = 0$. □

Lemma 9.7.10 *A module that is flat for each component of a direct sum of right modules is flat for the direct sum.*

Proof Say A is an R-module that is B_i-flat, and let C be any right submodule of $B \equiv \oplus B_i$. Any element of C will be non-zero in only finitely many components, so to see that $C \otimes_R A \to B \otimes_R A$ is injective, we can assume that the direct sum is finite. By induction we can therefore assume that $B = B_1 \oplus B_2$. Let $C_1 = C \cap B_1$ and let C_2 be the projection of C onto B_2. Then the diagram

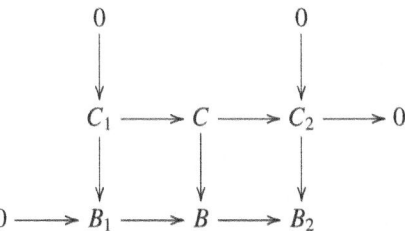

of right modules is obviously commutative and has exact rows. Tensoring with A we then get the commutative diagram

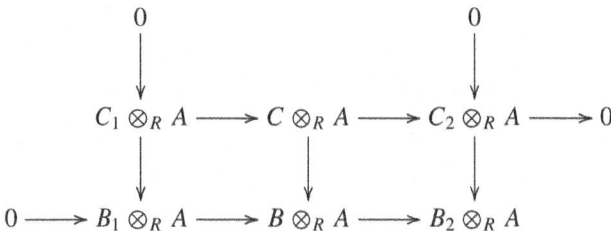

of additive groups with exact rows. The snake lemma shows that the kernel of the middle vertical arrow is trivial, so A is B-flat. □

Now suppose we have a module A over a unital ring R such that $I \otimes_R A \cong IA$ for all right ideals I of R. Then by the diagram consideration at the end of the proof of the theorem above, we know that $I \otimes_R A \to R \otimes_R A$ is an injection of additive groups, so A is R-flat. But then by the last lemma above, the module A is flat for any free right module, and thus it is flat for any right quotient module of a right free module by the lemma before. But any right R-module comes with a free presentation, and is therefore such a quotient, so A is flat for any right R-module, in other words, it is flat; and alternative proof, as promised.

Note that if I is an ideal, then $I \otimes_R A \cong IA$ is an isomorphism of R-modules. Also, when R is commutative, remember that a right ideal is an ideal.

We include a partial converse of Proposition 9.7.3.

Corollary 9.7.11 *Any torsion-free module over a PID is flat.*

Proof If A is a torsion-free module over a PID R, then any ideal of R is of the form $I = Ra$ for $a \in R$, so any element of $Ra \otimes_R A$ is of the form $b \otimes x$ for $b \in R$ and $x \in A$. If the map $I \otimes_R A \to IA$ kills such an element, then $bx = 0$, so $x = 0$ or $b = 0$ since A has no torsion. In either case $b \otimes x = 0$, so $I \otimes_R A \cong IA$, and A is flat by the theorem. □

This means that $\mathbb{Q}$ is flat as a $\mathbb{Z}$-module. However, by Corollary 9.9.6 we see that if it was projective, it would be free, and two distinct rational numbers are linear dependent over $\mathbb{Z}$ because $(cb^2d)a/b + (-abd^2)c/d = 0$. So it is not projective. Also, it cannot be finitely generated since if d is the product of the denominators of a finite collection of generators, then $1/n$ for $n > d$ does not belong to the $\mathbb{Z}$-span of the generator set. So it is certainly not finitely presented, which is also confirmed by the following result.

Theorem 9.7.12 *Finitely presented flat modules over unital rings are projective.*

Proof Suppose A is a finitely presented flat module over a unital ring R, and say we are given an epimorphism $B \to C$, so $\hat{C} \to \hat{B}$ is a monomorphism. Since A is flat, the top arrow of the commutative square

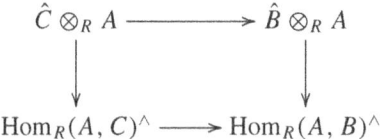

of additive groups is an injection, and the vertical arrows are isomorphisms by Proposition 9.6.8. Thus the bottom arrow is an injection, and then by Propositions 9.6.7 and 9.4.9, we deduce that $\mathrm{Hom}_R(A, B) \to \mathrm{Hom}_R(A, C)$ is a surjective map of additive groups. We conclude that A is projective. □

9.8 Duals

Definition 9.8.1 The *dual module* A^* of an RS-bimodule A is the SR-bimodule $\mathrm{Hom}_R(A, R)$. Regarded as an $R^{\mathrm{op}}S^{\mathrm{op}}$-bimodule, we define the *double dual* A^{**} of A as the $S^{\mathrm{op}}R^{\mathrm{op}}$-bimodule $\mathrm{Hom}_{R^{\mathrm{op}}}(A^*, R^{\mathrm{op}})$, which is the same thing as an RS-bimodule. Evidently the formula

$$(\eta(x))(f) = f(x)$$

defines an RS-bimodule map $\eta \colon A \to A^{**}$. The bimodule A is *reflexive* if η is an isomorphism.

Not all bimodules are reflexive. The dual bimodule can even be trivial.

Example 9.8.2 We have $\text{Hom}_\mathbb{Z}(\mathbb{Z}_n, \mathbb{Z}) = \{0\}$ since for $f \in \text{Hom}_\mathbb{Z}(\mathbb{Z}_n, \mathbb{Z})$, we have $nf([m]) = f([nm]) = f(0) = 0$, so $f([m]) = 0$. ◇

Normally the dual bimodule is larger.

Example 9.8.3 The dual of the countable $\mathbb{Q}$-module $A = \oplus_{i \in \mathbb{N}} \mathbb{Q}$ is the uncountable module $A^* = \prod_i \mathbb{Q}$. Letting B be the complement of the $\mathbb{Q}$-subspace A of A^*, we see that $A^{**} = (A \oplus B)^* \cong A^* \oplus B^*$ is also uncountable, so A is not reflexive. ◇

We have the following result with a straighforward proof and an obvious generalization.

Lemma 9.8.4 *Given two RS-bimodules A_i with projections $\pi_i \colon A_1 \oplus A_2 \to A_i$. Define $h \colon (A_1 \oplus A_2)^{**} \to A_1^{**} \oplus A_2^{**}$ by $h(x) = (x_1, x_2)$, where $x_i(f_i) = x(f_i \pi_i)$. Then h is an RS-bimodule isomorphism with inverse given by*

$$(h^{-1}(x_1, x_2))(f) = x_1(f\iota_1) + x_2(f\iota_2).$$

Moreover, we have $h\eta = (\eta_1, \eta_2)$, where $\eta_i \colon A_i \to A_i^{**}$ are the canonical maps.

Thus η is injective if and only if each η_i is injective. Also, we see that A is reflexive if and only if each A_i is reflexive.

Definition 9.8.5 Consider a finitely generated free module A over a unital ring with basis X. Define $x^* \in A^*$ for $x \in X$ by $x^*(y) = \delta_{x,y}$ for $y \in X$. Then $X^* \equiv \{x^* \mid x \in X\}$ is clearly a basis for the right module A^* called the *dual basis* of X.

Finitely generated free modules over unital rings are reflexive as $\eta(x) = x^{**}$ for any element x of a basis X. Alternatively, one can use the lemma above repeatedly.

The bimodule map $\eta \colon A \to A^{**}$ is always injective when A is a free module over a unital ring. To see this, write any non-zero $y \in A$ as a finite linear combination $y = \sum a_x x$ of a basis X with all $a_x \neq 0$ and observe that $(\eta(y))(x^*) = a_x$, where $x^*(y) = \delta_{x,y}$ for $y \in X$.

Proposition 9.8.6 *If an RS-bimodule A is a finitely generated projective module with R unital, then the SR-bimodule A^* is finitely generated and projective as a right module, and the bimodule A is reflexive.*

Proof This is immediate from the lemma above and the remark above on dual bases together with Propositions 9.3.8 and 9.3.6. □

Proposition 9.8.7 *Let A be an RS-bimodule and let B be an RT-bimodule with R unital. Then there is an ST-bimodule map $h \colon A^* \otimes_R B \to \text{Hom}_R(A, B)$ such that $h(f \otimes x) = f(\cdot)x$. Moreover, if A is a finitely generated projective module, then h is a bimodule isomorphism.*

9.8 Duals

Proof The map h comes from the balanced ST-bimodule map $(f, x) \mapsto f(\cdot)x$.

Next assume first that A is a free module with a finite basis X. Then any element of $A \otimes_S B$ is clearly of the form $\sum_{x \in X} x^* \otimes y_x$, and if such an element is zero under h, then evaluating at $x \in X$, we get $y_x = 0$, so the element itself is zero. Thus h is injective. To see that it is surjective, observe that $f \in \mathrm{Hom}_R(A, B)$ is the image under h of

$$\sum_{x \in X} x^* \otimes f(x) \in A^* \otimes_R B$$

as can be checked by evaluating at $x \in X$.

If A is a finitely generated projective module, then there is another module C such that $D \equiv A \oplus C$ is a finitely generated free module, see the proof of the proposition above. Let h_C and h_D be the maps in the proposition corresponding to $A = C$ and $A = D$, respectively. By the previous paragraph the map h_D is an isomorphism, and it can be checked that $h_D = h \oplus h_C$ under the natural isomorphisms

$$D^* \otimes_R B \cong (A^* \otimes_R B) \oplus (C^* \otimes_R B) \text{ and } \mathrm{Hom}_R(D, B) \cong \mathrm{Hom}_R(A, B) \oplus \mathrm{Hom}_R(C, B),$$

see we conclude that h is an isomorphism as well. $\square$

Corollary 9.8.8 *Let A be an RS-bimodule and let B be an TR-bimodule with R unital. If A is a finitely generated projective module, then the ST-bimodule map*

$$h \colon A^* \otimes_R B^* \to (B \otimes_R A)^* \, ; \, (h(f \otimes g))(x \otimes y) = f(y)g(x)$$

is an isomorphism.

Proof The map in question is given by the isomorphisms

$$A^* \otimes_R B^* \cong \mathrm{Hom}_R(A, B^*) \cong \mathrm{Hom}_R(B \otimes_R A, R)$$

provided by the proposition above and Proposition 9.5.15. $\square$

The following result shows that the tensor product of free modules is free if one of the rings is unital.

Proposition 9.8.9 *If A is an RS-bimodule with module basis X, and B is an ST-bimodule with module basis Y, then $X \otimes Y$ is a module basis for $A \otimes_S B$ if either S or R is unital.*

Proof By passing elements of S to the left of the tensor product symbol, we see that any element of $A \otimes_S B$ can be written as a finite sum of elements of the form $z \otimes y$ with $z \in A$ and $y \in Y$, so $X \otimes Y$ spans $A \otimes_S B$ over R.

If a finite sum $\sum a_{x,y} x \otimes y$ with $x \in X$ and $y \in Y$ and $a_{x,y} \in R$ is zero, and S is unital, then by applying $\iota \otimes y^*$, we get $\sum a_{x,y} x = 0$, so $a_{x,y} = 0$. Similarly, if R is unital, apply $x^* \otimes \iota$, to get $a_{x,y} = 0$. In either case $X \otimes Y$ is a basis. $\square$

9.9 Modules over PID's

We have the following fundamental result.

Theorem 9.9.1 *All bases of a free module over a commutative unital ring have the same cardinality, namely the rank of the module.*

Proof Given a free module A over a commutative unital ring R, we reduce to the vector space setting by first observing that the quotient R/I of R by a maximal ideal is a field by Corollary 6.11.4, and maximal ideals do exist by Proposition 6.11.7. Having picked any such ideal, we next observe that $R/I \otimes_R A$ is a vector space over R/I, and by Proposition 9.8.9 it has a linear basis $1 \otimes X$ for any R-linear basis X of A. By Theorem 3.2.5, if Y was another R-linear basis for A, then $1 \otimes Y$ and $1 \otimes X$ would have the same cardinality, so $|X| = |Y|$. Since a generator set of a vector space with minimal cardinality is automatically a linear basis, we conclude that $1 \otimes Z$ for a minimal generator set Z of A also has cardinality $|X|$, so rank $A = |X|$. □

By Proposition 9.7.7 we could alternatively have considered the vector space A/AI over the field R/I in the argument above, thus avoiding any use of tensor products. Alternatively one could adapt more directly the proof for vector spaces to modules.

Corollary 9.9.2 *Any two free modules over a PID are isomorphic if and only if they have the same rank.*

Proof This is immediate from the theorem above. □

The existence of a basis for a module cannot be adapted because for any non-zero element y in the complement of a free submodule B spanned by some maximal linear independent set X, the set $X \cup \{y\}$ is not linear independent. Setting off to show this, one can only deduce that $ay \in B$ for some non-zero element a of the ring. In the field case we could divide by a to get a contradiction, but for a general ring this is not possible. Indeed, the element a could be in the annihilator of y.

However, the following result carries over.

Theorem 9.9.3 *Any submodule of a free module over a PID is free, and never of greater rank than the larger module.*

Proof We give a proof only in the case of a submodule A of a module with finite basis $\{x_i\}_{i=1}^n$ over a PID R. Let A_m be the submodule $A \cap \langle\{x_i\}_{i=1}^m\rangle$. Since A_1 is a submodule of $\langle x_1 \rangle$, it is of the form $\langle ax_1 \rangle$ for some $a \in R$, so either it is trivial, or it is free with basis ax_1 as $bax_1 = 0$ implies $ba = 0$, so $b = 0$ as $a \neq 0$.

Assume inductively that A_m is free and of rank not greater than m, which as we have seen, holds for $m = 1$. Since R is a PID, the ideal I of all $a \in R$ such that $ax_{m+1} \in A + \sum_{i=1}^m Rx_i$ is principal, say with generator $c \in R$. If $c = 0$, then $A_{m+1} = A_m$, so the induction step holds. Thus $A = A_n$ is free and of rank not greater than n.

9.9 Modules over PID's

If $c \neq 0$, pick $y \in A$ such that $cx_{m+1} - y \in \sum_{i=1}^{m} Rx_i$. Now if $z \in A_{m+1}$, then its coefficient with respect to x_{m+1} belongs to I, and must be divisible by c. So there is $d \in R$ such that $z - dcx_{m+1} \in \sum_{i=1}^{m} Rx_i$. But then $z + dy \in A_m$, so $A_{m+1} = A_m + \langle y \rangle$, and $\{x_i\}_{i=1}^{m} \cup \{y\}$ is linear independent as y has the non-trivial coefficient c with respect to x_{m+1}. Again the induction step holds, and $A = A_n$ is free and of rank not greater than n. □

Corollary 9.9.4 *A submodule of a finitely generated module over a PID is finitely generated, and not of greater rank than the larger module.*

Proof By Proposition 9.3.6 we know that a finitely generated module A over a PID is the quotient of a free module B of not greater rank than that of A. If C is a submodule of A, then by the theorem above its inverse image in B is a free module of not greater rank than that of A. The image of a basis for this module will be a generator set for C with cardinality not greater than rank A. □

Corollary 9.9.5 *Any minimal generator set of a finitely generated free module over a PID is a basis.*

Proof Suppose $\{x_i\}_{i=1}^{n}$ is a minimal finite generator set of a free module A over a PID R. Let $\{e_i\}$ be the standard basis for R^n, and consider the module map $f \colon R^n \to A$ such that $f(e_i) = x_i$. Then as A is projective, we get the decomposition $R^n \cong A \oplus \ker f$ by Proposition 9.3.8. By the theorem above A is a free module of rank n, and $\ker f$ is a free module of rank $n - n = 0$, so $\ker f = \{0\}$. Thus f is an isomorphism and $\{x_i\}_{i=1}^{n}$ is a basis for A. □

Corollary 9.9.6 *Any projective module over a PID is free.*

Proof Combine the theorem above with Proposition 9.3.8. □

We have also seen that finitely generated projective modules over a PID are torsion-free, which is consistent with the corollary above and the following result.

Theorem 9.9.7 *Suppose A is a finitely generated module over a PID. Then A/A_τ is free and of finite rank, and*

$$A \cong A_\tau \oplus (A/A_\tau).$$

Any other free module B such that $A \cong A_\tau \oplus B$ is isomorphic to A/A_τ.

Proof By Proposition 9.1.17 the module A/A_τ is torsion-free, and obviously finitely generated, say with a finite generator set $\{x_i\}_{i=1}^{n}$. Among these generators pick a maximal subset of linear independent generators $\{y_i\}_{i=1}^{m}$. Then for each x_j there is a non-zero a_j in the PID such that $a_j x_j \in \langle y_1, \ldots, y_m \rangle$, otherwise x_j could be added to the assumed maximal collection. Hence $aA/A_\tau \in \langle y_1, \ldots, y_m \rangle$, where $a = a_1 \cdots a_n \neq 0$. Thus we have a map $A/A_\tau \to \langle y_1, \ldots, y_m \rangle$ given by $x \mapsto ax$. It is a monomorphism because A/A_τ is torsion-free. Since its image is a submodule of a free module, the theorem above tells us that this image module is free. But this image module is isomorphic to A/A_τ, so A/A_τ is free, and we get the decomposition by Proposition 9.3.8. □

Corollary 9.9.8 *A finitely generated flat module over a PID is free.*

Proof Combine the theorem with Corollary 9.7.3. □

The theorem is evidently also consistent with Corollary 9.7.11. More importantly, it reduces the study of finitely generated modules over PID's to finitely generated torsion modules.

9.10 Torsion Modules over PID's

Definition 9.10.1 Let A be a module over a PID with a prime element p. Denote by $A(p)$ the submodule of A that consists of all $x \in A$ such that $p^n x = 0$ for some $n \in \mathbb{N}$. For any element a of the PID, denote by A_a the submodule consisting of all $x \in A$ such that $ax = 0$.

Proposition 9.10.2 *Suppose A is a non-trivial finitely generated torsion module over a PID, and let P be a collection of representatives from each prime ideal of the PID. Then*
$$A = \oplus A(p),$$
where we sum over all $p \in P$ such that $A(p)$ is non-trivial.

Proof Pick $a \in R$ such that $ax = 0$ for all $x \in A$. Such an element exists because A is a finitely generated torsion module; we may pick the finite product of those elements that kill each generator. Now every element of the PID is up to units a product of powers of elements in P. Factorizing a this way, by induction it suffices to show that if $a = bc$ for relatively prime b and c, corresponding to powers of different prime elements, then $A = A_b \oplus A_c$. This follows because there are ring elements r, s such that $rb + sc = 1$, so for any $x \in A$, we have $x = rbx + scx$ with $rbx \in A_c$ and $scx \in A_b$ as $crbx = rax = 0 = sax = bscx$. Thus A is the sum of A_b and A_c, and if $y \in A_b \cap A_c$, then $y = rby + scy = 0 + 0 = 0$, so we have a direct sum. □

Next we decompose each $A(p)$ into cyclic modules.

Definition 9.10.3 Elements $x_1, \ldots, x_n$ of a module over a PID are *independent* if $\sum a_i x_i = 0$ implies $a_i x_i = 0$, or equivalently, if $\langle x_1, \ldots, x_n \rangle = \oplus \langle x_i \rangle$.

Lemma 9.10.4 *Suppose A is a module over a PID with $p^n A = 0$ for a prime element p and $n \in \mathbb{N}$. Assume we have $x_1 \in A$ with $\mathrm{Ann}(x_1) = (p^n)$ and independent elements $y_2, \ldots, y_n \in A/\langle x_1 \rangle$. Then there are representatives $x_i \in A$ of y_i with $\mathrm{Ann}(x_i) = \mathrm{Ann}(y_i)$. The elements are $x_1, \ldots, x_n$ automatically independent.*

Proof To get such representatives it suffices to show that a generator p^m of the annihilator of an element $y \in A/\langle x_1 \rangle$ kills some representative of y. The opposite

9.10 Torsion Modules over PID's

inclusion of annihilators is obvious. For any representative $z \in A$ of y, we have $p^m z \in \langle x_1 \rangle$. Write $p^m z = a p^k x_1$ for some $k \leq n$ and a not divisible by p. If $m > k$, then $a p^{n-(m-k)} x_1 = p^n z = 0$, an absurdity as p^n cannot divide $a p^{n-(m-k)}$. But then $z - a p^{k-m} x_1$ is a representative of y killed by p^m.

Therefore we can pick representatives x_i of y_i with no smaller annihilators. If $\sum_{i=1}^n a_i x_i = 0$, then $\sum_{i=2}^n a_i y_i = 0$. By assumption $a_i y_i = 0$, so $a_i x_i = 0$ for $i \geq 2$ by the paragraph above. But then also $a_1 x_1 = 0$. □

Observe that for any module A over a PID we have $\mathrm{Ann}(x) = (p)$ for $x \in A_p$ when p is a prime element of the PID. Recall that $R/(p)$ is a field, and note that A_p is a vector space over this field.

Lemma 9.10.5 *Suppose A is a module over a PID R with a prime element p, and say we have $x \in A$ with $\mathrm{Ann}(x) = (p^n)$ for some $n \in \mathbb{N}$. Then A_p has greater dimension than $(A/\langle x \rangle)_p$ as vector spaces over the field $R/(p)$.*

Proof Consider first $n = 1$. Say $y_2, \ldots, y_m$ is a basis for $(A/\langle x \rangle)_p$ over $R/(p)$. Then each $y_2, \ldots, y_m \in A/\langle x \rangle$ has annihilator (p). Moreover, they are independent over R because if $\sum a_i y_i = 0$ for $a_i \in R$, then $\sum (a_i + (p)) y_i = 0$ as $p y_i = 0$. So $a_i \in (p)$ by linear independence of $\{y_i\}$ over $R/(p)$. But then $a_i y_i = 0$ as $p y_i = 0$.

Since $p A_p = 0$ and $\mathrm{Ann}(x) = (p)$, the lemma above provides representatives $x_i \in A$ of y_i each with annihilator (p), so $x_i \in A_p$, and such that $x_1, \ldots, x_m$ with $x_1 = x$ are independent over R. To see that these elements in A_p are linearly independent over $R/(p)$, say that $\sum (a_i + (p)) x_i = 0$. Then as $p x_i = 0$, we get $\sum a_i x_i = 0$, so $a_i x_i = 0$ and $a_i \in (p)$. This settles the case $n = 1$.

Assume by induction that the lemma holds for n. To prove the lemma for $n + 1$ with $\mathrm{Ann}(x) = (p^{n+1})$, note that $\mathrm{Ann}(px) = (p^n)$. Thus A_p has greater dimension than $(A/\langle px \rangle)_p$, which again has greater dimension than $(A/\langle x \rangle)_p$ since the annihilator of $x + \langle px \rangle \in A/\langle px \rangle$ is (p), and the case $n = 1$ has already been proved. □

Theorem 9.10.6 *Suppose A is a finitely generated module over a PID R, and let p be a prime element of R. If $A(p)$ is non-trivial, then*

$$A(p) \cong \bigoplus_{i=1}^m R/(p^{n_i}),$$

where $m, n_i \in \mathbb{N}$ and $n_i \leq n_j$ for $i \leq j$. Moreover, the sequence $n_1, \ldots, n_m$ is uniquely determined by this.

Proof We prove the existence part of the theorem by induction on the dimension of the vector space $A(p)_p$ over the field $R/(p)$.

By Corollary 9.9.4 the submodule $A(p)$ of A is finitely generated, so we may pick an element $x_1 \in A(p)$ with annihilator (p^{r_1}) for a maximal r_1.

Suppose $A(p)_p$ is trivial. For $x \in A(p)$ we know that $p^n x = 0$ for some n, so $p^{n-1} x \in A(p)_p$ and $p^{n-1} x = 0$. But then $p^{n-2} x \in A(p)_p$ is zero. Continuing this we get $x = 0$, so $A(p)$ is trivial, which contradicts our assumption. So the dimension of $A(p)_p$ over $R/(p)$ is never zero.

If $\dim A(p)_p = 1$, then by the last lemma above $(A(p)/\langle x_1 \rangle)_p$ is trivial. If $x \in A(p)$, then $p^n x = 0$ for some n, so $p^{n-1} x + \langle x_1 \rangle \in (A(p)/\langle x_1 \rangle)_p$ and $p^{n-1} x$ belongs to $\langle x_1 \rangle$. Then $p^{n-2} x + \langle x_1 \rangle \in (A(p)/\langle x_1 \rangle)_p$ and $p^{n-2} x \in \langle x_1 \rangle$. Continuing this we get $x \in \langle x_1 \rangle$. Thus $A(p) = \langle x_1 \rangle$. The map $R \to A(p)$ given by $a \mapsto a x_1$ is an epimorphism with kernel (p^{r_1}), so the existence part of the theorem holds for the single element sequence r_1, and the induction can start.

Assume that the existence part of the theorem holds for each non-trivial module with prescribed vector space dimension less than $\dim A(p)_p$. In particular, by the last lemma it holds for the module $A(p)/\langle x_1 \rangle$. If this is non-trivial, then there are independent elements $y_2, \ldots, y_m \in A(p)/\langle x_1 \rangle$ with $\mathrm{Ann}(y_i) = (p^{r_i})$ and $r_2 \geq \cdots \geq r_m$. By the first lemma each y_i has a representative $x_i \in A(p)$ with the same annihilator, and such that $x_1, \ldots, x_m$ are independent. By choice of r_1 we have $r_1 \geq r_2$. Hence we get the desired decomposition with sequence $n_i = r_{m+1-i}$.

The uniqueness of the sequence will be clear from the next theorem. □

The following result is an obvious generalization of the Chinese remainder theorem to PID's. Of course it follows from the more general result Theorem 6.8.9, but we nevertheless include a separate proof of this important result adapted to this context.

Lemma 9.10.7 *Suppose $a = a_1 \cdots a_n$ in a PID R with the a_i's relatively prime. Given $b_i \in R$, there is a solution x of the system $x - b_i \in (a_i)$ uniquely determined up to addition by an element of (a). Moreover, the module map $f: R \to \bigoplus R/(a_i)$ given by $f(b) = (g_1(b), \ldots, g_n(b))$ with quotient maps $g_i: R \to R/(a_i)$, is surjective with kernel (a), so*

$$R/(a) \cong \bigoplus R/(a_i).$$

Proof Note that $c, d \in R$ are relatively prime if and only if $[c] \in R/(d)$ is a unit. If c_i are relatively prime to d, so is $c_1 \cdots c_n$ because $[c_1 \cdots c_n] = [c_1] \cdots [c_n]$ is a unit of $R/(d)$.

Thus $c_i = a/a_i$ is relatively prime to a_i, and there are $r_i, s_i \in R$ such that $r_i a_i + s_i c_i = 1$. Set $x = \sum b_j s_j c_j$. Then $x = b_i s_i c_i + (a_i) = b_i + (a_i)$.

If y is another solution, then $y - x \in (a_i)$ for all i, so $y - x = (a)$.

Surjectivity of f is now immediate. Its kernel consists of all $b \in R$ such that $b \in (a_i)$, which happens precisely when $b \in (a)$. □

When $R = \mathbb{Z}$ the surjectivity of f defined above is just a restatement of the Chinese remainder theorem. We could have defined a module map $R/(a) \to \bigoplus R/(a_i)$ directly, and in the case $R = \mathbb{Z}$, this map would be surjective exactly when it is injective because both sides are finite sets with the same cardinality.

Lemma 9.10.8 *If R is an integral domain, then*

$$aR/(ab) \cong R/(b)$$

for non-zero elements $a, b \in R$.

9.10 Torsion Modules over PID's

Proof The epimorphism $R \to aR/(ab)$ given by $c \mapsto ac + (ab)$ has kernel (b) because if $ac \in (ab)$, so $ac = abd$ for some $d \in R$, then $c = bd$. □

The following result is known as the *invariant factor theorem*.

Theorem 9.10.9 *Suppose A is a non-trivial finitely generated torsion module over a PID R. Then*
$$A \cong R/(a_1) \oplus \cdots \oplus R/(a_n)$$
for non-zero $a_i \in R$ such that a_i divides a_{i+1}. Moreover, the sequence $(a_1), \ldots, (a_n)$ of ideals is uniquely determined by this.

Proof By Proposition 9.10.2 we can write $A = \oplus_{i=1}^n A(p_i)$ for primes p_i. By the last theorem above
$$A(p_i) \cong \bigoplus R/(p_i^{m_{ij}})$$
for $m_{ij} \in \mathbb{N}$ with $m_{ij} \leq m_{i,j+1}$. Set $a_j = p_1^{m_{1j}} \cdots p_n^{m_{nj}}$. Clearly a_j divides a_{j+1}, and by Lemma 9.10.7 we have
$$R/(a_j) \cong R/(p_1^{m_{1j}}) \oplus \cdots \oplus R/(p_n^{m_{nj}}),$$
which gives the required decomposition.

To show uniqueness, first observe that given a decomposition as in the theorem, then
$$A_p \cong (R/(a_1))_p \oplus \cdots \oplus (R/(a_n))_p$$
for any prime p. Now $(R/(a))_p$ is a vector space over $R/(p)$ for any $a \in R$. We claim that its dimension is 1 if p divides a, and is otherwise zero. An element $b + (a)$ belongs to this space if $pb \in (a)$. If p does not divide a, then $b \in (a)$ and the space is trivial. On the other hand, if p divides a, so $a = pc$ for some c, then $pb \in (a)$ exactly when $b \in (c)$, again by unique factorization. So in this case $(R/(a))_p = cR/(cp) \cong R/(p)$ by the lemma above, and its dimension over $R/(p)$ is one. Hence the dimension of A_p equals the number of a_i's that has p as a factor.

Say we had another decomposition of A of the same type with respect to $b_1, \ldots, b_k$ such that b_i divides b_{i+1}. Pick a prime factor p of a_1. Then p divides every a_i as a_j divides a_{j+1}, so $\dim A_p = n$ by the preceding paragraph. But then p must also divide at least n of the b_i's, which is only possible if $k \geq n$. Similarly, picking a prime factor of b_1, we conclude that $n \geq k$. So $n = k$. Thus the number of components in the direct sum is uniquely determined.

The remaining uniqueness assertion certainly holds for $n = 1$. Assume by induction that this part holds for all decompositions into less than n components. Suppose we are given a length n decomposition as in the theorem.

If p is a prime factor of a_1, then there are $b_i \in R$ such that $a_i = pb_i$ for all i as a_j divides a_{j+1}. By the lemma above, we then get

$$pA \cong \bigoplus_{i=1}^{n} pR/(pb_i) \cong \bigoplus_{i=1}^{n} R/(b_i),$$

and evidently b_i divides b_{i+1}. Picking a possible prime factor for b_1 we can repeat this until the first component in the direct sum collapses to $\{0\}$, which happens when the generator of the corresponding ideal is a unit. We are then in a situation where

$$a_1 A \cong \bigoplus_{i=2}^{n} R/(c_i)$$

for some $c_i \in R$ with $a_i = a_1 c_i$. By our induction hypothesis the ideals $(c_2), \ldots, (c_n)$, and hence the ideals $(a_i) = a_1(c_i)$ for $i \geq 2$, are uniquely determined by $a_1 A$.

If we did the same for another length n decomposition of A with respect to d_i's, then since the length of a decomposition is invariant, we would get the first collapse at $a_1 A$, and consequently $(a_i) = (d_i)$ for $i \geq 2$. Returning to the two decompositions of A, we conclude that $R/(a_1) \cong R/(d_1)$ as the remaining components are isomorphic, so also $(a_1) = (d_1)$. □

Corollary 9.10.10 *Any finitely generated module over a PID R is isomorphic to*

$$R^m \oplus R/(a_1) \oplus \cdots \oplus R/(a_n)$$

for non-zero $a_i \in R$ such that a_i divides a_{i+1}. The number m and the sequence $(a_1), \ldots, (a_n)$ are uniquely determined by the isomorphism class of the module.

Proof This is immediate from Theorems 9.9.7 and 9.9.1 and the theorem above. □

This corollary and Theorem 9.10.6 are generalizations of the fundamental theorem for finitely generated abelian groups since such groups are finitely generated modules over the PID $\mathbb{Z}$ and $\mathbb{Z}/(n) = \mathbb{Z}_n$. The proof of the result for modules is a natural generalization of the one for groups, and the techniques used are similar to those used to prove the fundamental theorem of arithmetic. In the next section we give a different proof of the existence part, more in the spirit of linear algebra.

9.11 Smith Normal Form

Recall that we can multiply and add matrices with entries in a ring R just as we did for fields, and $M_n(R)$ becomes a ring under these operations, and it is unital if R is unital. We can also define the transpose of a matrix with entries in R as before, but the transpose of a product is no longer the opposite product of the transposes, nor is the inverse of a transpose the transpose of the inverse unless we for instance assume that the unital ring R is commutative. In the commutative case the theory of determinants works with the same definition. Of course division must be avoided

9.11 Smith Normal Form

when this is senseless. A matrix over a commutative unital ring is invertible if and only if its determinant is a unit.

Any $m \times n$-matrix over a ring R defines an R-module map $R^n \to R^m$ by matrix multiplication with column vectors. If A and B are free modules with finite bases $X = \{x_1, \ldots, x_n\}$ and $Y = \{y_1, \ldots, y_m\}$, respectively, and $f \in \text{Hom}_R(A, B)$, then the *associated matrix* $(f)_Y^X = (a_{ij}) \in M_{m,n}(R^{\text{op}})$ of f is given by $f(x_j) = \sum_i a_{ij} y_i$. If we let $(x)_X$ denote the column vector of the coordinates of $x \in A$ with respect to the basis X, then $(f(x))_Y = (f)_Y^X (x)_X$. It is also easy to see that if $g \in \text{Hom}_R(B, C)$ and C is free with finite basis Z, then

$$(gf)_Z^X = (g)_Z^Y (f)_Y^X.$$

Thus if $m = n$ and R is unital, then f is an isomorphism if and only if $(f)_Y^X$ is invertible. In general the map $f \mapsto (f)_X^X$ is trivially a ring isomorphism from $\text{End}_R(A)$ to $M_n(R^{\text{op}})$, which is isomorphic to $M_n(R)$ when R is commutative.

We will soon need two lemmas that do not need proofs.

Lemma 9.11.1 *If A and B are two modules with submodules C and D, respectively. Then $A \times B \to A/C \times B/D$ given by $(a, b) \mapsto (a + C, b + D)$ is an epimorphism with kernel $C \times D$, so*

$$(A \times B)/(C \times D) \cong A/C \times B/D.$$

Lemma 9.11.2 *If R is a unital ring with $A \in M_n(R)$ invertible, and if M is a submodule of R^n, then*
$$AR^n/AM \cong R^n/M$$

under the map given by $Ax + AM \mapsto x + M$.

Definition 9.11.3 Two $m \times n$-matrices A and B over a unital ring R are *equivalent* if $A = CBD$ for invertible matrices $C \in M_m(R)$ and $D \in M_n(R)$. They are *similar* if $n = m$ and $D = C^{-1}$.

Thus $A, B \in M_{m,n}(R)$ are equivalent if they are associated matrices of the same R-module map $R^n \to R^m$ but with respect to different bases of cardinality n and m.

When we performed Gauss-Jordan elimination for matrices over fields we introduced three elementary row operations, namely, interchanging two rows, multiplying a row by a unit, and multiplying a row by an element and adding this row to another. Given a unital ring R, let $P_{ij}, D_i(a), T_{ij}(b) \in M_n(R)$ be the matrices obtained from I_n respectively by interchanging row i and j, multiplying row i from the left by a unit $a \in R$, multiplying row j from the left with $b \in R$ and adding this row to row i. It is easy to see that

$$P_{ij} = I_n - E_{ii} - E_{jj} + E_{ij} + E_{ji}, \quad D_i(a) = I_n + (a-1)E_{ii}, \quad T_{ij}(b) = I_n + bE_{ij}$$

where E_{ij} are the matrix units. It is trivially verified that for any $A \in M_{m,n}(R)$, the matrix $P_{ij}A$ is the one obtained by performing the elementary row operation corresponding to P_{ij} on A. The same is true for the two other elementary matrices. Multiplication from the right by $n \times n$-elementary matrices produces the corresponding elementary column operations on A, then with multiplication by elements of the ring from the right. For instance, the matrix $AD_i(a)$ is the one obtained from A by multiplying its ith column from the right by the unit a. Also notice that the inverse of an elementary operation is an elementary operation of the same type.

A fourth row (column) operation, which in general cannot be obtained by repeated use of the three elementary ones, is to multiply A from the left (right) by an $m \times m$-matrix ($n \times n$-matrix) of the form

$$\begin{pmatrix} I_k & 0 & 0 \\ 0 & B & 0 \\ 0 & 0 & I_{m(n)-k-2} \end{pmatrix}$$

with invertible $B \in M_2(R)$.

Theorem 9.11.4 *Any matrix over a PID is equivalent to a matrix of the form*

$$\begin{pmatrix} E & 0 \\ 0 & 0 \end{pmatrix}$$

with $E = \mathrm{diag}(a_1, \ldots, a_k)$ having non-zero entries such that a_i divides a_{i+1}. The sequence $a_1, \ldots, a_k$ is unique up to multiplication by units.

Proof The *length of an element a* in the PID R is the number of prime factors in its decomposition, so the length of a unit is zero.

We may assume that the $m \times n$-matrix A over R is non-zero. Pick a non-zero entry with minimal length. Performing elementary row and column operations we can assume that it is a_{11}.

Two things can happen: Either every entry in the first row except a_{11} is divisible by a_{11}, and can therefore be eliminated using the third elementary column operation, producing a matrix with first row

$$\begin{pmatrix} d & 0 & \cdots & 0 \end{pmatrix}.$$

Or a_{11} does not divide some entry in the first row, which we by the first column operation may assume is a_{12}. Since a_{11} and a_{12} are relatively prime, their greatest common divisor d has smaller length than that of a_{11}, and there are $r, s, u, v \in R$ such that $d = ra_{11} + sa_{12}$ and $a_{11} = du$ and $a_{12} = dv$. Then $ru + sv = 1$, so

$$B = \begin{pmatrix} r & v \\ s & -u \end{pmatrix} \text{ is invertible with inverse } \begin{pmatrix} u & v \\ s & -r \end{pmatrix}.$$

Multiplying our matrix on the right by

9.11 Smith Normal Form

$$\begin{pmatrix} B & 0 \\ 0 & I_{n-2} \end{pmatrix}$$

we obtain a matrix with first row of the form

$$\begin{pmatrix} d & 0 & b_{13} & \cdots & b_{1n} \end{pmatrix}.$$

Next we turn attention to the first column. Two things can occur: Either all entries in the column other than the one in position (1, 1) is divisible by that entry, and we can eliminate them by the third elementary row operation. Or some entry in the column is not divisible by that entry, and we can, after having shifted it to the second row, multiply the matrix with

$$\begin{pmatrix} B & 0 \\ 0 & I_{m-2} \end{pmatrix}$$

from the left for an appropriate invertible B, and get a matrix with zeroes at the second place in the first column, whereas the (1, 1)-entry has length less than that of d.

Alternating this process between first rows and first columns, then since the length of the entry in position (1, 1) cannot reduce indefinitely below the length of a_{11}, sooner or later we will end up with an equivalent matrix $C_1 A D_1$ of the form

$$\begin{pmatrix} a_1 & 0 & \cdots & 0 \\ 0 & & & \\ \vdots & & A_1 & \\ 0 & & & \end{pmatrix}$$

for some $a_1 \in R$ and $A_1 \in M_{m-1,n-1}(R)$ and $C_1 \in M_m(R)$ and $D_1 \in M_n(R)$.

Doing the same thing for A_1, we can write

$$C'_2 A_1 D'_2 = \begin{pmatrix} a_2 & 0 & \cdots & 0 \\ 0 & & & \\ \vdots & & A_2 & \\ 0 & & & \end{pmatrix}$$

for $a_2 \in R$ and $A_2 \in M_{m-2,n-2}(R)$ and invertible matrices $C'_2 \in M_{m-1}(R)$ and $D'_2 \in M_{n-1}(R)$. Setting

$$C_2 = \begin{pmatrix} 1 & 0 \\ 0 & C'_2 \end{pmatrix} \text{ and } D_2 = \begin{pmatrix} 1 & 0 \\ 0 & D'_2 \end{pmatrix}$$

we get

$$C_2C_1AD_1D_2 = \begin{pmatrix} a_1 & 0 & 0 & \cdots & 0 \\ 0 & a_2 & 0 & \cdots & 0 \\ \cdot & \cdot & & & \\ \cdot & \cdot & & A_2 & \\ 0 & 0 & & & \end{pmatrix}.$$

Continuing this we get an equivalent matrix of A of the form in the theorem with $E = \mathrm{diag}(a_1, \ldots, a_k)$ for some $a_i \in R$. However, the division property between the a_i's might not hold.

If a_1 does not divide a_2, add the second row to the first, so that the first row becomes

$$\begin{pmatrix} a_1 & a_2 & 0 & \cdots & 0 \end{pmatrix}.$$

Now reduce the length of a_1 by resorting to the alternating procedure above. In the end, we get a new diagonal matrix, where the entry at $(1, 1)$ will divide the entry at $(2, 2)$, so we can assume that a_1 divides a_2. By the first elementary row and column operations we can get a new diagonal matrix where a_2 and a_3 have swopped places. Repeating the reduction of the length procedure above, we can assume that a_1 also divides a_3. Continuing this we can assume that a_1 divides $a_2, \ldots, a_k$.

Consider next the smaller matrix with a_2 in the upper left corner, and repeat the procedure. We can therefore assume that a_2 divides $a_3, \ldots, a_k$. Considering yet a smaller matrix with a_3 in the upper left corner, we can get an equivalent matrix where a_3 divides $a_4, \ldots, a_k$, till finally a_i divides a_{i+1} for all i.

Concerning uniqueness, say $A = CFD$, where F is as prescribed in the theorem and $C \in M_m(R)$ and $D \in M_n(R)$ are invertible. Then by the two lemmas above, we have the following chain

$$R^m/AR^n \cong C^{-1}R^m/FDR^n \cong R^m/FR^n \cong R^m/((a_1) \oplus \cdots \oplus (a_k))$$
$$\cong R/(a_1) \oplus \cdots \oplus R/(a_k) \oplus R^{m-k}$$

of isomorphisms. By Corollary 9.10.10 the sequence $a_1, \ldots, a_k$ is up to multiplication by units uniquely determined by the left hand side. □

The matrix above is said to be in *Smith normal form*, and the elements $a_1, \ldots, a_k$ are the *invariant factors of the matrix*.

The existence of a Smith normal form works also for matrices over UFD's since such rings also have greatest common divisors.

Corollary 9.11.5 *Two matrices of the same size over a PID are equivalent if and only if they have the same sequence of invariant factors up to multiplication by units.*

Proof If the matrices have the same invariant factors up to multiplication by units, they are both equivalent to the same matrix in Smith normal form, and therefore they are equivalent to each other.

Conversely, if the $m \times n$-matrices A and B are equivalent, then their Smith normal forms are equivalent. Now carry out the argument at the end of the proof of the

9.11 Smith Normal Form

theorem, to get

$$R/(a_1) \oplus \cdots \oplus R/(a_k) \oplus R^{m-k} \cong R/(b_1) \oplus \cdots \oplus R/(b_l) \oplus R^{m-l}.$$

By Corollary 9.10.10 their invariant factors a_i and b_i coincide up to multiplication by units. □

The proof of the theorem above provides an algorithm for how one to a given matrix $A \in M_{m,n}(R)$ over a PID R can find matrices $C \in M_m(R)$ and $D \in M_n(R)$ such that CAD is in Smith normal form. It also gives an alternative proof of the existence part of the decomposition in Corollary 9.10.10.

Namely, any finitely generated module M over R has a free presentation of finite rank, so by Theorem 9.9.3, it can be written as $R^m/f(R^n)$ for $n \leq m$ and a monomorphism $f: R^n \to R^m$. Let A be the associated matrix of f with respect to the standard bases. According to the theorem above $F = CAD$ is in Smith normal form with invariant factors a_i for some invertible matrices $C \in M_m(R)$ and $D \in M_n(R)$. Then

$$M \cong R^m/C^{-1}FD^{-1}R^n \cong R^m/FR^n \cong R/(a_1) \oplus \cdots \oplus R/(a_k) \oplus R^{m-k}$$

as claimed.

We can rephrase the theorem above in the following way.

Corollary 9.11.6 *Suppose M is a finitely generated submodule of a free module N over a PID R. Then there exists a basis $\{x_i\}$ for N and finitely many elements $a_i \in R$ such that a_i divides a_{i+1} and $\{a_1 x_1, \ldots, a_n x_n\}$ is a basis for M. Such a sequence $a_1, \ldots, a_n$ is unique up to multiplication by units.*

Proof Pick any basis $\{x_i\}$ for N. By Theorem 9.9.3 we know that M is a free module of finite rank say with basis $\{y_1, \ldots, y_n\}$. Let $A \in M_n(R)$ be the invertible matrix associated to the module map that sends x_i to y_i for $i = 1, \ldots, n$. By the theorem above there are invertible matrices $C, D \in M_n(R)$ such that $CAD = \text{diag}(a_1, \ldots, a_n)$ for some $a_i \in R$ such that a_i divides a_{i+1}. We may therefore assume that $\{a_1 x_1, \ldots, a_n x_n\}$ is a basis for M.

Uniqueness of the sequence $a_1 \ldots, a_n$ is clear from the corollary above. □

Definition 9.11.7 The *rank of an $m \times n$-matrix A over a PID R* is the rank of the free submodule AR^n of R^m.

Corollary 9.11.8 *The rank of a matrix over a PID coincides with the rank of its transpose.*

Proof Say A is an $m \times n$-matrix over a PID R having Smith normal form $F = CAD$ with $\text{diag}(a_1, \ldots, a_k)$. The rank of A coincides with the rank of F, which is k, and this is the same as the rank of F^T, which again is the rank of A^T. □

For Euclidean domains we have the following improvement of the theorem above.

Corollary 9.11.9 *Any matrix over a Euclidean domain can be brought to Smith normal form by finitely many elementary row and column operations.*

Proof In the proof of the theorem above we must be able to bring a matrix over a Euclidean domain R to Smith normal form without using the fourth operation. That operation involved multiplying from the left or right by a matrix that contained a 2×2-matrix B with determinant ± 1. If we can bring such a matrix to the identity matrix using only elementary row operations, we are done, since then B itself will be a product of such elementary matrices, which will result in elementary operations when we multiply the original matrix by the matrix containing B.

Say
$$B = \begin{pmatrix} a & b \\ c & d \end{pmatrix}$$

with $ad - bc = \pm 1$. In a Euclidean domain we have $\lambda \colon R \backslash \{0\} \to \mathbb{N} \cup \{0\}$ allowing for a division algorithm. If $c \neq 0$, then $a \neq 0$, so there are elements $q, r \in R$ such that $c = qa + r$ with $r = 0$ or $\lambda(r) < \lambda(a)$. Multiplying the first row of B by $-q$ and adding the resulting row to the second row of B, we get the row equivalent matrix

$$\begin{pmatrix} a & b \\ r & d - bq \end{pmatrix}$$

with determinant ± 1. If $r \neq 0$, interchange the two rows, thus changing the sign of the determinant, and repeat the procedure so that the new entry in position (2, 1) either is zero, or has λ-value less that the entry in position (1, 1). In the second case, interchange the rows again, changing the sign of the determinant, and repeat the procedure. After finitely many such steps we get an equivalent matrix of the form

$$\begin{pmatrix} a & b \\ 0 & d \end{pmatrix}$$

with $ad = \pm 1$. Multiplying the second row by $\mp ab$ and adding this to the first row gives $\mathrm{diag}(a, d)$ which is obviously equivalent to the unit matrix. □

A counterexample to this, which must necessarily be found among 2×2-matrices over PID that are not Euclidean domains, can be given. But for matrices over integers or polynomials in one indeterminant over fields, elementary row and column operations will bring a matrix to Smith normal form.

It is also immediate from the corollary above that any invertible matrix over a Euclidean domain can be inverted by a finite procedure involving elementary row and column operations.

Example 9.11.10 The Smith normal form of the matrix

$$\begin{pmatrix} 1 & 2 & 3 \\ 4 & 5 & 0 \end{pmatrix}$$

9.11 Smith Normal Form

with integral coefficients is after a couple of elementary row and column operations easily seen to be

$$\begin{pmatrix} 1 & 0 & 0 \\ 0 & 3 & 0 \end{pmatrix}$$

with invariant factors 1, 3 and rank two. ◇

Example 9.11.11 It is also easy to see that the Smith normal form of

$$\begin{pmatrix} -x & 4 & -2 \\ -3 & 8-x & 3 \\ 4 & -8 & -2-x \end{pmatrix} \in M_3(\mathbb{Q}[x])$$

after some elementary row and column operations is

$$\begin{pmatrix} 1 & 0 & 0 \\ 0 & 1 & 0 \\ 0 & 0 & (x-2)(x^2-4x+20) \end{pmatrix}$$

with invariant factors $1, 1, (x-2)(x^2-4x+20)$ and rank three. ◇

We can also use the theorem above to characterize modules over PID's given by generators and relations. The following result is immediate from the theorem.

Corollary 9.11.12 *If x_i are elements of a PID R subject to $\sum_j a_{ij} x_j = 0$ for $a_{ij} \in R$, then the module generated by $(x_1, \ldots, x_n) \in R^n$ is isomorphic to*

$$R^{n-k} \oplus R/(a_1) \oplus \cdots \oplus R/(a_k),$$

where the Smith normal form of $(a_{ij}) \in M_{m,n}(R)$ has invariant factors $a_1, \ldots, a_k$.

Example 9.11.13 Consider the additive group G generated by x_1, x_2, x_3 subject to $Ax = 0$, where $x = (x_1, x_2, x_3)^T$ and

$$A = \begin{pmatrix} 5 & 9 & 5 \\ 2 & 4 & 2 \\ 1 & 1 & -3 \end{pmatrix}.$$

Since the Smith normal form of $A \in M_3(\mathbb{Z})$ is easily seen to be diag(1, 2, 4), we get

$$G \cong \mathbb{Z}/(1) \times \mathbb{Z}/(2) \times \mathbb{Z}/(4) = \mathbb{Z}_2 \times \mathbb{Z}_4.$$

9.12 Applications to Linear Algebra

Suppose $A \in \text{End}(V)$ for a non-trivial finite dimensional vector space V over a field F. The evaluation homomorphism $f \in F[x] \mapsto f(A) \in \text{End}(V)$ turns V into a module over the PID $F[x]$. For dimension reasons the evaluation homomorphism has non-trivial kernel. The unique monic polynomial f_A that generates this ideal is called the *minimal polynomial* of A. Any F-linear basis of V clearly generate V as an $F[x]$-module, rendering it finitely generated. It is also a torsion module because for any $v \in V$, all vectors of the form $A^i v$ will be linear dependent, so there is a polynomial g such that $gv = 0$. The following result is therefore immediate from Theorem 9.10.9.

Theorem 9.12.1 *Suppose A is a linear map on a finite dimensional vector space V over a field F. Consider V as an $F[x]$-module under the action given by $fv = f(A)v$ for $v \in V$. Then we have a decomposition*

$$V \cong V_1 \oplus \cdots \oplus V_m$$

into cyclic modules V_i. If f_i is the minimal polynomial of A restricted to V_i, then f_i divides f_{i+1}, and $f_m = f_A$. Moreover, the sequence $f_1, \ldots, f_m$ is uniquely determined by A.

We refer to $f_1, \ldots, f_m$ above as the *invariants* of A. We can also describe the restriction of A to each cyclic submodule V_i. To simplify notation say from the outset that V is cyclic with generator v and minimal polynomial

$$f_A(x) = a_0 + a_1 x + \cdots + a_{n-1} x^{n-1} + x^n.$$

Then $v, Av, \ldots, A^{n-1} v$ constitute an F-basis for V. If they were linear dependent, some polynomial of smaller degree than f_A would annihilate V. They span V since to any polynomial g we have $g = q f_A + r$ with $\deg(r) < \deg(f_A)$, so $g(A) = r(A)$.

It is easy to see that the associated matrix of the endomorphism A with respect to the basis $\{v, Av, \ldots, A^{n-1} v\}$ is

$$\begin{pmatrix} 0 & 0 & 0 & \cdots & 0 & -a_0 \\ 1 & 0 & 0 & \cdots & 0 & -a_1 \\ 0 & 1 & 0 & \cdots & 0 & -a_2 \\ \cdot & \cdot & \cdot & \cdot & \cdot & \cdot \\ 0 & 0 & 0 & \cdots & 0 & -a_{n-2} \\ 0 & 0 & 0 & \cdots & 1 & -a_{n-1} \end{pmatrix}.$$

So the theorem above tells us that the associated matrix of A can be decomposed into matrices of this form; called its *rational canonical form*. Thus every matrix over a field is similar to one in rational canonical form with the components in the decomposition corresponding to matrix blocks along the diagonal.

9.12 Applications to Linear Algebra

Example 9.12.2 The rational canonical form of a matrix $A \in M_6(\mathbb{Q})$ associated to an endomorphism with invariants $x - 3$, $(x - 3)(x - 1)$, $(x - 3)(x - 1)^2$ is

$$\begin{pmatrix} 3 & 0 & 0 & 0 & 0 & 0 \\ 0 & 0 & -3 & 0 & 0 & 0 \\ 0 & 1 & 4 & 0 & 0 & 0 \\ 0 & 0 & 0 & 0 & 0 & 3 \\ 0 & 0 & 0 & 1 & 0 & -7 \\ 0 & 0 & 0 & 0 & 1 & 5 \end{pmatrix}.$$

◇

Definition 9.12.3 The *characteristic polynomial* of a linear map on a finite dimensional vector space over a field F is the polynomial $f \in F[x]$ of an associated matrix A given by $f(x) = \det(xI - A)$.

The characteristic polynomial of an endomorphism does not depend on the chosen associated matrix $A \in M_n(F)$ since

$$\det(xI - BAB^{-1}) = \det(B(xI - A)B^{-1}) = \det(xI - A)$$

for any invertible $B \in M_n(F)$.

The following result is known as the *Cayley–Hamilton theorem*.

Proposition 9.12.4 *If f is the characteristic polynomial of a linear map A, then $f(A) = 0$.*

Proof Let $B \in M_n(F[x])$ be the matrix associated to $x\iota - A$, where A acts on a vector space over F of dimension n. Let C be the quadratic matrix over the commutative unital ring $F[x]$ given by Cramer's rule, so $CB = fI$, and B would be invertible if f was a unit. Replace x by A in $CB = fI$ to get $f(A) = C(A)0 = 0$. □

The Cayley–Hamilton theorem is often used together with the division algorithm to calculate high powers of matrices.

If f is the characteristic polynomial of a linear map on a vector space with associated $n \times n$-matrix A, then clearly $\det(A)$ equals $(-1)^n$ times the constant term $f(0)$ of f, whereas the coefficient of x^{n-1} in $f(x)$ is easily seen to be $-\text{Tr}(A)$.

There is an important connection between the invariants of a linear map A and its characteristic polynomial. We approach this in two different ways. The present one is the more direct one, and it also reproves the theorem above using the Smith normal form of $xI - A$.

Proposition 9.12.5 *The characteristic polynomial of a linear map on a finite dimensional vector space is the product of its invariants.*

Proof Suppose $A = (a_{ij})$ is an associated matrix of a linear map on a vector space V over a field F of finite dimension n with respect to a basis $\{v_i\}$. Let $h: F[x]^n \to V$ be the $F[x]$-module map that sends the element e_i of the standard basis to v_i.

We claim that $\ker h = (xI - A)F[x]^n$. Recalling the module action on V a simple computation shows that $(xI - A)F[x]^n \subset \ker h$. In particular, we have

$$p_j(x) \equiv (xI - A)e_j = xe_j - \sum_i a_{ij}e_i \in \ker h.$$

If $y = \sum_j g_j e_j \in \ker h$, then repeated substitution of xe_j from the formula above with x's coming from $g_j \in F[x]$ shows that

$$y = \sum p_j h_j + \sum b_j e_j$$

for some $h_j \in F[x]$ and $b_j \in F$. The first sum belongs to the ideal $\ker h$, so $0 = h(y) = 0 + \sum b_j v_j$, which shows that $b_j = 0$. Thus

$$y = (xI - A)\begin{pmatrix} h_1(x) \\ \cdot \\ \cdot \\ h_n(x) \end{pmatrix} \in (xI - A)F[x]^n$$

establishes our claim.

By Theorem 9.11.4 there are invertible $C, D \in M_n(F[x])$ such that

$$C(xI - A)D = \mathrm{diag}(f_1(x), \ldots, f_n(x))$$

for unique monics $f_i \in F[x]$ such that f_i divides f_{i+1} for i less than some k.

Remembering that V is a torsion module, we get by the first isomorphism theorem

$$V \cong F[x]^n / \ker h \cong F[x]^n / \mathrm{diag}(f_1, \ldots, f_n) F[x]^n \cong F[x]/(f_1) \oplus \cdots \oplus F[x]/(f_k),$$

where we have also discarded possible constant polynomials among the first in the original sequence of f_i's. Thus $f_1, \ldots, f_k$ are the invariant factors of A.

Taking the determinant of $C(xI - A)D = \mathrm{diag}(f_1(x), \ldots, f_n(x))$, and using its multiplicative property of matrices over commutative rings, and the fact that the determinant of an invertible matrix over a commutative ring is a unit, we get $\det(xI - A) = f_1(x) \cdots f_k(x)$ by comparing coefficients of the highest power of x. $\square$

The proof of the theorem also tells us that the invariants of a linear map A are the non-unit invariant factors, that is, the non-unit diagonal elements of the Smith normal form of the associated matrix to $x\iota - A$.

In view of the proposition above the Cayley–Hamilton theorem follows immediately from the fact that the minimal polynomial of an endomorphism on a finite dimensional vector space annihilates the endomorphism.

Example 9.12.6 The rational canonical form of

$$A = \begin{pmatrix} -3 & 2 & 0 \\ 1 & 0 & 1 \\ 1 & -3 & -2 \end{pmatrix} \in M_3(\mathbb{Q}) \quad \text{is} \quad \begin{pmatrix} 0 & 0 & -3 \\ 1 & 0 & -7 \\ 0 & 1 & -5 \end{pmatrix}$$

because the Smith normal form of $xI - A \in M_3(\mathbb{Q}[x])$ is $\operatorname{diag}(1, 1, 3 + 7x + 5x^2 + x^3)$, so A has invariant $3 + 7x + 5x^2 + x^3 = \det(xI - A)$. ◇

9.13 Generalized Jordan Blocks

Let A be a linear map on a non-trivial finite dimensional vector space V over a field F. By Theorem 9.12.1 we can decompose the finitely generated torsion $F[x]$-module V as a direct sum of cyclic modules $V_i = F[x]/(f_i)$ with invariants $f_1, \ldots, f_m$. Each f_i can be written as a product of powers p^k of primes, and by the Chinese remainder theorem for PID's, the module V_i will then be a direct sum of the corresponding cyclic modules $F[x]/(p^k)$. The powers p^k occurring in each f_i are called the *elementary divisors* of A. Since every f_i divides the minimal polynomial $f_A = f_m$ of A, the powers p^k occurring in f_A are the highest possible. The direct sum of the cyclic modules corresponding to the elementary divisors of A that are powers of a fixed p is the module $V(p)$. The (in general non-cyclic) module $V(p)$ consists of the elements of V annihilated by the elementary divisor of type p^k occurring in the minimal polynomial.

Proposition 9.13.1 *Let F be a field and let $f(x) = (x - \lambda)^k \in F[x]$ with $k \in \mathbb{N}$. Then the cyclic $F[x]$-module $F[x]/(f)$ has an F-linear basis*

$$\{v, (x - \lambda)v, \ldots, (x - \lambda)^{k-1}v\}$$

with cyclic vector $v \equiv 1 + (f)$. The matrix of the module action of x on $F[x]/(f)$ with respect to this basis in order listet from the left is

$$\begin{pmatrix} \lambda & 0 & \cdots & \cdot & 0 \\ 1 & \lambda & \cdots & \cdot & 0 \\ \cdot & \cdot & \cdots & \cdot & \cdot \\ 0 & \cdot & \cdots & \cdot & 0 \\ 0 & 0 & \cdots & 1 & \lambda \end{pmatrix} \in M_k(F).$$

Proof The vectors in question must be linear independent, otherwise a polynomial of smaller degree than f would annihilate v, and f cannot generate such a polynomial. Using the division algorithm with steps determined by f, we see that there cannot be more than k linear independent vectors.

It is straightforward to check that the matrix above is the one associated to the action of x by componentwise multiplication with respect to the basis above. □

A matrix of the type in the proposition above is called a *Jordan block*.

Corollary 9.13.2 *Suppose A is a linear map on a finite dimensional vector space V over an algebraically closed field. Then V has a linear basis such that the associated matrix of A is a direct sum of Jordan blocks. Such a decomposition is unique up to reordering of Jordan blocks.*

Proof This is immediate from the proposition above and the discussion prior to it since prime elements are irreducible and thus of order one over an algebraically closed field. Each elementary divisor p^k of A will be of the type described in the proposition above, and A will act on the corresponding submodule of V just like x acts on $F[x]/(p^k)$, producing a Jordan block of size k.

The uniqueness statement is immediate from the uniqueness of the invariant factors of A. □

Another way of phrasing the corollary above, is to say that any quadratic matrix over an algebraically closed field is similar to a matrix with Jordan blocks along its diagonal, and otherwise has only zero entries. Such a matrix is said to be in *Jordan canonical form*. Strictly speaking we do not need an algebraically closed field, all we need is a field that contains the roots of the irreducible factors in the minimal polynomial.

Note also that a quadratic matrix over a field is diagonalizable if and only if its minimal polynomial has only distinct roots in the field, so that the elementary divisors all have degree one. By Proposition 9.12.5 the roots of the minimal polynomial are exactly the roots of the characteristic polynomial, but their multiplicity in the latter polynomial is in general larger. If there are as many distinct eigenvalues as the size of the matrix, then it is diagonalizable, and the minimal polynomial coincides with the characteristic polynomial.

Also, since the characteristic polynomial f of A above is the product of all the invariants of A, the dimension of the vector space $V(p)$, called the *generalized geometric multiplicity* of an eigenvalue λ of A that is a root of p, coincides with the multiplicity of λ in f, called the *algebraic multiplicity* of λ, if and only if $\deg(p) = 1$. This happens presicely when F contains all the roots of p, which is automatic when F is algebraically closed. Otherwise the geometric multiplicity of an eigenvalue that is a root of p is greater than its algebraic multiplicity.

Here is an alternative argument for why the characteristic polynomial of a linear map on a finite dimensional vector space V over a field F is the product of the invariants of the linear map. First notice that neither the invariants nor the characteristic polynomial alter by considering an extension field E of F with corresponding extension of the linear map to the vector space $E \otimes_F V$. The associated matrix of the extension will simply be the old matrix over F considered with entries in E. Therefore we can assume that F is algebraically closed. Next note that the determinant of a finite direct sum of endomorphisms is the product of the determinants of each endomorphism. The same will then be true for the characteristic polynomial. By the corollary above we therefore only need to prove the assertion for Jordan blocks, but clearly the characteristic polynomial of the Jordan block in the proposition above coincides with the polynomial in the proposition.

9.13 Generalized Jordan Blocks

Example 9.13.3 In Example 9.12.2 we had five elementary divisors

$$x - 3, x - 3, x - 3, x - 1, (x - 1)^2,$$

so the Jordan canonical form of the matrix is almost diagonal, except for a Jordan block of size 2 corresponding to the double root 1 of the minimal polynomial. Putting this block in the lower corner, we conclude that the original matrix is similar to

$$\begin{pmatrix} 3 & 0 & 0 & 0 & 0 & 0 \\ 0 & 3 & 0 & 0 & 0 & 0 \\ 0 & 0 & 3 & 0 & 0 & 0 \\ 0 & 0 & 0 & 1 & 0 & 0 \\ 0 & 0 & 0 & 0 & 1 & 0 \\ 0 & 0 & 0 & 0 & 1 & 1 \end{pmatrix}.$$

◇

Example 9.13.4 In Example 9.12.6 the elementary divisors of the linear map of the matrix is $x + 3, (x + 1)^2$ since the minimal polynomial is the only invariant, and its factorization into powers of primes is $3 + 7x + 5x^2 + x^3 = (x + 3)(x + 1)^2$. Thus its Jordan canonical forms are

$$\begin{pmatrix} -3 & 0 & 0 \\ 0 & -1 & 0 \\ 0 & 1 & -1 \end{pmatrix} \text{ and } \begin{pmatrix} -1 & 0 & 0 \\ 1 & -1 & 0 \\ 0 & 0 & -3 \end{pmatrix}$$

corresponding to the two possible positions of the Jordan blocks. ◇

For linear maps on finite dimensional vector spaces over general fields we need to generalize the proposition above to account for elementary divisors p^k with an irreducible monic p of possibly higher degree.

Proposition 9.13.5 *Let*

$$f(x) = a_0 + a_1 x + \cdots + a_{n-1} x^{n-1} + x^n$$

be a polynomial over a field F. Then the cyclic $F[x]$-module $F[x]/(f^k)$ for $k \in \mathbb{N}$ has an F-linear basis

$$\{v, xv, \ldots, x^{n-1}v\} \cup \{fv, fxv, \ldots, fx^{n-1}v\} \cup \cdots \cup \{f^{k-1}v, f^{k-1}xv, \ldots, f^{k-1}x^{n-1}v\}$$

with cyclic vector $v \equiv 1 + (f^k)$. The matrix of the module action of x on $F[x]/(f^k)$ with respect to this basis in order listet from the left is

$$J \equiv \begin{pmatrix} B & 0 & \cdots & \cdot & 0 \\ C & B & \cdots & \cdot & 0 \\ \cdot & \cdot & \cdots & \cdot & \cdot \\ 0 & \cdot & \cdots & \cdot & 0 \\ 0 & 0 & \cdots & C & B \end{pmatrix} \in M_{kn}(F),$$

where $C \in M_n(F)$ has 1 in the upper right corner and otherwise zeroes, and B is the familiar matrix

$$\begin{pmatrix} 0 & 0 & 0 & \cdots & 0 & -a_0 \\ 1 & 0 & 0 & \cdots & 0 & -a_1 \\ 0 & 1 & 0 & \cdots & 0 & -a_2 \\ \cdot & \cdot & \cdot & \cdot & \cdot & \cdot \\ 0 & 0 & 0 & \cdots & 0 & -a_{n-2} \\ 0 & 0 & 0 & \cdots & 1 & -a_{n-1} \end{pmatrix} \in M_n(F).$$

Proof The proof is like that of the previous proposition, with a slightly more tedious but straightforward identification of the matrix associated to the action of x on the module with respect to the listed basis. □

The matrix J in the proposition above is called a *generalized Jordan block*. We notice that such a block reduces to an ordinary Jordan block when f has degree one, so that $B = -a_0$ and $C = 1$.

As before we get the immediate corollary.

Corollary 9.13.6 *Suppose A is a linear map on a finite dimensional vector space V over any field. Then V has a linear basis such that the associated matrix of A is a direct sum of generalized Jordan blocks. Such a decomposition is unique up to reordering of generalized Jordan blocks.*

When working over real numbers we will occasionally get generalized Jordan blocks with $B, C \in M_2(\mathbb{R})$, but nothing larger than that since irreducible polynomial over $\mathbb{R}$ are maximally of second degree. An easy way to see this is to decompose an irreducible polynomial over $\mathbb{R}$ in $\mathbb{C}[x]$, and then notice that roots appear in conjugate pairs, so the product of the two corresponding linear factors is a second order real polynomial.

Write the minimal polynomial of an endomorphism A on a finite dimensional vector space V over a field F as

$$f_A = \prod_{i=1}^{r} p_i^{m_i}$$

for distinct prime factors p_i. Then as we know $V \cong \bigoplus_{i=1}^{r} V_i$, where

$$V_i \equiv \{v \in V \mid p_i^{m_i} v = 0\}$$

is a sum of cyclic modules each having a generalized Jordan block. Since the different $p_i^{n_i}$ are relatively prime we can find polynomials g_i over F such that

$$\sum_{i=1}^{r} g_i p_i^{m_i} = 1.$$

Set $P_i = \sum_{j \neq i} g_j(A) p_j^{m_j}(A)$. Then $P_i(v) = \delta_{ij} v$ for $v \in V_j$, so P_i is a projection onto V_i that is a polynomial in A. Note that although the projections P_i commute, they are not *orthogonal*, meaning $P_i P_j = \delta_{ij}$, when $r \geq 3$ because $\sum_i P_i = (r-1)\iota$.

9.14 The Jordan–Chevalley Decomposition

The deviation of a quadratic matrix over an algebraically closed field from being diagonalizable is measured by a *nilpotent matrix*, that is, a quadratic matrix with some vanishing positive integer power. This is so because we can write an $n \times n$-matrix in Jordan canonial form as $S + N$, where S is a diagonal matrix consisting of possibly repeated entries of the eigenvalues of the matrix, and N has non-zero entries only on the shorter diagonal just below the main diagonal, so it is nilpotent with $N^n = 0$. Obviously $SN = NS$. Any quadratic matrix over the field is similar to such a sum $S + N$, and the conjugation of S and N by the same invertible matrix clearly produces two matrices that are diagonalizable and nilpotent, respectively, and that commute with each other. We have proved the first existence part of the theorem below. Before we state it we include some preliminary results that are interesting in their own right.

Definition 9.14.1 An endomorphism on a finite dimensional vector space is *diagonalizable* if it has a basis of eigenvectors.

Note that the restriction of a diagonalizable linear map A on a finite dimensional vector space V to an invariant subspace is also diagonalizable. This is true because the minimal polynomial of the restricted endomorphism is obviously a factor of the minimal polynomial of A, which has only simple roots all belonging to the field. Note also that if $V = U \oplus W$ for A-invariant subspaces U and W, then the minimal polynomial of A is the least common multiplier of the minimal polynomials of A restricted to U and W. Only as a common multiplier can it annihilated both U and W, and it obviously has to be the least possible such.

Definition 9.14.2 A set of endomorphisms on a finite dimensional vector space are *simultaneously diagonalizable* if we can find a common basis of eigenvectors for them, in other words, if their associated matrices are all diagonal with respect to this basis.

The infinite dimensional version of the following result is of interest to quantum physicists who are keen to measure observables accurately simultaneously.

Proposition 9.14.3 *Two diagonalizable linear maps on a finite dimensional vector space are simultaneous diagonalizable if and only if they commute.*

Proof The forward implication is trivial.

If A and B are commuting linear maps on a finite dimensional space V, then any eigenvector subspace of A is B-invariant because if $Av = av$, then $ABv = BAv = aBv$. The restriction of a diagonalizable B to an eigenvector subspaces of A is as we have seen diagonalizable. Collecting the corresponding basis elements for B for each single eigenvector subspace of a diagonalizable A, we get the desired basis as the union of these. $\square$

Corollary 9.14.4 *A commutative subalgebra of* $\text{End}(V)$ *for a finite dimensional vector space V consists solely of diagonalizable linear maps if it is generated by diagonalizable ones.*

Proof Observe that the sum and product of two diagonalizable endomorphisms of the algebra are diagonalizable because they commute and can therefore be simultaneously diagonalized. Extending this result by induction to finite sums of finite products, we are done. $\square$

Definition 9.14.5 A linear map A on a vector space is *nilpotent* if $A^n = 0$ for some $n \in \mathbb{N}$.

Proposition 9.14.6 *A commutative subalgebra of* $\text{End}(V)$ *for a finite dimensional vector space V that is generated by nilpotent linear maps consists only of nilpotent linear maps.*

Proof If $A, B \in \text{End}(V)$ commute and satisfy $A^n = 0$ and $B^m = 0$, then $(AB)^{n+m} = A^n B^m = 0$ and $(A + B)^{n+m} = 0$ by the binomial formula. $\square$

Theorem 9.14.7 *If A is a linear map on a finite dimensional vector space over an algebraically closed field, then $A = A_s + A_n$ for unique diagonalizable and nilpotent endomorphisms A_s and A_n, respectively, such that $A_s A_n = A_n A_s$. In fact, there is a polynomial f with vanishing constant term over the field such that $A_s = f(A)$, so A_s and A_n commute with all linear maps that commute with A.*

Proof We have already proved existence of the decomposition, but for the uniqueness part we need the stronger existence result involving f.

Say the minimal polynomial of A is

$$\prod_i (x - \lambda_i)^{n_i}$$

for distinct eigenvalues λ_i. By the Chinese remainder theorem for PID's we can find a polynomial f over the field with zero constant term and such that

$$f(x) - \lambda_i \in ((x - \lambda_i)^{n_i}).$$

9.14 The Jordan–Chevalley Decomposition

Then $f(A)$ restricted to the subspace V_i of V annihilated by $(x - \lambda_i)^{n_i}$ acts as multiplication by λ_i. So V_i consists solely of eigenvectors for $f(A)$, and we know that $V = \oplus V_i$. Thus $A_s \equiv f(A)$ is semisimple.

But we can pick a basis for each V_i so that the associated matrix of A has Jordan blocks with λ_i along the diagonals. Thus the associated matrix of $A_n \equiv A - f(A)$ has non-zero entries only below the diagonal and is therefore nilpotent. Both A_n and A_s are polynomials in A, so they commute with each other and also with all endomorphisms that commute with A.

As for uniqueness, if also $A = S + N$ for semisimple and nilpotent linear maps S and N, respectively, that commute, then as both S and N commute with A, they also commute with A_s and A_n. By the last corollary and last proposition above, we know that $A_s - S$ is semisimple and $A_n - N$ is nilpotent. But $A_s - S = N - A_n$, and any eigenvalue of a nilpotent linear map is clearly zero, as otherwise no power of it evaluated at the corresponding eigenvector can vanish, so $A_s - S = 0$. Thus $A_s = S$ and $A_n = N$. □

One referes to the decomposition in the theorem above as the *Jordan–Chevalley decomposition* of A. We can strengthen this theorem slightly.

Definition 9.14.8 A linear map on a finite dimensional vector space is *semisimple* if every invariant subspace of it has an invariant complementary subspace. It is *simple* if the vector space has no invariant subspace apart from itself and the trivial subspace.

Proposition 9.14.9 *A linear map on a finite dimensional vector space is semisimple if and only if any prime factor in its minimal polynomial occur at most once.*

Proof By Theorem 9.15.5 a linear map A on a finite dimensional vector space V is semisimple if and only if V is a direct sum of invariant subspaces with simple restrictions of A.

Thus if A is semisimple, then it restricts to a simple endomorphism B on some subspace. Consider the generalized Jordan decomposition of the associated matrix to B. This matrix cannot have more than one generalized Jordan block, and moreover, considering such a generalized Jordan block of the type displayed in Proposition 9.13.5, we see that $k = 1$ is the only possibility. Otherwise the subspace consisting of all column vectors having zeroes everywhere except in the last n coordinates will be an invariant subspace, and this contradicts simplicity. By uniqueness up to reordering of the generalized Jordan decomposition of the associated matrix to A, we therefore conclude that there will be no such blocks with $k > 1$ in the decomposition. In other words, no prime factor in the minimal polynomial of A occur more than once.

Conversely, if no prime factor in the minimal polynomial of A occur more than once, the module V over the polynomials over the field F decomposes into a direct sum of submodules of the type $F[x]/(p)$, where p is prime. But such modules are simple by Proposition 9.15.2, so the restrictions of A to them are simple, and V is semisimple by the result recalled at the beginning of the proof. □

Corollary 9.14.10 *Suppose A is a linear map on a finite dimensional vector space V over a field F. Let E be an extension field of F, and let $\tilde{A}$ denote the extension*

$\iota \otimes A \in \operatorname{End}_E(E \otimes_F V)$ *of A to* $E \otimes_F V$. *Then A is semisimple whenever* $\tilde{A}$ *is semisimple. The converse holds if F is a perfect field. If A is diagonalizable, it is semisimple. The converse is true if F contains all the roots of the characteristic polynomial of A.*

Proof If A is diagonalizable, its minimal polynomial has only distinct roots, so no prime factor can occur more than once. By the proposition above A is semisimple. If F contains all the roots of the characteristic polynomial of a semisimple A, the minimal polynomial has all its roots in F, and they are distinct by the proposition above, so A is diagonalizable.

If $\tilde{A}$ is semisimple, then by the proposition above no prime factor polynomial over E in the minimal polynomial of $\tilde{A}$ occur more than once. Since A has the same minimal polynomial as $\tilde{A}$ both considered as polynomials over E, and as E is an extension field of F, these prime factors will be factors of the prime factor polynomials over F in the minimal polynomial of A. So the prime factors over F in the minimal polynomial of A cannot occur more than once. Thus A is semisimple by the proposition above.

Suppose F is a perfect field. By the proposition above, if A is semisimple, no prime factor in its characteristic polynomial can occur more than once. Prime elements are irreducible polynomials, and as F is perfect, each of these polynomials have distinct roots. As $\tilde{A}$ has the same minimal polynomial as A, no prime factor polynomial over E can thus occur more than once. □

In particular, we see that for an algebraically closed field, semisimplicity and diagonalizability are equivalent notions. Recall that perfect fields include finite fields and fields of characteristic zero, and of course, any algebraically closed field is perfect.

Proposition 9.14.11 *The theorem above holds for linear maps on finite dimensional vector spaces over perfect fields provided one replaces the word 'diagonalizable' by 'semisimple'.*

Proof Suppose we have a linear map A on a finite dimensional vector space V over a perfect field. Since F is perfect, it has a Galois extension E that contains all the roots of the characteristic polynomial of A. We use the same symbol for an endomorphism and an associated matrix with respect to a fixed basis $\{x_i\}$ of V, and also with respect to the basis $\{1 \otimes x_i\}$ of $E \otimes_F V$. The associated matrix of the extension $\tilde{A}$ of $A \in \operatorname{End}_F(V)$ to $E \otimes_F V$ is thus denoted A.

The theorem above obviously works with E instead of an algebraically closed field, so we have a unique Jordan–Chevalley decomposition $A = A_s + A_n$ of the matrix A into commuting diagonalizable and nilpotent matrices with entries possibly in E. By the corollary above we also know that the endomorphism of A_s is semisimple.

Now $\alpha \in G(E/F)$ applied to a matrix with entries in E means the matrix with α applied to each entry. Then α will obviously be an F-linear ring homomorphism on

9.14 The Jordan–Chevalley Decomposition

$M_n(E)$, and by the fundamental theorem of Galois theory, any matrix fixed under all such α's will belong to $M_n(F)$.

But $A = \alpha(A) = \alpha(A_s) + \alpha(A_n)$, and obviously $\alpha(A_s)$ is diagonalizable, whereas $\alpha(A_n)$ is nilpotent. By uniqueness of the decomposition as matrices over E, we conclude that $\alpha(A_s) = A_s$ and $\alpha(A_n) = A_n$. Since this holds for all $\alpha \in G(E/F)$, we deduce that A_s and A_n have entries in F, so both are associated matrices of extensions of endomorphism on V to $E \otimes_F V$. Thus by the corollary above the linear map A on V is a sum of a semisimple endomorphism A_s and a nilpotent endomorphism A_n that commute.

If $A = B_s + B_n$ is another such decomposition, then by the corollary above the extensions of B_s and B_n to $E \otimes_F V$ will be diagonalizable and nilpotent, respectively, and they will commute. By uniqueness of the Jordan–Chevalley decomposition on the level of E, we conclude that $B_s = A_s$ and $B_n = A_n$.

As for the existence of the polynomial, by the theorem above there is $f \in E[x]$ such that $f(A) = A_s$. By the division algorithm applied to f, we may assume that

$$A_s = a_0 + a_1 A + \cdots + a_m A^m,$$

where $a_i \in E$ and m is not greater than the degree of the minimal polynomial of $\tilde{A}$ over E. Then

$$A_s = \alpha(A_s) = \alpha(a_0) + \alpha(a_1)A + \cdots + \alpha(a_m)A^m$$

and since $\{A^i\}_{i=0}^m$ are linear independent over E, we conclude that $\alpha(a_i) = a_i$, so $a_i \in F$ and $f \in F[x]$. □

We have also the *multiplicative Jordan–Chevalley decomposition*.

Proposition 9.14.12 *Let A be an invertible linear map on a finite dimensional vector space over a perfect field. Then $A = A_s A_u$ for unique commuting linear maps with A_s semisimple and A_u unipotent, meaning that $A_u - \iota$ is nilpotent. Both are polynomials in A and they are invertible.*

Proof In the additive Jordan–Chevalley decomposition $A = A_s + A_n$, the endomorphism A_s is invertible because the associated matrix of an appropriate extension of it has the roots of the characteristic polynomial of A repeated along the diagonal, and these are non-zero as A is invertible. Clearly $A_u \equiv \iota + A_s^{-1} A_n$ commutes with A_s, is unipotent, and $A = A_s A_u$, so A_u is also invertible.

Uniqueness of the decomposition is more tricky. One first needs a polynomial for an appropriate extension of A_u in an extension of A. Proceed as in the proof of the theorem above, invoking the Chinese remainder theorem for PID's, and observe that the Jordan blocks of the extension of A_u are obtained from those of the extension of A by dividing all entries with the diagonal ones. This gives the required polynomial for the extension of A_u. Thus A_u will commute with all linear maps that commute with A.

Now suppose $A = B_s B_u$ for commuting endomorphisms with B_s semisimple and B_u unipotent. Both commute with A, and thus commute with A_u and A_s. Hence some integer power of

$$B_s - A_s = A_s(A_u - \iota) - B_s(B_u - \iota)$$

vanish as $A_u - \iota$ and $B_u - \iota$ are nilpotent. As appropriate extensions of A_s and B_s are simultaneously diagonalizable, we deduce that $A_s = B_s$, so $B_u = A_u$.

Combining this with the argument in the proof of the proposition above, involving the Galois group of the extension field, we conclude that the polynomial of the extension of A_u in the extension of A has coefficients in the non-extended field. So A_u is a polynomial in A over the original field. □

We have the following consequence of the second last proposition above.

Corollary 9.14.13 *Let A be a linear map on a finite dimensional vector space V over a perfect field. Define $\mathrm{ad}(B)\colon \mathrm{End}(V) \to \mathrm{End}(V)$ for $B \in \mathrm{End}(V)$ by $\mathrm{ad}(B)C = BC - CB$. Then $\mathrm{ad}(A_s)$ and $\mathrm{ad}(A_n)$ are the commuting semisimple and nilpotent parts, respectively, of the Jordan–Chevalley decomposition of $\mathrm{ad}(A)$.*

Proof There is a basis of eigenvectors x_i of an appropriate extension of A_s with eigenvalues λ_i. The corresponding matrix units E_{ij} is a basis of eigenvectors of the extension of $\mathrm{ad}(A_s)$ with eigenvalues $\lambda_i - \lambda_j$, so $\mathrm{ad}(A_s)$ is semisimple.

If $A_n^m = 0$, then simple arithmetic shows that $\mathrm{ad}(A_n)^{2m} = 0$, so $\mathrm{ad}(A_n)$ is nilpotent. But $\mathrm{ad}(A) = \mathrm{ad}(A_s) + \mathrm{ad}(A_n)$ is then the unique Jordan–Chevalley decomposition of $\mathrm{ad}(A)$ as the two terms obviously commute. □

In particular, the map ad preserves semisimplicity and nilpotency of linear maps on finite dimensional vector fields over perfect fields.

Beyond perfect fields there are counterexamples to the Jordan–Chevalley decomposition.

9.15 Semisimple Modules

Definition 9.15.1 A *simple module* A over a ring R is a module with no submodules other than itself and the trivial one, and such that $RA \neq \{0\}$.

The condition $RA \neq \{0\}$ holds for any non-trivial module A over a unital ring R. For a simple module A over any ring R, we have $RA = A$ as RA is a non-trivial submodule of A.

Irreducible representations of a finite group on a finite dimensional complex vector space are exactly the simple modules of the group ring.

Proposition 9.15.2 *For a module A over a unital ring R the following conditions are equivalent:*

9.15 Semisimple Modules

(i) *The module A is simple;*

(ii) *The module A is non-trivial and is generated by any non-zero element;*

(iii) *As modules $A \cong R/I$, where I is a maximal left ideal of R.*

Proof If A is a simple module, then $Rx = A$ for any non-zero $x \in A$ as Rx is a non-trivial submodule of A. So (i) implies (ii). Conversely, if (ii) holds, any non-trivial submodule B of A contains a non-zero element that necessarily generates A, so $A = B$.

If (ii) holds, then A is cyclic, so $A \cong R/I$ for some left ideal I of R, which by the correspondence theorem must be maximal as A is simple. So (iii) holds. That (iii) implies (i) is immediate from the correspondence theorem. □

Minimal left ideals of a unital ring are obviously simple. This is no longer true in the non-unital case which the additive group $\mathbb{Z}_p$ for p prime with trivial ring multiplication shows.

Example 9.15.3 The simple modules of a PID R are all of the form $R/(p)$ for p prime because the maximal ideals in R are of the form (p). ◇

Definition 9.15.4 A module is *semisimple* if it is a direct sum of simple modules.

A semisimple R-module A clearly satisfies $RA = A$.

Theorem 9.15.5 *A module over a unital ring is semisimple if and only if every submodule has a complementary submodule.*

Proof Suppose $A = \bigoplus_{x \in X} A_x$ with simple submodules A_x, and let B be a submodule of A. Let Y by Zorn's lemma be a maximal subset of X such that $B + \bigoplus_{y \in Y} A_y$ is a direct sum. Then A_x for any $x \in X$ will belong to this direct sum, because either $A_x \cap A_y = \{0\}$ for all $y \in Y$, and then the direct sum is not maximal, or $A_x \cap A_y \neq \{0\}$ for some $y \in Y$, and then $A_x = A_y$ since both modules are simple. Thus $\bigoplus_{y \in Y} A_y$ is a complementary submodule to B.

Conversely, assume that A is a non-trivial module such that every submodule has a complementary submodule. Then any non-trivial B submodule of A must have at least one simple submodule. To see this, pick a non-zero element $x \in B$, and consider its cyclic module R/I with left ideal I. Let J be a maximal left ideal of R that contains I. By the correspondence theorem $Jx \cong J/I$ is a maximal submodule of $Rx \cong R/I$. Let C be a complementary submodule to Jx in A. Then obviously

$$Rx = Jx \oplus (C \cap Rx)$$

and $C \cap Rx \cong Rx/Jx$ is a simple submodule of B.

If the direct sum of all the simple submodules of A is not all of A then a complementary submodule will by the previous paragraph contain a simple submodule, which is a contradiction. □

It is clear from the proof of the theorem above that a module is semisimple if and only if it is a sum of simple modules; the point beeing that the sum need not be direct.

Corollary 9.15.6 *Submodules and quotient modules of semisimple modules are semisimple.*

Proof Suppose B is a submodule of a semisimple module A. Let C be the sum of of all simple submodules of B, and let D be a submodule of A such that $A = C \oplus D$. Clearly $B = C \oplus (B \cap D)$, and $B \cap D$ must be trivial since according to the proof in the theorem above, it would otherwise have a simple submodule of B, which is impossible. Thus $B = C$ is semisimple.

Note that A/B can be identified with a submodule of A and is therefore semisimple. □

We could also have formulated this in terms of an exact sequence of modules.

For representations of finite groups on finite dimensional complex vector spaces we obtained a decomposition into irreducible subrepresentations with complementary invariant subspaces gotten from orthogonality with respect to an averaging of the inner product on $\mathbb{C}^n$.

Schur's lemma is valid also here.

Lemma 9.15.7 *Any non-zero homomorphism between simple modules is an isomorphism. If A is a simple R-module then $\mathrm{End}_R(A)$ is a division ring.*

Proof The kernel and image of a homomorphism between modules are submodules, so for a non-trivial homomorphism between simple modules, its kernel must be trivial and its image must be the whole module.

Any non-zero module map $A \to A$ must therefore have an inverse map. □

Any non-trivial module over a division ring has a basis, and two bases of a module have the same cardinality, which we call the *dimension of the module*. The reason why this is true is that the theorems stating this for vector spaces over fields, do not use commutativity of the field crucially in their proofs.

It is opportune to recall the following result.

Proposition 9.15.8 *Consider a direct sum $A = \bigoplus_{i=1}^n A_i$ of modules A_i. Let R_n be the unital ring of $n \times n$-matrices (f_{ij}) with entries $f_{ij} \in \mathrm{Hom}(A_j, A_i)$ under usual matrix addition and multiplication. Then the map $\mathrm{End}(A) \to R_n$ that sends f to (f_{ij}) with $f_{ij} = \pi_i \circ f \circ \iota_j$ is a ring isomorphism.*

Proof The map in question is a unital ring homomorphism because

$$\pi_i \circ (fg) \circ \iota_j = \sum_k (\pi_i \circ f \circ \iota_k)(\pi_k \circ g \circ \iota_j)$$

as $\sum \iota_k \pi_k = \iota$.

If $\pi_i \circ f \circ \iota_j = 0$ for $f \in \mathrm{End}(A)$ and all i, j, then $f = \sum_{ij} \pi_i \circ f \circ \iota_j = 0$, so the map is injective. To see that it is surjective, given $(f_{ij}) \in R_n$, define $f \in \mathrm{End}(A)$ by $f = \sum_{ij} \iota_i \circ f_{ij} \circ \pi_j$. Then $\pi_k \circ f \circ \iota_l = f_{kl}$ as $\pi_i \iota_j = \delta_{ij} \iota_{A_i}$. □

Definition 9.15.9 The number of times a simple module of a given isomorphism class occurs in the decomposition of a module A is called the *multiplicity of the simple module in A*. By nB we mean the direct sum $\bigoplus_{i=1}^{n} B$ of the module B.

The following result shows that the definition of multiplicity makes sense.

Theorem 9.15.10 *Say we have a finite decomposition $\bigoplus n_i A_i$ of a module A into pairwise non-isomorphic simple modules A_i. Then the A_i's are uniquely determined up to permutations and isomorphisms by the isomorphism class of A, with multiplicities n_i also uniquely determined. Moreover, the ring $\mathrm{End}(A)$ is isomorphic to the ring of matrices with blocks in $M_{n_i}(\mathrm{End}(A_i))$ along the diagonal and otherwise with zero entries.*

Proof The last statement is immediate from the proposition above combined with Schur's lemma.

Uniqueness of the A_i's up to permutations and isomorphisms is also immediate from Schur's lemma. The same lemma shows that we are done if we can show that $n = m$ whenever $nA \cong mA$ for a module A.

From the last statement of the theorem we know that $nA \cong mA$ implies $M_n(\mathrm{End}(A)) \cong M_m(\mathrm{End}(A))$ with dimensions $n^2 = m^2$ as modules over the division ring $\mathrm{End}(A)$; the division property is assured by Schur's lemma. $\square$

9.16 Density

Let A be a module over a ring R. Then A is also a module over the ring $R' \equiv \mathrm{End}_R(A)$ under the operation $fx = f(x)$. Let $R'' \equiv \mathrm{End}_{R'}(A)$. To $a \in R$ define a map $f_a \colon A \to A$ by $f_a(x) = ax$. Then $gf_a = f_a g$ for any $g \in R'$ because $gf_a(x) = g(ax) = ag(x) = f_a g(x)$. For this reason we call R' the *commutant* of R. As $f_a g = gf_a$ for $g \in R'$, by definition $f_a \in (R')'$, and $R'' \equiv (R')'$ is called the *bicommutant* of R. In fact, we see that $a \mapsto f_a$ is a ring homomorphism $R \to R''$. When this homomorphism is injective, we say that A is a *faithful module*. This means that $ax = 0$ for all $x \in A$ is possible only when $a = 0$. In this case the *natural map* $a \mapsto f_a$ embeds the ring R into the bicommutant R''. Remember that the commutants refer to a module although our notation suppresses this. To stress the module we sometimes talk about commutants in the module.

The following result, known as *Jacobson's density theorem*, says something about the denseness of the image of the natural map in the bicommutant.

Theorem 9.16.1 *If A is a semisimple module over a unital ring R with $x_1, \ldots, x_n \in A$ and $f \in R''$, then there is $a \in R$ such that $ax_i = f(x_i)$ for all i. Thus if A is finitely generated over R', the natural map $R \to R''$ is surjective.*

Proof Since A is semisimple, the submodule Rx_1 has a complementary submodule. Let $\pi \colon A \to A$ be the projection onto $Rx_1 \subset A$ with respect to this decomposition.

As $\pi \in R'$, we get $f(x_1) = f\pi(x_1) = \pi f(x_1)$, so $f(x_1) \in Rx_1$ and there is $b \in R$ such that $f(x_1) = bx_1$.

But we need an element in R that words for all x_i's simultaneously. To this end we employ a diagonal trick by von Neumann.

Assume first that A is simple, and consider the map $f^n \colon nA \to nA$ given by $f^n(y_1, \ldots, y_n) = (f(y_1), \ldots, f(y_n))$. Since $f \in R''$, we see that f^n is an $\text{End}_R(nA)$-module map as $\text{End}_R(nA) \cong M_n(R')$ by Theorem 9.15.10. Therefore there is $a \in R$ such that

$$(ax_1, \ldots, ax_n) = a(x_1, \ldots, x_n) = f^n(x_1, \ldots, x_n) = (f(x_1), \ldots, f(x_n))$$

by the first paragraph.

For semisimple A the proof is similar. Since each x_i lives in finitely many simple components in the decomposition of A, and since f commutes with the projection onto the direct sum of those simple modules where all the x_i's are non-zero, it suffices to repeat the argument above with A a finite direct sum of the type considered in Theorem 9.15.10. □

We get *Burnside's theorem* as a corollary.

Corollary 9.16.2 *Suppose V is a finite dimensional vector space over an algebraically closed field F, and that R is a unital subalgebra of $\text{End}(V)$. If V considered as an R-module is simple, then $R \cong \text{End}(V)$ as algebras.*

Proof By Schur's lemma we know that $R' = \text{End}_R(V)$ is a division ring. The field F obviously sits as a subring in the center of R'. We claim it is everything. Let $a \in R'$. Then the subring $F(a)$ of R' generated by a and F is a field, and since R' is finite dimensional as a vector space over F, we know that $F(a)$ is a finite extension of F, so it is an algebraic extension. Since F is algebraically closed, we conclude that $F(a) = F$. In particular, we have $a \in F$, so $R' = F$ as claimed.

Since V is finitely generated as a module over $R' = F$, the density theorem tells us that the natural map $R \to R'' = \text{End}(V)$ is surjective. It is also injective since by assumption $R \subset \text{End}(V)$. □

If we have an irreducible representation π of a finite group G on a finite dimensional vector space V over an algebraically closed field F, then V is a simple module over the ring $\pi(F[G])$. Thus the corollary tells us that

$$\pi(F[G]) \cong \text{End}(V)$$

as algebras.

As yet another immediate corollary of the density theorem we get *Wedderburn's theorem*.

Corollary 9.16.3 *Let A be a faithful, simple module over a unital ring R, and assume that A is finite dimensional over the division ring R'. Then $R = R''$.*

9.16 Density

Here is another application of the density theorem.

Corollary 9.16.4 *Let R be a unital algebra over a field. Let $\{V_i\}_{i=1}^n$ be finite dimensional vector spaces over the field, that are pairwise non-isomorphic as simple modules over R considered as a ring. Then there are elements $e_i \in R$ that act as the identity on V_i and such that $e_i V_j = \{0\}$ for $i \neq j$.*

Proof Apply the density theorem to a linear basis for $A = V_1 \oplus \cdots \oplus V_n$ and to the projection $\pi_i \in R''$ onto $V_i \subset A$. □

We then get yet another corollary.

Corollary 9.16.5 *Let R be a unital algebra over a field of characteristic zero. Suppose A and B are finite dimensional vector spaces that are semisimple modules over R as a ring. Let f_a and g_a be the linear maps on A and B, respectively, given by the action of $a \in R$. If $\mathrm{Tr}(f_a) = \mathrm{Tr}(g_a)$ for all $a \in R$, then $A \cong B$ as R-modules.*

Proof Both A and B are finite direct sums of simple modules, so we need only check that a simple module V occurs with the same multiplicity in both decompositions; say these from the outset are n and m, respectively. Pick by the corollary above an element $e \in R$ that acts as the identity on V and as the zero-map on the other simple modules in the decompositions of A and B. Then

$$n \dim(V) = \mathrm{Tr}(f_e) = \mathrm{Tr}(g_e) = m \dim(V),$$

so $n = m$ since we are in characteristic zero. □

In the language of representations we get the following statement: A representation of a finite group that is a finite direct sum of irreducible representations on finite dimensional vector spaces over a field of characteristic zero is uniquely determined up to equivalence by its character. To see this just extend the representation and its character to the group ring.

Decomposability of representations are provided by the following result by Maschke, which extends Theorem 5.4.3 to more general fields and rings. See the definition below for the notion of a semisimple ring.

Proposition 9.16.6 *Let G be a finite group and let R be a semisimple ring such that $|G|1$ is a unit in R, which for instance holds for fields with characteristic that does not divide $|G|$. Then $R[G]$ is a semisimple ring.*

Proof We show that any submodule B of a module A over $R[G]$ has a complementary submodule. Now B has a complementary R-submodule by semisimplicity of R, see the proposition below. Let $\pi: A \to B$ be the R-linear projection onto B with respect to this R-module decomposition. As $|G|1 \in R$ is a unit with inverse $1/|G|$, we can define a map $f: A \to B$ by

$$f(x) = \frac{1}{|G|} \sum_{a \in G} a\pi(a^{-1}x).$$

This is an $R[G]$-module map because

$$f(bx) = \frac{1}{|G|} \sum_{a \in G} a\pi(a^{-1}bx) = \frac{1}{|G|} \sum_{a \in G} b(b^{-1}a)\pi((b^{-1}a)^{-1}x) = bf(x),$$

where we in the last equality replaced a by $b^{-1}a$ in the summation over G. If $g \colon B \to A$ is the inclusion map of B in A, then $fg = \iota$ because $\pi g(a^{-1}x) = a^{-1}x$ for $x \in B$ and $a \in G$, so that

$$fg(x) = \frac{1}{|G|} \sum_{a \in G} a(a^{-1}x) = \frac{1}{|G|}|G|x = x.$$

Hence for $y \in A$, we see that $gf(y) \in B \subset A$ and $y - gf(y) \in \ker f$ with trivial $B \cap \ker f$, so $A \cong B \oplus \ker f$ as $R[G]$-modules. □

We again used the Haar integral crucially. The reader is encouraged to check which results in the representation theory of finite groups are valid for more general fields, especially those whose characteristic does not divide the order of the group.

Proposition 9.21.12 shows that the previous result is as good as it gets.

9.17 Semisimple Rings

Definition 9.17.1 A *semisimple ring* is a non-trivial unital ring that is semisimple as a left module over itself.

Proposition 9.17.2 *Every module over a semisimple ring is semisimple.*

Proof Every module over a unital ring has a free presentation. The result then follows because direct sums and quotients of semisimple modules are semisimple. □

The following result is immediate from the proposition above and the characterization of semisimple modules in terms of complementary submodules.

Corollary 9.17.3 *A non-trivial unital ring is semisimple if and only if every short exact sequence of modules over it splits.*

Corollary 9.17.4 *Modules over semisimple rings are projective. A non-trivial unital ring is semisimple if all its cyclic modules are projective.*

Proof The first statement is immediate from the proof of the proposition above and from the corollary above.

For the second statement observe that the quotient module of the ring R by a left ideal I is cyclic, and thus projective, so $R \cong I \oplus (R/I)$. Hence every submodule of the ring has a complementary submodule. □

9.17 Semisimple Rings

In particular, modules over semisimple rings are flat.

Corollary 9.17.5 *A non-trivial unital ring is semisimple if and only if every module over it is injective.*

Proof The forward direction is immediate from the second last corollary above and Proposition 9.4.4.

For the opposite direction one considers the short exact sequence given by a left ideal of the ring. By the same proposition it splits and provides a complementary submodule to the left ideal. □

In fact, if already the cyclic modules of a non-trivial unital ring are injective, the ring must be semisimple, but the proof of this is more involved.

We aim to reveal the structure of semisimple rings. Towards this goal it is natural to consider matrix rings.

Proposition 9.17.6 *Suppose R is a unital ring. Then any ideal of $M_n(R)$ is of the form $M_n(I)$ for an ideal I of R.*

Proof Let $E_{ij} \in M_n(R)$ be the matrix units, so $E_{ij}E_{kl} = \delta_{jk}E_{il}$ and thus $E_{ij}AE_{kl} = a_{jk}E_{il}$ for any $A = (a_{ij}) \in M_n(R)$.

Let J be an ideal of $M_n(R)$, and let I be the set consisting of all $a \in R$ such that $aE_{11} \in J$. Then I is an ideal of R because $abE_{11} = aE_{11}bE_{11}$ for $a, b \in R$. Furthermore, if $A = (a_{ij}) \in J$, then $a_{ij}E_{11} = E_{1i}AE_{j1} \in J$, so $J \subset M_n(I)$. But we also have $M_n(I) \subset J$ because if $a \in I$, then $aE_{ij} = E_{i1}(aE_{11})E_{1j} \in J$. □

Corollary 9.17.7 *Matrix rings over division rings are simple.*

Proof This is immediate from the proposition above as division rings are simple. □

However, as modules over themselves matrix rings over division rings are not simple. Indeed, let R be a division ring, and let A_i be the left ideal of $M_n(R)$ consisting of all matrices with non-zero entries only along the ith column. Then A_i is a simple $M_n(R)$-module. To see this, let B be a submodule, and say we have an element $\sum_r a_r E_{ri} \in B$ with $a_k \in R$ non-zero. Then $a_k^{-1} \in R$ and

$$E_{ji} = a_k^{-1} E_{jk} \sum_r a_r E_{ri} \in B,$$

so $B = A_i$. The $M_n(R)$-module A_i is obviously isomorphic to R^n, so

$$M_n(R) \cong A_1 \oplus \cdots \oplus A_n \cong nR^n$$

as $M_n(R)$-modules. By uniqueness of such decompositions, as stated in Theorem 9.15.10, all simple submodules of $M_n(R)$ are isomorphic to R^n.

Note that R^n as a product ring is not a division ring since elements with a zero coordinate are not invertible. Also, we stress that R^n is simple as an $M_n(R)$-module, not as an R^n-module.

In fact, for unital rings S and T, no non-trivial left ideal I of S will be isomorphic as an $S \times T$-module to a left ideal J of T. Indeed, say $f \colon I \to J$ is an $S \times T$-module map. Then with π_S and π_T the projections from $S \times T$ onto S and T, respectively, we have $\pi_S f = 0$ by assumption. For any non-zero $a \in I$, we have

$$(\pi_S f(a,0), 0) = (1,0) \cdot (\pi_S f(a,0), \pi_T f(a,0))$$
$$= (\pi_S f((1,0) \cdot (a,0)), \pi_T f((1,0) \cdot (a,0))) = (\pi_S f(a,0), \pi_T f(a,0)),$$

so $\pi_T f(a,0) = 0$, which combined with $\pi_S f(a,0) = 0$ from above, gives $f(a,0) = 0$. Thus f is manifestly not injective.

Another result which one should keep in mind in the discussion to follow, is the following.

Proposition 9.17.8 *Any finite direct product of semisimple rings is a semisimple ring.*

Proof Suppose R_i is a ring with a simple R_i-module A_i. Then $A_1 \times \cdots \times A_n$ is a simple $R_1 \times \cdots \times R_n$-module because $\pi_i(B)$ is an R_i-module for any $R_1 \times \cdots \times R_n$-module B.

If all R_i are semisimple rings, write each one of them as a direct sum of simple submodules. Then $R_1 \times \cdots \times R_n$ as a module over itself, will after plugging in the direct sums for R_i and expanding, be a direct sum of simple modules of the type described above. □

Suppose R_i are division rings, then by the proposition above, and by the discussion prior to it, the product ring

$$M_{n_1}(R_1) \times \cdots \times M_{n_m}(R_m)$$

is semisimple. As a module over itself it is a direct sum

$$n_1 R_1^{n_1} \oplus \cdots \oplus n_m R_m^{n_m}$$

of simple modules $R_i^{n_i}$ that occur with multiplicity n_i since those corresponding to different n_i's are not isomorphic as we pointed out above.

Below we shall see that any semisimple ring is of this form.

A unital ring cannot be an infinite direct sum of non-trivial subrings since any element of the ring can only have finitely many non-zero components by definition of direct sums. The projections of the identity of the ring will then be identities of the component rings in a finite decomposition.

The following lemma is straightforward.

Lemma 9.17.9 *For any ring R the transpose of a matrix with entries in R gives an isomorphism*
$$M_n(R)^{\mathrm{op}} \cong M_n(R^{\mathrm{op}})$$

of rings.

9.17 Semisimple Rings

Lemma 9.17.10 *If R is a unital ring, then $\operatorname{End}_R(R) \cong R^{\operatorname{op}}$ as rings.*

Proof The homomorphism $R^{\operatorname{op}} \to \operatorname{End}_R(R)$ given by $a \mapsto h_a$ with $h_a(x) = xa$ for $a, x \in R$ is evidently injective, and it is surjective because $h_{f(1)} = f$ for any $f \in \operatorname{End}_R(R)$. □

Theorem 9.17.11 *If R is a semisimple ring, then*

$$R \cong M_{n_1}(R_1) \times \cdots \times M_{n_m}(R_m)$$

as rings, where the R_i's are division rings uniquely determined up to isomorphisms and permutations along with the numbers n_i.

Proof By the remark above we can write R as a finite direct sum of simple submodules. Consider the R-endomorphism ring of this direct sum, and use the last lemma above in combination with Theorem 9.15.10. Then one obtains an expression for R^{op} of the type in the theorem with division rings S_i. Take opposites, use that the opposite of a direct product of rings is the direct product of the opposite rings, together with the second last lemma above, to get R as a product of matrix rings over S_i^{op}. It remains to observe that the opposite of a division ring is again a division ring, so let $R_i = S_i^{\operatorname{op}}$. □

Note that the matrix rings $M_{n_i}(R_i)$ in the theorem above are two-sided ideals of R; not merely left ideals. Any right module over a left semisimple ring can be decomposed into a direct sum of simple right modules. Indeed, for a semisimple ring R with a decomposition as in the theorem, these simple right modules are the right ideals of $M_{n_i}(R_i)$ having non-zero entries only along a given row (rather than a column), and they occur with multiplicity n_i in R, being all isomorphic to the right module R^{n_i} of row vectors acted upon from the right by elements of $M_{n_i}(R_i)$.

By the *dimension of an algebra* we mean its dimension as a vector space. When we talk about *modules over algebras* we mean modules over them considered as rings. When we write down properties of algebras that we have defined for rings, we mean that they have these properties as rings.

Corollary 9.17.12 *Any semisimple algebra over an algebraically closed field is isomorphic to a finite direct sum of matrix algebras over the field, and these components are unique up to permutations.*

Proof From the first part of the proof of Burnside's theorem, we see that the division rings R_i occurring in the proof of the theorem above, then as endomorphism rings of simple modules, are all isomorphic to the field. □

Proposition 9.17.13 *There exists no finite dimensional division algebra over an algebraically closed field other than the field itself.*

Proof Say R is such an algebra over a field F, and let $a \in R$. Since R is finite dimensional as a vector space over F, there is a polynomial f over F of minimal degree such that $f(a) = 0$. As F is algebraically closed, we may write $f(x) = g(x)(x - b)$ for $b \in F$ and $g \in F[x]$. Then $g(a)(a - b) = 0$, and $g(a) \neq 0$ by minimality of f. Hence $a = b \in F$ since $g(a)$ has an inverse in the division ring R. □

There is perhaps a more intrinsic approach to the submodule structure of a semisimple ring, which we now present. We refer to left ideals of a ring that are simple as modules as *simple left ideals*.

Lemma 9.17.14 *Let I be a simple left ideal of a unital ring R, and let A be a simple module over R. Then either $I \cong A$ as modules, or $IA = \{0\}$.*

Proof Clearly IA is a submodule of A, so by simplicity of A, either IA is trivial, or $IA = A$. Assume $IA = A$. As A is non-trivial, we can pick $x \in A$ such that Ix is non-trivial. But Ix is a left ideal of A, and so again by simplicity of A, we get $Ix = A$. Thus the map $a \mapsto ax$ from I to A is an epimorphism. Its kernel, which is a left ideal of the simple module I, must be trivial since it cannot be all of I. So the map is an isomorphism. □

Theorem 9.17.15 *Let R be a semisimple ring. Then R has only finitely many pairwise non-isomorphic simple left ideals $I_1, \ldots, I_m$. There is only a finite number n_i of simple left ideals of R isomorphic to I_i, and the sum S_i of these left ideals is a two-sided ideal of R. The ring R is isomorphic to the direct product $S_1 \times \cdots \times S_m$ of the unital rings S_i.*

Proof Assume from the outset that R has arbitrary many pairwise non-isomorphic simple left ideals I_i with associated S_i.

By the lemma above $S_i S_j = \{0\}$ for $i \neq j$. In particular, as R is unital and $R = \sum_i S_i$, we get $S_i \subset S_i R = S_i S_i \subset S_i$, so S_i is a two-sided ideal of R.

We can obviously write the identity element 1 of R as

$$1 = e_1 + \cdots + e_m$$

for a finite m and non-zero $e_i \in S_i$. For any $x \in R$, we have $x = x1 = xe_1 + \cdots + xe_m$ with $xe_i \in S_i$. Hence the total number of S_i's is m, showing that there are only m pairwise non-isomorphic left ideals I_i of R.

Let $x_i \in S_i$. Since $S_i S_j = \{0\}$ for $i \neq j$, we get $e_i x_i = 1 x_i = x_i$ and $x_i e_i = x_i 1 = x_i$, so e_i is an identity element for the ring S_i. Also if $x_1 + \cdots + x_m = 0$, then

$$x_i = e_i x_i = e_i(x_1 + \cdots + x_m) = e_i 0 = 0,$$

so $R \cong S_1 \times \cdots \times S_m$.

Since R is a direct sum of simple left ideals, the identity element e_i of S_i will be a finite sum of non-zero elements from simple left ideals of R isomorphic to I_i. Again we see that every element of S_i has non-zero coordinates only in these simple left ideals, so say there are n_i simple left ideals of R isomorphic to I_i. □

Of course, in this theorem $S_i \cong M_{n_i}(R_i^{\mathrm{op}})$ and $R_i \cong \mathrm{End}_R(I_i)$. The identity elements e_i of these matrix rings from an orthogonal family of idempotents.

9.17 Semisimple Rings

Theorem 9.17.16 *Every simple module of a semisimple ring R is isomorphic to one of the finitely many pairwise non-isomorphic simple left ideals I_i of R. Every non-trivial module of R is a direct sum of simple submodules, each of which is isomorphic to some I_i.*

Proof By Proposition 9.17.2 we know that any non-trivial module A can be written as a direct sum of simple submodules.

Suppose B is a simple submodule of A. If $IB = \{0\}$ for every simple left ideal I of R, then $B = RB$ is trivial. Since this cannot be the case, by the lemma above we conclude that $B \cong I$ for some simple left ideal I of R. $\square$

Suppose R is a semisimple ring, say with an exhaustive family $I_1, \ldots, I_m$ of pairwise non-isomorphic simple left ideals. Let S_i be the sum of the simple left ideals of R isomorphic to I_i. We claim that for a non-trivial R-module A, we have

$$A \cong S_1 A \oplus \cdots \oplus S_m A,$$

and that $S_i A$ is the sum of the simple submodules of A isomorphic to I_i. To verify this, observe that $A = RA = \sum S_i A$. And if B is a simple submodule of $S_i A$, we have $B = RB = S_i B$ as $S_i S_j = \delta_{ij} S_i$, so $B \cong I_i$ by the lemma above.

From the discussion earlier in this section we record the following corollary.

Corollary 9.17.17 *For a division ring R every simple $M_n(R)$-module is isomorphic to R^n. In particular, they are all faithful.*

Be aware that unital simple rings need not be semisimple. According to Theorem 9.17.11 the semisimple simple rings are matrix rings over division rings. Of course, commutative unital simple rings are semisimple as left ideals are then two sided ideals. As any matrix ring over a ring with a non-trivial product, like non-trivial unital rings, are noncommutative, except the case of 1×1-matrices, we conclude that commutative semisimple rings are finite products of fields.

The converse of the following result is obvious.

Proposition 9.17.18 *If R is a non-trivial unital subalgebra of $\mathrm{End}(V)$ for a finite dimensional vector space V that is semisimple as an R-module, then R is semisimple.*

Proof We can obviously find vectors $v_i \in V$ such that $V \cong Rv_1 \oplus \cdots \oplus Rv_n$ as R-modules. Then $a \mapsto (av_1, \ldots, av_n)$ is an injective R-module map from R to V. Thus R is semisimple, being a submodule of a semisimple module. $\square$

Suppose we have an endomorphism A acting on a finite dimensional vector space V over a field F. Then obviously A is semisimple if and only if V is semisimple as a module over the unital subalgebra $F[A]$ of $\mathrm{End}(V)$ generated by A. By the proposition above this happens if and only if the ring $F[A]$ is semisimple. Since this ring is commutative, we deduce from the discussion prior to the proposition above, that the algebra $F[A]$ is isomorphic to a finite direct product of fields. Let $f_A \in F[x]$ be the minimal polynomial of A. Then

$$F[A] \cong F[x]/(f_A).$$

If
$$f_A = \prod_i p_i^{m_i}$$

is a factorization of f_A into primes p_i, then the Chinese remainder theorem for PID's tells us that
$$F[A] \cong \bigoplus_i F[x]/(p_i^{m_i})$$

as modules. By uniqueness of decompositions of semisimple modules, we see that all $F[x]/(p_i^{m_i})$ must be fields, and this can only happen if all $m_i = 1$. This is consistent with Proposition 9.14.9.

Proposition 9.17.19 *Let I be a non-trivial left ideal of a simple unital ring R. Then the natural map $R \to R''$ is an isomorphism.*

Proof The kernel of the natural map $f \colon a \mapsto f_a$ is a proper ideal of R, so f is injective.

As IR is a non-trivial ideal of R, we must have $IR = R$ and $f(I)f(R) = f(R)$. If $g \in R''$ and $x, y \in I$, then
$$g f_x(y) = g(xy) = g(x)y = f_{g(x)}(y),$$

where we in the second step used that the map $x \mapsto xy$ belongs to R' and that $g \in R''$. Thus $g f_x = f_{g(x)}$, so $R'' f(I) \subset f(I)$. Therefore as the rings $f(R)$ and R'' are unital, we get
$$R'' = R'' f(R) = R'' f(I) f(R) = f(I) f(R) = f(R).$$

□

We can generalize this proposition.

Definition 9.17.20 A module A is a *generator* if every module is the homomorphic image of a (possibly infinite) direct sum of A with itself. A module over a unital ring R is *balanced* if the natural map $R \to R''$ is an isomorphism.

A unital ring as a module over itself is a generator because every module over a unital ring has a free presentation

The left ideal I in the proposition above is a generator for modules over R because $R = IR$, so we can write the identity $1 \in R$ as a finite sum
$$1 = x_1 a_1 + \cdots + x_n a_n,$$

which shows that $(y_1, \ldots, y_n) \mapsto y_1 a_1 + \cdots + y_n a_n$ is an R-module map from $I \oplus \cdots \oplus I$ onto R. Then we use that R is a generator.

Clearly, if a module is a generator, then finitely many copies of it will do to cover by a homomorphism any finitely generated module.

Theorem 9.17.21 *If a module over a unital ring R is a generator, then it is balanced and finitely generated projective as a module over R'.*

Proof Suppose we have an R-module A that is a generator. Then there is some $n \in \mathbb{N}$ such that R is the homomorpic image of nA. As R is free, there is a module B such that $nA \cong R \oplus B$. This direct sum is in fact balanced. The presence of R with its identity element assures that the natural map here is injective. This map is also surjective since for g in the bicommutant of R in $R \oplus B$ we have for $a \in R$ and $x \in B$, that

$$\pi_1 g(a+x) = g\pi_1(a+x) = g(a) = g(1 \cdot a) = g(1)a$$

and

$$\pi_2 g(a+x) = g\pi_2(a+x) = g(x) = g(1 \cdot x) = g(1)x$$

as the maps $b + y \mapsto ab + y$ and $b + y \mapsto bx$ and the projections π_i belong to the commutant of R in $R \oplus B$. Thus $g(a+x) = g(1) \cdot (a+x)$, which shows that g belongs to the image of the natural map. Hence nA is balanced.

We can now show that the natural map $f \colon R \to R''$ is an isomorphism. Suppose $h \in R''$. Then its diagonal amplification $h^n \colon nA \to nA$ belongs to the bicommutant of R in nA since the commutant of R in nA consists of the $n \times n$-matrices with entries in R'. Since nA is balanced, there is $a \in R$ such that $h^n = f_a^n$, so $h = f_a$. But f is also injective because if $ax = 0$ for all $x \in A$, then also $ab = 0$ for all $b \in R$ as A is a generator, so $a = 0$.

The last part is immediate from the usual isomorphisms

$$nR' \cong \mathrm{Hom}_R(R, A) \oplus \mathrm{Hom}_R(B, A) \cong A \oplus \mathrm{Hom}_R(B, A)$$

as additive groups, since these are actually R'-isomorphisms. □

The reader will have noticed that a lot of the ingredients from the density theorem are found in the previous proof. The converse of the theorem is also true, and we leave this as an exercise.

9.18 Noetherian and Artinian Modules

Definition 9.18.1 A module is *noetherian* if it has the *ascending chain condition* for submodules, that is, any ascending chain

$$A_1 \subset A_2 \subset \cdots$$

of *submodules stabilizes* in the sense that $A_n = A_{n+1} = \cdots$ for some n. A module is *artinian* if it satisfies the same property with inclusions reversed, and we then talk about a *descending chain condition*. We can also talk about noetherianess (artinianess) for bimodules, and also with the prefix left or right if we look at chains of such submodules. A ring is a *noetherian (artinian) ring* if it is noetherian (artinian) as a left module over itself; if it is so as a bimodule we rather speak of left and right noetherian (artinian).

Remember that the submodules of a ring are the left ideals.

By Lemma 6.14.4, or the theorem below, any PID is noetherian. A PID that is not a field cannot be artinian. To see this pick a prime element p of the PID. Then the descending chain

$$(p) \supset (p^2) \supset \cdots$$

of ideals cannot break off, because if $(p^n) = (p^{n+1})$ for some n, then $p^n = ap^{n+1}$ for some element a of the ring, so p would be a unit. In fact, replacing p by any non-zero element in the argument above, we see that an artinian integral domain is a field.

Proposition 9.18.2 *In an artinian commutative ring prime ideals are maximal.*

Proof If I is a prime ideal in an artinian commutative ring R, then R/I is an integral domain, and so is a field. Thus I is a maximal ideal. □

Theorem 9.18.3 *For a module A the following conditions are equivalent:*

(i) *The module A is noetherian;*

(ii) *Each submodule of A is finitely generated;*

(iii) *Any non-empty family of proper submodules of A has a maximal element.*

Proof If (i) holds and A has a submodule B that is not finitely generated, we can inductively pick elements $a_i \in B$ such that

$$(a_1) \subset (a_1, a_2) \subset \cdots$$

with only proper inclusions, which is a contradiction. So (ii) must hold.

Conversely, if we assume that (ii) holds, and $A_1 \subset A_2 \subset \cdots$ is an ascending chain of submodules of A. Then $\cup A_i$ is a submodule of A, which by assumption is finitely generated, by say X. Since X is finite it will be contained in A_n for some n, and then $A_n = A_{n+1} = \cdots$, so (i) holds.

Assume that (i) holds. If we have a non-empty family of proper submodules of A with no maximal submodule, it must contain a submodule A_1. Since this is not a maximal element of the family, it is properly contained in another member of the family, and then in another member, and so on, so (i) is violated, which is a contradiction. Thus (iii) must hold.

9.18 Noetherian and Artinian Modules

Conversely, if we assume that (iii) holds, then the family of proper submodules in any ascending chain must have a maximal element, which stabilizes the chain. □

Corollary 9.18.4 *In a noetherian ring each ideal contains a finite product of prime ideals.*

Proof Let $\mathcal{F}$ be the family of ideals in R that do not contain any finite product of prime ideals. If $\mathcal{F}$ is non-empty, then by the theorem above it has a maximal element A which is certainly not a prime ideal. So there are ideals B and C that are not contained in A but such that $BC \subset A$. By maximality of A both $B + A$ and $C + A$ will contain a finite product of prime ideals, and $(B + A)(C + A) \subset A$, so A also contains a finite product of prime ideals, which is absurd. □

Definition 9.18.5 A module is *finitely cogenerated* if each family of submodules with trivial intersection has a finite subcollection with trivial intersection.

The proof of the following result is 'dual' to the proof in the theorem above, and is left out.

Theorem 9.18.6 *For a module A the following conditions are equivalent:*

(i) *The module A is artinian;*

(ii) *Each quotient module of A is finitely cogenerated;*

(iii) *Any non-empty family of non-trivial submodules of A has a minimal element.*

The following result is immediate from the two theorems above.

Corollary 9.18.7 *Submodules and homomorphic images of noetherian (artinian) modules are noetherian (artinian).*

Proposition 9.18.8 *Let $0 \to A \to B \to C \to 0$ be an exact sequence of modules. Then B is noetherian (artinian) if and only if both A and C are noetherian (artinian).*

Proof The forward implication is the corollary above.

For the opposite direction assume A is a noetherian submodule of B with a noetherian quotient module $C = B/A$.

If D is a submodule of B, then the module $D/(A \cap D) \cong (D + A)/A$ is a finitely generated submodule of B/A by Theorem 9.18.3, so there are elements $x_i \in D$ such that
$$D = (x_1) + \cdots + (x_n) + A \cap D.$$

But since by Theorem 9.18.3, we know that $A \cap D$ is a finitely generated submodule of A, say with generators $y_1, \ldots, y_m$. Then $\{x_i\} \cup \{y_j\}$ will generate D. Thus B is noetherian by Theorem 9.18.3.

The proof in the artinian case is similar. Alternatively one can use the five lemma. □

Corollary 9.18.9 *A finite direct sum of modules is noetherian (artinian) if and only if every component in the sum is noetherian (artinian). Any finitely generated module over a noetherian (artinian) unital ring is noetherian (artinian).*

Proof The first statement is by the proposition above true for a direct sum of two modules, and thus holds by induction for finite direct sums. A finitely generated module over a noetherian (artinian) unital ring R is a quotient module of the noetherian (artinian) ring R^n for some n. Thus by the proposition above it is noetherian (artinian). □

Definition 9.18.10 A module is *completely reducible* if it is a finite direct sum of simple modules.

Completely reducible modules are obviously semisimple, and semisimple rings are completely reducible.

Corollary 9.18.11 *Completely reducible modules are noetherian and artinian. A semisimple module is completely reducible if it is noetherian or artinian.*

Proof Any ascending or descending chain with proper inclusions in a simple module cannot have more than one inclusion, so the first statement in the proposition follows from the corollary above.

In a countable direct sum $\oplus_{i=1}^{\infty} A_i$ of simple modules A_i we have the infinite ascending and descending chains $A_1 \subset A_1 \oplus A_2 \subset \cdots$ and $\oplus_{i=1}^{\infty} A_i \supset \oplus_{i=2}^{\infty} A_i \supset \cdots$ of submodules with only proper inclusions. □

Proposition 9.18.12 *In a noetherian ring every one-sided inverse is a two-sided inverse.*

Proof Suppose we have $ab = 1$ but $ba \neq 1$ in a noetherian ring R. Then

$$e_{ij} \equiv b^{i-1}a^{j-1} - b^i a^j \neq 0$$

and satisfies

$$e_{ij}e_{kl} = \delta_{jk}e_{il}$$

for $i, j, k, l \in \mathbb{N}$. Hence R contains the submodule $\bigoplus_i Re_{ii}$ and has an infinite chain of submodules with only proper inclusions. □

Corollary 9.18.13 *Given a module A and a finite chain*

$$\{0\} \equiv A_0 \subset A_1 \subset \cdots \subset A_n \equiv A$$

of submodules. Then A is noetherian (artinian) if and only if every quotient module A_{i+1}/A_i is noetherian (artinian).

9.18 Noetherian and Artinian Modules

Proof Apply the second last proposition above inductively to the exact sequence

$$0 \longrightarrow A_i \longrightarrow A_{i+1} \longrightarrow A_{i+1}/A_i \longrightarrow 0$$

for $i \geq 0$. □

Example 9.18.14 Any finite dimensional vector space is both noetherian and artinian because there are no ascending or descending chains with more proper inclusions than the dimension of the vector space. Clearly finite dimensionality of the vector space is equivalent to noetherianess or artinianess. Despite the fact that $F[x]$ over a non-trivial field F is not an artinian ring, any proper quotient ring $F[x]/(f)$ of it is artinian because $F[x]/(f)$ is a finite dimensional vector space with dimension $\deg(f)$, and every left ideal of it is a vector subspace. More generally, finite dimensional algebras are noetherian and artinian because subalgebras are subspaces. ◇

Example 9.18.15 Pick a prime number p. Let R be the set of rational numbers m/p^n for all non-negative integers m and n such that $m/p^n \in [0, 1)$. We turn R into a commutative non-unital ring by trivial multiplication and with addition defined modulo integers.

Let I be a non-trivial proper ideal in R. Remember that any integer times an element of I will belong to I since I is closed under addition and subtraction.

Denote by k the least natural number such that $l/p^k \notin I$ for some integer l.

Consider natural numbers m and n such that $n \geq k$ and $\gcd(m, p) = 1$. If $m/p^n \in I$, then $m/p^k = p^{n-k}m/p^n \in I$. And $1/p^{k-1} \in I$ by definition of k. There are $a, b \in \mathbb{Z}$ such that $am + bp = 1$. Thus $l/p^k = lam/p^k + lb/p^{k-1} \in I$, which is impossible. Therefore $m/p^n \notin I$. Hence

$$I_{k-1} \equiv I = \{0, 1/p^{k-1}, 2/p^{k-1}, \ldots, (p^{k-1} - 1)/p^{k-1}\}.$$

Since all proper ideals are finite the ring R is obviously artinian, but it is not noetherian because $I_1 \subset I_2 \subset \cdots$ is an infinite ascending chain of ideals with only proper inclusions. ◇

Proposition 9.18.16 *A surjective (injective) endomorphism of a noetherian (artinian) module is an isomorphism.*

Proof Say f is a surjective endomorphism of a noetherian module A. The ascending chain $\ker f \subset \ker f^2 \subset \cdots$ of submodules must stabilize, say from $\ker f^n$ onward. Write $x \in \ker f$ as $x = f^n(y)$. Then $y \in \ker f^{n+1} = \ker f^n$, so $x = 0$.

The argument in the artinian case is similar. □

Corollary 9.18.17 *All bases of a finitely generated free module over a noetherian or artinian ring have the same cardinality.*

Proof Let $\{x_i\}_{i=1}^n$ and $\{y_i\}_{i=1}^m$ be two bases with $n \geq m$ of a finitely generated free module A over a noetherian ring. Let f be the surjective endomorphism of A that sends x_i to y_i for $i \leq m$ and the remaining x_i to zero. By the proposition above the kernel of f is zero, so there cannot be any remaining x_i's.

The proof in the artinian case is similar. □

Definition 9.18.18 The *length of a series*

$$\{0\} = A_0 \subset \cdots \subset A_n = A$$

of submodules A_i of a module A is the number of proper inclusion, so the length might be less than $n \in \mathbb{N}$. The series is a *composition series* of A if each *factor* A_{i+1}/A_i is a simple module. The length $l(A)$ of A is the length of a composition series with least length among all composition series of A; if A has no (finite) composition series we set $l(A) = \infty$.

A *refinement of a series* is a series obtained by inserting a finite number of additional submodules in the original series. Two series are *equivalent series* if there is a bijection between their non-trivial factors such that the corresponding factors are isomorphic as modules. Obviously a composition series is maximal among all possible refinements of a series.

We have the following straightforward generalization of the results of Schreier and Jordan–Hölder for groups. The proof would involve a butterfly lemma adapted to the context of modules.

Theorem 9.18.19 *Any two series of a module have equivalent refinements. In particular, all composition series of a module are pairwise equivalent.*

Since the length of a composition series of a module is the number of factors of the series, all composition series of a module have the same length, which is then the length of the module.

Proposition 9.18.20 *A module has finite length if and only if it is both noetherian and artinian.*

Proof If a module A has finite length it has a composition series with simple, hence noetherian and artinian factors, so A is both noetherian and artinian by Corollary 9.18.13.

Suppose A from the outset is both noetherian and artinian. If A is non-trivial, then by Theorem 9.18.6 it has a minimal non-trivial submodule A_1, which obviously is simple. If $A_1 \neq A$, then by the same theorem there is a submodule A_2 that is minimal among all submodules of A that properly contains A_1, so A_2/A_1 is a simple module. If $A_2 \neq A$, there is an A_3 such that A_3/A_2 is simple, and so on. Since A is noetherian, the ascending chain $\{0\} \equiv A_0 \subset A_1 \subset A_2 \cdots$ must terminate with A after finitely many steps. The result is a composition series, so A has finite length. □

It is easily verified that $l(B) = l(A) + l(C)$ for the modules in Proposition 9.18.8.

9.18 Noetherian and Artinian Modules

Example 9.18.21 Subrings of artinian rings need not be artinian because $\mathbb{Q}$ is artinian while the subring $\mathbb{Z}$ is not artinian as the infinite descending chain $\mathbb{Z} \supset 2\mathbb{Z} \supset 2^2\mathbb{Z} \supset 2^3\mathbb{Z} \supset \cdots$ has only proper inclusions.

Also, subrings of noetherian rings need not be noetherian. For instance, the ring $M_2(\mathbb{Q})$ is noetherian because every submodule of it is also a vector space over $\mathbb{Q}$ under the change of ring homomorphism $\mathbb{Q} \to M_2(\mathbb{Q})$ that sends $a \in \mathbb{Q}$ to $\text{diag}(a, a)$. However, the subring

$$R = \begin{pmatrix} \mathbb{Z} & \mathbb{Q} \\ 0 & \mathbb{Q} \end{pmatrix} \equiv \{ \begin{pmatrix} a & b \\ 0 & c \end{pmatrix} \mid a \in \mathbb{Z} \text{ and } b, c \in \mathbb{Q} \}$$

of $M_2(\mathbb{Q})$ is not noetherian because

$$I_n = \{ \begin{pmatrix} 0 & m/2^n \\ 0 & 0 \end{pmatrix} \mid m \in \mathbb{Z} \}$$

form an infinite ascending chain $I_1 \subset I_2 \subset \cdots$ of left ideals of R with only proper inclusions.

So R is not (left) noetherian, but it is actually right noetherian. To see this, consider the right ideal I of R consisting of matrices having rational numbers in the right lower corner and otherwise only zero entries. The right R-module I has clearly no non-trivial proper right R-submodules, so it is right noetherian. The additive group R/I is a right R-module and is isomorphic to

$$A \equiv \begin{pmatrix} \mathbb{Z} & \mathbb{Q} \\ 0 & 0 \end{pmatrix}$$

as a right R-module. Let B be a non-trivial right R-submodule of A. Then

$$J = \{ a \in \mathbb{Z} \mid \begin{pmatrix} a & b \\ 0 & 0 \end{pmatrix} \in B \text{ for some } b \in \mathbb{Q} \}$$

is an ideal in the PID $\mathbb{Z}$, so $J = a\mathbb{Z}$ for some $a \in \mathbb{Z}$. Evidently

$$B \subset \begin{pmatrix} a\mathbb{Z} & \mathbb{Q} \\ 0 & 0 \end{pmatrix}.$$

Suppose $b \in \mathbb{Q}$ is such that

$$\begin{pmatrix} a & b \\ 0 & 0 \end{pmatrix} \in B.$$

If $a \neq 0$, then

$$\begin{pmatrix} ac & d \\ 0 & 0 \end{pmatrix} = \begin{pmatrix} a & b \\ 0 & 0 \end{pmatrix} \begin{pmatrix} c & d/a \\ 0 & 0 \end{pmatrix} \in B,$$

for any $c \in \mathbb{Z}$ and $d \in \mathbb{Q}$, so
$$B = \begin{pmatrix} a\mathbb{Z} & \mathbb{Q} \\ 0 & 0 \end{pmatrix}.$$

If $a = 0$, one checks that
$$B = \begin{pmatrix} 0 & \mathbb{Q} \\ 0 & 0 \end{pmatrix}.$$

Thus any ascending chain $A_1 \subset A_2 \subset \cdots$ of right R-submodules of A with only proper inclusions produce an ascending chain $J_1 \subset J_2 \subset \cdots$ of ideals of $\mathbb{Z}$ with only proper inclusions except perhaps in the first step. But $\mathbb{Z}$ is noetherian, so A is right noetherian. By the right-module-version of Proposition 9.18.8, we conclude that also R is right noetherian.

Consider the ring
$$S = \begin{pmatrix} \mathbb{Q} & \mathbb{Q} \\ 0 & 0 \end{pmatrix}$$

and the ideal
$$K = \begin{pmatrix} 0 & \mathbb{Q} \\ 0 & 0 \end{pmatrix}.$$

Then both K and S/K are artinian as S-modules, so the ring S is artinian. But S is not right artinian because we have an infinite descending chain
$$\begin{pmatrix} 0 & 2\mathbb{Z} \\ 0 & 0 \end{pmatrix} \supset \begin{pmatrix} 0 & 2^2\mathbb{Z} \\ 0 & 0 \end{pmatrix} \supset \begin{pmatrix} 0 & 2^3\mathbb{Z} \\ 0 & 0 \end{pmatrix} \supset \cdots$$

of right S-submodules with only proper inclusions. ◇

Definition 9.18.22 A *boolean ring* is a unital ring that consists entirely of idempotents.

Boolean rings are commutative and have characteristic two if they are non-trivial. To see this, consider elements a and b in a boolean ring. Then $2a = (2a)^2 = 4a$, so $2a = 0$ and $a = -a$. Thus $a + b = (a+b)^2$ gives $ab = ba$.

Example 9.18.23 The arctypical example of a boolean ring is the power set $P(X)$ of a set X with addition and multiplication defined as
$$a + b = (a \cup b) - (a \cap b) \quad \text{and} \quad ab = a \cap b$$

for $a, b \in P(X)$. Here X is the identity element and the empty set is the zero element. Unital subrings of $P(X)$ are also boolean rings. ◇

Proposition 9.18.24 *In a boolean ring any prime ideal is maximal. Any boolean noetherian ring is a finite direct product of the field $\mathbb{Z}_2$ with itself, so its cardinality is a power of two.*

9.18 Noetherian and Artinian Modules

Proof If P is a proper prime ideal of a boolean ring R, then R/P is a boolean ring and an integral domain. The only idempotents in an integral domain are 0 and the identity element, so R/P is the field $\mathbb{Z}_2$ and P is maximal.

If a non-trivial boolean ring R is noetherian, then by Corollary 9.18.4 the zero ideal contains a finite product $P_1 \cdots P_n$ of distinct proper prime ideals P_i. Then

$$R \cong R/\{0\} \cong R/P_1 \times \cdots \times R/P_n \cong \mathbb{Z}_2^n$$

by Proposition 6.11.8 and Theorem 6.8.9. Here $|R| = 2^n$. □

Here is a change of rings result which is often applied to quotient maps for rings.

Proposition 9.18.25 *Let $f: R \to S$ be a ring homomorphism, and let A be an S-module. If A is noetherian (artinian) as an R-module via f, then A is noetherian (artinian) as an S-module. The converse holds if f is surjective.*

Proof Any S-submodule of A is an R-submodule via f, which proves the first statement. When f is surjective, then every additive subgroup of A that is invariant under $f(R)$ is obviously also an S-submodule, which proves the second statement.
□

The following result is known as the *Hilbert basis theorem*.

Theorem 9.18.26 *Polynomial rings over noetherian unital rings are noetherian.*

Proof Let $\mathcal{E}$ and $\mathcal{F}$ be the families of left ideals in a noetherian ring R and in the polynomial ring $R[x]$, respectively. For each non-negative integer n define $f_n: \mathcal{F} \to \mathcal{E}$ by letting $f_n(A)$ consist of the zero-element and of the non-zero leading coefficients of all nth degree polynomials in the left ideal A of $R[x]$.

We claim that if $f_n(A) = f_n(B)$ for $A \subset B$ and all n, then $A = B$. To see this suppose $g \in B$ has degree n. By assumption there is $g_n \in A$ with the same leading coefficient as g. Either $g - g_n = 0$, and we are done, or $g - g_n$ has degree less than n. Since $g - g_n \in B$, we can repeat the argument and find $g_{n-1} \in A$ such that $g - g_n - g_{n-1} \in B$ either is zero or has degree less than $n - 1$. After at most n steps the only option is zero, and then g is a finite sum of elements $g_i \in A$. Thence $g \in A$ and $A = B$.

Let $A_1 \subset A_2 \subset \cdots$ be an ascending chain of left ideals in $R[x]$. For each n we get an ascending chain

$$f_n(A_1) \subset f_n(A_2) \subset \cdots$$

of left ideals in R. Since R is noetherian, there are integers $m(n)$ such that the above chain stabilizes at $f_n(A_{m(n)})$.

Again because R is noetherian the collection $\{f_i(A_j)\}$ has a maximal element, say $f_p(A_q)$. For any left ideal C of $R[x]$ we have $f_i(C) \subset f_j(C)$ when $j \geq i$ because we can multiply a polynomial in C of degree i by $x^{j-i} \in R[x]$, and get a polynomial in C of degree j that has the same leading coefficient. Hence we get the equalities

$$f_p(A_q) = f_n(A_q) = f_n(A_j)$$

for $n \geq p$ and $j \geq q$. Set $r = m(1)m(2) \cdots m(p-1)q$. Then

$$f_n(A_r) = f_n(A_{r+1}) = \cdots$$

for all non-negative n. By the first paragraph in the proof we get $A_r = A_{r+1} = \cdots$. In other words, the ring $R[x]$ is noetherian. □

Theorem 9.18.27 *Formal power series rings over noetherian unital rings are noetherian.*

Proof Let R be noetherian, and consider a left ideal A in the formal power series ring $R[[x]]$. For any non-negative n let B_n denote the left ideal of R generated by the leading coefficients of the elements in the left ideal $A \cap (x^n)$ of $R[[x]]$. Since R is noetherian, the ascending chain $B_0 \subset B_1 \subset \cdots$ stabilizes, say at B_m, and by Theorem 9.18.3 we can moreover assume that each B_n is finitely generated, say with finitely many generators a_{ni}. Choose $f_{ni} \in A \cap (x^n)$ having a_{ni} as leading coefficient. Again by Theorem 9.18.3 we have completed the proof if we can show that the finitely many elements f_{ni} for $n \leq m$ generate A as a left ideal.

Take any $g \in A$. Then we can certainly find an R-linear combination f_0 of f_{0i}'s such that $g - f_0 \in A \cap (x)$. Proceeding like this we can find an R-linear combination f_j of f_{ji}'s such that

$$g - f_0 - \cdots - f_j \in A \cap (x^{j+1}).$$

For $j \geq m$ we may write $f_j = \sum_i b_{ji} x^{j-m} f_{mi}$ with $b_{ji} \in R$ because $B_m = B_{m+1} = \cdots$ and then

$$g = f_0 + \cdots + f_{m-1} + \sum_{j=m}^{\infty} \sum_i b_{ji} x^{j-m} f_{mi} = f_0 + \cdots + f_{m-1} + \sum_i \left(\sum_{j=m}^{\infty} b_{ji} x^{j-m} \right) f_{mi}.$$

□

Corollary 9.18.28 *In a noetherian unital ring R both $R[x_1, \ldots, x_n]$ and $R[[x_1, \ldots, x_n]]$ are noetherian.*

9.19 Nilpotence

Definition 9.19.1 An element a of a ring is *nilpotent* if $a^n = 0$ for some $n \in \mathbb{N}$. A *nil ideal of a ring* is an ideal consisting only of nilpotent elements. An ideal I of a ring is *nilpotent* if $I^n = \{0\}$ for some $n \in \mathbb{N}$. Left and right nil and nilpotent ideals are defined similarly.

If e is a nilpotent idempotent in a ring with $e^n = 0$ for some n, then $e = e^n = 0$.

9.19 Nilpotence

If a is a nilpotent element of a unital ring with $a^n = 0$, then $1 - a$ is invertible with
$$(1-a)^{-1} = 1 + a + \cdots + a^n.$$

The ideal $I = \{0, 2\}$ of $\mathbb{Z}_4$ is nilpotent with $I^2 = \{0\}$.
Here is a non-commutative example.

Example 9.19.2 If I is any ideal of a ring R, then
$$J = \begin{pmatrix} 0 & I \\ 0 & 0 \end{pmatrix}$$
is a nilpotent ideal of
$$\begin{pmatrix} R & R \\ 0 & R \end{pmatrix}$$
with $J^2 = \{0\}$. ◇

The off diagonal matrix units of a matrix ring over a division ring are nilpotent. Yet the matrix ring is simple, so it has no non-trivial nil ideals. It has also no non-trivial nilpotent left ideals due to the following result.

Proposition 9.19.3 *Simple unital rings have no non-trivial left or right nilpotent ideals.*

Proof Suppose I is a left ideal of a simple unital non-trivial ring R with $I^n = \{0\}$ for some $n \in \mathbb{N}$. Then IR is a nilpotent ideal with $(IR)^n \subset I^n R = \{0\}$, and since R is unital and non-trivial we must have $IR \neq R$. As R is simple, we conclude that $IR = \{0\}$, and again since R is unital, the left ideal I must be trivial. □

Clearly any nilpotent ideal is a nil ideal, but as the next example shows, the converse is not in general true, although in this case the nilpotent elements in the ring form an ideal.

Example 9.19.4 Let p be a prime number and consider the ring $\bigoplus_{n=1}^{\infty} \mathbb{Z}/(p^n)$. The subset of nilpotent elements form an ideal I because the ring is commutative. But this nil ideal in not nilpotent because if $I^m = \{0\}$ for some m, then the element a of the ring that has $p + (p^{m+1})$ as its $(m+1)$th coordinate and otherwise has only zero coordinates is nilpotent with $a^{m+1} = 0$, so $a \in I$, but $a^m \neq 0$, which is a contradiction. ◇

However, we will see that nil ideals are nilpotent when the ring is artinian or noetherian.

Proposition 9.19.5 *Nil left ideals of artinian rings are nilpotent.*

Proof Suppose A is a nil left ideal of an artinian ring and that A is not nilpotent. Since the ring is artinian, the family $\{A^n \mid n \in \mathbb{N}\}$ has a minimal element, say $B \equiv A^n$. Then $B^2 = B$ as $A^{2n} \subset A^n$.

Let $\mathcal{F}$ be the family of left ideals C of the ring that are contained in B and such that BC is non-trivial. Since $B^2 = B \neq \{0\}$, the family $\mathcal{F}$ in non-empty, and as the ring is artinian, the family has a minimal element C. Thus there is $a \in C$ such that Ba is non-trivial. But $Ba \subset C$ and $B(Ba) = B^2 a = Ba \neq \{0\}$, so $Ba \in \mathcal{F}$. By minimality of C we get $Ba = C$. Thus there is $b \in B$ such that $ba = a$. But b is nilpotent as A is nil, say with $b^m = 0$. Thus $a = b^m a = 0 a = 0$, which is a contradiction, so A must be nilpotent. $\square$

Lemma 9.19.6 *In noetherian rings sums of nilpotent ideals are nilpotent.*

Proof Let $\sum_i I_i$ be an arbitrary sum of nilpotent ideals in a noetherian ring. Then as a left ideal the sum is finitely generated. But each generator lies in finitely many I_i's and therefore the sum is contained in a sum of finitely many I_i's. So we may assume that the initial sum above is finite.

By induction it therefore suffices to show that $I + J$ is nilpotent for ideals I and J with $I^n = J^m = \{0\}$ for some n and m. But $(I + J)^m \subset I + J^m = I$ and $(I + J)^{n+m} = I^n = \{0\}$. $\square$

Proposition 9.19.7 *A noetherian ring with no non-trivial nilpotent ideals has no non-trivial nil ideals.*

Proof Let I be a non-trivial nil ideal of a noetherian ring R. Since R is noetherian, the family $\mathcal{F}$ of annihilators of non-zero elements of I has a maximal left ideal $\text{Ann}(a)$ for some $a \in I$. Let $b \in R$. Then $ab \in I$. Either $ab = 0$, and we can jump to the next paragraph of this proof, or $ab \neq 0$. Then, and as I is nil, there exists a least $n \in \mathbb{N}$ such that $(ab)^n = 0$. Since $(ab)^{n-1}$ is a non-zero element of I, we have $\text{Ann}((ab)^{n-1}) \in \mathcal{F}$ which clearly contains $\text{Ann}(a)$. By maximality of the latter left ideal we conclude that $\text{Ann}((ab)^{n-1}) = \text{Ann}(a)$. But $(ab)^n = 0$ shows that $ab \in \text{Ann}((ab)^{n-1})$, so $aba = 0$ by definition of $\text{Ann}(a)$. Thus $(RaR)^2 = RaRRaR = \{0\}$ and since R has no non-trivial nilpotent ideals, we deduce $RaR = \{0\}$.

If R is unital, we therefore get $a = 0$, which is a contradiction. If R is non-unital, the ideal (a) generated by a is nilpotent since $RaR = \{0\}$, so $(a) = \{0\}$ and $a = 0$. $\square$

Proposition 9.19.8 *Nil ideals in noetherian rings are nilpotent.*

Proof Let I be a nil ideal in a noetherian ring R.

If J is the sum of all nilpotent ideals in R, then R/J has no non-trivial nilpotent ideals because by the correspondence theorem any such ideal is of the form K/J for an ideal K of R, and if $(K/J)^n = \{0\}$ for some n, then $K^n \subset J$. But $J^m = \{0\}$ for some m by the lemma above, so $K^{m+n} = \{0\}$, and the nilpotent ideal K must belong to J, which entails that K/J is trivial.

Hence the nil ideal I/J in R/J is trivial by the proposition above. So $I \subset J$ and I must be nilpotent. $\square$

9.19 Nilpotence

We know that matrix rings over divisions rings are noetherian and right noetherian as well as artinian and right artinian, so the same is true for semisimple unital rings since these are finite directs sums of such matrix rings.

In Example 9.18.21 we exhibited rings that where only one-sidedly noetherian or artinian.

Lemma 9.19.9 *If A is a simple left ideal in a ring R, then either A^2 is trivial or $A = Re$ for an idempotent e.*

Proof Suppose A^2 is non-trivial. Then there is $a \in A$ such that Aa is non-trivial. Since Aa is a left ideal contained in A, we must have $Aa = A$ because A is simple. So there is $e \in A$ such that $ea = a$. Then the left ideal $\text{Ann}(a)$ is strictly contained in A because e does not belong to it, so $\text{Ann}(a)$ is trivial since A is simple. As $(e^2 - e)a = a - a = 0$, we thus get $e^2 - e = 0$. But Re is a left ideal contained in A that is non-trivial because $0 \neq e = e^2 \in Re$. Hence $A = Re$ by simplicity of A. □

Theorem 9.19.10 *We have the following equivalent statements for a unital ring:*

(i) *It is artinian and has no non-trivial nilpotent ideals;*

(ii) *Each left ideal is generated by an idempotent;*

(iii) *It is completely reducible;*

(iv) *It is a finite direct product of matrix rings over division rings.*

Proof Let R be a unital ring.

Assume that (i) holds for R, and let A be a non-trivial left ideal of R. Since R is artinian, the left ideal A contains a minimal non-trivial left ideal B, which is obviously a simple left ideal, so by the lemma above, either B^2 is trivial, or $B = Ra$ for an idempotent $a \in R$.

If B^2 is trivial, then BR is an ideal of R and $(BR)^2 \subset B^2R = \{0\}$. By assumption BR must therefore be trivial, so B is trivial as R is unital, and this is absurd. So $B = Ra$ and A contains a non-zero idempotent.

Hence the family $\mathcal{F}$ consisting of all left ideals of the form $R(1-a) \cap A$ for some non-zero idempotent $a \in A$ is non-empty, and since R is artinian, it must have a minimal element, say $R(1-a) \cap A$. We claim that this left ideal is trivial.

Otherwise $R(1-a) \cap A$ contains a non-zero idempotent, say b. Notice that $ba = 0$. Let $c = a + b - ab$. Then c is an idempotent of A that is non-zero because $bc = b \neq 0$. Also, since $1 - c = (1-a) - (1-a)b$ and $b \in R(1-a)$, we see that $R(1-c) \cap A \subset R(1-a) \cap A$. The inclusion is proper because $b \notin R(1-c) \cap A$ as $bc \neq 0$, and this contradicts minimality of $R(1-a) \cap A$.

Let $d \in A$. Then $d(1-a) \in R(1-a) \cap A = \{0\}$, so $d = da$ and $A = Ra$. Thus (ii) holds.

To see that (ii) implies (iii), note that if A is a submodule of R, then $A = Re$ for some idempotent $e \in R$. But then $R(1-e)$ is a complementary submodule of A in R, so R is by Theorem 9.15.5 semisimple as a module over itself, and thus completely reducible by Theorem 9.17.15 since R is unital.

The implication (iii) to (iv) is Theorem 9.17.11.

The implication (iv) to (i) is clear since a matrix ring over a division ring is artinian as it is a finite dimensional algebra over the division ring, and the finite direct product of artinian rings is obviously artinian. As matrix rings over division rings are simple, the only ideals in R are direct subproducts of such matrix rings. But they have units so any non-trivial ideal of R has non-trivial idempotents and thus cannot be nilpotent. □

The step (iv) $\Rightarrow$ (iii) above is immediate from the fact that $M_n(R) \cong nR^n$ as modules over $M_n(R)$, whereas the step (iii) $\Rightarrow$ (ii) is Proposition 9.2.6.

9.20 The Jacobson Radical

Recall that a maximal submodule of a module is a proper submodule that is not strictly contained in another proper submodule.

Definition 9.20.1 The *Jacobson radical* rad(A) of a module A is the intersection of the maximal submodules of A. If there are no maximal submodules, we set rad(A) = A. The Jacobson radical of a ring is the Jacobson radical of the ring considered as a module over itself.

Note that rad(A) is a submodule of A. When A is noetherian it has a maximal submodule, so in this case rad(A) $\neq A$.

Example 9.20.2 If $p_1^{m_1} \cdots p_n^{m_n}$ is the prime number factorization of $a \in \mathbb{Z}$, then

$$\mathrm{rad}(\mathbb{Z}/a\mathbb{Z}) = p_1 \cdots p_n \mathbb{Z}/a\mathbb{Z}$$

by Proposition 6.11.8 since $p_i \mathbb{Z}/a\mathbb{Z}$ are the maximal submodules of the $\mathbb{Z}$-module $\mathbb{Z}/a\mathbb{Z}$. ◇

Proposition 9.20.3 *The Jacobson radical of a semisimple module is trivial.*

Proof The maximal submodules of a semisimple module are the direct sums obtained by setting one simple component to zero. The intersection of all these is obviously trivial. □

Proposition 9.20.4 *The Jacobson radical of a direct product of rings is the direct product of the Jacobson radical of each ring.*

9.20 The Jacobson Radical

Proof The result is immediate from the trivial observation that a left ideal A of a direct product of rings is maximal if and only if the projection of A on each ring is maximal. $\square$

Say A_i are the maximal submodules of a module A, so $\text{rad}(A) = \cap A_i$. Then the map $A \to \prod A/A_i$ which sends $a \in A$ to $i \mapsto a + A_i$ has kernel $\cap A_i$, so

$$A/\text{rad}(A) \subset \prod A/A_i$$

and each A/A_i is a simple module.

Proposition 9.20.5 *When the Jacobson radical of a module A is a finite intersection of maximal submodules, then $A/\text{rad}(A)$ is completely reducible. Any artinian module with vanishing Jacobson radical is completely reducible.*

Proof The first statement is immediate from the discussion above since submodules of completely reducible modules are completely reducible.

The second statement follows from the first provided we show that the radical of an artinian module is a finite intersection of maximal submodules. To this end note that intersections of an increasing collection of maximal submodules form a descending chain, which in an artinian module must stabilize. $\square$

The kernel of a non-zero module map from A to any simple module is clearly a maximal submodule of A. Thus we get the following result.

Proposition 9.20.6 *The radical of a module A is given by*

$$\text{rad}(A) = \cap \ker f,$$

where the intersection is taken over all module maps f from A to simple modules.

Corollary 9.20.7 *If $g: A \to B$ is a module map, then $g(\text{rad}(A)) \subset \text{rad}(B)$. Thus the radical $\text{rad}(A)$ is invariant under the action of any module map $A \to A$. The Jacobson radical of a ring is an ideal.*

Proof If $f: B \to C$ is a module map into a simple module, then $fg: A \to C$ is also a module map into C, so $fg(\text{rad}(A)) = 0$ by the proposition above, which by the same proposition gives the first result.

For a ring R the left ideal $\text{rad}(R)$ is invariant under the action of the module map $a \mapsto ab$ on R for any $b \in R$. Thus $\text{rad}(R)$ is an ideal. $\square$

Corollary 9.20.8 *For any R-module A we have $\text{rad}(R)A \subset \text{rad}(A)$.*

Proof Let x be an element of a simple R-module B, and consider the module map $f: R \to B$ given by $f(a) = ax$. Then $f(\text{rad}(R)) = 0$ by the proposition above, so $\text{rad}(R)B$ vanishes. If $g: A \to B$ is a module map, then

$$g(\text{rad}(R)A) \subset \text{rad}(R)B = \{0\},$$

and the result follows from the proposition above. $\square$

Corollary 9.20.9 *An element a of a ring belongs to $\mathrm{rad}(R)$ if and only if aA vanishes for every simple R-module A.*

Proof The proof is immediate from the proof of the corollary above, and by considering quotients of R. □

Corollary 9.20.10 *If R is a ring and A is a module over R with composition series of length not greater than n, then $(\mathrm{rad}(R))^n A$ is trivial.*

Proof Say $\{A_i\}$ is a composition series of A of length n. For $x \in A_i$ write $[x] \in A_i/A_{i-1}$. Let $a_i \in \mathrm{rad}(R)$. Apply the corollary above successively to A_i/A_{i-1} starting with $y \in A = A_n$. Then $a_1[y] = [0]$, so $a_1 y \in A_{n-1}$. Thus $a_2[a_1 y] = [0]$, so $a_2 a_1 y \in A_{n-2}$. Thus $a_3[a_2 a_1 y] = [0]$, so $a_3 a_2 a_1 y \in A_{n-3}$, arriving finally at $a_n \cdots a_1 y \in \{0\}$. If the length was less than n we would arrive at zero even faster. □

The following result comes with a trivial proof.

Proposition 9.20.11 *Let A be a module with a submodule $B \subset \mathrm{rad}(A)$, and let $f: A \to A/B$ be the quotient map. The correspondence theorem provides a one-to-one correspondence $C \to f(C)$ between maximal submodules of A and maximal submodules of A/B. Thus $\mathrm{rad}(A/B) = f(\mathrm{rad}(A))$. In particular, the radical of $A/\mathrm{rad}(A)$ vanishes.*

The following consequence is immediate.

Corollary 9.20.12 *The simple R-modules correspond to quotients by maximal ideals I of R, and R/I are also the simple $R/\mathrm{rad}(R)$-modules.*

Proposition 9.20.13 *For an artinian ring R the ring $R/\mathrm{rad}(R)$ is completely reducible.*

Proof Note that $\mathrm{rad}(R)$ is an ideal, so $R/\mathrm{rad}(R)$ is indeed a ring. It is artinian as an R-module since it is a quotient module of an artinian ring. By Proposition 9.18.25 it is therefore also artinian as a module over itself, so $R/\mathrm{rad}(R)$ is artinian as a ring. But the radical of $R/\mathrm{rad}(R)$ vanishes by the previous proposition. By Proposition 9.20.5 it is therefore completely reducible. □

Corollary 9.20.14 *For any module A over an artinian ring R, we have $\mathrm{rad}(A) = \mathrm{rad}(R)A$. The module A is semisimple if and only if $\mathrm{rad}(R)A$ vanishes.*

Proof By Corollary 9.20.8 we have $\mathrm{rad}(R)A \subset \mathrm{rad}(A)$. For the opposite inclusion observe that $A/\mathrm{rad}(R)A$ is a module over $R/\mathrm{rad}(R)$ which is semisimple by the proposition above. Thus $A/\mathrm{rad}(R)A$ is a semisimple module over $R/\mathrm{rad}(R)$ and thus also over R by the complementary submodule property of semisimple modules. So its Jacobson radical as an R-module vanishes by Proposition 9.20.3. But then $\mathrm{rad}(A) \subset \mathrm{rad}(R)A$ by Corollary 9.20.7 applied to the quotient R-module map $A \to A/\mathrm{rad}(R)A$. □

There is a slightly different way of looking at the Jacobson radical.

9.20 The Jacobson Radical

Definition 9.20.15 An element x of a module A is a *non-generator* of A if whenever A is generated by x and any $X \subset A$, then X alone will generate A.

We write $\langle X, x \rangle$ for the submodule generated by a subset X and an element x of a module.

Lemma 9.20.16 *Suppose A is a module generated by a subset X and an element x, but that $\langle X \rangle \neq A$. Then A has a maximal submodule that contains X but not x.*

Proof Consider the family of submodules of A that contain X but not x. This family is not empty since $\langle X \rangle$ belongs there. Ordered under inclusion by Zorn's lemma it has a maximal member. This is a maximal submodule since any submodule strictly larger than it must contain x, and would therefore have to be all of A. □

Proposition 9.20.17 *The subset of non-generators of a module is the Jacobson radical of the module.*

Proof Any non-generator x of a module A must belong to any maximal submodule B otherwise $\langle B, x \rangle = A$ and yet $B \neq A$. Thus $x \in \text{rad}(A)$.

Conversely, if x is not a non-generator of A, then there is some subset X of A such that $\langle X, x \rangle = A$ but $\langle X \rangle \neq A$. By the lemma above there is a maximal submodule B that contains X but not x. So $x \notin \text{rad}(A)$. □

The following useful result is known as *Nakayama's lemma*.

Theorem 9.20.18 *If A is a finitely generated module with a submodule B such that $A = \text{rad}(A) + B$, then $A = B$.*

Proof If A is generated by $\{a_1 + b_1, \cdots, a_n + b_n\}$ with $a_i \in \text{rad}(A)$ and $b_i \in B$, then

$$A = \langle b_1, \ldots, b_n, a_1, \ldots, a_n \rangle = \langle b_1, \ldots, b_n, a_1, \ldots, a_{n-1} \rangle = \cdots = \langle b_1, \ldots, b_n \rangle \subset B$$

by the characterization of the Jacobson radical in the proposition above. □

Corollary 9.20.19 *Let A be a finitely generated module with a submodule C. Then $C \subset \text{rad}(A)$ if and only if $A = B$ whenever B is a submodule of A such that $A = C + B$.*

Proof The forward implication is immediate from the theorem.

For the opposite implication observe that if B is a maximal submodule of A, then the submodule $C + B$ cannot be A, as B would then have to be A. So $C + B = B$, again by maximality of B. Thus $C \subset B$ and $C \subset \text{rad}(A)$. □

Corollary 9.20.20 *If A is a finitely generated R-module with a submodule B such that $A = \text{rad}(R)A + B$, then $A = B$. A finitely generated module A vanishes if $A = \text{rad}(R)A$.*

Proof By Corollary 9.20.8 we have $\text{rad}(R)A \subset \text{rad}(A)$, so $A = \text{rad}(A) + B$ and $A = B$ by the theorem above. This proves the first statement.

The second statement follows from the first by setting $B = \{0\}$. □

As above we can obviously replace $\text{rad}(R)$ in this corollary by any left ideal C of R contained in $\text{rad}(R)$. The case $A = CA + B \Rightarrow A = B$ can be obtained from the special case $A = CA \Rightarrow A = \{0\}$ by considering the module A/B.

If R is a unital ring considered as a finitely generated module over itself, then the inclusion $R \subset \text{rad}(R)$ is only possible when R is trivial, which is consistent with the fact that any non-trivial ring contains maximal ideals.

We can also relate the Jacobson radical of a unital ring to invertible elements.

Notice that if R is a unital ring with a left ideal A, then the subset $1 + A$ is closed under multiplication and inverses when they exist. To see this, suppose $b \in R$ is a left inverse of $1 + a$ with $a \in A$. Then $b(1 + a) = 1$, so $b = 1 + (-b)a \in 1 + A$.

Proposition 9.20.21 *Let A be a left ideal in a unital ring R. Then $1 + A$ is a subgroup of $U(R)$ if and only if $A \subset \text{rad}(R)$.*

Proof By the discussion prior to the proposition we need only show that the elements of $1 + A$ are invertile if and only if $A \subset \text{rad}(R)$.

For the forward implication, if A is not contained in some maximal left ideal B, then $R = A + B$ by maximality of B, so $1 = a + b$ for some $a \in A$ and $b \in B$. By assumption $b = 1 - a$ is invertible, and thus $B = A$, which is a contradiction. Hence $A \subset \text{rad}(R)$.

Conversely, consider $c \equiv 1 + a$ with $a \in A \subset \text{rad}(R)$. Then $R = A + Rc$. By the corollary above, we get $R = Rc$, and $1 = dc$ for some $d \in R$. But $d = 1 - da$ and $-da \in A$, so d like c, has also a left inverse, which necessarily must be c. Thus c has both a left and a right inverse. □

The following result displays a left-right symmetry in the Jacobson radical of a unital ring.

Corollary 9.20.22 *The Jacobson radical of a unital ring is the intersection of the maximal right ideals in the ring.*

Proof By Corollary 9.20.7 the Jacobson radical of a unital ring R is an ideal. Thus $\text{rad}(R)$ is a left ideal of the opposite unital ring R^{op}. By the proposition above $1 + \text{rad}(R)$ is a subgroup of $U(R) = U(R^{\text{op}})$. By the same proposition we get $\text{rad}(R) \subset \text{rad}(R^{\text{op}})$. Replacing R by R^{op} we get $\text{rad}(R^{\text{op}}) = \text{rad}(R)$. □

Thus for instance the previous proposition holds also for right ideals.

Proposition 9.20.23 *Let R be a unital ring, and let Q be the set of all $a \in R$ such that $1 + ba$ has a left inverse for every $b \in R$. Then $Q = \text{rad}(R)$.*

9.20 The Jacobson Radical

Proof If $a \in \mathrm{rad}(R)$ and $1 + a$ has no left inverse, then $R(1 + a) \neq R$, so $R(1 + a)$ is contained in a maximal left ideal B. But since the radical is the intersection of all such ideals, and $a \in \mathrm{rad}(R)$, we know that $a \in B$, and clearly $1 + a \in B$. But then also $1 \in B$, which is impossible. So $1 + a$ has a left inverse. Since Q obviously contains any left ideal that consists only of elements a such that $1 + a$ has a left inverse, we conclude that $\mathrm{rad}(R) \subset Q$.

The opposite inclusion follows by the proposition above and the remark prior to it provided we can show that Q is a left ideal, and we only need to show that $a + b \in Q$ for $a, b \in Q$ since closedness under multiplication with ring elements from the left is built into the definition of Q.

Since $a \in Q$, then to $c \in R$ we can find an element $d \in R$ such that $d(1 + ca) = 1$. Then $d(1 + c(a + b)) = 1 + dcb$ has a left inverse as $b \in Q$. But then also $1 + c(a + b)$ has a left inverse, so $a + b \in Q$. □

Note that Q above is the largest left ideal consisting of elements $a \in R$ such that $1 + a$ has a left inverse.

Similarly, by left-right symmetry, the Jacobson radical of a unital ring R coincides with the set of all $a \in R$ such that $1 + ab$ has a right inverse for every $b \in R$.

Proposition 9.20.24 *The Jacobson radical of a unital ring R vanishes if $U(R) \cup \{0\}$ is a ring.*

Proof Note that $S = U(R) \cup \{0\}$ is a division ring if it is a ring, and that $S \cap \mathrm{rad}(R)$ is a proper ideal in S if R is non-trivial. Thus $S \cap \mathrm{rad}(R)$ is trivial.

Now if $a \in \mathrm{rad}(R)$, then $1 + a \in U(R)$ by Proposition 9.20.21, so $1, 1 + a \in S$, and therefore $a \in S \cap \mathrm{rad}(R) = \{0\}$. □

Corollary 9.20.25 *Any polynomial ring $R[x_1, \ldots, x_n]$ over a division ring R has trivial Jacobson radical.*

Proof Arguing by degrees it is clear that an element of $R[x_1, \ldots, x_n]$ is invertible if and only if it is a non-zero element of R, so $U(R[x_1, \ldots, x_n]) \cup \{0\} = R$, and the result follows from the proposition above. □

In the next section we will generalize this result.

Proposition 9.20.26 *We have*

$$\mathrm{rad}(M_n(R)) = M_n(\mathrm{rad}(R))$$

for any unital ring R.

Proof Since the radical of a ring is an ideal we may by Proposition 9.17.6 write $\mathrm{rad}(M_n(R)) = M_n(J)$ for an ideal J of R. Let $a \in J$. By Proposition 9.20.21 the element $(1 + a)I_n = I_n + aI_n$ is invertible in $M_n(R)$, so $1 + a$ is invertible in R, and by the same proposition $a \in \mathrm{rad}(R)$. Thus $\mathrm{rad}(M_n(R)) \subset M_n(\mathrm{rad}(R))$.

For the opposite inclusion it it enough to show that $aE_{ij} \in \mathrm{rad}(M_n(R))$ for any $a \in \mathrm{rad}(R)$ and any matrix unit E_{ij}. By Proposition 9.20.21 this amounts to showing that $I_n + aE_{ij}$ is invertible. Let S be the commutative unital subring of R generated by a. The determinant of $I_n + aE_{ij} \in M_n(S)$ makes sense. It is 1 if $i \neq j$ and $1+a$ if $i = j$. By Cramer's rule for matrices over commutative rings $I_n + aE_{ij}$ is invertible since both 1 and $1+a$ are units by Proposition 9.20.21. □

9.21 The Wedderburn Radical

The set Nil(R) consisting of all nilpotent elements of a ring R is an ideal, known as the *Wedderburn radical*, when R is commutative, but otherwise this is not always true as we have seen. The following result shows that for an artinian unital commutative ring R we have Nil(R) = rad(R). This suggests that the natural generalization of the Wedderburn radical to non-commutative artinian unital rings is the Jacobson radical.

Proposition 9.21.1 *The Jacobson radical of a unital ring contains all left and right nil ideals. If in addition the ring is artinian, the Jacobson radical is a nilpotent ideal.*

Proof Any element a of a nil left ideal A in a unital ring R is nilpotent, so $1+a \in U(R)$ and thus $A \subset \text{rad}(R)$ by Proposition 9.20.21. The right ideal case holds by the comment below Corollary 9.20.22.

Assume in addition that the ring R is artinian. Then the descending chain rad(R) $\supset$ (rad(R))$^2 \supset \cdots$ of left ideals stabilizes, say at some power I of the Jacobson radical.

If I is non-trivial, then among all left ideals B such that IB is non-trivial, there is again by artinianess, a minimal one, say B. Pick $b \in B$ such that Ib is non-trivial. As $I^2b = Ib$, we get $Ib = B$ by minimality of B. Hence $b = cb$ for some $c \in I$. Then $(1-c)b = 0$ implies the absurdity $b = 0$ because $1 - c \in U(R)$ by Proposition 9.20.21. □

The proposition above tells us that left (right) nil ideals of artinian rings are left (right) nilpotent; a fact we have already established in Proposition 9.19.5.

The following characterization of artinian unital rings goes under the name of the *Hopkins-Levitzki theorem*, and is an exercise in change of rings.

Theorem 9.21.2 *A unital ring R is artinian if and only if it is noetherian and has nilpotent Jacobson radical and such that $R/\text{rad}(R)$ is completely reducible.*

Proof By Propositions 9.21.1 and 9.20.13 it suffices to show that R is artinian if and only if it is noetherian under the common assumption that rad(R) is nilpotent and $R/\text{rad}(R)$ is completely reducible. We prove this by induction on the smallest $n \in \mathbb{N}$ such that $(\text{rad}(R))^n$ is trivial.

Consider $S = R/(\text{rad}(R))^{n-1}$. Then $(\text{rad}(S))^{n-1} = \{0\}$ by Proposition 9.20.11. Clearly the natural composition $R/\text{rad}(R) \to S \to S/\text{rad}(S)$ is a well-defined ring epimorphism. Now $S/\text{rad}(S)$ is completely reducible as an $R/\text{rad}(R)$-module since the ring $R/\text{rad}(R)$ is artinian and semisimple, and thus $S/\text{rad}(S)$ is completely reducible as a ring over itself.

9.21 The Wedderburn Radical

By our induction hypothesis the ring S is artinian precisely when it is noetherian. By Proposition 9.18.25 this is equivalent to the assertion that S as an R-module is artinian precisely when it is noetherian. Thus by Proposition 9.18.8 it is enough to show that $(\text{rad}(R))^{n-1}$ as an R-module is artinian precisely when it is noetherian.

But to show this observe that $(\text{rad}(R))^{n-1}$ is an $R/\text{rad}(R)$-module because $(\text{rad}(R))^n$ is trivial. By Proposition 9.18.25 we only need to show that $(\text{rad}(R))^{n-1}$ as an $R/\text{rad}(R)$-module is artinian precisely when it is noetherian. By Corollary 9.18.11 this is true since the ring $R/\text{rad}(R)$ is semisimple. $\square$

Proposition 9.21.3 *The Jacobson radical of the upper triangular $n \times n$-matrices over a division ring consists of the matrices that are zero on the diagonal.*

Proof Let R be the upper triangular $n \times n$-matrices over a division ring S. Let J be the nilpotent ideal of R consisting of the matrices that are zero on the diagonal. Then by the previous proposition $J \subset \text{rad}(R)$. By Proposition 9.20.11 we have $\text{rad}(R)/J = \text{rad}(R/J) = \{0\}$ as $R/J \cong S^n$ is semisimple. $\square$

Definition 9.21.4 An element in an algebra over a field F is *algebraic over F* if it is a root of a polynomial with coefficients in F. The algebra is *algebraic* if every element of it is algebraic over F.

Finite dimensional algebras over fields are clearly algebraic, and any algebraic algebra is a union of such algebras since any subalgebra generated by one element is finite dimensional. Infinite algebraic field extensions of a field F are algebraic algebras over F. The group ring over a field of a torsion group is algebraic. The same is true for *locally finite groups*, that is, groups such that their finitely generated subgroups are finite.

Proposition 9.21.5 *Suppose a belongs to the Jacobson radical of a unital algebra over a field F. Then a is nilpotent if and only if a is algebraic over F.*

Proof The forward implication is obvious. For the opposite direction, suppose a is an element in the Jacobson radical of the algebra that satisfies

$$a^m + b_1 a^{m+1} + \cdots b_n a^{m+n} = 0$$

for $b_i \in F$. By Proposition 9.20.21 we know that $1 + b_1 a + \cdots b_n a^n$ is invertible, so $a^m = 0$. $\square$

The following result is now immediate.

Corollary 9.21.6 *The Jacobson radical of a unital algebraic algebra is the largest nil ideal.*

Theorem 9.21.7 *If the vector space dimension of a unital algebra over a field F has cardinality less than $|F|$, then the Jacobson radical of the algebra is the largest nil ideal.*

Proof By Proposition 9.21.1 it suffices to show that the Jacobson radical of the algebra R in question is a nil ideal. The same proposition tells us that the theorem holds when F is finite since then R is finite and is certainly artinian.

So by the proposition above it is enough to show that every $a \in \mathrm{rad}(R)$ is algebraic over an infinite field F. By Proposition 9.20.21 we see that $b - a = b(1 - b^{-1}a) \in U(R)$ for any non-zero $b \in F$. By our cardinality assumption the elements $b - a$ as b varies over F_* cannot be linear independent over F. So there are distinct elements $b_1, \ldots, b_n \in F_*$ such that
$$\sum_{i=1}^n c_i(b_i - a)^{-1} = 0$$
with $c_i \in F$ not all zero. Clearing denominators we see that a is a root of a polynomial f over F which is non-zero as $f(b_i) = c_i \prod_{j \ne i}(b_j - b_i) \ne 0$ for some i. □

Corollary 9.21.8 *For countably generated unital algebras over uncountable fields the Jacobson radical is the largest nil ideal.*

Proof Any algebra of the type in question obviously has a countable linear basis as a vector space over the field, so the theorem above applies. □

The corollary above is useful when one studies group rings of countable groups over real and complex fields.

Definition 9.21.9 An element a of a ring R is *von Neumann regular* if $a \in aRa$, and the ring R is a *von Neumann regular ring* if this holds for every element.

Clearly quotients and arbitrary direct products of von Neumann regular rings are von Neumann regular. Any unital ring in which every element a satisfies $a^n = a$ for some $n \geq 2$ is von Neumann regular, like Boolean rings for instance.

Proposition 9.21.10 *For a unital ring the following statements are equivalent:*

(i) *It is von Neumann regular;*

(ii) *Every principal left ideal of it is generated by an idempotent;*

(iii) *Every finitely generated left ideal of it is generated by an idempotent.*

Proof If (i) holds for a unital ring R, then given a principal ideal Ra, we can pick $b \in R$ such that $aba = a$. Then $e = ba$ is an idempotent such that $Re = Ra$, so (ii) holds.

Conversely, if (ii) is assumed and $a \in R$, then $Ra = Re$ for some idempotent $e \in R$. Thus $e = ca$ and $a = de$ for $c, d \in R$, and $aca = dee = a$, so (i) holds.

It only remains to check that (ii) implies (iii) since the converse is obvious. By induction it suffices to show that for any idempotents $a, b \in R$, the left ideal $A = Ra + Rb$ is generated by an idempotent. Write $A = Ra + Rb(1-a)$ with $Rb(1-$

9.21 The Wedderburn Radical

a) $= Rc$ for some idempotent c such that $ca \in Rb(1-a)a = \{0\}$. Then $A = Ra + Rc = R(a+c)$ because $Rc = Rc(a+c)$ as $c = c(a+c)$ and $Ra \subset R(a+c)$ as $a = (1-c)(a+c)$. So A is principal and is generated by an idempotent. $\square$

Due to the left-right symmetry in the definition of von Neumann regular rings the proposition above is also true for right ideals.

Corollary 9.21.11 *A unital semisimple ring is von Neumann regular. The radical of a von Neumann regular unital ring is trivial.*

Proof The first assertion is the last proposition and Theorem 9.19.10.

As for the second assertion, note that if $a = aba$ in a unital ring R, then $a(1-ba) = 0$. If $a \in \mathrm{rad}(R)$, then $1 - ba \in U(R)$ by Proposition 9.20.21, and therefore $a = 0$. $\square$

We have the following converse result of Maschke's theorem.

Proposition 9.21.12 *Let G be a finite group and let R be a unital ring such that $R[G]$ is a semisimple ring. Then R is semisimple and $|G|1$ is a unit in R.*

Proof The ring R is the image of the ring homomorphism $f \colon R[G] \to R$ given by $\sum a_s s \mapsto \sum a_s$, so R is semisimple since the homomorphic image of any semisimple ring is semisimple.

To prove that $|G|1$ is a unit, by the fundamental theorem of arithmetic it is enough to show that $p1$ is a unit in R for any prime number p that divides $|G|$. By Cauchy's theorem the group G has an element s of order p. By the corollary above the ring $R[G]$ is von Neumann regular, so there is an element $a \in R[G]$ such that $(1-s)a(1-s) = 1 - s$, or $(1 - (1-s)a)(1-s) = 0$.

We claim that if $b(1-s) = 0$ for $b \in R[G]$, then $b = c(1 + s + \cdots + s^{p-1})$ for some $c \in R[G]$. We prove this by induction on the number n of non-zero b_s in $b = \sum b_s s$. It is certainly true for $n = 0$. Assume it is true for all natural numbers less than n. Pick any t with $b_t \neq 0$. If no such t exists there is nothing to prove. Since $b1 = bs = bs^2 = \cdots$, the elements $t, ts, \ldots, ts^{p-1}$ all have the coefficient b_t in b. Thus
$$b = b_t t(1 + s + \cdots s^{p-1}) + d$$

for some $d \in R[G]$ that obviously has less non-zero coefficients than b. The claim follows now by the induction hypothesis.

Setting $b = 1 - (1-s)a$, we get $1 - (1-s)a = c(1 + s + \cdots + s^{p-1})$ for some $c \in R[G]$. Applying f to this identity yields $1 = f(c)(1 + \cdots + 1) = f(c)p$. $\square$

Proposition 9.21.13 *A unital ring is completely reducible if and only if it is noetherian and von Neumann regular.*

Proof The forward implication is immediate from Corollary 9.18.11 and the corollary above.

Since every left ideal in a noetherian ring is finitely generated, the second last proposition above and Theorem 9.19.10 give the backward implication. $\square$

Together with Corollary 9.18.11 the proposition above tells us that noetherian von Neumann regular unital rings are artinian.

Proposition 9.21.14 *The ring of endomorphisms of a semisimple module is von Neumann regular.*

Proof Let f be an endomorphism of a semisimple module A. Let B and C be complementary submodules of $\ker f$ and $\operatorname{im} f$, respectively. Define an endomorphism g of A to be zero on C and the inverse of $f|B$ on $\operatorname{im} f$. Then $fgf = f$. □

Of course any module over a division ring is semisimple, so matrix rings over division rings are von Neumann regular. Any product of such matrix algebras is also von Neumann regular, and this includes completely reducible rings in accordance with Proposition 9.21.13.

9.22 Radicals Under Change of Rings

We want to relate the Jacobson radical of a ring to that of a subring.

Proposition 9.22.1 *Let R be a subring of a unital ring S sharing the same identity. If R is a direct summand of S considered as an R-module, then $R \cap \operatorname{rad}(S) \subset \operatorname{rad}(R)$. The same conclusion holds if there is a group G of automorphisms of the ring S such that R consists of the elements in S fixed under the action of G.*

Proof Assume first that R is a direct summand of S, so that we have a complementary R-submodule A of R in S. Let a belong to the left ideal $R \cap \operatorname{rad}(S)$ of R. Then $1 + a \in U(S)$ by Proposition 9.20.21, and by the same proposition it suffices to show $1 + a \in U(R)$, which is accomplished by showing that $1 + a$ has a right inverse in R. Pick $b + c$ with $b \in R$ and $c \in A$ such that $1 = (1 + a)(b + c)$. Since $1 \in R$ and $(1 + a)b \in R$, whereas $(1 + a)c \in A$, we get $1 = (1 + a)b$.

If $a \in R \cap \operatorname{rad}(S)$, then by Proposition 9.20.21 we can write $(1 + a)d = 1$ for some $d \in S$. Let $\alpha \in G$. Then $1 = (1 + a)\alpha(d)$. Now $1 + a$ has a left inverse in S. Using this we get $\alpha(d) = d$, so $d \in R$, and thus $d \in \operatorname{rad}(R)$. □

The following result generalizes the fact that $f(\operatorname{rad}(R)) \subset (\operatorname{rad}(S))$ for any ring epimorphism $f : R \to S$, in the case when the rings are unital.

Proposition 9.22.2 *Suppose we have a ring homomorphism $f : R \to S$ and finitely many elements $a_i \in S$ that commute with every element of $f(R)$ and such that*

$$S = \sum f(R)a_i.$$

Then $f(\operatorname{rad}(R)) \subset \operatorname{rad}(S)$.

Proof By Corollary 9.20.9 it is enough to show that $f(\mathrm{rad}(R))A$ vanishes for every simple S-module A. Write $A = Sx$ for some $x \in A$. Then $A = \sum f(R)a_i x$ shows that A is finitely generated as an R-module via f. Since A is non-trivial we have a proper inclusion $f(\mathrm{rad}(R))A \subset A$ by Corollary 9.20.20.

At the same time $f(\mathrm{rad}(R))A$ is an S-module because

$$a_i f(\mathrm{rad}(R))A = f(\mathrm{rad}(R))a_i A \subset f(\mathrm{rad}(R))A,$$

and since A is simple as an S-module, the proper submodule $f(\mathrm{rad}(R))A$ must vanish. $\square$

From the proof above it is clear that we can dispense with the assumption that a_i should commute with $f(R)$ as long as $a_i f(\mathrm{rad}(R)) \subset f(\mathrm{rad}(R))S$.

One often talks about algebras over commutative rings rather than just fields.

Definition 9.22.3 By an *R-algebra* S over a commutative ring R we mean a ring S together with a homomorphism $R \to S$ into the center of S.

Normally we suppress the map $R \to S$ in the definition above. Note that S becomes an R-module, and that the ring multiplication in S is R-bilinear.

Corollary 9.22.4 *Let S be an R-algebra that is finitely generated as an R-module. Then $\mathrm{rad}(R)S \subset S$.*

If F is a field with an extension field E and R is a an algebra over F, then the extension of scalar ring $E \otimes_F R$ is an algebra over E under multiplication uniquely determined by $(a \otimes b)(c \otimes d) = ac \otimes bd$. We write R^E for the algebra $E \otimes_F R$ and will regard R as a subring of R^E under the identification $R \cong F \otimes_F R$.

Proposition 9.22.5 *Suppose F is a field with an extension field E and R is a unital algebra over F. Then $R \cap \mathrm{rad}(R^E) \subset \mathrm{rad}(R)$. Equality is achieved if R is a finite dimensional algebra or if E is an algebraic extension of F. If E is a finite extension of F then*

$$(\mathrm{rad}(R^E))^{[E:F]} \subset E \otimes_F \mathrm{rad}(R).$$

Proof Picking a linear basis $\{x_i\}$ for the vector space E over F, we see that $R^E \cong \bigoplus x_i R$ and $x_i R \cong R$ as R-modules. So R is a direct summand of R^E and $R \cap \mathrm{rad}(R^E) \subset \mathrm{rad}(R)$ by Proposition 9.22.1.

If R is finite dimensional as a vector space over F, then it is artinian, so $\mathrm{rad}(R)$ is nilpotent by Proposition 9.21.1. But then $E \otimes_F \mathrm{rad}(R)$ is a nil ideal of R^E, and by the same proposition, we get $E \otimes_F \mathrm{rad}(R) \subset \mathrm{rad}(R^E)$. Thus $\mathrm{rad}(R) \subset R \cap \mathrm{rad}(R^E)$, and equality is obtained by the first part of the proof.

Suppose E is a finite extension of F. Then $\{x_i\}$ is finite and each x_i commutes with $R \subset R^E$, so $\mathrm{rad}(R) \subset \mathrm{rad}(R^E)$ by Proposition 9.22.2.

If E is an algebraic extension field of F, write $E = \cup_X F(X)$, where the union is taken over all finite subsets X of E and $F(X)$ is the subfield of E generated by

X and F. By the previous paragraph we have $\operatorname{rad}(R) \subset \operatorname{rad}(R^{F(X)})$ for every finite subset X of E.

Now $A \equiv \cup_X \operatorname{rad}(R^{F(X)})$ is a left ideal of R^E because if $a \in \operatorname{rad}(R^{F(X)})$ and $b \in \operatorname{rad}(R^{F(Y)})$, then $a + b \in \operatorname{rad}(R^{F(X \cup Y)})$ by the previous observation as $F(X \cup Y)$ is a finite extension of $F(X)$ and of $F(Y)$. And if $c \in R^E$, say $c \in R^{F(Z)}$, then $ca \in \operatorname{rad}(R^{F(X \cup Z)})$ as $c \in R^{F(X \cup Z)}$ and $a \in \operatorname{rad}(R^{F(X \cup Z)})$.

By Proposition 9.20.21 we have $A \subset \operatorname{rad}(R^E)$ if $1 + a \in U(R^E)$, but the same proposition gives $1 + a \in U(R^{F(X)})$. Hence $\operatorname{rad}(R) \subset A \subset \operatorname{rad}(R^E)$, and thus $\operatorname{rad}(R) = R \cap \operatorname{rad}(R^E)$ by the first paragraph in the proof.

To prove the last assertion, let A be a simple R-module. Then the R^E-module $A^E \equiv E \otimes_F A$ has a composition series of length $[E : F]$ viewed as an R-module. So A^E viewed as an R^E-module, cannot have a composition series of length greater than $[E : F]$. By Corollary 9.20.10 we have $aA^E = \{0\}$ for any $a \in (\operatorname{rad}(R^E))^{[E:F]}$. Say $a = \sum x_i \otimes b_i$ with $b_i \in R$. Then for any $y \in A$ we have

$$0 = \left(\sum x_i \otimes b_i\right)(1 \otimes y) = \sum x_i \otimes b_i y \Rightarrow b_i y = 0,$$

which by Corollary 9.20.9 tells us that $b_i \in \operatorname{rad}(R)$. Thus $a \in E \otimes_F \operatorname{rad}(R)$. □

The following result shows that the inclusion $\operatorname{rad}(R) \subset R \cap \operatorname{rad}(R^E)$ does not hold if $\operatorname{rad}(R)$ is not a nilpotent ideal of R, which of course can happen for a non-algebraic extension E of F.

Proposition 9.22.6 *Let E be a non-algebraic extension of a field F. Then $R \cap \operatorname{rad}(R^E)$ is a nil ideal of R for any unital algebra R over F.*

Proof Let $a \in R \cap \operatorname{rad}(R^E)$ and pick $u \in E$ which is transcendental over F. By the first part of the previous proposition applied to the extension $F(u) \subset E$, we get $a \in R \cap \operatorname{rad}(R^{F(u)})$, so to prove that a is nilpotent we might as well assume that $E = F(u)$ from the beginning. Since $\operatorname{rad}(R^E)$ is an ideal in R^E and $u \in R^E$, we get $1 - au \in U(R^E)$ by Proposition 9.20.21. So there are

$$f(u) = b_0 + \cdots + b_n u^n \in R[u] \quad \text{and} \quad g(u) = c_0 + \cdots + c_{n+1} u^{n+1} \in F[u]$$

with $b_n \neq 0$ and such that $(1 - au)f(u)/g(u) = 1$.

Hence $c_i = b_i - ab_{i-1}$ upon comparing coefficients of powers of u. Solving this gives $b_i = a^i c_0 + a^{i-1} c_1 + \cdots + c_i$. Thus

$$0 = a^{n+1} c_0 + a^n c_1 + \cdots + c_{n+1}$$

and as $b_n \neq 0$, the c_i's are not all zero, so a is algebraic over F and certainly over E. Thus a is nilpotent by Proposition 9.21.5. □

Proposition 9.22.7 *Let E be a separable algebraic extension of a field F, and let R be a unital algebra over F. Then $\operatorname{rad}(R^E) = \{0\}$ if $\operatorname{rad}(R) = \{0\}$.*

9.22 Radicals Under Change of Rings

Proof Any $a \in \mathrm{rad}(R^E)$ is a finite sum of elementary tensors with the first factors in E. Let K be the subfield of E generated by F and these factors. Then K is a finite field extension of F, and obviously $a \in R^K$, so $a \in \mathrm{rad}(R^K)$ by the first part of Proposition 9.22.5. Since $K \subset E$, the field extension K of F is evidently also separable. By Proposition 7.8.2 we know that K is a simple extension of F generated by a separable element z. The splitting field $L = F(z_1, \ldots, z_n)$ over F of the minimal polynomial of $z_1 = z$ is a Galois extension of F; it is a separable extension of F by repeated application of Proposition 7.8.4 since the minimal polynomial of z_i over $F(z_1, \ldots, z_{i-1})$ divides the minimal polynomial of z_i over F, and the latter is a separable polynomial. Since L is a finite (and thus an algebraic) extension of K, we have

$$\mathrm{rad}(R^K) \subset \mathrm{rad}((R^K)^L)$$

by Proposition 9.22.5. Obviously $(R^K)^L = R^L$, so we are done if we can show that $\mathrm{rad}(R^L)$ vanishes. Therefore we may from the outset assume that E is a Galois extension of F.

Define an F-linear map $f \colon E \to F$ by $f(x) = \sum \alpha(x)$, where we are summing over all $\alpha \in G(E/F)$. That $f(x) \in F$ is clear from the fundamental theorem of Galois theory because $\sum \beta\alpha = \sum \alpha$ for any $\beta \in G(E/F)$. The F-linearity is immediate from the definition of $G(E/F)$. By Lemma 8.1.5 it is clear that f is non-zero, say with $f(x) \neq 0$. In fact, it is non-degenerate in the sense that $z = 0$ whenever $f(yz) = 0$ for all y otherwise we may pick $y = xz^{-1}$. Since F is a field the map f is also surjective. In fact, the map $E \to E^*$ given by $y \mapsto f(\cdot y)$ is a vector space isomorphism because its kernel is trivial, and the dual space E^* has the same dimension as E. Thus to any basis $\{x_i\}$ of E there are elements $y_i \in E$ such that $f(y_i x_j) = \delta_{ij}$.

The Galois group acts on R^E by $\alpha \otimes \iota$. Write $a = \sum x_j \otimes a_j$. Then

$$1 \otimes a_i = 1 \otimes \sum f(y_i x_j) a_j = \sum f(y_i x_j) \otimes a_j = \sum (\alpha \otimes \iota)((y_i \otimes 1)a) \in R \cap \mathrm{rad}(R^E)$$

since $\mathrm{rad}(R^E)$ is an ideal of R^E invariant under all automorphisms of R^E. Thus $a_i \in \mathrm{rad}(R)$ by the first part of Proposition 9.22.5, so $a_i = 0$ and $a = 0$. □

Theorem 9.22.8 *If E is a separable algebraic extension of a field F, then $\mathrm{rad}(R^E) = E \otimes_F \mathrm{rad}(R)$ for any unital algebra R over F.*

Proof The inclusion $\mathrm{rad}(R) \subset \mathrm{rad}(R^E)$ is clear from Proposition 9.22.5, and since $\mathrm{rad}(R^E)$ is an ideal in R^E, we get

$$E \otimes_F \mathrm{rad}(R) \subset (E \otimes 1)\mathrm{rad}(R) \subset (E \otimes 1)\mathrm{rad}(R^E) \subset \mathrm{rad}(R^E).$$

We evidently have a ring isomorphism

$$R^E / (E \otimes_F \mathrm{rad}(R)) \cong E \otimes_F (R/\mathrm{rad}(R)),$$

and $\mathrm{rad}(E \otimes_F (R/\mathrm{rad}(R))) = \{0\}$ by the proposition above as $\mathrm{rad}(R/\mathrm{rad}(R))$ vanishes by Proposition 9.20.11. Thus $\mathrm{rad}(R^E) \subset E \otimes_F \mathrm{rad}(R)$ by the same proposition. $\square$

The following example shows that separability is essential in the previous result.

Example 9.22.9 Let F be a field of prime characteristic p that has an element not in F^p. Let a be a pth root of this element. The minimal polynomial of a divides $x^p - a^p$, and $x^p - a^p = (x - a)^p$ in characteristic p. Hence the finite extension $F(a)$ of F is normal but not separable.

Now the Jacobson radical of the algebra $F(a)$ over F vanishes as $F(a)$ is a vector space over F and is therefore semisimple. However, the Jacobson radical of $F(a) \otimes_F F(a)$ is not trivial. As

$$F(a) \otimes_F F(a) \cong F(a) \otimes_F F[x]/(x^p - a^p) \cong F(a)[x]/(x^p - a^p) = F(a)[x]/(x - a)^p$$

we see that $\mathrm{rad}(F(a) \otimes_F F(a))$ is the nilpotent ideal generated by $(x - a)/(x - a)^p$.
$\diamondsuit$

The previous proposition implies that if R is a finite dimensional semisimple algebra over a field F, then the algebra $R^{\overline{F}}$ over the algebraic closure of F is semisimple when F is perfect, which guarantees that $\overline{F}$ is a separable extension of F. This should be compared with Corollary 9.14.10, where the algebra is generated by a single endomorphism, and is studied from the point of view of Jordan canonical forms.

Definition 9.22.10 An algebra over a field is *absolutely semisimple* if its extension by scalars to the algebraic closure is semisimple.

Proposition 9.22.11 *If R and S are finite dimensional absolutely semisimple algebras over a field F, then $R \otimes_F S$ is semisimple.*

Proof By assumption the algebras $R^{\overline{F}}$ and $S^{\overline{F}}$ are semisimple, so they are direct products of matrix algebras, and so is evidently $R^{\overline{F}} \otimes_{\overline{F}} S^{\overline{F}}$. Thus its Jacobson radical vanishes. But

$$R^{\overline{F}} \otimes_{\overline{F}} S^{\overline{F}} \cong \overline{F} \otimes_F (R \otimes_F S),$$

so $\mathrm{rad}(R \otimes_F S)$ vanishes by Proposition 9.22.5. $\square$

9.23 Radicals of Polynomial Rings

The following result is useful.

Proposition 9.23.1 *Let S be a multiplicatively closed subset of a commutative ring R. Then any ideal of R which is maximal among the ideals that do not intersect S is a prime ideal. In particular, such an ideal exists when $S \neq R$ by Zorn's lemma.*

9.23 Radicals of Polynomial Rings

Proof Let $a, b \notin I$ for an ideal I maximal among all ideals that do not intersect S. By maximality of I both $I + (a)$ and $I + (b)$ intersect S. If $ab \in I$ then by commutativity, some element of $(I + (a))(I + (b))$ will belong to both the multiplicative set S and the ideal I, which is absurd. □

Definition 9.23.2 Say R is a ring with an ideal I. Let $\mathrm{rad}_I(R)$ denote the intersection of the maximal ideals of R that contain I, and let $\mathrm{Nil}_I(R)$ be the set of $a \in R$ such that $a^n \in I$ for some natural number n.

Clearly $\mathrm{rad}_{\{0\}}(R) = \mathrm{rad}(R)$ and $\mathrm{Nil}_{\{0\}}(R) = \mathrm{Nil}(R)$. While $\mathrm{rad}_I(R)$ by Zorn's lemma is always an ideal for I proper, the set $\mathrm{Nil}_I(R)$ need not be an ideal in general; it is of course when R is commutative. In unital rings maximal ideals are prime, so for a proper ideal I in a unital ring R we have $\cap P \subset \mathrm{rad}_I(R)$, where the intersection is over the prime ideals that contain I.

Definition 9.23.3 A commutative ring is a *Hilbert ring* if every prime ideal is an intersection of maximal ideals.

Proposition 9.23.4 *If R is a unital commutative ring with a proper ideal I, then*

$$\mathrm{Nil}_I(R) = \cap P \subset \mathrm{rad}_I(R),$$

where we intersect over all prime ideals that contain I. The last inclusion is evidently an equality when R is a unital Hilbert ring.

Proof If $a \in \mathrm{Nil}_I(R)$, then $a^n \in I$ for some $n \in \mathbb{N}$. Thus $a^n \in P$ for any prime ideal P that contains I. By repeated use of the definition of primeness in the commutative case we get $a \in P$, so $\mathrm{Nil}_I(R) \subset \cap P$.

If $a \notin \mathrm{Nil}_I(R)$, then by the proposition above there is a prime ideal P that contains I and does not meet the multiplicatively closed subset $\{a^n \mid n \in \mathbb{N}\}$ of R, so $a^1 \notin P$. Thus $\mathrm{Nil}_I(R) = \cap P$. □

Corollary 9.23.5 *If R is a commutative ring, then $\mathrm{Nil}(R)$ is the intersection of all prime ideals of R.*

Theorem 9.23.6 *Let $R[X]$ be the polynomial ring over a commutative unital ring R with indeterminants in a non-empty set X. Then*

$$\mathrm{rad}(R[X]) = \mathrm{Nil}(R[X]) = \mathrm{Nil}(R)[X].$$

In particular, the Jacobson radical of $R[X]$ vanishes if and only if R has no non-zero nilpotent elements.

Proof Clearly $\mathrm{Nil}(R)[X] \subset \mathrm{Nil}(R[X])$. As $R/\mathrm{Nil}(R)$ has no non-zero nilpotent members, neither has $R[X]/\mathrm{Nil}(R)[X] \cong (R/\mathrm{Nil}(R))[X]$, so $\mathrm{Nil}(R)[X] = \mathrm{Nil}(R[X])$.

Now $\mathrm{Nil}(R[X]) \subset \mathrm{rad}(R[X])$ by Proposition 9.21.1. To prove the opposite inclusion, let $f \in \mathrm{rad}(R[X])$, and write

$$f = a_0 + a_1 s_1 + \cdots + a_n s_n,$$

where $a_i \in R$ and s_i are members of the free abelian monoid generated by X. Pick $x \in X$. Then $1 + xf \in U(R[X])$ by Proposition 9.20.21. Let P be a prime ideal of R with quotient map $h \colon R \to R/P$. Extend this to a unital ring homomorphism $h \colon R[X] \to (R/P)[X]$ which fixes the monoid elements. Then

$$1 + h(a_0)x + h(a_1)xs_1 + \cdots + h(a_n)xs_n = h(1+xf) \in U((R/P)[x]),$$

and since R/P is an integral domain, this is only possible if all $h(a_i) = 0$. So $a_i \in P$, and $a_i \in \mathrm{Nil}(R)$ by the corollary above. Thus $f \in \mathrm{Nil}(R)[X]$. □

Here is a kind of generalization to non-commutative rings.

Theorem 9.23.7 *Consider the polynomial ring $R[X]$ over a unital ring R with indeterminants in a non-empty set X. Then $R \cap \mathrm{rad}(R[X])$ is a nil ideal in R and $\mathrm{rad}(R[X]) = (R \cap \mathrm{rad}(R[X]))[X]$. So $R[X]$ has vanishing Jacobson radical if R has no non-trivial nil ideals.*

Proof Let $a \in R \cap \mathrm{rad}(R[X])$ and pick $x \in X$. Then $1 - ax \in U(R[X])$ by Proposition 9.20.21, say with inverse f, so $(1-ax)f = 1$. Applying to this identity the unital ring homomorphism that evaluates all the indeterminants different from x at zero, we get

$$(1-ax)(a_0 + a_1 x + \cdots + a_n x^n) = 1$$

for some $a_i \in R$ and n. Comparing coefficients of x gives

$$a_0 = 1, \quad a_1 = aa_0 = a, \quad \ldots, \quad a_n = aa_{n-1} = a^n, \quad 0 = aa_n = a^{n+1},$$

so $R \cap \mathrm{rad}(R[X])$ is a nil ideal in R.

The inclusion $(R \cap \mathrm{rad}(R[X]))[X] \subset \mathrm{rad}(R[X])$ is clear as $\mathrm{rad}(R[X])$ is an ideal of $R[X]$. The opposite inclusion, which remains to be proven, means that if $f \in \mathrm{rad}(R[X])$, then all its coefficients must also belong to $\mathrm{rad}(R[X])$. We prove this by induction on the finite number m of indeterminants appearing in f. The case $m = 0$ is obvious. Suppose we have proved such a statement for the polynomial ring in one variable over any unital ring. Picking any $y \in X$ that appears in f, and writing $f = \sum f_i y^i$ for $f_i \in R[X \setminus \{y\}]$, we conclude that all f_i belong to the Jacobson radical of $(R[X \setminus \{y\}])[y] = R[X]$. The number of variables appearing in f_i is less than m. Hence, by the induction hypothesis, the coefficients in R of f_i belong to $\mathrm{rad}(R[X])$, and so do the coefficients of f.

We can therefore assume that X consists of a single element x. We assert that if $f(x) = b_0 + \cdots + b_n x^n \in \mathrm{rad}(R[x])$, then all $b_i x^i \in \mathrm{rad}(R[x])$. This will suffice because

$$b_i + ib_i x + \cdots + b_i x^i = b_i(1+x)^i \in \mathrm{rad}(R[x])$$

by Corollary 9.20.7 as $x \mapsto x + 1$ defines a module automorphism on $R[x]$, and then by the assertion applied to this new polynomial, we get $b_i = b_i x^0 \in \mathrm{rad}(R[x])$.

9.23 Radicals of Polynomial Rings

We prove the assertion by induction on n. It is trivially true for $n = 0$, and we assume that it is true for all polynomials of degree less than n over any unital ring.

Pick any prime number p larger than n, and consider the quotient ring

$$S = R[u]/(1 + \cdots + u^{p-1}).$$

Then $u^p - 1 = (u - 1)(1 + \cdots + u^{p-1}) = 0$ in S, where u also denotes the equivalence class of $u \in R[u]$ in S. For any positive integer j less than p we have $[u]^j = 1$ in the quotient ring $S/(u^j - 1)$. This entails $[u] = 1$ for the unit $[u]$ because if $j > 1$, then j and p are relatively prime, so there are integers r, s such that $1 = rj + sp$, and thus $[u] = [u]^{rj+sp} = 1$. Hence $p1 = 1 + [u] + \cdots + [u]^{p-1} = [1 + \cdots + u^{p-1}] = 0$, so $p1$ belongs to the ideal $(u^j - 1)$ in S.

Since

$$S[x] = R[x] \oplus uR[x] \oplus \cdots \oplus u^{p-2}R[x]$$

and u is central in $S[x]$, we get $R[x] \cap \mathrm{rad}(S[x]) = \mathrm{rad}(R[x])$ by Propositions 9.22.1 and 9.22.2. By acting on $f(x) \in \mathrm{rad}(S[x])$ with the module automorphism on $S[x]$ given by $x \mapsto ux$, we get $f(ux) \in \mathrm{rad}(S[x])$ by Corollary 9.20.7. Hence

$$b_0(u^n - 1) + b_1(u^n - u)x + \cdots + b_{n-1}(u^n - u^{n-1})x^{n-1} = u^n f(x) - f(ux) \in \mathrm{rad}(S[x])$$

as $\mathrm{rad}(S[x])$ is an ideal. By the induction hypothesis applied to the above polynomial over S, we get $b_i(u^n - u^i)x^i \in \mathrm{rad}(S[x])$ for $i < n$. Since $\mathrm{rad}(S[x])$ is an ideal, we also get $b_i(u^{n-i} - 1)x^i = u^{-i}b_i(u^n - u^i)x^i \in \mathrm{rad}(S[x])$. Now $p1 \in (u^{n-i} - 1)$ by the previous paragraph, so $pb_i x^i$ belongs to the ideal $\mathrm{rad}(S[x])$.

Thus $qb_i x^i \in \mathrm{rad}(S[x])$ also for a prime number $q > n$ different from p. As p and q are relatively prime, there are integers k, l such that $1 = kp + lq$. Again by the ideal property we see that $b_i x^i = kpb_i x^i + lqb_i x^i \in \mathrm{rad}(S[x])$ for $i < n$. But since $f \in \mathrm{rad}(S[x])$, we also get $b_n x^n \in \mathrm{rad}(S[x])$. We have proved the induction step, so the assertion is verified, and the theorem holds. □

From the previous result we know that $R \cap \mathrm{rad}(R[X])$ is a nil ideal of R. A natural question is whether it is the largest nil ideal in R, that is, the sum of all nil ideals in R. In other words, will $N[X] \subset \mathrm{rad}(R[X])$ for every nil ideal N of R? When N is nilpotent this is true because if N^n is trivial, then so is $(N[X])^n$, and the inclusion holds by Proposition 9.21.1. But in general this is an open problem.

We have the following analogue of the theorem above when the ring is an algebra.

Theorem 9.23.8 *Let R be a unital algebra over a field F. Let $F(X)$ be the field of fractions of the polynomial ring $F[X]$ over F with indeterminants in a non-empty set X. Consider the unital algebra $R(X) \equiv F(X) \otimes_F R$ over $F(X)$. Then $R \cap \mathrm{rad}(R(X))$ is a nil ideal in R and $\mathrm{rad}(R(X)) = F(X) \otimes_F (R \cap \mathrm{rad}(R(X)))$. Thus the Jacobson radical of $R(X)$ vanishes if R has no non-trivial nil ideals.*

Proof That $R \cap \mathrm{rad}(R(X))$ is a nil ideal is immediate from Proposition 9.22.6 since $F(X)$ is a non-algebraic extension of F.

The proof of $\operatorname{rad}(R(X)) = F(X) \otimes_F (R \cap \operatorname{rad}(R(X)))$ in the one-variable case is analogous to that of the theorem above. So we will take for granted that whenever $a_i \in R$ and $\sum a_i x^i \in \operatorname{rad}(R(x))$, then $a_i \in \operatorname{rad}(R(x))$.

As for the many variable case consider a typical element $f/g \in \operatorname{rad}(R(X))$ with $f \in R[X]$ and a non-zero $g \in F[X]$. We must show that the coefficients of f/g are in $\operatorname{rad}(R(X))$, but since $f \in g \operatorname{rad}(R(X)) \subset \operatorname{rad}(R(X))$, it suffices to show this for f. As in the proof of the previous theorem we prove this by induction on the number of variables appearing in any such f. Write $f = \sum f_i x^i$ for an $x \in X$ appearing in the expression for f, where all f_i belong to the algebra $R(Y)$ over $F(Y)$ with $Y = X \backslash \{x\}$. Since

$$R(X) \cong F(X) \otimes_{F(Y)} F(Y) \otimes_F R \cong F(Y)(x) \otimes_{F(Y)} R(Y) \cong R(Y)(x),$$

the paragraph above yields $f_i \in \operatorname{rad}(R(X))$. As each f_i have one variable less than f, the induction hypothesis implies that their coefficients belong to $\operatorname{rad}(R(X))$. □

9.24 Radicals of Groups Rings

By Maschke's theorem we know that any group ring of a finite group over a field with characteristic not dividing the order of the group is semisimple. For infinite groups this breaks down.

Proposition 9.24.1 *The group ring of any infinite group over a unital non-trivial ring is never semisimple.*

Proof Let G be an infinite group over a unital non-trivial ring R. Consider the unital ring homomorphism $f: R[G] \to R$ given by $\sum a_s s \mapsto \sum a_s$.

Suppose $R[G]$ is semisimple. Then the ideal $\ker f$ of $R[G]$ has a complementary left ideal A with a such that $a_t \neq 0$ for some t. As $(\ker f) a \subset A \cap \ker f = \{0\}$ and $1 - s \in \ker f$ for all $s \in G$, we get $(1 - s) a = 0$, so $sa = a$. Hence $a_{st} = a_t \neq 0$ for infinitely many $s \in G$, rendering an element a that does not belong to the group ring. □

Here we shall study the weaker property of vanishing Jacobson radical. Let us first include an elementary result of a more general nature.

Proposition 9.24.2 *Let R be a unital ring and let H be a subgroup of a group G. Then $R[H] \cap \operatorname{rad}(R[G]) \subset \operatorname{rad}(R[H])$. If $R[H]$ has vanishing Jacobson radical for any finitely generated subgroup H of G, then $\operatorname{rad}(R[G])$ vanishes.*

Proof Suppose $a \in R[H] \cap \operatorname{rad}(R[G])$. By Proposition 9.20.21 there is $b \in R[G]$ such that $b(1 + a) = 1$. By the same proposition we aim to remove terms from b till $b \in R[H]$ and in such a way that we still have $b(1 + a) = 1$. Write $a = \sum_{s \in H} a_s s$ and $b = \sum_{t \in G} b_t t$, where only finitely many a_s's and b_t's are non-zero. Since the cosets of H form a partition of G, we have

9.24 Radicals of Groups Rings

$$e = b(1+a) = \sum_{r \in G/H} \sum_{t \in rH} b_t(t + \sum_{s \in H} a_s ts).$$

Thus, keeping only the terms in b with b_t corresponding to $r = e$, we get the desired inclusion.

For the last statement, if $c \in \text{rad}(R[G])$, then $c \in R[H]$, where H is the subgroup of G generated by the finitely many group elements with non-zero coefficients in c. Assuming that $\text{rad}(R[H])$ vanishes, we get $c \in R[H] \cap \text{rad}(R[G]) \subset \text{rad}(R[H]) = \{0\}$. □

Definition 9.24.3 An *involution on a ring* R is an additive anti-multiplicative map $* : R \to R$ such that $a^{**} = a$.

The identity map on a commutative ring is an involution. The transpose in a matrix ring over a commutative ring is an involution, and so is the complex conjugate on the field of complex numbers.

Proposition 9.24.4 *Suppose R is a ring with an involution such that $a_i = 0$ whenever $\sum a_i^* a_i = 0$. Then the group ring $R[G]$ of any group G has only trivial nil left ideals. Thus group rings over rational, real and complex numbers have no non-trivial nil left ideals.*

Proof Define the trace $\text{tr} \colon R[G] \to R$ to be the R-linear map that picks out the coefficient of the unit element of the group. Define an involution on $R[G]$ by $f^*(s) = f(s^{-1})^*$ for $f \in R[G]$ and $s \in G$. Then $\text{tr}(f^*f) = \sum f(s)^*f(s)$, so by assumption $f = 0$ if $\text{tr}(f^*f) = 0$.

Now if $R[G]$ has a non-trivial nil left ideal A, say with non-zero $f \in A$, then $g \equiv f^*f \in A$ is also non-zero by the first paragraph. Pick the greatest $n \in \mathbb{N}$ such that g^n is non-zero. Then $(g^n)^*g^n = g^{2n} = 0$, so $g^n = 0$ by the first paragraph, which is impossible. □

Definition 9.24.5 The *spectrum* of an element a in a unital algebra over a field F is the subset $\text{sp}(a)$ of F consisting of all scalars x such that $x1 - a$ is not invertible.

The spectrum of an element in $M_n(F)$ consists of all its eigenvalues, which is non-empty if the field F is algebraically closed. In the direct product F^X over a set X considered as a commutative algebra over a field F, the spectrum of an element $f \colon X \to F$ is its image.

Keep in mind that the spectrum of an element referes to the algebra it is a member of. If R is a unital subalgebra of a unital algebra S, meaning that they have the same identity, then the spectrum of $a \in R$ contains the spectrum of $a \in S$ since it is easier to find an inverse of $x1 - a$ in S than in the smaller algebra R.

Proposition 9.24.6 *If R is a unital algebra over a field, then $\text{sp}(a)$ vanishes for $a \in \text{rad}(R)$.*

Proof Since $\text{rad}(R)$ is a ideal in R, which is proper by the definition of maximal ideals, none of its elements can be invertible, so $0 \in \text{sp}(a)$.

If x is a non-zero scalar, then $x1 - a$ is invertible with inverse $x^{-1}(1 - x^{-1}a)^{-1}$, which exists by Proposition 9.20.21. □

Remark 9.24.7 This is a digression for readers with background in functional analysis. A unital complex C*-algebra is a complex involutive unital algebra which is a Banach space for a submultiplicative norm that satisfies $\|a\|^2 = \|a^*a\|$ for every element. By complex function theory it follows that $\|a\|^2 = \sup \text{sp}(a^*a)$.

The Jacobson radical of a complex involutive unital subalgebra R of a unital C*-algebra vanishes. Indeed, if $a \in \text{rad}(R)$, then $a^*a \in \text{rad}(R)$ since R is involutive and the Jacobson radical is an ideal. But then by the proposition above and the remark prior to it, the spectrum of a^*a in the C*-algebra is trivial, so $\|a\| = 0$ and $a = 0$. This covers a fairly broad class of algebras.

Using the Haar integral with respect to the counting measure, one may complete the group ring $\mathbb{C}[G]$ with involution $f^*(s) = \overline{f(s^{-1})}$ to a unital C*-algebra. Hence the Jacobson radical of $\mathbb{C}[G]$ vanishes, which incidentally implies the last part of Proposition 9.24.4. ◇

We have seen that the Jacobson radical of the complex group ring of any group vanishes. This can be generalized.

Theorem 9.24.8 *Let G be a group, and let F be a non-algebraic field extension of $\mathbb{Q}$. Then the Jacobson radical of the group ring $F[G]$ vanishes.*

Proof By Zorn's lemma there is a maximal subset X of F such that no $x \in X$ is algebraic over the subfield of F generated by $\mathbb{Q}$ and $X \setminus \{x\}$. Clearly the subfield of F generated by $\mathbb{Q}$ and X is then isomorphic to the field $\mathbb{Q}(X)$ of fractions of the polynomial ring $\mathbb{Q}[X]$, and F is an algebraic extension of $\mathbb{Q}(X)$.

By Theorem 9.23.8 we know that $\mathbb{Q}[G] \cap \text{rad}(\mathbb{Q}(X) \otimes_\mathbb{Q} \mathbb{Q}[G])$ is a nil ideal of the group ring $\mathbb{Q}[G]$, and that

$$\text{rad}(\mathbb{Q}(X) \otimes_\mathbb{Q} \mathbb{Q}[G]) = \mathbb{Q}(X) \otimes_\mathbb{Q} (\mathbb{Q}[G] \cap \text{rad}(\mathbb{Q}(X) \otimes_\mathbb{Q} \mathbb{Q}[G])).$$

Since $\mathbb{Q}[G]$ has no nil ideals by Proposition 9.24.4, the Jacobson radical of the ring $\mathbb{Q}(X) \otimes_\mathbb{Q} \mathbb{Q}[G]$ must therefore vanish.

Since we are in characteristic zero, the algebraic extension F of $\mathbb{Q}(X)$ is separable, and therefore by Proposition 9.22.7, the Jacobson radical of the group ring $F[G] \cong F \otimes_{\mathbb{Q}(X)} (\mathbb{Q}(X) \otimes_\mathbb{Q} \mathbb{Q}[G])$ must also vanish. □

One might conjecture that the result above holds for the group ring over any field of characteristic zero, that is, also for algebraic extensions of $\mathbb{Q}$. Since any algebraic extension of a field of characteristic zero is automatically separable, the conjecture would follow from Proposition 9.22.7, if one could prove that the Jacobson radical of $\mathbb{Q}[G]$ vanishes for any group G. However, this has not yet been verified.

We consider now the case when the characteristic of the field is not zero.

9.24 Radicals of Groups Rings

Definition 9.24.9 A *p'-group* is a group that has no elements of prime number order p.

By Corollary 4.20.3 and Lagrange's theorem, a group is a p'-group presicely when the prime number p does not divide its order.

Proposition 9.24.10 *Let F be a field of prime characteristic p. Then the group ring of a p'-group over F has no non-trivial nil left ideals.*

Proof Say G is a p'-group and assume that $F[G]$ has a nil left ideal with a non-zero element $a = \sum a_s s$. By multiplying from left with a suitable group element we may assume that $\mathrm{tr}(a) = a_e \neq 0$. Then

$$\mathrm{tr}(a^p) = \sum a_{s_1} \cdots a_{s_p},$$

where we sum over the set X of all tuples $(s_1, \ldots, s_p)$ such that $s_1 \cdots s_p = e$.

The cyclic group $\langle f \rangle$ of order p acts on X by $f \cdot (s_1, \ldots, s_p) = (s_p, s_1, \ldots s_{p-1})$ and partitions X into orbits with 1 or p elements. The tuples in an orbit with p elements contribute to $\mathrm{tr}(a^p)$ with the same scalar, so each such orbit produces a scalar p times, which in characteristic p vanishes. A singular orbit consists of a tuple with $s_1 = \cdots = s_p \equiv s$. Thus $s^p = e$. As p is prime, we get $s = e$ since we are in a p'-group. So we actually have only one single orbit, which then contributes with a_e^p. Thus $\mathrm{tr}(a^p) = \mathrm{tr}(a)^p$.

Repeated use of this identity gives $\mathrm{tr}(a^{p^n}) = \mathrm{tr}(a)^{p^n}$ for all $n \in \mathbb{N}$. Eventually the left hand side will vanish, whereas the right hand side will never vanish, and this is absurd. □

Proposition 9.24.11 *Suppose E is an algebraic extension of a field F of prime characteristic p. If G is a p'-group, then $\mathrm{rad}(E[G])$ vanishes whenever $\mathrm{rad}(F[G])$ is trivial.*

Proof By Proposition 9.22.5, if E is a finite extension of F then

$$(\mathrm{rad}(E[G]))^{[E:F]} \subset E \otimes_F \mathrm{rad}(F[G]) = \{0\},$$

and the nilpotent ideal $\mathrm{rad}(E[G])$ must vanish by the proposition above.

If E is not a finite extension of F, any $a \in \mathrm{rad}(E[G])$ will belong to $K[G]$ for the subfield K of E generated by F and the coefficients of a. By Corollary 7.2.12 the field K is a finite extension of F, and by Proposition 9.22.5 we have $a \in K[G] \cap \mathrm{rad}(E[G]) \subset \mathrm{rad}(K[G])$, and $\mathrm{rad}(K[G])$ is trivial by what we have already proved. □

Theorem 9.24.12 *Suppose F is a non-algebraic extension field of $\mathbb{Z}_p$. Then the group ring over F of any p'-group has trivial Jacobson radical.*

Proof Let G be a p'-group. By Proposition 9.24.10 we know that $\mathbb{Z}_p[G]$ has no non-trivial nil left ideals. Repeating the first two paragraphs in the proof of the previous theorem with $\mathbb{Q}$ replaced by $\mathbb{Z}_p$, we conclude that the Jacobson radical of $\mathbb{Z}_p(X) \otimes_{\mathbb{Z}_p} \mathbb{Z}_p[G]$ vanishes. Applying the proposition above to the algebraic extension F of $\mathbb{Z}_p(X)$, we get the desired result. □

One might conjecture that the result above holds for any field of prime characteristic p, that is, also for algebraic extensions of $\mathbb{Z}_p$. Since any algebraic extension of $\mathbb{Z}_p$ is automatically separable, the conjecture would hold by Proposition 9.22.7, if the Jacobson radical of $\mathbb{Z}_p[G]$ vanishes for any p'-group G; but this is unsettled despite various results pointing to such a property.

9.25 Units in Group Rings

It is easy to see that the units of the monoid ring over a unital ring R of a free monoid are the units of R. For group rings the situation is more complicated since group elements are obviously also units.

Definition 9.25.1 In the group ring $R[G]$ of a group G over a unital ring R the *trivial units* are the elements as with $a \in U(R)$ and $s \in G$.

Example 9.25.2 If $\langle x \rangle$ is the cyclic group of order 5, then $f = 1 - x^2 - x^3$ and $g = 1 - x - x^4$ are non-trivial units in the group ring of $\langle x \rangle$ over $\mathbb{Z}$ with $fg = 1$. ◇

The presence of torsion in a group also prevents the group ring even over a field from being a domain because if a is a group element of order n, then

$$(a-1)(a^{n-1} + \cdots + 1) = a^n - 1 = 0.$$

Proposition 9.25.3 *Group rings over domains of torsion-free abelian groups are domains with only trivial units.*

Proof Since one or two elements in a group ring has only finitely many non-zero coefficients, we may by restricting to the subgroup generated by the corresponding group elements, assume that the group in question is finitely generated. Since there is no torsion we may assume that it is a finitely generated free abelian group, and then its group ring is the ring of Laurent polynomials in finitely many indeterminants over a domain. By induction we can reduce to one indeterminant, and in this case a simple degree-argument will do. □

By Theorem 4.34.8 we know that among abelian groups the torsion-free groups are exactly the ordered groups. The following result is therefore a generalization of the proposition above.

Proposition 9.25.4 *The group ring over a domain of an ordered group has only trivial units and is a domain.*

9.25 Units in Group Rings

Proof Let a and b be two non-zero elements of the group ring in question, say with s and t the least group elements with non-zero coefficients a_s and b_t in a and b, respectively. Since the coefficients belong to a domain, in ab the least group element with non-zero coefficient $a_s b_t$ is st. So $ab \neq 0$ and the group ring is a domain. Also, if $ab = 1 = ba$, we can only have one non-zero term in a and in b, and $st = e = ts$ and $a_s b_t = 1 = b_t a_s$. So a and b are trivial units. $\square$

Obviously the Jacobson radical of any domain vanishes and there are no non-trivial nilpotent elements in a domain. The following converse result is less obvious.

Proposition 9.25.5 *Suppose the group ring $R[G]$ of a non-trivial group over a non-trivial unital ring has only trivial units. Then $R[G]$ has no non-zero nilpotent elements provided R has no non-zero nilpotent elements and G has no elements of order two. And $R[G]$ has vanishing Jacobson radical unless $|G| = |R| = 2$.*

Proof Let $a \in R[G]$ with $a^2 = 0$. The first claim follows if we can show that $a = 0$. Since $1 - a$ is a unit with inverse $1 + a$, there are by assumption $s \in G$ and $b \in U(R)$ such that $1 - a = bs$. If $bs \neq 1$, then as $0 = a^2 = 1 - 2bs + b^2 s^2$ and $s^2 \neq 1$ by assumption, we get a contradiction. Thus $a \in R$ and $a = 0$ as R has no non-zero nilpotent elements.

For the second claim, first note that if $|G| = |R| = 2$ with a generator $s \in G$, then $R[G] = \{0, 1, s, s + 1\}$ and $\text{rad}(R[G]) = \{0, s + 1\}$.

So assume that either $|G|$ or $|R|$ is greater than two. Let $a \in \text{rad}(R[G])$. By Proposition 9.20.21 we know that $1 - a \in U(R[G])$, so by assumption $1 - a = bs$ for some $s \in G$ and $b \in U(R)$.

If $|R| \geq 3$, there is a non-zero $c \in R$ with $c \neq 1$. Again by Proposition 9.20.21 the element $1 - ca = 1 - c + cbs$ is a unit, and it will be non-trivial unless $s = 1$.

If $|G| \geq 3$, there is $t \in G$ with $t \neq 1$ and $t \neq s^{-1}$. Now $1 - at = 1 - t + bst$ is a non-trivial unit unless $s = 1$.

In either case $a \in R$. But then for any $r \in G$ with $r \neq 1$, the unit $1 - ar$ must be trivial, so $a = 0$. $\square$

Corollary 9.25.6 *The group ring over a domain of a non-trivial ordered group has vanishing Jacobson radical.*

Proof This is evident from the previous two propositions. $\square$

We have the following satisfactory situation for abelian groups.

Theorem 9.25.7 *Let F be a field and let G be an abelian group. Then $F[G]$ has vanishing Jacobson radical when F has characteristic zero. When F has prime characteristic p, then $F[G]$ has vanishing Jacobson radical if and only if G is a p'-group.*

Proof If the characteristic of F is p and $F[G]$ has vanishing Jacobson radical, then G must be a p'-group, because if G has an element s of order p, then by abelianess of G, the left ideal $F[G](s - 1)$ is nilpotent as

$$(s-1)^p = (s^p - 1)(1 + \cdots + s^{p-1}) = 0.$$

Thus $\{0\} \neq F[G](s-1) \subset \mathrm{rad}(F[G])$ by Proposition 9.21.1.

Assume next that F has characteristic zero, or that it has characteristic p and that G is a p'-group. By Proposition 9.24.2 we may assume that G is finitely generated. Then by the fundamental theorem for finitely generated abelian groups, we can write $G = K \times H$, where K is a finite torsion group and H is a free abelian group of finite rank. Since the characteristic of F does not divide the order of K, we know that $R = F[K]$ is semisimple by Maschke's theorem. Thus $R \cong F_1 \times \cdots \times F_n$ for suitable fields F_i as R is commutative. Evidently $F[G] \cong R[H]$, and

$$F_1[H] \times \cdots \times F_n[H] \to R[H]; \quad (a_{s_1}s_1, \ldots, a_{s_n}s_n) \mapsto a_{s_1}\cdots a_{s_n}s_1\cdots s_n$$

is an isomorphism. Hence by Proposition 9.20.4 it suffices to see that the Jacobson radical of each $F_i[H]$ vanishes, and this is clear from Proposition 9.25.3. □

We will presently show that if the group ring of a torsion free-group over a domain has only trivial units, then the group ring itself is a domain.

Let H be a subgroup of a group G, and let R be a unital ring. Under left- and right multiplication both $R[G]$ and $R[H]$ are obviously $R[H]R[H]$-bimodules, and the projection map $f \colon R[G] \to R[H]$ given by

$$f\left(\sum_{s \in G} a_s s\right) = \sum_{s \in H} a_s s$$

is clearly an $R[H]R[H]$-bimodule map.

Lemma 9.25.8 *Let notation be as in the previous paragraph. If A is a left ideal of $R[G]$, then $A \subset R[G]f(A)$. Thus $f(A)$ is non-trivial when A is non-trivial.*

Proof Write $G = \cup_{x \in X} xH$ for a set X of coset representatives. Thus for $a \in A$ there are $a_x \in R[H]$ such that $a = \sum_{x \in X} xa_x$. Multiplying this from left by x^{-1} and using the definition of f, we get $a_x = f(x^{-1}a) \in f(A)$ since A is a left ideal. □

Proposition 9.25.9 *Let R be a unital ring, let G be a group and let H be the subgroup of all elements having finitely many distinct conjugates. Let f be the R-linear projection map $R[G] \to R[H]$ that fixes the elements of H and kills those in $G \backslash H$. Then $f(a)f(b) = 0$ for any $a, b \in R[G]$ with $aR[G]b$ trivial.*

Proof Since $f(a)f(b) = f(f(a)b)$ by the bimodule property of f, let us assume $f(a)b \neq 0$ and derive a contradiction. Fix $s \in G$ having a non-zero coefficient in $f(a)b$. Pick a finite set F of elements in G implementing a conjugation whenever there is one between group elements having non-zero coefficients in $f(a) - a$ and s times the inverse of those with non-zero coefficients in b.

Form the intersection $K = \cap N(t)$ over the finitely many $t \in H$ having a non-zero coefficient in $f(a)$. Then $[G : K] < \infty$ by the assumption on H. Let $u \in K$. Since

9.25 Units in Group Rings

$aR[G]b = \{0\}$, we have $f(a)b = u(f(a) - a)u^{-1}b$, so $s = uvu^{-1}w$ for some v and w having non-zero coefficients in $f(a) - a$ and b, respectively. So $sw^{-1} = xvx^{-1}$ for some $x \in F$. Hence $u = x(x^{-1}u) \in xN(v)$ and $K \subset \cup xN(v)$, where the union is over $x \in F$ and the finitely many v with non-zero coefficients in $f(a) - a$. As G is a finite union of K-cosets, we see that G is finite union of $N(v)$'s each having infinite index in G as $v \in G\setminus H$, and this contradicts Proposition 4.4.4. □

Theorem 9.25.10 *The group ring over a domain of a torsion-free group is a domain if it has no non-zero nilpotent elements.*

Proof Let R be a domain and G a torsion-free group such that $R[G]$ has no non-zero nilpotent elements. If $R[G]$ is not a domain, there are non-zero elements $a, b \in R[G]$ such that $ba = 0$. Then for any $c \in R[G]$ we have $(acb)^2 = 0$, so $acb = 0$ since $R[G]$ has no non-zero nilpotent elements.

Let H be the subgroup of G of all elements having finitely many distinct conjugates, and let f be the R-linear projection map $R[G] \to R[H]$ that fixes the elements of H and kills those in $G\setminus H$. Note that H is abelian by Corollary 4.29.5 and hence $R[H]$ is a domain by Proposition 9.25.3.

Since $b \ne 0$ we have $f(R[G]b) \ne 0$ by the previous lemma, so there is $b' \in R[G]$ with $f(b'b) \ne 0$. Similarly, there is $a' \in R[G]$ with $f(aa') \ne 0$. Yet $f(aa')f(b'b) = 0$ by the proposition above as $aa'b'b = 0$ by the first paragraph. This contradicts the fact that $R[H]$ is a domain. □

Corollary 9.25.11 *The group ring over a domain of a torsion-free group is a domain if it has only trivial units.*

Proof Since domains cannot have non-zero nilpotent elements, the corollary is immediate from the theorem above and Proposition 9.25.5. □

Remark 9.25.12 Here we will use elementary functional analysis and complex function theory to conclude that the Jacobson radical of any complex unital algebra $\mathcal{A}$ is trivial if it is a subalgebra of a unital Banach algebra A.

Suppose we have a non-zero $a \in \text{rad}(\mathcal{A})$, so $1 - xa \in U(\mathcal{A})$ for any $x \in \mathbb{C}$ by Proposition 9.20.21. For any norm bounded functional g on A we can therefore define a complex function $f \colon \mathbb{C} \to \mathbb{C}$ by $f(x) = g((1-xa)^{-1})$.

This function is entire. To see this first note that $\|a(1-xa)^{-1}\| \ne 0$ since

$$0 < \|a\| = \|a(1-xa)^{-1}(1-xa)\| \le \|a(1-xa)^{-1}\| \cdot \|(1-xa)\|$$

by submultiplicativity of the norm. Then for any $y \in \mathbb{C}$ with $|y| < \|a(1-xa)^{-1}\|^{-1}$ we have by recognizing a Neumann series, that

$$(1-(x+y)a)^{-1} = ((1-xa)(1-ya(1-xa)^{-1}))^{-1} = (1-xa)^{-1} \sum_{n=0}^{\infty} (ya(1-xa)^{-1})^n$$

with convergence in norm. Hence

$$f(x+y) = \sum_{n=0}^{\infty} g(a^n(1-xa)^{-n-1}) y^n$$

by linearity and continuity of g. This shows that f is holomorphic on the entire complex plane.

When $|x| < \|a\|^{-1}$, then by recognizing more Neumann series, we have

$$|xf(x)| \le |x| \cdot \|g\| \cdot \|\sum_{n=0}^{\infty}(xa)^n\| \le |x| \cdot \|g\| \sum_{n=0}^{\infty} \|xa\|^n = |x| \cdot \|g\| \,(1 - |x| \cdot \|a\|)^{-1},$$

which is zero for $x = 0$. So we have an entire complex function $x \mapsto xf(x)$ that is identically zero by Liouville's theorem, which is a contradiction due to the richness of bounded functionals supplied by the Hahn–Banach theorem. $\diamond$

9.26 Division Rings

In view of Theorem 9.17.11 it is important to understand division rings well. The theory of division rings is rich and uses both field theory and group theory. This is due to the facts that the center $Z(R)$ of a division ring R is a field, and that its non-zero elements R_* is a group under multiplication. The following beautiful result illustrates this interplay.

Theorem 9.26.1 *Finite division rings are fields. In particular, finite unital subrings of division rings are fields.*

Proof The center of a finite non-trivial division ring R is a finite field, say with cardinality $m \ge 2$ as it must be a power of its characteristic. Let n be the dimension of R as a vector space over its center. Suppose $n \ge 2$.

The class formula for the finite group R_* is

$$|R_*| = Z(R_*) + \sum_{a \in D}[R_* : N(a)],$$

where D is a subset of R_* that contains exactly one element from each non-single conjugacy class. Let $k(a)$ be the dimension of $N(a) \cup \{0\}$ as a vector space over the center of R, so $k(a)$ is less than n and divides n. The class formula becomes

$$m^n - 1 = m - 1 + \sum_{a}(m^n - 1)/(m^{k(a)} - 1).$$

Since $k(a)$ divides n, we have

$$x^n - 1 = (x^{k(a)} - 1) f_a(x) \Phi_n(x)$$

9.26 Division Rings

for some $f_a \in \mathbb{Z}[x]$. Thus $m - 1$ is divisible by the cyclotomic polynomial Φ_n evaluated at m. So

$$|\Phi_n(m)| = \prod |m - b| \leq m - 1,$$

where b ranges over all primitive roots of $x^n - 1$. As $m \geq 2$, geometrically it is easy to see that $|m - b| > m - 1 \geq 1$ for any complex number b on the unit circle different from 1. But for $n \geq 2$ no primitive roots are 1, which contradicts the last formula above. Hence n must be one, so R coincides with its center and is a field.

For the last statement in the theorem, if $a \neq 0$ belongs to a finite subring S of R, we may assume that $a^n = a^m$ for some natural numbers $n > m$. Then $a^{-1} = a^{n-m-1} \in S$, so S is a finite division ring, and is therefore a field. □

Corollary 9.26.2 *Given a division ring R of positive characteristic, then any finite subgroup of R_* is cyclic.*

Proof The subring S of R generated by the subgroup G in question and the prime field of R is finite as the characteristic of R is positive. By the theorem above S is a field. Then G, being a subgroup of S_*, must be cyclic. □

The corollary breaks down for division rings R of characteristic zero, but the possible finite subgroups of R_* can be characterized also in this case.

Example 9.26.3 The quaternions H is isomorphic to $\mathbb{R}1 \oplus \mathbb{R}i \oplus \mathbb{R}j \oplus \mathbb{R}ij$ as rings, where

$$1 = \begin{pmatrix} 1 & 0 \\ 0 & 1 \end{pmatrix}, \quad i = \begin{pmatrix} i & 0 \\ 0 & -i \end{pmatrix}, \quad j = \begin{pmatrix} 0 & -1 \\ 1 & 0 \end{pmatrix},$$

so $i^2 = -1 = j^2$ and $ij = -ji$. It is a division ring of characteristic zero, in fact, a real division algebra over its center, and H_* has the finite non-cyclic subgroup $\{\pm 1, \pm i, \pm j, \pm ij\}$. ◊

Theorem 9.26.4 *Any real finite dimensional division algebra is either isomorphic to $\mathbb{R}$ or $\mathbb{C}$ or H.*

Proof Suppose R is such a division algebra. We may assume that its dimension as a vector space over $\mathbb{R}$ is greater than one. Since this dimension is finite, the elements of R are algebraic over $\mathbb{R}$ in the sense that the field $\mathbb{R}[a]$ is an algebraic extension of $\mathbb{R}$ for any $a \in R$.

Pick $a \in R \backslash \mathbb{R}$. Then $\mathbb{R}[a]$ is a proper algebraic extension of $\mathbb{R}$, and since $\mathbb{C}$ is algebraically closed, it must be a subfield of $\mathbb{C}$. But $[\mathbb{C} : \mathbb{R}] = 2$, so $\mathbb{R}[a] \cong \mathbb{C}$. Fix a copy of $\mathbb{C}$ with an i in R, and view R as a vector space over $\mathbb{C}$.

Let $R^+ = \{b \in R \mid bi = ib\}$ and $R^- = \{b \in R \mid bi = -ib\}$. Now $R^+ \cap R^- = \{0\}$ and

$$b = (2i)^{-1}(ib + bi) + (2i)^{-1}(ib - bi) \in R^+ + R^-$$

for any $b \in R$, so $R = R^+ \oplus R^-$ as complex vector spaces. By definition $\mathbb{C} \subset R^+$. Any $b \in R^+$ is a root of a real polynomial, so $\mathbb{C}[b]$ is an algebraic extension field of $\mathbb{C}$. Thus $\mathbb{C}[b] = \mathbb{C}$ as $\mathbb{C}$ is algebraically closed, and $R^+ = \mathbb{C}$.

We may assume that R^- is non-trivial. Fix a non-zero element $c \in R^-$. As $d \mapsto dc$ is a $\mathbb{C}$-linear injection from R^- to R^+, the dimension of R^- is 1 as a complex vector space. So R has real dimension four.

Since c is algebraic over $\mathbb{R}$, we must have $c^2 \in \mathbb{R} + \mathbb{R}c$, and since also $c^2 \in \mathbb{C}$, we get $c^2 \in \mathbb{R}$. If $c^2 > 0$, then $c^2 = r^2$ for $r \in \mathbb{R}$. But then $c = \pm r \in \mathbb{R}$, which is impossible. So $c^2 < 0$. Then $c^2 = -r^2$ for some $r \in \mathbb{R}_*$. Put $j = c/r$. Then $j^2 = -1 = i^2$ and $ij = -ji$, so R is isomorphic as an algebra to $\mathbb{R}1 \oplus \mathbb{R}i \oplus \mathbb{R}j \oplus \mathbb{R}ij$. □

The division ring in the above theorem need not be finite dimensional over $\mathbb{R}$, all we needed was that its elements were algebraic over $\mathbb{R}$. Recall that an algebra is *algebraic* if its elements are algebraic over the underlying field. Note that an algebraic unital algebra with no zero-divisors is automatically a division algebra. Indeed, any non-zero element a of such an algebra over a field F will be a root of a polynomial. Cancel a sufficiently many times so that $af(a)$ is a non-zero element $b \in F$ for a polynomial $f \in F[x]$. Then $b^{-1}f(a)$ belongs to the algebra and will be the inverse of a, so the algebra is a division ring.

Proposition 9.26.5 *Elements of a division ring that commute with all commutators of the ring belong to its center. In particular, if the commutators are central, then the division ring is a field. A non-commutative division ring is generated as a division ring by its center and its commutators.*

Proof Let R be a division ring. If $a \notin Z(R)$, then $ab \neq ba$ for some $b \in R$. The identity
$$a(ab) - (ab)a = a(ab - ba)$$
contradicts the first statement in the proposition, and it verifies the third statement. □

Definition 9.26.6 An additive map δ on a ring R is a *derivation of the ring* if $\delta(ab) = \delta(a)b + a\delta(b)$. The map $\delta_a \colon R \to R$ given by $\delta_a(b) = ab - ba$ is a derivation called an *inner derivation*.

Proposition 9.26.7 *Suppose R is a division ring with characteristic different from 2. If S is a proper division subring of R that is invariant under the inner derivations of R, then $S \subset Z(R)$.*

Proof We claim that any element a in the complement of S must commute with any $b \in S$. To see this, note that
$$2a\delta_a(b) = \delta_a^2(b) + \delta_{a^2}(b) \in S.$$

If $\delta_a(b)$ is non-zero, then twice this element can be inverted in S, so $a \in S$, which is impossible.

If $c \in S_*$, then both a and ac are in the complement of S, so they must commute with b by the previous argument. Hence b commutes with $c = a^{-1}ac$. All in all $b \in Z(R)$, so S is in the center of R. □

9.26 Division Rings

Lemma 9.26.8 *Suppose a is a non-central torsion element of a division ring R with positive characteristic. Then there exists $c \in R_*$ such that $cac^{-1} = a^n \neq a$ for some $n \in \mathbb{N}$. This can moreover be done with a commutator c in R_*.*

Proof Let F be the field generated by a, so $|F| = p^m$ for some prime number p, and $a^{p^m} = a$. Since a is not central the inner projection δ_a is non-zero on R, but zero on F, so it is F-linear on R considered as a vector space over F. We will show that δ_a has an eigenvector.

Write $\delta_a = \lambda - \rho$, where $\lambda, \rho \colon R \to R$ are F-linear maps given by $\lambda(b) = ab$ and $\rho(b) = ba$. Since λ and ρ commute, we get in characteristic p that

$$\delta_a^{p^m}(b) = (\lambda - \rho)^{p^m}(b) = \lambda^{p^m}(b) - \rho^{p^m}(b) = a^{p^m}b - ba^{p^m} = ab - ba = \delta_a(b).$$

Consider the factorization

$$x^{p^m} - x = \prod_{b \in F}(x - b),$$

which holds by Proposition 7.7.2. This gives

$$0 = \delta_a^{p^m} - \delta_a = \prod_{b \in F_*}(\delta_a - b) \cdot \delta_a$$

and as $\delta_a \neq 0$, the map $\delta_a - b$ is not injective for some $b \in F_*$. So for some $c \in R_*$ we have $\delta_a(c) = bc$, which says that c is an eigenvector of δ_a with eigenvalue $b \in F_*$.

As $ac - ca = bc$, we get $cac^{-1} = a - b \in F\setminus\{a\}$. Now a and cac^{-1} have the same order in the finite cyclic group F_*, so by Proposition 4.7.4 they generate the same cyclic subgroup. Hence $cac^{-1} = a^n$ for some $n \in \mathbb{N}$.

Replacing c by the commutator $ac - ca \neq 0$ we still get $cac^{-1} = a^n \neq a$. □

Theorem 9.26.9 *A division ring is a field if every non-zero commutator has finite order.*

Proof If the center of such a division ring R is not all of R, there is by Proposition 9.26.5 a commutator $a \notin Z(R)$. If R does not have characteristic 2, then as both a and $2a$ are non-zero commutators, by assumption they have a common finite order m, and

$$1 = (2a)^m = 2^m a^m = 2^m.$$

This shows that R has positive characteristic. Since a is a non-central torsion element, the lemma above yields a commutator $c \in R_*$ such that $cac^{-1} = a^n \neq a$ for some $n \in \mathbb{N}$. Since c is a non-zero commutator, by assumption it has finite order. Thus a and c generate a finite subgroup of R_*, which by Corollary 9.26.2 is cyclic, and this contradicts $cac^{-1} \neq a$. □

Corollary 9.26.10 *If R is an infinite division ring with center F, then the division subring generated by F and an element a of R is contained in an infinite subfield of R. In particular, the centralizer $\{b \in R \mid ab = ba\}$ of a in R is infinite.*

Proof We may assume that $F \neq R$, that $a \notin F$, and that the subring $F(a)$ generated by F and a is finite. By the lemma above there is $c \in R_*$ such that $aca^{-1} = a^n \neq a$ for some $n \in \mathbb{N}$. Since a has finite order the element c^n commutes with a for some $n \in \mathbb{N}$. The division ring generated by a and c^n and F is therefore a field E which clearly contains $F(a)$. And E is infinite because if c had finite order, then c and a would generate a finite subgroup of R_*, and such a group would be cyclic by Corollary 9.26.2. This contradicts $cac^{-1} \neq a$. □

Corollary 9.26.11 *Any algebraic division algebra over a finite field F is an algebraic field extension of F.*

Proof The field generated by F and any non-zero element a of the division algebra is finite, so a has finite order. Since this holds for all non-zero elements of the division ring, it must certainly hold for non-zero commutators. By the theorem above the division ring is commutative, and is thus an algebraic field extension of F. □

Thus algebraic division algebras over finite fields and over $\mathbb{R}$ are well understood. Finite dimensional algebraic division algebras over $\mathbb{Q}$ have been classified, whereas the situation for infinite dimensional ones is more open.

Definition 9.26.12 Elements of the type $a^{-1}b^{-1}ab$ for invertible elements a, b of a ring are called *multiplicative commutators of the ring*.

Proposition 9.26.13 *Let R be a division ring. The center of R consists of all elements that commute with the multiplicative commutators of the ring. Thus if the multiplicative commutators of R belong to $Z(R)$, then R is a field. If S is a proper division subring of R such that S_* is a normal subgroup of R_*, then $S \subset Z(R)$. Any division ring is generated by the conjugates of any non-central element. If R is non-commutative, it is generated as a division ring by its multiplicative commutators.*

Proof Any $c \in R$ that commutes with all multiplicative commutators of R and satisfies $ca \neq ac$ for some $a \in R$ contradicts the identity

$$a(a^{-1}cac^{-1} - b^{-1}cbc^{-1}) = 1 - b^{-1}cbc^{-1}$$

with $b = a - 1$. This proves the first two claims.

Any $a \in R \backslash S$ and $c \in S$ that do not commute will contradict the identity

$$a(a^{-1}ca - b^{-1}cb) = c - b^{-1}cb$$

with $b = a - 1$. So such a and c must commute. If $d \in S_*$, then as $ad \notin S$, we conclude from the previous argument that $d = a^{-1}ad$ must commute with c. So $c \in Z(R)$ and $S \subset Z(R)$.

If T is the division subring of R generated by the conjugates of a non-central element r of R, then T_* is a normal subgroup of R_* since for any $s \in R_*$, the set $s^{-1}Ts$ is a division ring that will contain all conjugates of r, so $T \subset s^{-1}Ts$ and $sTs^{-1} \subset T$. Since $r \notin Z(R)$, the previous paragraph tells us that $T = R$.

9.26 Division Rings

Suppose R is non-commutative. The group of non-zero elements of a division subring of R that is generated by the multiplicative commutators of R is obviously normal in R_*. The subring must be all of R since R is non-commutative, so there are non-central multiplicative commutators by the second statement in the proposition, and then the third statement forbids the subring to be proper. □

Theorem 9.26.14 *A division ring is a field if and only if its multiplicative group is nilpotent.*

Proof If the division ring R is a field, then R_* is abelian and nilpotent.

Conversely, if R_* is nilpotent but non-abelian, then we may assume that there is $c \notin Z(R)$ such that $d^{-1}cdc^{-1} \in Z(R_*)$ for every $d \in R_*$. Since $c \notin Z(R)$, there is $a \in R$ such that $ca \neq ac$. This contradicts the first identity in the proof of the proposition above since that identity forces a to be in the center of R. □

Proposition 9.26.15 *If S is a proper division subring of a division ring R, then R is finite if and only if R_*/S_* is finite.*

Proof Let u, v be right S-independent vectors in R, and define $f: S \to R_*/S_*$ by

$$f(a) = (u + va)S_*.$$

This map is injective because if $(u + va)S = (u + vb)S$, then $u + va = (u + vb)c$ for some $c \in S$, so $c = 1$ and $a = bc = b$. If R_*/S_* is finite, then S and hence R must therefore be finite. □

Definition 9.26.16 If S is a division subring of a division ring R, then an *S-conjugate* of $a \in R$ is an element in R of the form bab^{-1} for $b \in S_*$.

Now such a ring S acts by conjugation on the set of S-conjugates of a, and the isotropy subgroup of this transitive action is T_*, where T is the division subring of S consisting of elements that commute with a. Thus the set of S-conjugates of a is in one-to-one correspondence with S_*/T_*. The following result is then immediate.

Corollary 9.26.17 *If a division subring S of a division ring R is infinite, then either there is only one S-conjugate of $a \in R$, or there are infinitely many.*

Corollary 9.26.18 *A non-central element of a division ring has infinitely many conjugates.*

Proof A division ring R with a non-central element a is infinite by Theorem 9.26.1, and there cannot be only one R-conjugate of a. By the corollary above a must have infinitely many conjugates. □

Chapter 10
Appendix

For cultural reasons, out of pure curiosity, we include here some classical results from number theory that are a little bit beside the main focus of the book, such as some analytic number theory, and the transcendentality of e and π, plus two results by Liouville and Thue.

10.1 The Function $\pi(x)$ for Large x

Obviously we want to study $\pi(x)$ for large x. The infinity of the number of primes can be expressed as $\lim_{x \to \infty} \pi(x) = \infty$, but there are more subtle ways of investigating the behavior of a function at infinity; you can e.g. compare it to known functions.

To get an idea of this behavior, let us provide a probabilistic argument. Looking for primes not greater than x, half of them are odd, and then $2/3$ of these are not divisible by 3 etc., so we arrive at the approximation formula

$$\pi(x) \approx x \prod_{n=1}^{\pi(x^{1/2})} (1 - 1/p_n),$$

which should be compared with the exact formula of Legendre.

Example 10.1.1

$$\pi(25) \approx 25 \left(1 - \frac{1}{2}\right)\left(1 - \frac{1}{3}\right)\left(1 - \frac{1}{5}\right) \approx 6,7$$

and

$$25\left(1-\frac{1}{2}\right)\left(1-\frac{1}{3}\right)\left(1-\frac{1}{5}\right) = 25 - \frac{25}{2} - \frac{25}{3} - \frac{25}{5} + \frac{25}{2 \cdot 3} + \frac{25}{2 \cdot 5} + \frac{25}{3 \cdot 5} - \frac{25}{2 \cdot 3 \cdot 5}.$$

Most notably, we no longer truncate with the greatest integer function.

For large x we actually get a reasonable approximation. For instance, we know that $\pi(100) = 25$ and according to tables $\pi(1000) = 168$, while the formula gives respectively 23 and 153 up to the closest integers.

Since

$$\frac{1}{2}\ln x = \int_1^{x^{1/2}} \frac{1}{t} dt \approx \sum_{n=1}^{[\sqrt{x}]} \frac{1}{n}$$

and each $n \leq \sqrt{x}$ can, by the fundamental theorem of arithmetic, be written uniquely as a product of primes not exceeding $\sqrt{x}$, we see that for large x, the number $\frac{1}{2}\ln x$ is roughly less than

$$\sum_{s_1 = \cdots = s_u = 0}^{\infty} \frac{1}{p_1^{s_1} \cdots p_u^{s_u}} = \prod_{k=1}^{u}(1 - 1/p_k)^{-1} \approx x/\pi(x),$$

where $u = \pi(\sqrt{x})$, and we have used the summation formula for geometric series together with rearrangement of terms under absolute convergence. Thus we see that $\pi(x)$ is approximately less than $2x/\ln x$, if we buy the somewhat fishy probabilistic argument.

Chebyshev proved the following estimates by honest methods.

Theorem 10.1.2 *There exist constants $0 < A < 1 < B$ such that*

$$A\frac{x}{\ln x} < \pi(x) < B\frac{x}{\ln x}$$

for all x sufficiently large.

This result was soon subsumed in the famous *prime number theorem*.

Theorem 10.1.3

$$\lim_{x \to \infty} \frac{\pi(x)}{x/\ln x} = 1.$$

So asymptotically there are $x/\ln x$ primes less than x.

For example, $100/\ln 100 = 21{,}714...$ and $1000/\ln 1000 = 144{,}764...$, which is not too bad. The approximation does of course improve with larger x, but not impressively fast. According to tables $\pi(100000000) = 5761455$ which should be compared with $100000000/\ln 100000000 = 5428681{,}024....$

Another way of phrasing the prime number theorem is to say that statistically, among the first n natural numbers with n large, only 1 in $\ln n$ numbers is a prime

10.1 The Function $\pi(x)$ for Large x

number. This means roughly speaking that the probability of a randomly picked natural number being prime, is inverse proportional to its number of digits.

So prime numbers occur less and less frequently as we wander along the path to infinity. This seems intuitively correct since there are more ways large numbers can be decomposed into an increasing number of building blocks.

Since by the prime number theorem $\frac{\pi(x)}{x} \ln x$ tends to 1 as $x \to \infty$, the fraction $\frac{\pi(x)}{x}$ must tend to zero equally fast as $\ln x$ tends to infinity. We will presently prove the following much softer result:

Theorem 10.1.4
$$\lim_{x \to \infty} \frac{\pi(x)}{x} = 0.$$

Or put differently, the prime numbers are so widely spaced at large that the average number of them within an interval stretching from the origin tends to zero as the length of the interval goes to infinity.

One can prove this theorem by a careful inspection of Legendre's formula, but we rather give a proof using techniques of Chebyshev.

Proof Consider an integer $n > 1$. Observe that a prime number p divides $n!$ if and only if $p \leq n$. So any prime number $p \in \langle 2^{n-1}, 2^n]$ divides $2^n!$ and cannot divide $2^{n-1}!$. But then the binomial coefficient

$$\binom{2^n}{2^{n-1}} = \frac{2^n!}{2^{n-1}!(2^n - 2^{n-1})!} = \frac{2^n!}{2^{n-1}!2^{n-1}!}$$

will also be divisible by p as this prime number sits as a factor in the nominator and not in the denominator. Therefore the product of all distinct primes $p \in \langle 2^{n-1}, 2^n]$ has to divide this binomial coefficient, which obviously means that this product is smaller than the binomial coefficient itself. The number of factors in this product is $\pi(2^n) - \pi(2^{n-1})$, and the product will be larger than the one where all the primes are replaced by their lower bound 2^{n-1}. At the other extreme the binomial coefficient under consideration occurs in Newton's expansion formula for $(1 + 1)^{2^n}$ and hence must be smaller than the latter. In conclusion we get the inequality

$$(2^{n-1})^{\pi(2^n) - \pi(2^{n-1})} < 2^{2^n},$$

which for the exponents of 2 on each side means that

$$\pi(2^n) - \pi(2^{n-1}) < \frac{2^n}{n-1}.$$

Hence

$$\pi(2^{2m}) - \pi(2^2) = \sum_{n=3}^{2m} (\pi(2^n) - \pi(2^{n-1})) < \sum_{n=3}^{2m} \frac{2^n}{n-1},$$

and trivially $\pi(2^2) < 2^2$, so

$$\pi(2^{2m}) < \sum_{n=2}^{2m} \frac{2^n}{n-1} = \sum_{n=2}^{m} \frac{2^n}{n-1} + \sum_{n=m+1}^{2m} \frac{2^n}{n-1} \le \sum_{n=2}^{m} 2^n + \sum_{n=m+1}^{2m} \frac{2^n}{m},$$

having replaced the denominators in the last two sums by the smallest ones, namely 1 and m, occurring in the sums. Since $\sum_{r=0}^{k} 2^r = 2^{k+1} - 1 < 2^{k+1}$, and $2^{m+1} < 2^{2m+1}/m$ as $m < 2^m$, we therefore get

$$\pi(2^{2m}) < 2^{m+1} + 2^{2m+1}/m < 2 \cdot 2^{2m+1}/m = 2^{2m} \cdot 4/m.$$

Now every real number $x \ge 2$ obviously belongs to some $\langle 2^{2m-2}, 2^{2m}]$ for a unique natural number m since such increasingly larger intervals partition the segment of the real axis stretching from 1 to infinity. But then

$$\pi(x)/x < \pi(2^{2m})/2^{2m-2} < 16/m,$$

which clearly tends to zero as x and subsequently m goes to infinity. $\square$

The following result, known as *Bertrand's postulate*, is obtained using similar techniques.

Theorem 10.1.5 *For any integer n larger than 1, the interval $\langle n, 2n \rangle$ will always contain a prime number.*

In the discussion prior to Chebyshev's result we incidentally proved, resorting to the divergence of the harmonic series, that the product $\prod (1 - 1/p)^{-1}$ over the primes is infinite. This gives the following result by Euler:

Proposition 10.1.6 *The sum $\sum 1/p$ over all primes diverges.*

Proof Let $S = \sum 1/p$. If we take the logarithm of the corresponding product $\prod (1 - 1/p)^{-1}$ and Taylor expand around 1, we get

$$\ln \prod_p (1 - 1/p)^{-1} = -\sum_p \ln(1 - 1/p) = S + \sum_p \sum_{n=2}^{\infty} 1/np^n$$

$$< S + \sum_p \sum_{n=2}^{\infty} 1/p^n < S + \sum_{m=1}^{\infty} 1/m^2$$

as

$$\sum_{n=2}^{\infty} 1/p^n = p^{-2}(1 - 1/p)^{-1} = \frac{1}{p(p-1)} < \frac{1}{(1-p)^2}.$$

This is only possible if S is infinite because $\sum_{m=1}^{\infty} 1/m^2$ is finite. $\square$

The divergence of $\sum 1/p$ not only shows in yet another way that there are infinitely many primes (there exist more than a dozen of proofs of both results), but this divergence also suggests that the primes are denser than the squares.

It is in fact conjectured that for any natural number $n > 1$, there exists a prime number between n^2 and $(n + 1)^2$.

The following result by Brun shows that there are infinitely fewer twins than primes.

Theorem 10.1.7 *The sum of the reciprocals of all twins is finite.*

10.2 The Riemann Zeta Function

Much of what has been said is tied in with the *Riemann zeta function*

$$\zeta(s) = \sum_{n=0}^{\infty} 1/n^s.$$

Euler considered the zeta function as a function of a real variable, and got the famous product formula

$$\zeta(s) = \prod (1 - 1/p^s)^{-1}$$

relating the zeta function to the prime numbers by the following simple procedure: Divide $\zeta(s)$ by 2^s and subtract this from $\zeta(s)$ to obtain

$$(1 - 1/2^s)\zeta(s) = 1 + 1/3^s + 1/5^s + \cdots.$$

Next, dividing this identity by 3^s and subtracting from the same identity gives

$$(1 - 1/2^s)(1 - 1/3^s)\zeta(s) = 1 + 1/5^s + 1/7^s + 1/11^s + \cdots.$$

Continuing this way there will eventually be no terms left on the right hand side except 1, and we arrive at the desired formula

$$\prod (1 - 1/p^s)\zeta(s) = 1.$$

Euler did not let reservations about convergence get in the way for clever arguments and great results. He has of course been credited many results which he proved only formally, i.e. by disregarding problems with limits, a notion that was not properly pinned down at the time, so he could hardly be blamed. Also formal calculations can sometimes be an advantage because they bring out the essence in an argument, and are often valid in much greater generality.

It should be noted that Euler's argument for the product formula reminds a lot of Eratosthenes sieve, in that first all the terms of the zeta function with n divisible by 2 are discarded, then those terms with n divisible by 3 are removed, and so on. It is also striking that this derivation does not use directly the fundamental theorem of arithmetic, which is perhaps the most common way of deducing the product formula: Every n will appear exactly once in a typical term $1/n^s$ of the infinite product

$$\prod(1-1/p^s)^{-1} = \prod(1 + 1/p^s + 1/p^{2s} + 1/p^{3s} + \cdots),$$

as is readily seen by multiplying out the latter product.

10.3 Bernoulli Numbers

Euler solved Basel's problem by showing that

$$\sum_{m=1}^{\infty} 1/m^2 = \pi^2/6.$$

In fact, he calculated the zeta function $\zeta(s)$ for every even natural number, not only for $s = 2$. His calculation was rather formal. He argued that just like any monic polynomial in x can be decomposed into its irreducibles $x - \alpha$ formed by its complex roots α, so can an analytic function like

$$\sin x = x - x^3/3! + x^5/5! - \cdots$$

be factored into similar bits, being so to speak a polynomial of infinite degree having infinitely many roots. Now $\sin x$ equals zero exactly when x is an integer multiple of π, so we therefore get the decomposition

$$\sin x = x \prod_{n=1}^{\infty} (1 - (x/n\pi)^2).$$

Taking the logarithm of this infinite product and differentiating termwise, we get

$$\cot x = (\ln(\sin x))' = \left(\ln x + \sum_{n=1}^{\infty} \ln(1 - (x/n\pi)^2)\right)' = \frac{1}{x} - \frac{2x}{\pi^2} \sum_{n=1}^{\infty} \frac{1}{n^2}(1 - (x/n\pi)^2)^{-1}.$$

Inserting $(1 - (x/n\pi)^2)^{-1} = \sum_{m=0}^{\infty} (x/n\pi)^{2m}$ and swapping summation under absolute convergence yields

10.3 Bernoulli Numbers

$$x \cot x = 1 - 2 \sum_{m=1}^{\infty} \left(\sum_{n=1}^{\infty} 1/n^{2m} \right) x^{2m}/\pi^{2m}.$$

On the other hand

$$x \cot x = ix \frac{e^{ix} + e^{-ix}}{e^{ix} - e^{-ix}} = ix + \frac{2ix}{e^{2ix} - 1} = ix + \sum_{m=0}^{\infty} B_n(2ix)^n/n!,$$

where B_n are the *Bernoulli numbers*.

Definition 10.3.1 Let B_n be the numbers determined by the identity

$$\sum_{n=0}^{\infty} B_n z^n/n! = z/(e^z - 1).$$

Comparing coefficients of powers of x in the two expressions above for $x \cot x$ gives Euler's formula, which has later been proved more rigorously:

Proposition 10.3.2 *The formula*

$$\sum_{n=1}^{\infty} 1/n^{2m} = \frac{(-1)^{m+1} 2^{2m-1} \pi^{2m}}{(2m)!} B_{2m}$$

holds for every natural number m.

Further comparison gives $B_0 = 1$, $B_1 = -1/2$ and $B_m = 0$ for all odd integers $m > 1$. Now all the Bernoulli numbers can be computed inductively from the formula

$$B_m = -(m+1)^{-1} \sum_{k=0}^{m-1} \binom{m+1}{k} B_k,$$

which is derived by comparing coefficients of powers of z after multiplying up $e^z - 1$ and expanding the exponential function. This gives the recursive identities

$$0 = 1 + 2B_1 = 1 + 3B_1 + 3B_2 = 1 + 4B_1 + 6B_2 + 4B_3 = 1 + 5B_1 + 10B_2 + 10B_3 + 5B_4 = \ldots$$

and $B_2 = 1/6$, $B_4 = -1/30$, $B_6 = 1/42, \ldots$.
Hence

$$\sum_{n=1}^{\infty} 1/n^4 = \pi^4/90 \text{ and } \sum_{n=1}^{\infty} 1/n^6 = \pi^6/945,$$

etc.

Bernoulli who failed to determine any of these sums, introduced his numbers to find an efficient expression for the sum of the m-th power of the first $n-1$ natural numbers.

Proposition 10.3.3 *The formula*

$$\sum_{k=1}^{n-1} k^m = (m+1)^{-1} \sum_{k=0}^{m} \binom{m+1}{k} B_k n^{m+1-k}.$$

is valid for any natural numbers n and m.

For $m=1$ we recover the formula in Sect. 1.1, which was verified by induction, and for $m=2$, we get

$$1^2 + \cdots + (n-1)^2 = n(n-1)(2n-1)/6.$$

Proof A short proof of the general formula consists of comparing coefficients of the powers of x in the first and last expressions of

$$\sum_{m=0}^{\infty}(\sum_{k=1}^{n-1} k^m) x^m/m! = \sum_{k=0}^{n-1} e^{kx} = \frac{e^{nx}-1}{x} \cdot \frac{x}{e^x-1} = \sum_{k=1}^{\infty} n^k \frac{x^{k-1}}{k!} \sum_{i=0}^{\infty} B_i \frac{x^i}{i!},$$

and then use Newton's binomial formula. $\square$

Remark 10.3.4 It was Riemann who first considered zeta as a function of a complex variable and extended its domain by analytic continuation. This turned out to be an extremely fruitful idea. By pushing the whole area into the realm of complex analysis stunning results could be obtained thanks to the powerful tools suddenly at hand, including Cauchy's theorem.

Riemann's eighth page short memoir with its ingenuity and radical ideas paved the way for the proofs by Hadamard and de la vallee Poussin of the prime number theorem which had defied elementary approaches of Chebyshev and others.

It therefore came as a surprise when Selberg and Erdös fifty years further down the road proved the same theorem by so called elementary means, i.e. by avoiding complex analysis. Still there is no reason to doubt the utility of the zeta function in any investigation of the distribution of primes. It suffices to mention the sharp estimates obtained relating to the error in the prime number theorem, with the sharpest such possible being realized if and only if the Riemann hypothesis holds.

All of this concerns the location in the complex plane of the zeros of the extended Riemann zeta function. Riemann's main contribution was to set up an amazing correspondence between these zeros and the distribution of the primes. The more you know about the location of the zeros, the more you know about the primes. Riemann hypothesized that all the zeros of the zeta function (except some obvious ones) lie on the vertical line one-half to the right of the imaginary axis. This conjecture has withstood the efforts of the brightest minds ever since and is by now the most famous and also regarded as the most important unsolved problem in mathematics. The prime number theorem amounts to verifying that none of the zeros are located on the vertical line at distance one to the right of the imaginary axis.

Generalizations of the Riemann zeta function play a crucial role in e.g. class field theory and in Dirichlet's theorem on primes in arithmetic progression. The branch of mathematics centered around this function and which freely uses complex analysis is called analytic number theory.

10.4 Transcendentality of e and π

We include Hilbert's proof of the following result.

Theorem 10.4.1 *The number e is transcendental.*

Proof Suppose there are integers a_i with $a_0 + a_1 e + \cdots + a_n e^n = 0$ and $a_0 \neq 0$.

Let $f(x) = x^m((x-1)\cdots(x-n))^{m+1} e^{-x}$ for any natural number m. Multiplying the equation above with $\int_0^\infty f(x)dx/m!$, and splitting up the integral, gives $A + B = 0$, where

$$A = a_0 \int_0^\infty f(x)dx/m! + a_1 \int_1^\infty f(x)dx/m! + \cdots + a_n e^n \int_n^\infty f(x)dx/m!$$

and

$$B = a_1 \int_0^1 f(x)dx/m! + \cdots + a_n e^n \int_0^n f(x)dx/m!.$$

Using the identity $\int_0^\infty x^l e^{-x} dx = l!$, we see that for $k \geq 1$, the number

$$e^k \int_k^\infty f(x)dx = \int_0^\infty f(y+k)dy$$

is an integer divisible by $(m+1)!$. Similarly, the first term in A is also an integer, with the lowest power of x in $e^x f(x)$ contributing to $\int_0^\infty f(x)dx$ with $((-1)^n n!)^{m+1} m!$, and where the other terms are divisible by $m+1$. Hence A is an integer and $A \equiv a_0((-1)^n n!)^{m+1} \pmod{(m+1)}$. When m is a prime number greater than both a_0 and n, we thus see that A is non-zero.

On the other hand, using the mean value theorem for integrals, we see that

$$a_k \int_0^k f(x)dx/m! = a_k f(c)k/m!$$

for some $c \in [0, k]$. As $|f(c)| \leq k^m n^{m+1}$, the absolute value of $a_k f(c)k/m!$ can be made arbitrary small for large enough m, so $|B| < 1$ for large enough m. Since there are infinitely many primes, this contradicts $A + B = 0$ as A is a non-zero integer for a large enough prime m. □

The transcendentality of π is harder to check, and is a consequence of the following theorem by Lindemann and Weierstrass.

Theorem 10.4.2 *The exponentials $e^{a_1}, \ldots, e^{a_n}$ of pairwise distinct algebraic numbers a_i are linear independent over the algebraic numbers.*

Proof Suppose we have a vanishing linear combination of the complex numbers e^{a_i} with algebraic numbers as coefficients which are not all zero. Pick a Galois extension $E \subset \mathbb{C}$ of $\mathbb{Q}$ containing these coefficients and the a_i's. We may also assume that all the a_i's are non-zero. Otherwise we could multiply the vanishing linear combination with e^{-a_j} for some $a_j \neq 0$ when $n \geq 2$ to get another vanishing linear combination of the same form where all $a_i \neq 0$.

Let $E[E]$ be the group ring over E of the group E considered as an additive group, so the elements of $E[E]$ are finitely supported E-valued functions with pointwise addition and convolution product. The formula $\Phi(f) = \sum_{a \in E} f(a) e^a$ defines a unital homomorphism $\Phi \colon E[E] \to \mathbb{C}$. Its kernel is an ideal of $E[E]$ which by assumption contains a non-trivial element $f \in \ker \Phi$.

Using the lexicographic order on $\mathbb{C}$ considered as a set of ordered pairs of real numbers, one sees that $E[E]$ is an integral domain. Therefore the finite product

$$g = \prod_{\sigma, \eta \in G(E/\mathbb{Q})} \eta f \sigma \in \ker \Phi$$

is a non-zero function. By the fundamental theorem in Galois theory it takes values in $\mathbb{Q}$ since $\eta g = g$ for all $\eta \in G(E/\mathbb{Q})$. Multiplying g by a large enough integer, we may assume that it is integer valued.

Pick $a \in E$ such that $g(a) \neq 0$, and define $h \in E[E]$ with $h(-b)$ to be $g(a)$ when $b = \sigma(a)$ for some $\sigma \in G(E/\mathbb{Q})$, and otherwise zero. Then $gh \in \ker \Phi$ is integer-valued and $(gh)(0) = \sum_{b+c=0} g(b) h(c) = kg(a)^2 \neq 0$, where k is the number of elements in the $G(E/\mathbb{Q})$-orbit of a. Also, since $g\sigma = g$ and $h\sigma = h$, we see that $(gh) \circ \sigma = gh$, so the function gh assigns the same values to $\sigma(a)$ for all $\sigma \in G(E/\mathbb{Q})$. In other words, we have a vanishing sum

$$b_0 + b_1 \sum_\sigma e^{\sigma(a_1)} + \cdots + b_n \sum_\sigma e^{\sigma(a_n)} = \Phi(gh) = 0,$$

where b_i are integers with $b_0 \neq 0$, and where σ runs over $G(E/\mathbb{Q})$ in each sum.

Take any polynomial $u(x)$ with integer coefficients such that $u(\sigma(a_i)) = 0 \neq u(0)$ for all $\sigma \in G(E/\mathbb{Q})$ and i. For any prime number p define a polynomial by $v(x) = x^{p-1} u(x)^p / (p-1)!$, so for any integer $q \geq p$, its q-th derivative $v^{(q)}(x)$ is a polynomial with integer coefficients divisible by p. The finite sum $w(x) = v(x) + v'(x) + v''(x) + \ldots$ defines a polynomial w with the property that $w(\sigma(a_i)) = v^{(p)}(\sigma(a_i)) + v^{(p+1)}(\sigma(a_i)) + \ldots$ and such that $w(0) = v^{(p-1)}(0) + v^{(p)}(0) + \ldots$

10.4 Transcendentality of e and π

is an integer not divisible by sufficiently large p. Moreover, for any $a \in E$ of the form $\sigma(a_i)$ and with r the maximum of all $|\sigma(a_i)|$, we have

$$|w(a) - e^a w(0)| = |e^a| \cdot |e^{-a} w(a) - e^{-0} w(0)| \leq re^r \max\{|(e^{-x} w(x))'| \mid |x| \leq r\}$$
$$\leq re^r \max\{|-e^{-x} v(x)| \mid |x| \leq r\}.$$

Thus there is a constant c independent of p such that

$$|w(\sigma(a_i)) - e^{\sigma(a_i)} w(0)| \leq c^p/(p-1)!$$

for all σ and i.

Pick a natural number $m \neq p$ such that ma_i are all algebraic integers. This is always possible. Just let m be larger than the product of all the denominators of the rational coefficients of a monic polynomial h for which all the a_i's are roots. Then $h_1(x) = m^{\deg(h)} h(x/m)$ is a monic polynomial with integer coefficients and $h_1(ma_i) = 0$ for all i.

Let $d = \deg(u)$ and define

$$A = m^{dp-1}(w(0)b_0 + b_1 \sum_\sigma w(\sigma(a_1)) + \cdots + b_n \sum_\sigma w(\sigma(a_n))).$$

Since $m^{dp-1} \sum_\sigma w(\sigma(a_i))$ is a $G(E/\mathbb{Q})$-invariant element of E, it is a rational number. But it is also an algebraic integer since it is an integer linear combination of powers of ma_i and the algebraic integers form a ring much in the same way as the algebraic numbers form a field. By Proposition 7.1.6 it is therefore an integer. This is true for all i, so A is an integer. Since the integers $m^{dp-1} \sum_\sigma w(\sigma(a_i))$ are all divisible by p, whereas $m^{dp-1} w(0) b_0$ is not for sufficiently large p, we see that $A \neq 0$, so $|A| \geq 1$ for such p.

On the other hand, subtracting

$$w(0)m^{dp-1}(b_0 + b_1 \sum_\sigma e^{\sigma(a_1)} + \cdots + b_n \sum_\sigma e^{\sigma(a_n)}) = 0$$

from A, gives

$$|A| \leq m^{dp-1} \sum_i |b_i| \sum_\sigma |w(\sigma(a_i)) - w(0)e^{\sigma(a_i)}|$$
$$\leq n \max_i |b_i| \cdot |G(E/\mathbb{Q})| m^{dp-1} c^p/(p-1)! < 1$$

for large enough p, which is a contradiction. □

Corollary 10.4.3 *The number π is transcendental.*

Proof Since $e^{i\pi} + e^0 = 0$, the number π cannot be algebraic. □

We state the following deep result bu Gelfond and Schneider without proof.

Theorem 10.4.4 *If a and b are algebraic numbers such that $\log a$ and $\log b$ are linear independent over $\mathbb{Q}$, then they are also linear independent over the algebraic numbers.*

In particular, if p and q are distinct primes, then $\log p$ and $\log q$ are linear independent over the algbraic numbers because whenever $a \log p + b \log q = 0$ for $a, b \in \mathbb{Q}$, then $p^a q^b = 1$, so $a = b = 0$ by the fundamental theorem of arithmetic.

A more striking consequence of the theorem is that Hilbert's α^β-conjecture follows:

Corollary 10.4.5 *If α and β are algebraic numbers with α different from 0 and 1 and with β irrational, then α^β is transcendental.*

Proof If $a \log \alpha + b \log(\alpha^\beta) = 0$ for $a, b \in \mathbb{Q}$ with $b \neq 0$, then

$$\beta = \log(\alpha^\beta)/\log \alpha = -a/b \in \mathbb{Q},$$

which is absurd. So $\log \alpha$ and $\log(\alpha^\beta)$ are linear independent over $\mathbb{Q}$. If α^β was an algebraic number, then $0 \neq \beta \log \alpha - \log \alpha^\beta = 0$, which is impossible. □

In particular, the numbers $2^{\sqrt{2}}$ and $e^\pi = (e^{-i\pi})^i = (-1)^i$ are transcendental.
Let us also mention a result by Apery which says that $\zeta(3)$ is irrational.

10.5 Proof of Liouville's Theorem

We prove the following theorem.

Theorem 10.5.1 *For any irrational algebraic number a of degree $n > 0$ there is a real number $c > 0$ such that $|a - \frac{p}{q}| > \frac{c}{q^n}$ for all integers p, q with $q > 0$.*

Proof Let $f(x)$ be the minimal polynomial of a over $\mathbb{Q}$, so f has no rational roots, while $f(a) = 0$. If $|a - \frac{p}{q}| < 1$, let $c > 0$ be less than the minimal value of the continuous function $x \mapsto |1/f'(x)|$ over the compact interval $[a-1, a+1]$. Then by the mean value theorem, there is some b between a and p/q such that $|f(a) - f(p/q)| = |f'(b)(a - p/q)|$. Hence

$$|a - p/q| = |f(p/q)/f'(b)| > c|f(p/q)| = (c/q^n)|q^n f(p/q)| \geq c/q^n$$

as $q^n f(p/q)$ is a non-zero integer. □

10.6 Thue's Theorem

The following remarkable result is due to Axel Thue.

Theorem 10.6.1 *Let $f(z) = a_n z^n + \cdots + a_1 z + a_0$ be an irreducible polynomial over $\mathbb{Q}$ with $n \geq 3$ and integer coefficients a_i with $a_n > 0$. If for any real number $\varepsilon \in \langle 0, n-2 \rangle$ and any complex root a of f, there is a real constant $c > 0$ such that $|a - p/q| > c/q^{n-\varepsilon}$ for all integers p, q with $q > 0$, then the homogeneous equation $y^n f(x/y) = m$ has only finitely many integer solutions (x, y) for each integer m.*

Proof We factorize f over $\mathbb{C}$, say

$$a_m(x/y - b_1) \cdots (x/y - b_n) = m/y^n,$$

and let A be the minimal distance between any two distinct roots b_i and b_j. For any solution (x, y) with $y \neq 0$ we can have $|x/y - b_i| < A/2$ for at most one root b_i since otherwise the distance between any two such roots would be less than A by the triangle inequality. If moreover, both x and y are integers, then by the assumed estimate, we get

$$|m/y^n| > a_m (A/2)^{n-1} c/|y^{n-\varepsilon}|.$$

Hence $|y|$ is bounded, and for each fixed y there are only finitely many possibilities for x. $\square$

Already Thue's own improvements on the lowering of the exponential in Liouville's theorem gives the following immediate corollary, where we finally use irreducibility of f.

Corollary 10.6.2 *The Diophantine equation*

$$a_n x^n + a_{n-1} x^{n-1} y + \cdots + a_1 x y^{n-1} + a_0 y^n = m$$

given by the theorem above has only finitely many solutions.

So for instance, the equation $x^3 - 2y^3 = 1$ has only finitely many integer solutions, while the Pell equation $x^2 - 2y^2 = 1$ has infinitely many.

Thue's theorem with its method of proof sparked what soon became heavy investigations on the relationship between Diophantine equations and Diophantine approximations.

The most spectacular result in this direction is perhaps the one by Baker, which says that the integer solutions (x, y) of the Diophantine equation in the corollary above satisfy $\max\{|x|, |y|\} < e^d$, where $d = (n \max\{|m|, |a_1|, \ldots, |a_n|\})^{10^5}$. This huge bound has turned out to be effective in ruling out possible integer solutions of certain classical Diophantine equations.

Bibliography

1. C. Adams, *The Knot Book* (AMS, 2004)
2. V. Chari, A. Pressley, *Quantum Groups* (Cambridge University Press, 1994)
3. A. Connes, *Noncommutative Geometry* (Academic, 1994)
4. F. Diamond, J. Shurman, *A First Course in Modular Forms*, GTM, vol. 228 (Springer, 2005)
5. C.T.J. Dodson, P.E. Parker, *A User's Guide to Algebraic Topology*, Mathematics and its applications, vol. 387 (Kluwer Academic Publisher, 1997)
6. A. Einstein, *The Principle of Relativity* (Dover Publications, 1952)
7. D. Eisenbud, *Commutative Algebra*, GTM, vol. 150 (Springer, 1991)
8. W. Fulton, J. Harris, *Representation Theory*, GTM, vol. 129 (Springer, 2000)
9. G.H. Hardy, E.M. Wright, *The Theory of Numbers* (Oxford University Press, 1938)
10. K. Ireland, M. Rosen, *A Classical Introduction to Modern Number Theory*, GTM, vol. 84 (Springer, 2000)
11. C. Kassel, *Quantum Groups*, GTM, vol. 155 (Springer, 1991)
12. A.W. Knapp, *Lie Groups Beyond an Introduction*, Progress in Mathematics, vol. 140 (Birkhäuser, 1996)
13. T.Y. Lam, *A First Course in Noncommutative Rings*, GTM, vol. 131 (Springer, 2000)
14. T.Y. Lam, *Lectures on Modules and Rings*, GTM, vol. 189 (Springer, 1999)
15. S. Lang, *Algebra*, GTM, vol. 211 (Springer, 2000)
16. S. Lang, *Algebraic Number Theory*, GTM, vol. 110 (Springer, 1994)
17. H.B. Lawson, M-L. Michelsohn, *Spin Geometry* (Princeton University Press, 1989)
18. J.M. Lee, *Introduction to Smooth Manifolds*, GTM, vol. 218 (Springer, 2012)
19. S. Mac Lane, *Categories for the Working Mathematician*, GTM, vol. 5 (Springer, 1978)
20. F. Mandl, G. Shaw, *Quantum Field Theory* (Wiley, 1984)
21. S. Neshveyev, L. Tuset, *Compact Quantum Groups and Their Representation Categories*, CS, vol. 20 (SMF, 2013)
22. L. Tuset, *Analysis and Quantum Groups* (Springer, 2022)
23. S. Willard, *General Topology* (Addison-Wesley Publishing Company, 1968)
24. R.A. Wilson, *The Finite Simple Groups*, GTM, vol. 251 (Springer, 2009)

Index

A
Abelian, 134
Absolutely semisimple, 410
Absolute value, 76
Action of a group, 154
Additive group, 134
Adjoint A^*, 103
Algebra $\mathcal{A}$ over a field, 99
Algebraic, 261, 403, 423, 424
Algebraic closure, 75, 265
Algebraic equation, 70
Algebraic extension, 261
Algebraic multiplicity, 362
Algebraic numbers, 267
Algebraic over F, 403
Algebraic product of the vector spaces, 92
Algebraic real number, 70
Algebraic version of Mackey's theorem, 226
Alternating, 111
Alternating group, 145
Annihilator of a subset X of a module, 311
Antiendomorphism, 309
Antisymmetric, xi
Archimedean property, 7
Archimedian value, 69
Arcwise connected, 191
$a \in R$ divides $b \in R$, 250
Arithmetic function, 22
Arithmetic progression, 16
Artinian, 384
Ascending chain condition, 252, 383
Associate, 250
Associated eigenvalue, 118
Associated matrix, 351
Associated norm, 128

Associative, 134
At most countable, 81
Augmented matrix, 104
Automorphism, 138
Axiom of choice, xi

B
Baer's criterion, 321
Balanced, 324, 382
Basis, 316
Bernoulli numbers, 435
Bertrand's postulate, 432
B-flat, 339
Bicommutant, 373
Bidual of an abelian group, 206
Bijective, x
Bimodule map, 325
Binary operation on a set, x
Binomial coefficients, 12
Boolean ring, 390
Braid group, 195
Burnside's theorem, 156, 374
Butterfly lemma, 170

C
Cancellation property, 51, 135, 233
Cardinality, 81
Cardinal number, 85
(Cartesian) product, ix
Cauche complete, 62
Cauchy-Schwarz inequality, 128
Cauchy sequence, 60
Cauchy's theorem, 161

© The Editor(s) (if applicable) and The Author(s), under exclusive license to Springer Nature Switzerland AG 2025
L. Tuset, *Abstract Algebra via Numbers*,
https://doi.org/10.1007/978-3-031-74623-9

Cayley–Hamilton theorem, 359
Cayley-Hamilton theorem, 123
Cayley's theorem, 145
Center of a group, 142
Center of a ring, 233
Centralizer, 142
Central series, 174
Chain in a partially ordered set, xi
Chain of generalized eigenvectors of A of length n, 121
Change of the base ring, 339
Characteristic equation, 120
Characteristic function, xi
Characteristic of a ring, 234
Characteristic polynomial, 359
Characteristic subgroup, 178
Characteristic zero, 56
Character of a group, 205
Character of π, 212
Chinese remainder theorem, 21
Choice function, xi, xii
Class formula, 156
Class function, 213
Classical Gauss sum, 217
Class of nilpotency of the group, 173
Cofactor, 115
Cokernel, 330
Column vector, 104
Commutant, 373
Commutative, 134, 232
Commutative algebra, 99
Commutative ring with identity, 53
Commutators, 171
Complement, ix
Complemented in the module, 311
Complete group, 147
Completely reducible, 386
Complex conjugate, 75
Complex conjugate $\bar{A}$, 103
Complexification of a real vector space, 327
Complex numbers, 75
Complex symplectic group, 153
Composite number, 4
Composition series, 167, 388
Congruence (or residue) classes, 19, 135
Conjugacy problem, 188
Conjugate, 147, 156, 319
Constructable, 271
Content, 254
Continuum Hypothesis, 84
Contragredient representation, 201
Convolution product, 214, 236
Correspondence theorem, 143

Correspondence theorem for rings, 242
Coset, 137
Countable, 81
Coxeter presentation, 188
Cramer's rule, 114
Cycle, 109
Cycles, 193
Cyclic extension of F, 294
Cyclic group, 136
Cyclic module, 310
Cyclotomic polynomial, 291

D

Defect, 123
Degree $\deg p$ of a non-zero polynomial p over a ring, 248
deMorgan's laws, ix
Derivation of the ring, 424
Derivative, 274
Derived group, 171
Descending chain condition, 384
Determinant, 112
Determinant of a linear operator, 114
Diagonal, 118
Diagonalizable, 118, 365
Diamond isomorphism theorem, 143
Dihedral group, 146
Dimension, 96
Dimension of an algebra, 379
Dimension of the module, 372
Direct product of groups, 139
Direct sum, 92
Direct summand of the module, 311
Direct sum of representations, 201
Dirichlet's theorem, 16
Discriminant, 304
Disjoint cycles, 109
Distance, 127
Divisible, 322
Division algorithm, 8
Division ring, 232
Domain, x
Double cosets, 225
Double dual, 341
Dual basis, 127, 342
Dual group, 205
Dual module, 341
Dual space, 127

E

Eigenvector, 118
Element, ix

Index   447

Elementary divisors, 361
Elementary matrix, 108
Elementary row operations, 104
Elementary symmetric function, 300
Elementary tensor, 130
Embedding, 138, 233
Endomorphism, 138
Enough injectives, 322
Enough projectives, 322
Entry, 99
Epimorphism, 138, 233
Equivalence class, xi
Equivalence relation, xi
Equivalent, 351
Equivalent representations, 199
Equivalent series, 168, 388
Equivariant, 154
Euclidean algorithm, 10
Euclidean domain, 248
Euclid's lemma, 8
Euler's criterion, 30
Euler's phi-function, 23
Evaluation, 238
Even permutation, 111
Exact, 312
Exact at A_i, 312
Expansion of $\det(A)$ according to the ith row, 115
Extension field, 260
Extension of a field F by radicals, 296
Extension of scalars, 327

F

Factors, 167, 388
(Faithfully) flat, 335
Faithful module, 373
Faithful (or effective) action, 155
Faithful representation, 199
Fermat number, 37
Fermat prime, 37
Fermat's last theorem, 38
Fermat's little theorem, 20
Fibonacci numbers, 44
Field, 56, 232
Finite, 81
Finite dimensional vector space, 96
Finitely cogenerated, 385
Finitely generated, 240, 261
Finitely generated group, 136
Finitely generated module, 310
Finitely presented module, 332
Finite presentation, 187

Finite simple continued fraction, 57
First isomorphism theorem, 140
First principle of induction, 5
First version of Frobenius reciprocity, 223
Five-lemma, 333
Fixed field of G, 283
Fixed point, 154
Flat for B, 339
Flip, 130
Flip map, 326
Formal Laurent series, 239
Formal power series, 239
Fourier coefficients, 129, 207
Fourier inversion formula, 207
Fourier transform, 207
Fractional linear transformations, 185
Free abelian group, 184
Free action, 155
Free group generated by a set, 182
Free module, 316
Free presentation, 317
Free product of the groups, 182
Free ring, 244
Frobenius endomorphism, 278
Function, ix
Fundamental group, 191
Fundamental homomorphism theorem for rings, 242
Fundamental theorem for finitely generated abelian groups, 179
Fundamental theorem of algebra, 74
Fundamental theorem of arithmetic, 4

G

Galois extension, 287
Galois group of a polynomial, 286
Gaussian integers, 249
Gauss-Jordan elimination, 105
Gauss' lemma, 31
Gauss' quadratic reciprocity law, 33
Gauss sum, 216
Generalized geometric multiplicity, 362
Generalized Jordan block, 364
Generalized polyhedron, 191
General linear group, 150
Generator, 233, 310, 382
Generators of G, 136
Geometric series, 64
Geometric version of Mackey's theorem, 224
G-invariants, 212
Goldbach conjecture, 17

Gram-Schmidt orthonormalization, 129
Graph, x
Grassmannian, 158
Greatest common divisor, 8, 251
Group, 134
Group algebra, 214
Group ring, 236
G-spaces, 154, 199

H

Haar integral, 203
Hilbert basis theorem, 391
Hilbert ring, 411
Homogeneous space, 154
Homogeneous system, 103
Homology group, 193
Homomorphism, 138, 233
Homotopy, 191
Hopf-fibration, 159
Hopkins-Levitzki theorem, 402

I

Ideal, 240
Ideal generated by a subset, 240
Idempotent in a ring, 313
Identity, 232
Identity matrix, 101
Image, 98
Imaginary axis, 75
Independent, 346
Independent transcendental elements, 301
Index, 126, 137
Index of a relative to a primitive root b, 28
Induced representation, 221
Infinite descent, 39
Infinite dihedral group, 187
Infinite dimensional, 96
Infinite order, 137
Infinite simple continued fraction, 66
Infinum, 63
Inhomogeneous system, 103
Injective module, 320
Inner automorphism, 147
Inner derivation, 424
Inner product, 127
Integral domain, 55, 232
Integral operator, 215
Integral root test, 259
Internal direct sum, 98
Intersection, ix
Intertwiner, 199

Invariant factors of the matrix, 354
Invariant factor theorem, 349
Invariant inner product, 202
Invariants, 358
Invariant subset of a G-space, 154
Inverse element, 134
Inverse image, xi
Inverse map, x
Invertible, 99
Involution on a ring, 414, 415
Irreducible, 199, 250
Isomorphic vector spaces, 97
Isomorphism, 138, 233
Isomorphism of G-spaces, 154
Isomorphism problem, 188
Isotropy (stabilizer) group, 155

J

Jacobson radical, 396
Jacobson's density theorem, 373
Jordan block, 361
Jordan blocks, 122
Jordan canonical form, 122, 362
Jordan–Chevalley decomposition, 367
Jordan-Chevalley decomposition of A, 126

K

Kernel, 98, 138, 215
Kernel of a homomorphism, 242
kth convergent, 57

L

Lagrange's theorem, 137
Laurent polynomials, 239
Leading coefficient, 248
Leading columns, 105
Leading entry in a row, 105
Least upper bound property, 63
Left inverse, 134
Left (or right) ideal, 240
Left (right) zero divisor, 233
Left unit, 134
Legendre's formula, 23
Legendre symbol (a/p), 30
Length of a series, 388
Length of an element, 352
Lexicographical order, 189
Linear basis, 93
Linear combination, 93
Linear dependent, 93, 316
Linear Diophantine equation, 10

Index 449

Linear independent, 93, 316
Linear isomorphism, 97
Linear operator, 97
Linear transformation, 97
Liouville number, 70
Liouvilles's constant, 71
Local field, 81
Locally finite groups, 403
Loop, 191
Lower central series, 174

M

Map, ix, x
Matrix, 99
Matrix coefficients, 200
Matrix of $A \in \text{End}(V, W)$ with respect to bases, 100
(Matrix) product, 101
Matrix representation, 200
Matrix unit, 102
Maximal, 143
Maximal ideal in a ring, 246
Member, ix
Mersenne prime, 14, 36
Metric, 127
Minimal polynomial, 262, 358
Möbius function, 22
Möbius inversion formula, 23
Modular group, 185
Module, 309
Module homomorphism, 309
Module map, 309
Module over a ring, 309
Modules over algebras, 379
Monic, 248
Monoid, 134
Monomorphism, 138, 233
Morphism of G-spaces, 154
Morphisms, x
Multilinear or m-linear, 111
Multiplication, 232
Multiplicative, 22
Multiplicative commutators of the ring, 426
Multiplicative Jordan–Chevalley decomposition, 369
Multiplicity, 203, 274
Multiplicity of the simple module in A, 373

N

Nakayama's lemma, 399
Natural map, 373
Newton's binomial formula, 12

n-factorial, 12
Nil ideal of a ring, 392
Nilpotent, 126, 173, 366, 392
Nilpotent matrix, 365
Noetherian, 383
Noetherian (artinian) ring, 384
Non-archimedean, 76
Non-generator, 398, 399
Norm, 127
Normal extension, 272
Normal form, 75
Normalizer, 142
Normal series, 167
Normal subgroup, 139
n-th center of a group, 172
n-tuples, x

O

Octic group, 146
Odd permutation, 111
Opposite product, 309
Orbit, 154
Orbit decomposition, 155
Orbit decomposition formula, 156
Order, xi
Ordered domain, 54
Ordered field, 56
Ordered pairs, ix
Order homomorphism, 189
Order ideal, 311
Order of a group, 134
Order of a modulo b, 25
Order of an element, 137
Order $\text{ord}_p a$ of the prime element p, 254
Orientation, 193
Orthogonal, 128, 313, 365
Orthogonal group, 152
Orthogonal matrix, 152
Orthogonal projection, 205
Orthonormal basis, 129
Orthonormal k-frames, 158
Outer conjugacy classes, 147

P

p-adic absolute value, 77
p-adic expansion, 79
p-adic integers, 80
Parallelogram law, 128
Parseval identity, 129
Partial denominators, 57
(Partially) ordered, xi, 189

Partition, xi
Partition of a natural number, 181
Pascal's rule, 12
Pascal's triangle, 12
Pell's equation, 71
Perfect, 35, 278
Permutation group, 144
Permutation representation, 215
Permutations, 109, 144
p-group, 160
p'-group, 416, 417
Plancherel's formula, 207
Poisson summation formula, 208
Polarization identity, 128
Polynomial ring, 244
Polynomial ring in n indeterminates x_i with coefficients in R, 237
Pontryagin dual, 322
Pontryagin's duality theorem, 206
Positive cone for a group, 189
Power set, xi
Presentation for a group, 187
Prime, 250
Prime number, 4
Prime number theorem, 430
Prime subfield, 234
Primitive, 38, 254
Primitive nth root of unity in a field, 291
Primitive root, 26
Principal ideal, 240
Principal ideal domain, 248
Product, x
Projection map, 139
Projections, 319
Projective group, 150, 153
Projective module, 318
Proper, 233
Proper ideal, 240
Proper subgroup, 135
Proper submodule, 309
Pure braid group, 195
Pushout of two module maps, 320
Pythagoras' identity, 129
Pythagorean triangle, 39
Pythagorean triple, 38

Q
Quadratic congruence, 21
Quadratic residue (non-residue) of an odd prime, 30
Quaternions, 235
Quotient group, 139

Quotient map, 140, 241
Quotient module, 310
Quotient ring, 241
Quotient set, xi

R
R-algebra, 407
Rank-nullity theorem, 98
Rank of an $m \times n$-matrix A over a PID, 355
Rank of the group, 180
Rank of the module, 310
Rational canonical form, 358
Rational functions in n indeterminates, 238
Rational integers, 80
Real and complex projective spaces, 159
Real and imaginary parts, 75
Real axis, 75
Real numbers, 61
Reduced echelon form, 105
Reduced form, 183
Reducible, 250
Reducible over a field, 258
Reduction map, 339
Refinement of a normal series, 170
Refinement of a series, 388
Reflexive, 341
Regular function, 200
Regular n-polygon, 141
Regular representation, 204
Relation between the generators, 187
Relation on a set, ix
Relatively prime, 8, 251
Representation, 199
Representation of the algebra $F[G]$, 215
Resolvent cubic, 303
Riemann sphere S^2, 159
Riemann zeta function, 433
Right actions, 157
Right cosets, 137
Right inverse, 134
Right regular representation, 204
Right R-module, 309
Right unit, 134
Ring, 232
Ring over R generated by the elements of X subject to the relations Y, 244
R-linear combination, 316
R-module, 309
R-module map, 309
Root, 258
Row equivalent, 104
RS-bimodule, 325
R-submodule, 309

Index 451

S

Scalar product, 101
Schur's lemma, 204
S-conjugate, 427
Second principle of induction, 6
Second version of Frobenius reciprocity, 223
Semidirect product, 148
Semigroup, 134
Semisimple, 367, 371
Semisimple ring, 376
Separable, 278
Separable element, 278
Separable extension, 278
Sequence, x
Set, ix
Short extact, 312
Sieve of Eratosthenes, 18
Sign of a permutation, 110
Similar, 351
Similar matrices, 101
Simple, 367
Simple extension fields, 261
Simple group, 140
Simple left ideals, 380
Simple module, 370
Simple ring, 246
Simple root, 274
Simplicial complex, 191
Simply connected, 191
Simultaneously diagonalizable, 365
Smith normal form, 354
Snake diagram, 331
Snake lemma, 331
Solvable, 171
Solvable by radicals, 296
Span of a subset of a vector space, 93
Spans, 93
Special linear group, 151
Special orthogonal group, 152
Special unitary group, 152
Spectrum, 415
Split exact, 312
Split the sequence, 313
Splitting field, 271
Splitting field of a family, 272
Standard basis, 95
Standard inner product, 128, 204
Standard representation, 221
Stiefel manifold, 158
Subgroup, 135
Subgroup generated by X, 136
Submodule $\langle X \rangle$ generated by X, 310
Submodules stabilizes, 384

Subrepresentation, 199
Subring, 233
Subring generated by, 233
Subset, ix
Subspaces, 91
Supremum, 63
Surjective, ix
Sylow p-subgroup, 163
Sylow's first theorem, 162
Symmetric, xi
Symmetric function, 300
Symmetric group, 145
Symmetry, 146
Symmetry group, 146
Symplectic bilinear form, 152
Symplectic group, 153
System of m linear equations in n unknowns x_i with coefficients a_{ij} in a field F, 103

T

Tensor product, 324
Tensor product of representations, 201
Tensor product of two vector spaces, 129
Third (or double quotient) isomorphism theorem, 144
Torsion element, 312
Torsion-free group, 178
Torsion-free module, 312
Torsion module, 312
Torsion part of the group, 180
Torsion submodule, 312
Trace, 103
Trace formula, 215
Transcendental field, 268
Transcendental number, 268
Transcendental over a field, 268
Transcendental real numbers, 70
Transfer homomorphism, 177
Transitive, 154, 298
Transpose, 102
Transposition, 109
Transversal, 176
Triangular numbers $n(n + 1)/2$, 40
Triangulation, 191
Trivial, 233, 240
Trivial absolute value, 76
Trivial homomorphism, 139
Trivial representation, 200
Trivial subgroup, 135
Trivial units, 418
Trivial vector space, 96

Twins, 15
Twisted group ring, 245
Type of the group, 180
Type of the torsion part, 180

U
Uncertainty principle, 207
Union, ix
Unipotent, 369
Unique factorization domain, 254
Unit, 250
Unital, 232
Unital algebra, 99
Unitary group, 151
Unitary matrix, 151
Unit element, 134
Universal property, 325
Upper central series, 173
Upper (lower) triangular matrix, 236

V
Valuation, 76

Vandermonde determinant, 117
Variable, 238
Vectors, 90
Vector space, 90
Von Neumann regular, 404
Von Neumann regular ring, 404

W
Waring's problem, 43
Wedderburn radical, 402
Wedderburn's theorem, 374
Well-ordering principle, 7
Weyl algebra, 244
Wilson's theorem, 21
Word of length n, 183
Word problem, 188
Wreath product, 149

Z
Zero divisor, 233
Zorn's lemma, xi, xii

SPRINGER NATURE

GPSR Compliance

The European Union's (EU) General Product Safety Regulation (GPSR) is a set of rules that requires consumer products to be safe and our obligations to ensure this.

If you have any concerns about our products, you can contact us on ProductSafety@springernature.com

In case Publisher is established outside the EU, the EU authorized representative is:

Springer Nature Customer Service Center GmbH
Europaplatz 3
69115 Heidelberg, Germany

The manufacturer's authorised representative in the EU is Springer Nature Customer Service Centre GmbH, Europaplatz 3, 69115 Heidelberg, Germany. If you have any concerns regarding our products, please contact ProductSafety@springernature.com

Printed and bound by CPI Group (UK) Ltd, Croydon, CR0 4YY

25/03/2026

02078174-0013